OPTICAL INSTRUMENTS AND TECHNIQUES 1969

OPTICAL INSTRUMENTS AND TECHNIQUES 1969

Proceedings of the Conference held at the University of Reading during 14*th*–19*th July,* 1969, *under the auspices of the International Commission for Optics and arranged by the Optical Sub-Committee of the British National Committee for Physics.*

Edited by

J. HOME DICKSON

ORIEL PRESS

First published 1970

ISBN 0 85362 065 2
Library of Congress Catalog Card No. 71–112717

Published by
ORIEL PRESS LIMITED
32 Ridley Place,
Newcastle upon Tyne, NE1 8LH, England.

Text set in 10 on 12 point Times Roman and
Printed by Neill and Company Ltd.,
Edinburgh, Scotland.

PREFACE

The International Commission for Optics normally meets at three-year intervals and a scientific conference on some aspect of optics is associated with each meeting. In 1951 and in 1960 the Commission met in London and the associated conferences were devoted to optical instruments. This volume reports the Proceedings of the Conference on optical instruments and Techniques held at Reading, U.K., in 1969 in association with the eighth meeting of the commission. The emphasis on 'instruments and techniques' rather than on 'optical instruments' is significant. It is now generally recognised that the concept of systems design must be applied to optical instrumentation as a whole. Mechanical design must always be closely related to optical design and usually the success of an instrument also involves elaborate and ingenious electronics design. It may happen that, in a radically new system, the purely optical design of lens or mirror systems is the least novel part. All aspects of design must be considered together from the beginning in order to find the most efficient and economical way of satisfying the requirements of a problem.

I wish, on behalf of the Organising Committee, to thank the following organisations for their sponsorship and generous financial support:

The International Commission for Optics
The International Union of Pure and Applied Physics
(and, through the Union, UNESCO)
The Royal Society
The Institute of Physics and Physical Society
The Scientific Instrument Manufacturers Association of Great Britain.

I also wish to thank the Editor of these proceedings for the care and skill he has devoted to the work, and the authors, a high proportion of whom delivered their manuscripts in good time.

R. W. DITCHBURN.

EDITOR'S NOTE

In preparing these Proceedings of the Conference I have aimed primarily at an early date of publication. The papers have been arranged as far as possible in groups of cognate subjects, but there are some exceptions to this general rule. Abstracts have been added, these being kept as short as possible as in many cases the importance of the paper is such that an abstract is misleading; English abstracts have been given for the French language papers. A list of delegates to the Conference has been added and there is a fairly comprehensive index, as well as a table of contents with extended descriptions.

I cannot express too warmly my gratitude to Dr. Barry Shurlock, Mrs Byass and the staff of Oriel Press for their great help throughout, their patience and forbearance and the excellence of their sub-editing. I have also had great help with the proof reading from the authors, who have been most prompt in returning their corrected proofs, well and clearly marked; in fact their cooperation throughout was invaluable.

J. H. D.

OFFICERS OF THE CONFERENCE

President

G. Toraldo di Francia

Organising Committee

Professor R. W. Ditchburn (*Chairman*), University of Reading
Professor P. B. Fellgett, University of Reading
Dr. K. M. Greenland, British Scientific Instrument Research Association
Mr. K. J. Habell, National Physical Laboratory
Professor H. H. Hopkins, University of Reading
Mr. J. R. Stansfield, Hilger and Watts Limited
Professor W. D. Wright, Imperial College, London
Dr. D. A. Thetford (*Secretary*), University of Reading

The Sub-Committee was appointed by the British National Committee for Optics which is a sub-committee of the British National Committee for Physics, The Royal Society, 6 Carlton House Terrace, London S.W.1.

CONTENTS

Part I. New techniques and instruments for spectroscopy, including modern interferometric methods, stimulated emission and non-linear phenomena.

Part II. Recent developments in optical production techniques, including the use of new materials.

Part III. Optical metrology and optical processing of data, including coherent light techniques.

Part IV. Advances in assessment and specification of performance of optical instruments.

Part V. Image forming systems of essentially novel design.

Part VI. Systems design of astronomical instruments.

Part I.

New techniques and instruments for spectroscopy, including modern interferometric methods, stimulated emission and non-linear phenomena.

High Resolution Fourier Spectroscopy*

Pierre Connes

Laboratoire Aimé Cotton, C.N.R.S.
Faculté des Sciences,
ORSAY, 91, *France*

The technique of Fourier spectroscopy at first appeared suitable only for very weak sources and low resolving powers; most of the first applications were limited to the far infrared range. This situation arose from several factors: the difficulty of controlling the path difference is much less in the far infrared; also the performances of grating spectrometers are poor and easy to improve upon.

The work at the Laboratoire Aimé Cotton has been directed from the start towards high resolution in the *near* infrared. The most recent results clearly demonstrate that a Fourier interferometer is better than a grating spectrometer for resolution, a Fabry-Pérot étalon for spectral range and both for accuracy of instrumental line shape and wavenumber measurement. These factors are independent of (and additive to) the now classical advantages which come from the multiplexing ability and the large light gathering power. Altogether they make Fourier spectroscopy a most powerful tool which can be used in any spectral range for the solution of spectroscopic problems not considered possible a few years ago.

The technical problems which had to be solved are of several kinds. First, the path difference must be controlled with very great accuracy. A servo system using monochromatic fringes as a reference ensures that for each interferogram sample the path difference error is of the order of 1 Å (small compared with the irregularities in the optical surfaces of classical spectrometers (J. Connes & P. Connes 1966)).

Second the interferogram from a wide-band spectrum contains mostly ripples which are small compared to the mean intensity and these hold all the relevant information. However, even in the case of a rapidly fluctuating light source their *slopes* can be measured accurately enough, using the path difference modulation scheme first proposed by L. Mertz (1958).

* A résumé by the author.

The first interferometer built according to these principles (J. Connes & P. Connes 1966; J. Connes, P. Connes & J. P. Maillard 1967), has a 5 cm carriage motion, i.e. 10 cm maximum path difference and 0·1 cm^{-1} resolution. The sampling speed is limited to 5 samples per second, which means that 60,000 samples can be recorded in approximately 3 hours; this turned out to be the practical limit. The spectral range which can be computed from one interferogram is then 3000 cm^{-1} when working at 0·1 cm^{-1} resolution. This instrument has been used mainly for recording spectra of planets in the near infrared. Scientific results so far include the detection of CO, HCl and HF in the Venus atmosphere (J. Connes, P. Connes, W. S. Benedict & L. D. Kaplan 1967, 1968), and of CO in Mars (J. Connes, P. Connes & L. D. Kaplan 1969). Figure 1 shows that a gain in resolving power by a factor of 100 has been achieved in the Venus case compared to the best spectra recorded by classical means, i.e. grating spectrometers. The complete spectra of Venus, Mars, Jupiter and Saturn, from 1·1 to 2·5 μ (9000 to 4000 cm^{-1}) have been published in atlas form (J. Connes, P. Connes & J. P. Maillard 1969). Figure 2 shows a fraction of a sample page which compares Venus, Mars and solar spectra; the resolution is uniform and equal to 0·08 cm^{-1}. Wavenumbers can be directly measured with a vernier grid on the atlas pages within 10^{-2} cm^{-1}; this accuracy can be improved by a factor of 10 by treating the intensity samples themselves with computer programs. Some spectra of cool stars have also been recorded (P. Connes *et al.* 1968).

A second generation interferometer has been built by Jacques Pinard (1967, 1969). The maximum path difference has been increased by a factor of 20 to 200 cm; the resolution is thus 5.10^{-3} cm^{-1}. Since the speed of operation and maximum number of samples are about the same as for the first instrument, the spectral range covered within one interferogram is 20 times smaller, or about 150 cm^{-1} and it must therefore be isolated with a suitable monochromator.

This interferometer is being used for recording emission and absorption laboratory spectra. Figures 3 and 4, (J. Pinard 1969), show the improvement in resolving power and the even more striking one in wavenumber accuracy when compared with the best grating recorded spectra.

The computer program for performing the transforms has been developed by J. Connes at the Observatoire de Meudon. Using a 7040 IBM computer, a transform time of about 4 hours was needed for 64,000 samples, and this proved a practical limit. Recent advances in programming, however, plus the availability of large computers have changed the situation entirely; 64,000 points can now be transformed in 30 seconds of actual central processor time on a 360/75 computer. It is hoped that transforming 10^6 samples will prove entirely feasible.

In order to make full use of this new capacity a third generation interferometer with improved speed of operation, (50 to 100 samples sec^{-1}), is being built; it will give the same resolution as the second instrument but across a much wider spectral range. Used on emission line spectra

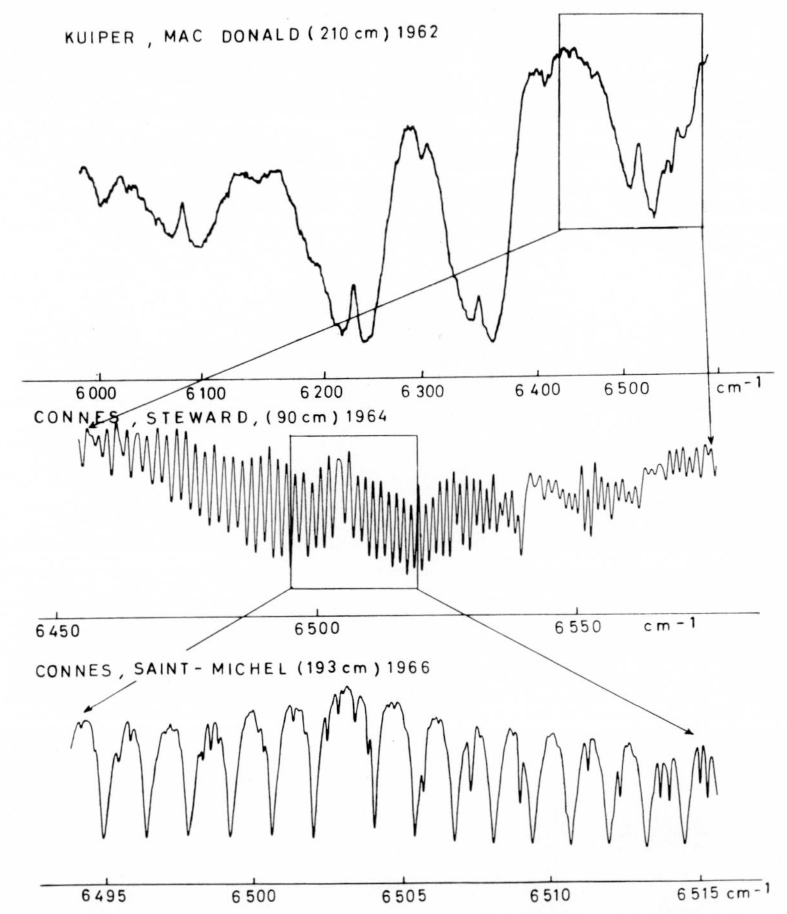

FIG. 1. *Upper trace.* A portion of the best previously available near infrared spectrum of Venus, recorded with a scanning, grating spectrometer (G. P. Kuiper 1962). The resolution is about 8 cm^{-1}; the trace shows mainly 4 unresolved CO_2 bands.
Middle trace. First results given in 1964 by Fourier spectroscopy with a resolution of 1 cm^{-1}; the rotational structure is resolved.
Lower trace. Latest results (1966), with 0·08 cm^{-1} resolution; many weaker lines appear.

it should be able to provide from just one interferogram, within the range of a given IR receiver, (e.g. lead sulphide, 3000 to 10,000 cm^{-1}), essentially *all* the available information on such features as the hyperfine structure, isotope shift, Zeeman patterns and line profiles, together with accurate absolute wavenumber calibration. For complex spectra such

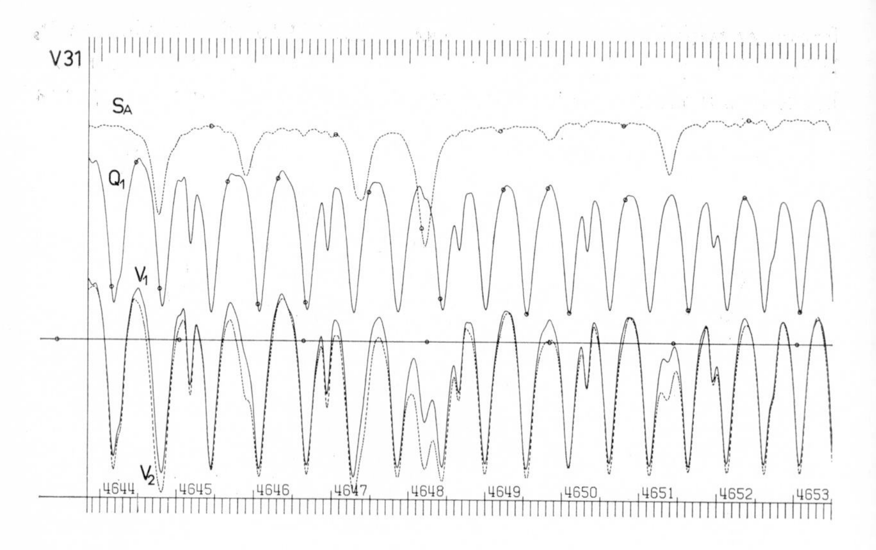

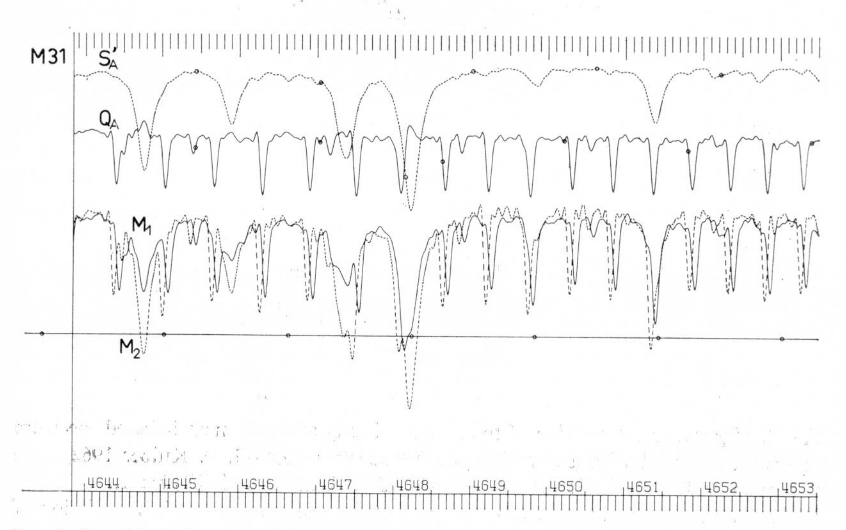

FIG. 2. One-third of a page of the Planetary Atlas (J. Connes, P. Connes & J. P. Maillard 1969). S_A and S'_A are solar spectra which show solar and telluric lines, V_1 and V_2 are Venus spectra taken at different elevations, which show identical Venus lines and telluric lines of different intensities. Q_1 is the ratio V_1/S_A and shows near cancellation of telluric lines. M_1 and M_2 are Mars spectra taken at different elevations, but also with different Doppler shifts; hence the Martian lines are displaced. Q_A is the ratio M_1/S_A. The planetary band is in both cases due to CO_2; the amount is much greater on Venus. The scale is in vacuum wavenumbers.

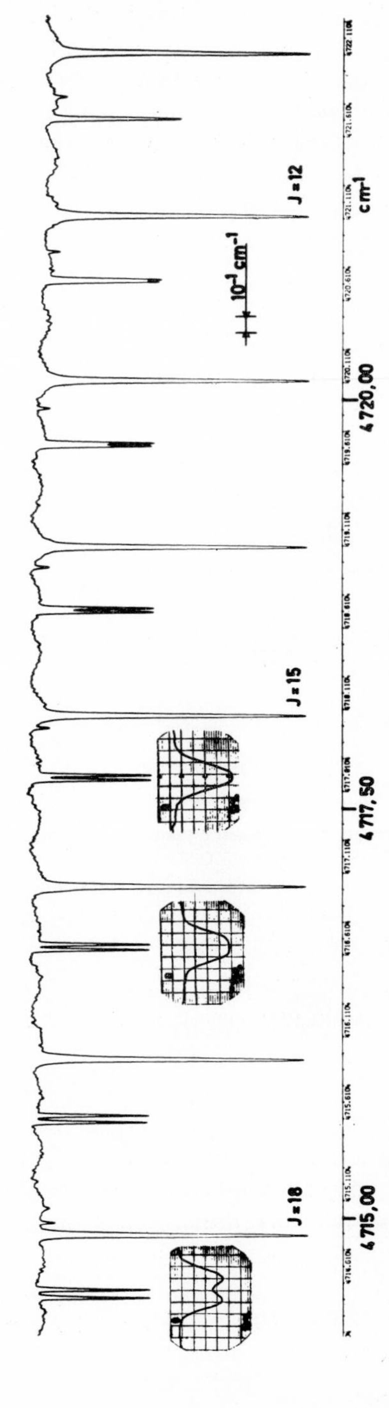

FIG. 3. A small portion of an N_2O band recorded with the high resolution interferometer (J. Pinard 1969). The inserts show some of the double lines recorded with the spectrometer (equipped with a double-passed 25-cm grating) which is currently producing the highest resolution spectra in the same range (Henry 1969).

as those of the rare earths for example, this kind of work has so far required hundreds of separate Fabry Pérot recordings.*

In the near IR, visible or UV ranges where photo-emissive receivers and photographic plates can be used the multiplexing ability would not give any signal to noise ratio gain; however we can nevertheless expect Fourier spectroscopy to produce useful results since these other advantages will be retained.

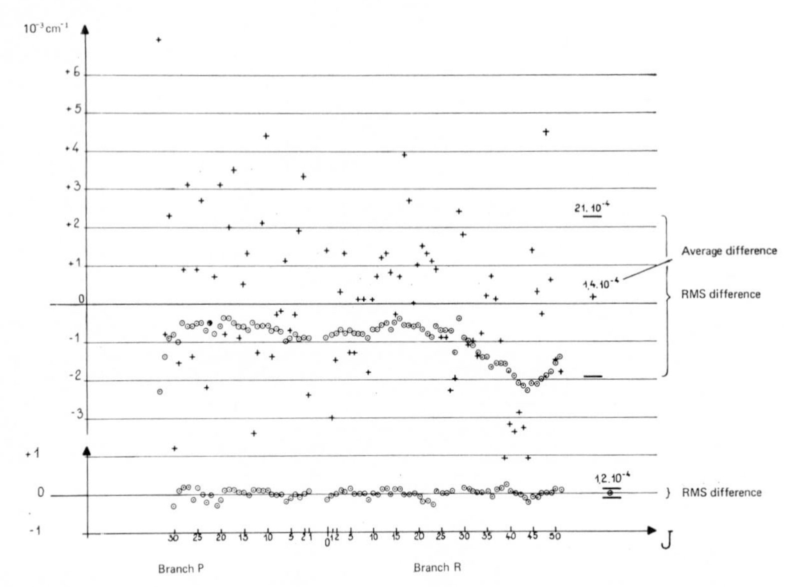

FIG. 4. Scatter of computed minus measured line positions for the same N_2O band in three different cases (Pinard 1969).
(1) *Upper part* (+): measurements made with a grating scanning spectrometer and computed positions from known band constants (Rao, Humphreys & Rank 1966).
(2) *Upper part* (⊙): same computed positions; measurements with the high resolution Fourier interferometer.
(3) *Lower part* (⊙): same measurements by Fourier spectroscopy; computed positions from improved band constants deduced from the experimental data.

The displacement of the carriage is 1 m, in both the second and third interferometers. However, it should be pointed out that the techniques involved (use of cat's eye retro-reflectors insensitive to tilt; oil-film bearings; two-phase linear motor) could be extrapolated to any reasonable displacement – say 20 m – without difficulty, if the need arose. Thus the system represents – for non-laser sources anyhow – a final step in the quest for

* *Note added in proof*: This interferometer has now been completed and satisfactory Fourier spectra computed from 10^6 sample interferograms. An account is being published in *Revue d'Optique* (1970).

ever increasing resolving power. This was only theoretically true of the Fabry-Pérot interferometer, since increasing the plate separation above a few centimetres reduced the free spectral range to uselessly small values.

REFERENCES

CONNES, J. and CONNES, P., 1966, *J. Opt. Soc. Am.*, **56,** 896.
CONNES, J., CONNES, P., BENEDICT, W. S. and KAPLAN, L. D., 1967, *Astrophys. Jour.*, **147**, 1230.
CONNES, J., CONNES, P., BENEDICT, W. S. and KAPLAN, L. D., 1968, *Astrophys. Jour.*, **152,** 731.
CONNES, J., CONNES, P. and MAILLARD, J. P., 1967, *Jour. Phys.*, **28,** 120.
CONNES, J., CONNES, P. and MAILLARD, J. P., 1969, *Atlas des Spectres dans le proche infrarouge de Venus, Mars, Jupiter et Saturne*, Editions du CNRS, 15 quai Anatole, France, Paris.
CONNES P., CONNES, J., BOUIGUE, R., QUERCY, M. and CHAUVILLE, J., 1968, *Ann. Astrophys*, **31,** 485.
HENRY, L., 1969, private communication.
KAPLAN L. D., CONNES, J. and CONNES, P. 1969, *Astrophys. Jour.*, **157,** L. 188.
KUIPER, G. P., 1962, *Comm. Lun. and Plan. Laboratory*, University of Arizona, **1,** 83.
MERTZ. L., 1958, *Jour. Phys.*, **19,** 233.
PINARD. J., 1967, *Jour. Phys.*, **28,** Ch. 2, 136.
PINARD, J., 1969, Thèse, Annales de Physique.
RAO, K. NARAHAKI, HUMPHREYS, C. J. and RANK, D. H., 1966, *Wavelength Standards in the Infrared*, Academic Press,

Photon Correlations as a Technique for Studying Spectral Intensity Profiles*

L. Mandel

Department of Physics and Astronomy,
The University of Rochester,
Rochester, New York 14627,
U.S.A.

Abstract—The principle of the correlation technique rests on the fact that, when a light beam falls on a photodetector, the probability distribution of the photoelectric emission times is closely related to certain correlation functions of the optical field. In practice, the distribution of time intervals between two photoelectric emissions is measured with an electronic time-to-amplitude converter and the data are stored in a pulse height analyser, which displays the form of the correlation directly. The apparatus and procedure used for two- and three-channel correlation measurements are described. Some recent experimental results obtained with this technique, showing the transition of a laser through its threshold of oscillation, are presented.

INTRODUCTION

The subject of optical correlation measurements has its origin in some ingenious and now classic experiments of Brown and Twiss (1956; Twiss, Little & Brown 1957) and Little (1956, 1957), who showed that there exist correlations in the photoelectric detection of photons in two mutually coherent light beams. The effect was the subject of some controversy at first, but is now recognised to be a consequence of the non-random emission of photons by the light source (Dicke 1954), which leads to non-random emission of photoelectrons at the detector (Purcell 1956; Hanbury Brown & Twiss 1957; Mandel 1958).

Now any departure from randomness is a potential source of information, and it was soon realized that a study of the (non-random) photoelectric fluctuations should yield information about the spatial coherence pf the radiation field (Hanbury Brown & Twiss 1957), the state of oolarization (Wolf 1960), the temporal coherence or spectral distribution

* This work was supported by the Air Force Office of Scientific Research, Office of Aerospace Research.

(Mandel 1959, 1963*a*, *b*; Goldberger, Lewis & Watson 1963; Gamo 1963; Wolf 1965) and the higher order correlations and spectral densities (Glauber 1963; Mandel 1964; Mehta & Mandel 1967).

In the following we consider the time dependence of the photoelectric correlations, which determine the spectral densities of the light. We discuss the technique of using time-to-amplitude converters in order to measure correlation times, and describe some recent work on the temporal coherence or spectral properties of laser beams, particularly in the neighbourhood of the threshold of oscillation.

THEORY OF CORRELATION MEASUREMENTS

It is not difficult to show from general arguments that, when a light beam falls on a photodetector, the photoelectrons will not usually be emitted at random. While the proof is perhaps most natural in terms of the quantum theory of electromagnetic fields, it is possible to obtain the main conclusions from a simple, semiclassical treatment of the photodetection problem (Mandel, Sudarshan & Wolf 1964; see also Mandel & Wolf 1966).

Let us suppose that the optical field is polarized and represented by a complex analytical signal (Born & Wolf 1965) $\mathbf{V}(\mathbf{x}, t)$ at the space-time point $(\mathbf{x}, t)$. Let the corresponding instantaneous light intensity be $I(\mathbf{x}, t) \equiv \mathbf{V}^*(\mathbf{x}, t).\mathbf{V}(\mathbf{x}, t)$. In practice all radiation fields fluctuate; they therefore have to be described by random functions of space and time, or by an ensemble of possible realizations. For the moment, however, we shall focus attention on one particular realization $\mathbf{V}(\mathbf{x}, t)$. It has been shown (Mandel *et al.* 1964) that for polarized, quasi-monochromatic fields, the probability of photoemission at time t within some time interval Δt which is many optical periods long, at a detector located at $\mathbf{x}$, is proportional to $I(\mathbf{x}, t)\Delta t$. The constant of proportionality involves both the atomic structure and the geometry of the detector. However, for a light beam which is effectively plane over the surface ΔS of the detector and is normally incident on it, we may write $\alpha cSI(\mathbf{x}, t)$ for the photoemission probability density per unit time $P_1(\mathbf{x}, t)$. If we express the light intensity in units of photons per unit volume, then α is a dimensionless parameter describing the quantum efficiency of the detector.

Let us now consider two photoemission events at $\mathbf{x}_1$, t_1 and $\mathbf{x}_2$, t_2, registered by two photodetectors, or by the same photodetector at different times. If we make the reasonable assumption that one photoemission does not influence the other, then the joint probability density $P_2(\mathbf{x}_1, t_1; \mathbf{x}_2, t_2)$ for the two photoemissions will be the product $\alpha_1 cS_1 I(\mathbf{x}_1, t_1)\alpha_2 cS_2 I(\mathbf{x}_2, t_2)$, and a corresponding expression can be written down for any other number of photoemissions. All these conclusions hold for one particular realization of the radiation field. However, since the field is not generally describable by a well-defined function of space and time, but only by an ensemble of functions, corresponding to an ensemble of realizations, we have to average over the ensemble in order to arrive at meaningful photoemission probability densities. Thus, in place of the foregoing we now have:

$$P_1(\mathbf{x}, t) = \alpha c \Delta S \langle I(\mathbf{x}, t) \rangle, \tag{1}$$

$$P_2(\mathbf{x}_1, t_1; \mathbf{x}_2, t_2) = \alpha_1 c \Delta S_1 \alpha_2 c \Delta S_2 \langle I(\mathbf{x}_1, t_1) I(\mathbf{x}_2, t_2) \rangle, \tag{2}$$

and similarly for any higher order probability density. It is interesting to note that the probabilities for two or more photoemissions now involve correlation functions of the light intensities, by virtue of the fluctuations of the radiation field.

Equations (1) and (2) have an immediate consequence. Since, in general,

$$\langle I(\mathbf{x}_1, t_1) I(\mathbf{x}_2, t_2) \rangle \neq \langle I(\mathbf{x}_1, t_1) \rangle \langle I(\mathbf{x}_2, t_2) \rangle,$$

it follows at once that

$$P_2(\mathbf{x}_1, t_1; \mathbf{x}_2, t_2) \neq P_1(\mathbf{x}_1, t_1) P_1(\mathbf{x}_2, t_2),$$

which means that the two photoemissions are not statistically independent. Because of the fluctuations of the radiation field, we see that information about one photoemission influences our assessment of the probability for the other. These conclusions hold also for successive photoemissions at the same detector.

For a stationary radiation field, whose statistical properties do not change with time, the expectation value $\langle I(\mathbf{x}, t) \rangle$ is independent of t and the correlation function in Equation (2) is a function only of the difference $t_2 - t_1 \equiv \tau$. Moreover, for all the problems we shall be discussing here, the points $\mathbf{x}_1$ and $\mathbf{x}_2$ may be taken to be effectively coincident. This is so either because only one photodetector is actually used, or because the light beam is split into two mutually coherent beams by a partly silvered mirror, and the light intensity functions $I(\mathbf{x}_1, t)$ and $I(\mathbf{x}_2, t)$ at the two photodetectors are then similar functions of time. In either case the correlation functions which appear in Equation (2) are essentially auto-correlations. On introducing the intensity deviation

$$\Delta I(\mathbf{x}, t) \equiv I(\mathbf{x}, t) - \langle I(\mathbf{x}) \rangle, \tag{3}$$

we can re-express Equation (2) in the form

$$\begin{aligned} P_2(\mathbf{x}_1, t_1; \mathbf{x}_2, t_1 + \tau) &= \alpha_1 c \Delta S_1 \alpha_2 c \Delta S_2 \langle I(\mathbf{x}_1) \rangle \langle I(\mathbf{x}_2) \rangle [1 + \lambda(\tau)] \\ &= P_1(\mathbf{x}_1, t_1) P_1(\mathbf{x}_2, t_1 + \tau) [1 + \lambda(\tau)], \end{aligned} \tag{4}$$

where $\lambda(\tau)$ is a normalized intensity correlation function defined by $\lambda(\tau) \equiv \langle \Delta I(\mathbf{x}, t) \Delta I(\mathbf{x}, t + \tau) \rangle / \langle I(\mathbf{x}) \rangle^2$. This shows that $P_2(\mathbf{x}_1, t_1; \mathbf{x}_2, t_2)$

can be regarded as a sum of two contributions, the second of which involves the correlation of the intensity fluctuations of the light. For sufficiently large τ we expect $\lambda(\tau)$ to vanish for any realizable optical field, and under those conditions $P_2(\mathbf{x}_1, t; \mathbf{x}_2, t+\tau) = P_1(\mathbf{x}_1, t)P_1(\mathbf{x}_2, t+\tau)$, which implies that the photoemissions are independent. However, for smaller τ for which $\lambda(\tau)$ is non-zero, the appearance of the first photoemission affects the observation probability of the second.

The departure of $P_2(\mathbf{x}_1, t; \mathbf{x}_2, t+\tau)$ from $P_1(\mathbf{x}_1, t)P_1(\mathbf{x}_2, t+\tau)$ due to non-zero $\lambda(\tau)$ has sometimes been attributed to the natural tendency of bosons – in this case photons – to form clusters or bunches. While this point of view is valid for thermal sources of light, in which case the distribution of photon numbers is of the Bose-Einstein form (Mandel 1963*a*) and $P_2(\mathbf{x}_1, t; \mathbf{x}_2, t+\tau) \geqslant P_1(\mathbf{x}_1, t)P_1(\mathbf{x}_2, t+\tau)$, it is not valid in general, for there is no fundamental reason why $\lambda(\tau)$ should always be positive. We shall shortly encounter an example of negative $\lambda(\tau)$.

The expression for the joint photoemission probability density may readily be generalized to three successive photodetections. We then have the three-fold photoemission probability density:

$$P_3(\mathbf{x}_1, t_1; \mathbf{x}_2, t_2; \mathbf{x}_3, t_3) = \alpha_1 c\Delta S_1 \alpha_2 c\Delta S_2 \alpha_3 c\Delta S_3 \langle I(\mathbf{x}_1, t_1) I(\mathbf{x}_2, t_2) I(\mathbf{x}_3, t_3)\rangle, \quad (5)$$

and if $t_2 - t_1 \equiv \tau_1$, $t_3 - t_2 \equiv \tau_2$, and if we introduce the intensity deviations $\Delta I(\mathbf{x}, t)$ given by Equation (3) as before, we may write

$$P_3(\mathbf{x}_1, t_1; \mathbf{x}_2, t_1+\tau_1; \mathbf{x}_3, t_1+\tau_1+\tau_2) = P_1(\mathbf{x}_1, t_1)P_1(\mathbf{x}_2, t_1+\tau_1)P_1(\mathbf{x}_3, t_1+\tau_1+\tau_2)$$
$$\times [1 + \lambda(\tau_1) + \lambda(\tau_2) + \lambda(\tau_1+\tau_2) + \mu(\tau_1, \tau_2)]. \quad (6)$$

Here $\mu(\tau_1, \tau_2)$ is a third order intensity correlation function defined by $\mu(\tau_1, \tau_2) \equiv \langle \Delta I(\mathbf{x}, t)\Delta I(\mathbf{x}, t+\tau_1)\Delta I(\mathbf{x}, t+\tau_1+\tau_2)\rangle / \langle I(\mathbf{x})\rangle^3$, which depends on two time parameters.

The intensity correlation function $\lambda(\tau)$ which appears in Equation (4), may be associated with a spectral density $\psi(v)$ in the usual way, viz:

$$\lambda(\tau) = \int_{-\infty}^{\infty} \psi(v)\exp(-2\pi i v\tau)\mathrm{d}v. \quad (7)$$

However, this spectral density of the intensity fluctuations must be clearly distinguished from the more usual spectral density $\phi(v)$ associated with the fluctuations of the complex field amplitude, which would be obtained from measurements with an optical spectrometer. If $\gamma(\tau)$ is the normalized amplitude auto-correlation function of a polarized field, defined by $\gamma(\tau) = \langle \mathbf{V}^*(\mathbf{x}, t)\cdot \mathbf{V}(\mathbf{x}, t+\tau)\rangle / \langle I(\mathbf{x})\rangle$ (Born & Wolf 1965), then $\phi(v)$ is the Fourier transform of $\gamma(\tau)$,

$$\gamma(\tau) = \int_{0}^{\infty} \phi(v)\exp(-2\pi i v\tau)\mathrm{d}v. \quad (8)$$

Similarly, the higher order correlation $\mu(\tau_1, \tau_2)$ may be associated with a higher order spectral density via a multiple Fourier transform (Mehta & Mandel 1967; Mandel & Mehta 1969). The higher order spectral densities contain information on the coupling between different Fourier components of the optical field, but we shall not deal with them here. In general there is no obvious relationship among $\gamma(\tau)$, $\lambda(\tau)$ and $\mu(\tau_1, \tau_2)$, or among the various spectral densities, which refer to different fluctuating processes. While the phase fluctuations of the optical field usually determine $\phi(v)$ and $\gamma(\tau)$, they do not influence $\psi(v)$, $\lambda(\tau)$ and $\mu(\tau_1, \tau_2)$ at all.

However, in the special case of a light beam produced by a thermal type source, for which the complex wave amplitude $\mathbf{V}(\mathbf{x}, t)$ behaves as a Gaussian random process, the different correlations are simply related. It may be shown with the help of the Gaussian moment theorem (see, for example, Mehta 1965) that for polarized, thermal light

$$\lambda(\tau) = |\gamma(\tau)|^2, \tag{9}$$

which implies that

$$\psi(v) = \int_0^\infty \phi(v')\phi(v'+v)\mathrm{d}v' \tag{10}$$

while

$$\mu(\tau_1, \tau_2) = \gamma(\tau_1)\gamma(\tau_2)\gamma^*(\tau_1+\tau_2) + \text{c.c.} \tag{11}$$

Photoelectric correlation measurements can therefore be used to study the spectral profile $\phi(v)$ of thermal light, as was pointed out some time ago (Mandel 1963*b*; Goldberger *et al.* 1963; Wolf 1965), and several such measurements have been reported (Morgan & Mandel 1966; Scarl 1966; Phillips, Kleiman & Davis 1967). There is some ambiguity in the reconstruction of $\phi(v)$ from $\psi(v)$ via Equation (10), which is connected with the fact that $\lambda(\tau)$ does not determine the phase of $\gamma(\tau)$, but some procedures for overcoming this problem have been suggested (Gamo 1963; Mehta 1968).

For laser beams, with which most of our recent work has dealt, no simple relations connecting $\gamma(\tau)$ and $\lambda(\tau)$ hold in general. However, for the special case of a laser which is oscillating well above threshold in two independent modes of equal intensities and spectral widths Δv_1 and Δv_2 centred on frequencies v_1 and v_2, it may be shown (see, for example, Mandel 1967) that $\lambda(\tau)$ is of the form

$$\lambda(\tau) = \exp[-\tfrac{1}{2}(\Delta v_1 + \Delta v_2)|\tau|]\cos[2\pi(v_2 - v_1)\tau] \tag{12}$$

Thus, the spectral density $\phi(v)$ of the light may again be determined from photoelectric correlation measurement via Equation (4), and this has been confirmed experimentally (Arecchi, Gatti & Sona 1966).

USE OF TIME-TO-AMPLITUDE CONVERTERS

Time-resolved photoelectric correlation measurements are most easily carried out with the help of a combination of a time-to-amplitude converter and pulse height analyser. The light beam to be studied is split into two or more parts by partly silvered mirrors, and the beams fall on separate phototubes. In the case of a two-channel experiment, the photoelectric pulses from the two detectors are fed to the *start* and *stop* inputs of the time-to-amplitude converter (TAC). This is an electronic device in which the *start* pulse initiates a ramp wave form which is cut off when the *stop* pulse is received. The output from the TAC is therefore a pulse whose amplitude is proportional to the time interval between the two input pulses, and the various output pulses are sorted into appropriate channels by a multi-channel pulse height analyser. The distribution of pulse heights registered by the analyser is therefore a measure of the distribution of time intervals between photoelectric pulses.

Nevertheless, this probability distribution cannot be immediately identified with the expression for $P_2(\mathbf{x}_1, t; \mathbf{x}_2, t+\tau)$ given by Equation (2). In order that an event corresponding to the time interval t to $t+\tau$ shall be registered by the TAC, it is necessary not only that a *start* pulse is received at time t and a *stop* pulse at time $t+\tau$, but also that no other *stop* pulse appears in between. The probability density for this event is given not by Equation (2), but by the more complicated expression

$$\tilde{P}_2(\mathbf{x}_1, t; \mathbf{x}_2, t+\tau) = \alpha_1 c\Delta S_1 \alpha_2 c\Delta S_2 < I(\mathbf{x}_1, t) I(\mathbf{x}_2, t+\tau) \exp[-\alpha_2 c\Delta S_2 \int_t^{t+\tau} I(\mathbf{x}_2, t')\mathrm{d}t'] >. \quad (13)$$

Evidently, if the mean number of photoelectric counts $\alpha_2 c\Delta S_2 < I(\mathbf{x}_2) > \tau$ registered in the *stop* channel during the interval t to $t+\tau$ is much less than unity, then the effect of the exponential factor in Equation (13) is small, and $\tilde{P}_2(\mathbf{x}_1, t; \mathbf{x}_2, t+\tau)$ does not differ appreciably from $P_2(\mathbf{x}_1, t; \mathbf{x}_2, t+\tau)$. However, in general, corrections have to be applied to $\tilde{P}_2(\mathbf{x}_1, t; \mathbf{x}_2, t+\tau)$ if the correlation function $\lambda(\tau)$ is to be determined from measurements with a TAC analyser combination. Even in the absence of intensity correlations, when $\lambda(\tau)=0$, $\tilde{P}_2(\mathbf{x}_1, t; \mathbf{x}_2, t+\tau)$ will have some non-zero value, which we denote by $\tilde{P}_2^{(R)}(\mathbf{x}_1, t; \mathbf{x}_2, t+\tau)$, due to random photoemissions. Evidently

$$\tilde{P}_2^{(R)}(\mathbf{x}_1, t; \mathbf{x}_2, t+\tau) = \alpha_1 c\Delta S_1 \alpha_2 c\Delta S_2 < I(\mathbf{x}_1) > < I(\mathbf{x}_2) > \exp[-\alpha_2 c\Delta S_2 < I(\mathbf{x}_2) > \tau]. \quad (14)$$

If the 'excess' of $\tilde{P}_2(\mathbf{x}_1, t; \mathbf{x}_2, t+\tau)$ over $\tilde{P}_2^{(R)}(\mathbf{x}_1, t; \mathbf{x}_2, t+\tau)$ divided by $\tilde{P}_2^{(R)}(\mathbf{x}_1, t; \mathbf{x}_2, t+\tau)$ is denoted by $\mathscr{E}(\tau)$, and if the counting rates are sufficiently low that only the first two terms in the power series expansion of the exponential need to be retained, we find from Equations (13) and (14):

$$\mathscr{E}(\tau) = \lambda(\tau) - 2\alpha_2 c\Delta S_2 < I(\mathbf{x}_2) > \int_0^\tau \lambda(t')\mathrm{d}t'. \tag{15}$$

This is an integral equation for $\lambda(\tau)$ in terms of the observed $\mathscr{E}(\tau)$ which can be solved by iteration. In practice, the situation is usually more complicated and other dead-time effects, as well as the contribution of photoelectric dark currents, have to be considered also. The problem has been treated in some detail recently (Davidson & Mandel 1968*a*; Davidson 1969) and it was found that $\lambda(\tau)$ is still derivable from $\mathscr{E}(\tau)$ via an integral equation, although the equation is more complicated than Equation (15). A corresponding procedure is also required in the analysis of three-channel correlation experiments.

EXPERIMENTAL PROCEDURE AND RESULTS

The light beam under study was produced by a single mode He:Ne laser, whose optical cavity could be controllably detuned by a feedback arrange-

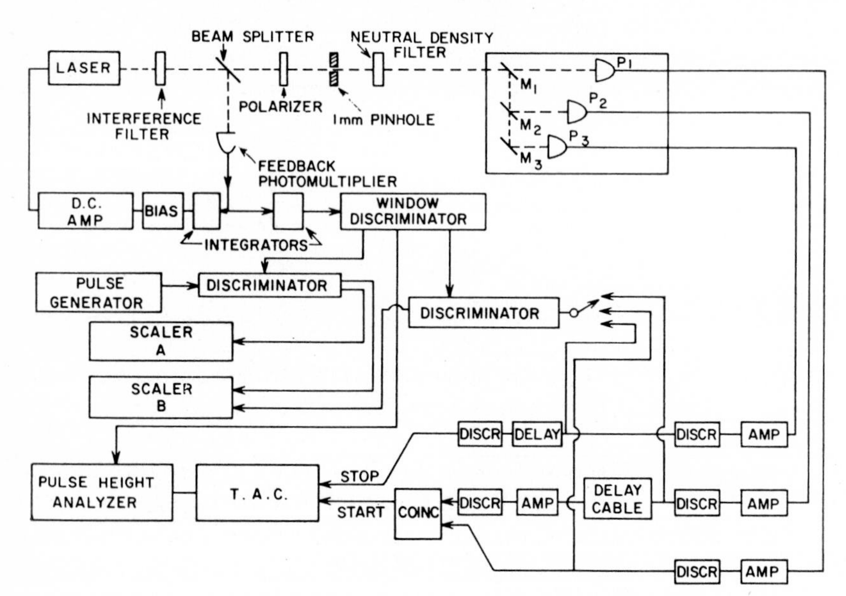

FIG. 1. Outline of the apparatus used for two- and three-channel correlation measurements of a laser beam near threshold.

ment (see Figure 1). By this means the laser could be operated in a reasonably stable manner anywhere in the neighbourhood of the threshold of oscillation. As a precaution against instability, a 'window discriminator'

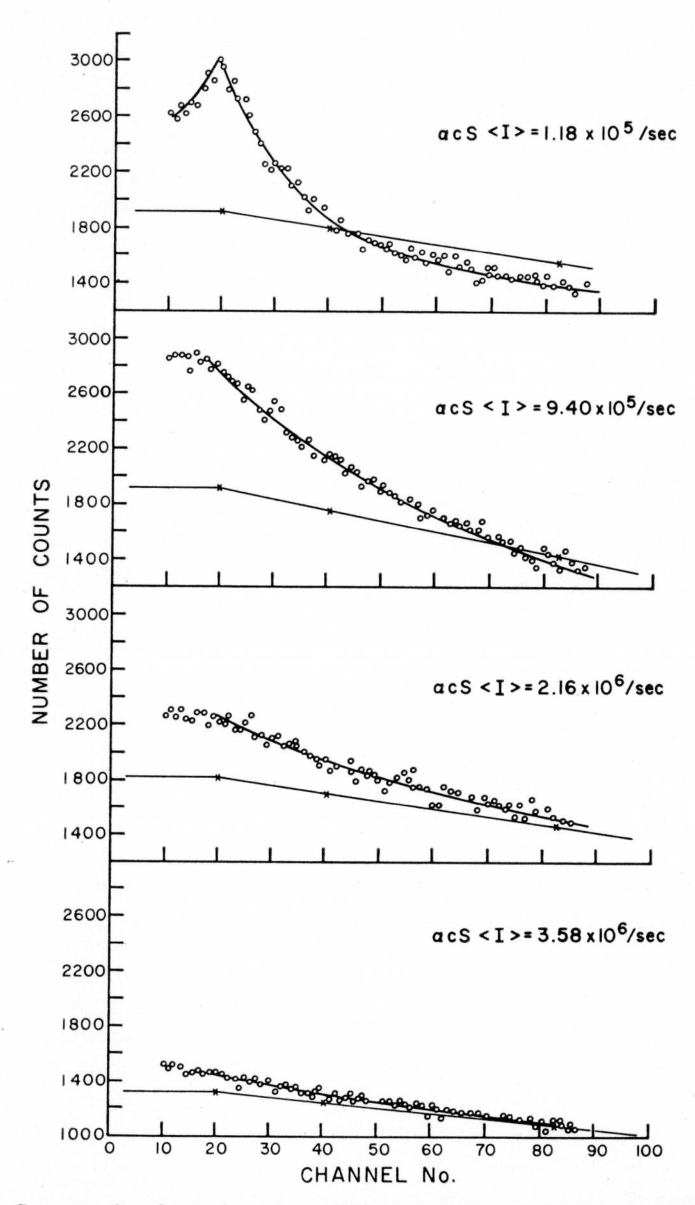

FIG. 2. Some results of two-channel correlation measurements for various settings of the laser. The numbers of counts are those recorded in the pulse height analyser. The channel number is proportional to the time interval between pulses, with zero interval corresponding to channel no. 20.

was used to monitor the light intensity continuously, and the counting circuits were gated open only when the light intensity fell within a narrow predetermined range.

The laser beam passed through an interference filter, which excluded extraneous light from the gas discharge, and was then split into three beams of approximately equal intensities, which fell on three photomultiplier tubes, as shown. The tubes were so arranged that the path differences to the three photocathodes were very nearly equal. The photoelectric pulses from each of the phototubes were fed to amplifiers and

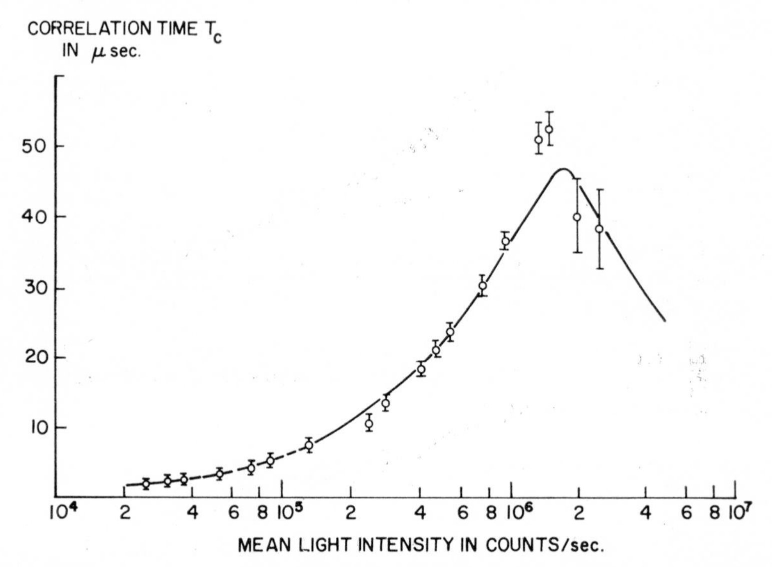

FIG. 3. The measured two-channel correlation times T_c for various settings of the laser. The full curve is based on the rotating-wave van der Pol oscillator model of the laser (Lax & Louisell 1967).

discriminators. In the two-channel experiments, the pulses in two channels, with a delay in one channel, provided the *start* and *stop* inputs for a time-to-amplitude converter. In the three-channel experiments, two of the pulses, with a delay in one channel, were fed to a high resolution coincidence circuit, whose output provided the *start* input to the TAC, while the third channel provided the *stop* input (see Figure 1). In all cases, the data were accumulated in a 100-channel pulse height analyser.

Figure 2 shows the results of some typical correlation measurements in a two-channel experiment, for various settings of the laser below and above threshold. Also shown as full lines are the contributions due to random photoemissions alone. Each set of data determines a function $\mathscr{E}(\tau)$, and $\lambda(\tau)$ can then be derived by solution of the integral equation.

The functions $\lambda(\tau)$ were found to be nearly exponential in all cases, of the form

$$\lambda(\tau) = \exp(-|\tau|/T_c), \tag{16}$$

so that the spectral density associated with the intensity fluctuations is nearly Lorentzian. The corresponding correlation times T_c are shown as a

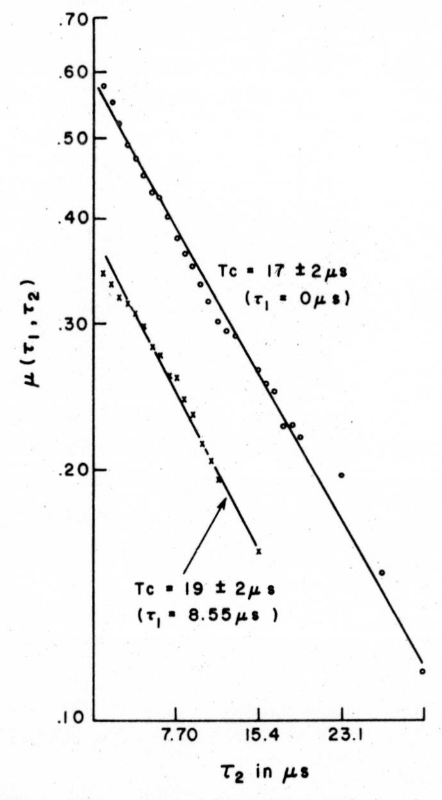

FIG. 4. The dependence of the measured intensity correlation function $\mu(\tau_1, \tau_2)$ on τ_2 for two fixed values of τ_1.

function of light intensity in Figure 3 (see Davidson & Mandel 1967), in which the full curve corresponds to the theoretical prediction based on the rotating-wave van der Pol oscillator model of a laser (Lax & Louisell 1967). It will be seen that the correlation time passes through a maximum at oscillation threshold, in a manner that is usually characteristic of a phase transition. This suggests that the condensation of the radiation field (almost) to a coherent state above threshold, may be regarded as a phase transition. The correlation times are essentially the reciprocals of the fluctuation bandwidths which were measured in the experiments of Freed and Haus (1966). It is worth noting that these spectral widths, corresponding to intensity fluctuations, are at least an order of magnitude narrower than the spectral widths associated with phase fluctuations of the optical field.

Some typical results for $\mu(\tau_1, \tau_2)$ obtained from the three-channel correlation experiments are shown in Figure 4. It will be seen that the τ_2-dependence of $\mu(\tau_1, \tau_2)$ is nearly exponential, and, moreover, that the time constant for the τ_2-decay is nearly the same for two very different

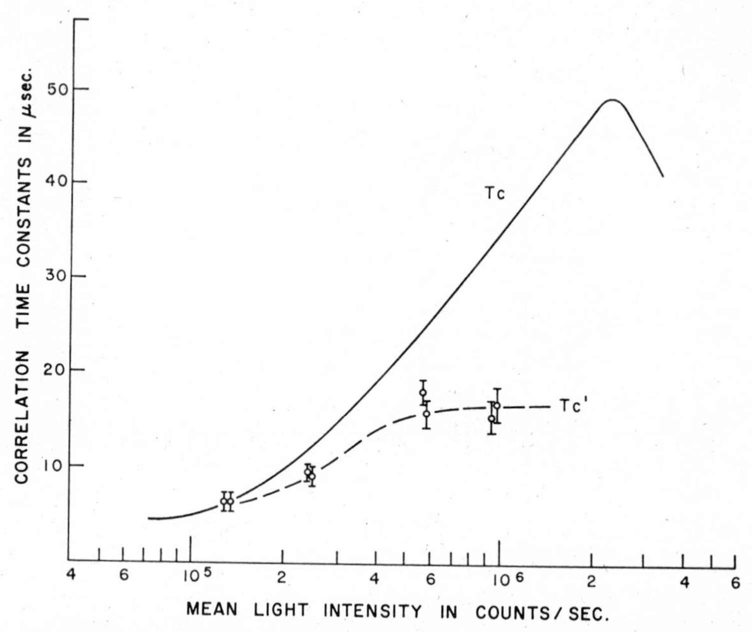

FIG. 5. The measured three-channel correlation times T_c' for various settings of the laser. The full curve shows the theoretically predicted two-channel correlation times T_c, for comparison.

values of τ_1. Indeed from a number of such results it appears that $\mu(\tau_1, \tau_2)$ is of the form

$$\mu(\tau_1, \tau_2) = 2\exp[-(|\tau_1| + |\tau_2|)/T_c'], \tag{17}$$

with similar time constants for the τ_1- and τ_2-decay. However, the values of the correlation times T_c' are different in general from the T_c in Equation (16). This is shown in Figure 5 where the experimental values of T_c' are plotted as a function of light intensity (see also Davidson & Mandel 1968*b*; Davidson 1969). Also shown for comparison is the curve of T_c versus light intensity. It will be seen that T_c' is appreciably smaller than T_c, except far below threshold where the two coincide. This coincidence is no accident, for the radiation has thermal characteristics far below threshold, so that Equations (9) and (11) hold. If $\lambda(\tau)$ is given by Equation (16), $\gamma(\tau)$ will be of the form

$$\gamma(\tau) = \exp[-\tfrac{1}{2}|\tau|/T_c - 2\pi i v_0 \tau], \tag{18}$$

and Equation (11) then requires that

$$\mu(\tau_1, \tau_2) = 2\exp[-(|\tau_1| + |\tau_2|)/T_c], \tag{19}$$

so that $T_c' = T_c$. However, once we leave the thermal region and approach the oscillation threshold, there is no longer any simple relation between the correlation times T_c' and T_c. It is interesting to note that the spectral width associated with the higher order fluctuations of the intensity is greater than that corresponding to $\lambda(\tau)$, which appears to be a minimum.

These results show that there are spectral features of the optical field which are not measurable with any of the traditional optical methods, but are readily studied by correlation techniques.

Acknowledgment

The author gladly acknowledges that the experimental work described herein was carried out in collaboration with, and largely, by F. Davidson, and formed part of the subject of his Ph.D. thesis.

REFERENCES

ARECCHI, F. T., GATTI, E. and SONA, A., 1966, *Phys. Lett.*, **20**, 27.
BORN, M. and WOLF, E., 1965, *Principles of Optics*, Oxford, Pergamon Press, 3rd ed., Chapter 10.
BROWN, R. HANBURY and TWISS, R. Q., 1956, *Nature*, **177**, 27.
BROWN, R. HANBURY and TWISS, R. Q., 1957, *Proc. Roy. Soc.*, **A242**, 300.
DAVIDSON, F. 1969, *Phys. Rev.*, **185**, 446.
DAVIDSON, F. and MANDEL, L., 1967, *Phys. Lett.*, **25A**, 700.
DAVIDSON, F. and MANDEL, L., 1968*a*, *J. Appl. Phys.*, **39**, 62.
DAVIDSON, F. and MANDEL, L., 1968*b*, *Phys. Lett.*, **27A**, 579.
DICKE, R. H., 1954, *Phys. Rev.*, **93**, 99.
FREED, C. and HAUS, H. A., 1966, *Phys. Rev.* **141**, 287.
GAMO, H., 1963, *J. Appl. Phys.*, **34**, 875.
GLAUBER, R. J., 1963, *Phys. Rev.*, **130**, 2529.
GOLDBERGER, M. L., LEWIS, H. W. and WATSON, K. M., 1963, *Phys. Rev.*, **132**, 2764.
LAX, M. and LOUISELL, W. H., **1967**, *J. Quant. Electronics*, **QE-3**, 47.
MANDEL, L., 1958, *Proc. Phys. Soc.*, **72**, 1037.
MANDEL, L., 1959, *Proc. Phys. Soc.*, **74**, 233.
MANDEL, L., 1963*a*, *Progress in Optics*, *Vol. II*, ed. E. Wolf, Amsterdam, North-Holland Publishing Co., p. 181.
MANDEL, L., 1963*b*, *Electromagnetic Theory and Antennas*, ed. E. C. Jordan, Oxford, Pergamon Press, p. 811.
MANDEL, L., 1964, *Quantum Electronics III*, ed. N. Bloembergen and P. Grivet, Paris, Dunod et Cie; New York, Columbia University Press, p. 101.
MANDEL, L., 1967, *Modern Optics*, ed. J. Fox, Brooklyn, N.Y., Polytechnic Press of the Polytechnic Institute, p. 143.
MANDEL, L. and MEHTA, C. L., 1969, *Il Nuovo Cimento*, 149.
MANDEL, L., SUDARSHAN, E. C. G. and WOLF, E., 1964, *Proc. Phys. Soc.*, **84**, 435.

Mandel, L. and Wolf, E., 1966, *Phys. Rev.*, **149**, 1033.
Mehta, C. L., 1965, *Lectures in Theoretical Physics*, Vol. 7C, ed. W. E. Britten, Boulder, University of Colarado Press, p. 345.
Mehta, C. L., 1968, *J. Opt. Soc. Am.*, **58**, 1233.
Mehta, C. L. and Mandel, L., 1967, *Electromagnetic Wave Theory*, Oxford, Pergamon Press, p. 1069.
Morgan, B. L. and Mandel, L., 1966, *Phys. Rev. Lett.*, **16**, 1012.
Phillips, D. T., Kleiman, H. and Davis, Sumner P., 1967, *Phys. Rev.*, **153**, 113.
Purcell, E. M., 1956, *Nature*, **178**, 1449.
Scarl, D. B., 1966, *Phys. Rev. Lett.*, **17**, 663.
Twiss, R. Q., Little, A. G. and Brown, R. Hanbury, 1957, *Nature*, **180**, 324.
Wolf, E., 1960, *Proc. Phys. Soc.*, **76**, 424.
Wolf, E., 1965, *Jap. J. Appl. Phys.*, **4**, (Suppl. 1), 1.

Field-Compensated Michelson Spectrometers

J. W. Schofield and J. Ring

Imperial College, London

Abstract—The luminosity-resolution product of a Michelson interferometer used as a spectrometer may be substantially increased by compensating the field of the instrument. There are many ways of compensating the field, each having different aberrations and degrees of mechanical and optical complexity. The most advantageous method depends on the maximum resolution, the spectral region of interest, and the spectral range required. Examples of the gains obtainable are given for two spectrometers under construction in the authors' laboratory.

The Resolution-Luminosity Products of Various Spectrometers

Any system which resolves the various wavelengths of electromagnetic radiation incident upon it can be said to have a resolution-luminosity product (RL), where

$$\text{Resolution, } R = \lambda/\delta\lambda, \text{ and}$$

$$\text{Luminosity, } L = A\omega,$$

where λ is the mean wavelength of two wavelengths a distance $\delta\lambda$ apart which the instrument is just capable of separating, and A is the area of some pupil or stop in the system which receives radiation from a solid angle ω.

Jaquinot (1954) has shown that the resolution-luminosity product for a given spectrometer of the grating, prism or classical interferometric type is constant with changing resolution, and that interferometers have a larger RL-product, for the same aperture, than prism or grating spectrometers by a factor of about 100.

The RL-products of field-compensated Michelson spectrometers and spherical Fabry-Pérot spectrometers are not constant, and for certain values of resolution may be much greater than their classical counterparts (Figure 1). In the case of field-compensated Michelson spectrometers, this advantage is gained at the cost of greater instrumental complexity than that obtaining in a classical instrument.

Field Compensation

In a classical Michelson interferometer which is adjusted for zero path difference, the image of one reflector in the beam-splitter is superimposed on the other reflector, and vice versa. Since the beam-splitter is a plane mirror, and therefore forms perfect optical images, application of Fermat's

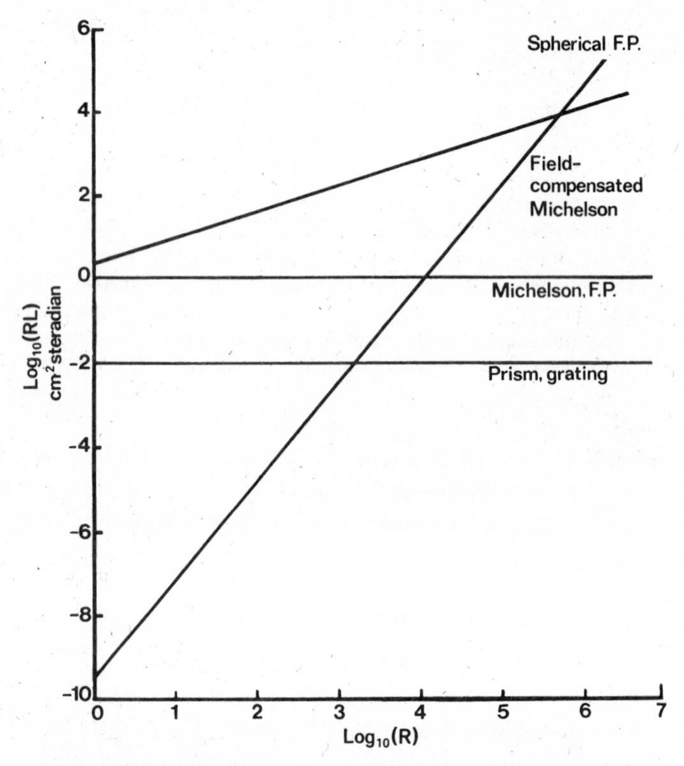

FIG. 1. Resolution-luminosity products of spectrometers as a function of resolution for a common aperture of $(2\pi)^{-1}$ cm^2.

Principle shows that all optical paths joining the two reflectors are equal, the path difference is zero for all rays, and the field is infinite (Figure 2*a*).

If one mirror is displaced from the image of the other to form an interferogram, the condition of mutual imaging no longer applies, and the path difference is no longer similar for all rays passing through the system (Figure 2*b*), but is given by

$$\Delta = \Delta_0 \cos i \tag{1}$$

Where Δ_0 is the path difference for the axial ray, and Δ is the path difference for a ray making an angle i with the axial ray. This dependence of the path difference on i limits the *RL*-product (Jaquinot 1954) to 2π steradians, for an instrument of unit aperture.

Field compensation consists of the introduction, into the interfering beams, of optical systems which are arranged to maintain the mutual imaging of the reflectors for all values of the path difference employed to obtain interferograms. The field is limited by the aberrations of the optical systems (Figure 2*c*).

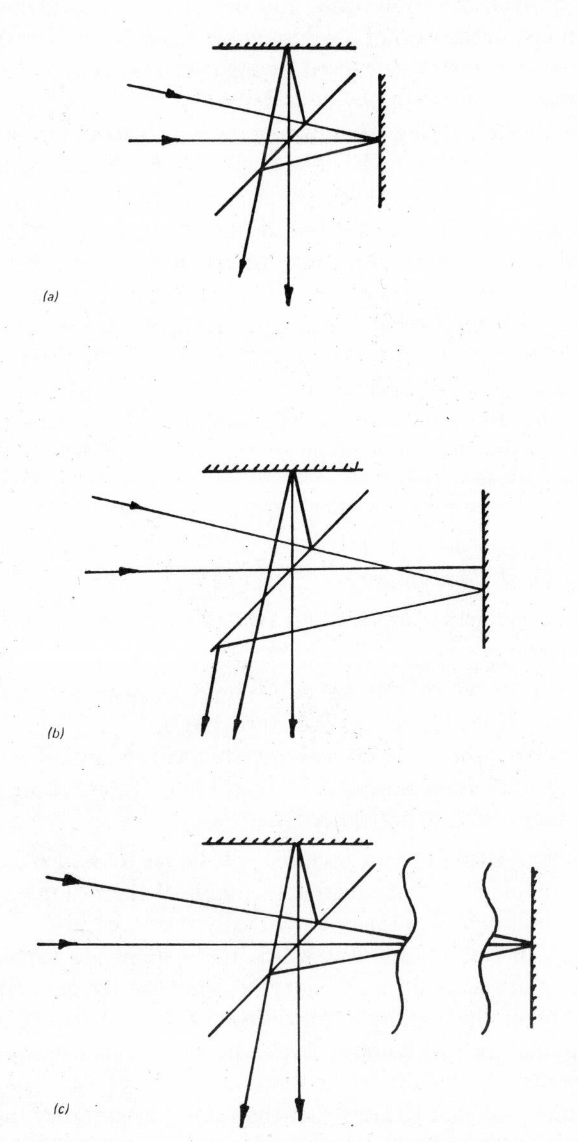

FIG. 2. Field compensation: (*a*) classical Michelson at zero path difference; (*b*) classical Michelson with path difference Δ; (*c*) field-compensated Michelson.

Advantages and Usefulness of Field-Compensated Spectrometers

The photometric accuracy of spectra obtained from spectrometers in a given time depends on the signal-to-noise ratio obtainable at the detector. Increase of the luminosity of a spectrometer at a given resolution may improve the signal-to-noise ratio by making a larger signal available at the detector, both in the shot-noise- and detector-noise-limited cases. There may, however, in the case of a field-compensated Michelson spectrometer, be an increase in the noise caused by instrumental defects if the design of the instrument is not carefully considered.

A spectrometer of high luminosity is only useful for increasing the signal-to-noise ratio if it can be filled with light. For example, in astronomy, stellar spectrophotometry with the largest telescopes available can be efficiently accomplished by a classical interferometer of small aperture at all attainable resolutions. However, for extended objects, such as gaseous nebulae, the luminosity required to obtain a spectrum utilizing all the light gathered by a 1 m telescope at a resolution of $\sim 2 \cdot 10^4$ would require a classical interferometer of 1 m aperture, whereas a field-compensated interferometer of 0·1 m aperture is sufficient.

In the laboratory the use of field-compensated instruments may be advantageous for the study of large objects of low brightness, i.e. very low pressure gases.

Systems of Field Compensation

The systems of field compensation available may be conveniently divided into two types:

(i) Systems in which the optical system consists of curved surfaces (Connes 1956; Cuisienier & Pinard 1967).

(ii) Systems in which the optical system consists only of plane surfaces (Mertz 1965, Bouchairene & Connes 1963; Hillard & Shepherd 1966; Shepherd 1969; Schofield & Ring 1969).

Some of the systems in category (ii) are liable to astigmatism, and all systems are liable in varying degrees to spherical aberration and chromatic aberration, although these may be partially corrected in some systems.

The range of resolutions over which the systems are capable of being operated is more limited for category (ii) than for category (i) because of the limitation of the available thicknesses of solid dielectric materials of sufficient quality. Mme. Duboin (1969), however, is constructing a system utilizing water as the dielectric material. Systems of category (i) are generally more difficult to construct than those of category (ii).

Below we briefly discuss two systems from category (ii), examples of both of which are under construction in the authors' laboratory.

The Mark 1 *Spectrometer*

This instrument is designed to obtain spectra of extended astronomical sources in the wavelength region 1 to 2·5μm at resolutions up to 10^4.

The compensation system employed is that due to Mertz (1965). The thickness of a plane parallel piece of glass in one of the interfering beams is increased as the reflector in the same beam is moved away from the beam-splitter to form the interferogram. The optical layout is illustrated in Figure 3. Calculations show that the tolerances, both on synchronism

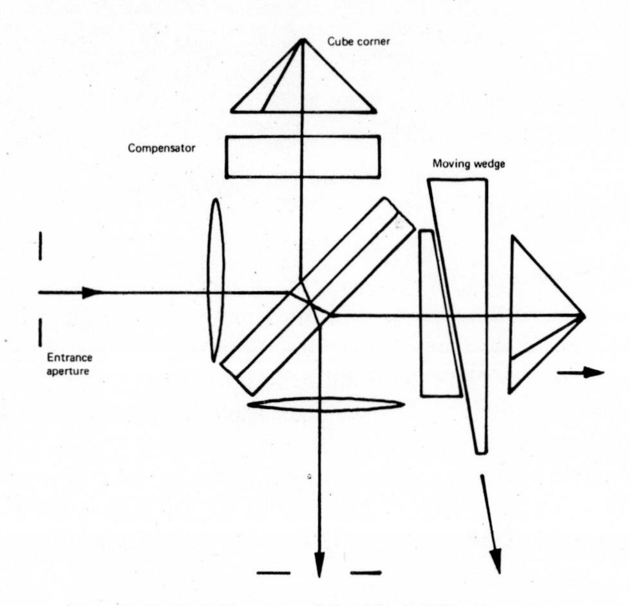

FIG. 3. Optical diagram of the Mark I Spectrometer.

between the prism and the reflector motions and the prism alignment are easily achieved, provided that the overall path difference in the system is measured and/or controlled by reference fringes.

Typically the glass thickness can vary by 25μ and the glass angle by 0′·5 arc from the calculated values with negligible effect on the instrumental profile. Figure 4 illustrates the instrument as of July 1969, and shows the two synchronized kinematic slides with prism and corner cube, together with the fixed plane parallel piece of glass in the other beam, which is necessary to obtain the correct conditions at zero path difference.

The instrument, which has an aperture of 0·1 m, has an *RL*-product of 1·88 m^2 steradian, which is equivalent to that of a classical interferometer with a 1 m aperture.

FIG. 4. The Mark I Spectrometer.

The Mark II *Spectrometer*

This instrument is based on the third modification of the four-prism system (Schofield & Ring 1969); the optical layout is illustrated in Figure 5. In each of the interfering beams are located three prisms, two of which are oppositely wedged to form a plane parallel block, the third pıoviding the correct zero path difference conditions.

Both outside prisms are mounted on a common carriage, which translates them along their largest faces to change the path difference. By a suitable choice of glass types the system can be made achromatic. Con-

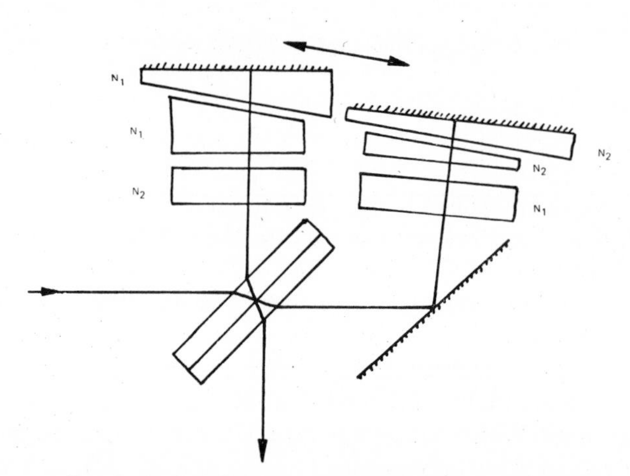

FIG. 5. Optical diagram of the Mark II Spectrometer.

struction has just begun on this instrument, which should have the following characteristics:

Resolution: 10^5 at 0·5μM.

Resolution-luminosity product: 13·8 M^2 steradian.

Spectral range: 0·4–1·0μM.

REFERENCES

BOUCHAIRENE, P. and CONNES, P., 1963, *J. Rad. Phys.*, **24,** 134.
CONNES, P., 1956, *Rev. D'Optique*, **35,** 37.
CUISENIER, M. and PINARD, J., 1967, *J. de Physique*, **28,** C2, 97.
DUBOIN, MME. L., 1969, private communication.
HILLARD, R. L. and SHEPHERD, G. C., 1966, *J. Opt. Soc. Amer.*, **56,** 362.
JAQUINOT, P., 1954, *J. Opt. Soc. Amer.*, **44,** 761.
MERTZ, L., 1965, *Transforms in Optics*, Wiley, New York, p. 18.
SCHOFIELD, J. W. and RING, J., 1969, paper in preparation.
SHEPHERD, G. C., 1969, private communication.

A Double-beam Far-infrared Interferometric Fourier Spectrometer System

L. W. Thorpe, R. C. Milward, G. C. Hayward and J. D. Yewen

Research & Industrial Instruments Company,
Worsley Bridge Road, London, S.E.26, England

Abstract—This paper describes an experimental far-infrared interferometric Fourier spectrometer which has been developed to operate in a 'double-beam' mode, in an effort to overcome the limitations in photometric accuracy which are inherent in any 'single-beam' interferometric spectrometer. The 'sum and difference' method of double-beaming employed is discussed in detail, and the requirements of the special electronic system devised for this instrument and the suitability of two types of infrared detector for this method of operation are critically assessed. The operational performance of the system has been evaluated, and has been found to yield a photometric accuracy and reproducibility of better than ±2 per cent full scale for far-infrared transmittance measurements over a wide spectral range.

INTRODUCTION

The technique of interferometric Fourier transform spectroscopy (Fellgett 1951, 1958; Jacquinot 1954) has received considerable application for far-infrared studies (cf. Richards 1964) in the last decade, due principally to the high efficiency of operation and broad-band capabilities of interferometric Fourier spectrophotometers. Existing commercial instrumentation of this type (cf. Milward 1968) has allowed the entire far-infrared region 3 to 500 cm^{-1} to be routinely covered, with resolutions as high as 0·1 cm^{-1}. These interferometers operate in a conventional 'single-beam' mode however, and in most instances have not been used for serious measurements of photometric intensities, although under normal conditions they readily allow far-infrared transmittance or reflectance measurements to be made to an accuracy of ±5 per cent full scale in regions of the spectrum where the signal-to-noise ratio is high.

The main limitations in obtaining higher photometric accuracies with these instruments are inherent in their 'single-beam' mode of operation, and arise from two distinct causes. First, effects of long-term instrumental drift which lead to a gradually changing signal output from the detector between or during successive interferogram scans, and cause the true

100 per cent level of the transmittance measurements to be altered. Such drifts are usually caused by changes in source output intensity and/or detector sensitivity. The Golay detectors (Golay 1947a, b; 1949) which are commonly used in commercial far-infrared interferometers are known to be particularly temperature sensitive ($\sim$1·5 per cent decrease in output signal per temperature rise of 1°C, at room temperature), and can exhibit considerable drift over a day's operation unless special care is taken to water-cool their casings. A subsidiary effect of instrumental drift during an interferogram scan is to introduce spurious low-frequency components in the transformed spectrum, which, however, usually lie well below the range of experimental interest (<1 cm^{-1}).

The second source of intensity error which may arise, is due to the 'phase errors' introduced in recorded interferograms by failure to digitize exactly at the zero path interferogram position. Unless care is taken to mathematically correct interferograms for these phase errors (cf. Connes 1961), then considerable amplitude distortions can occur in the computed spectra. It is thus possible that spectra computed from interferograms having widely different phase errors might exhibit slight intensity differences if the phase correction procedures employed are not 100 per cent efficient in each case.

For improved accuracy, therefore, it seems desirable to resort to a double-beam mode of operation, in which both 'sample' and 'reference' interferograms are simultaneously scanned, and received by the same detector under identical conditions of instrumental vacuum, source brightness, etc., so that effects of drift should cancel when double-beam spectra are computed. Furthermore, if both interferograms are sampled at exactly the same points of optical path difference, by means of a stepping drive arrangement, then intensity errors due to inadequate phase error correction will be minimized. The purpose of this article therefore is to discuss a method by which a commercial Michelson interferometer has been converted to this particular double-beam mode of operation, and to assess the performance of this system in operation, when Golay and other forms of infrared detector are utilised.

METHOD

The fundamental practical equation of optical Fourier transform spectroscopy relates the oscillatory interferogram intensities $I(n\Delta x)$ with their transformed frequency power spectrum $G(K)$ by a summation of the form:

$$G(K) = \sum_{n=0}^{n=N} I(n\Delta x)\cos(2\pi Kn\Delta x)\Delta x \qquad (1)$$

where the interferogram optical path difference is measured at regular intervals Δx up to some maximum value $N\Delta x$ and the frequency K can take an arbitrary range of values within the theoretical domain allowed by the choice of sampling interval Δx.

As shown by Equation (1), it is not mathematically possible to obtain conventional double-beam spectra by ratioing interferogram intensities directly, and it is therefore necessary to record two separate complete interferograms, and compute the power spectrum of each before taking a ratio $G_B(K)/G_A(K)$, for a range of values of K.

In order to achieve this, the standard single-beam Michelson interferometer shown in Figure 1 was modified to take a special unit which allowed the interferometer output beam to be switched through two different optical

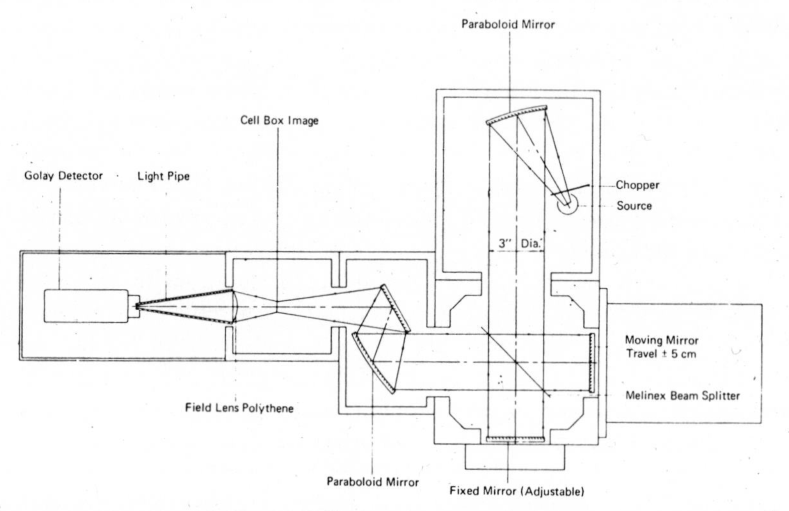

FIG. 1. Optical ray diagram of the R.I.I.C. FS-720 Michelson Interferometer for the 10–500 cm^{-1} spectral region.

'channels' and subsequently combined at the same detector. The beam-switching unit replaces the normal sample optics of the interferometer, and is shown in Figure 2.

This unit was inserted in the interferometer sample optics section rather than in the source optics prior to the interferometer, for reasons of greater convenience and to avoid any spectral mismatch which might arise due to inability to direct two separate optical beams along exactly the same path through the interferometer. The only disadvantage to this approach is that the detector can receive some radiative signal directly from the chopper blade itself, although in the instrument described here this contribution was usually very small compared to the intensity of the 'true' optical signals and tended to balance out for each channel. Furthermore, such 'false' radiation only raises the background levels of interferograms by a constant amount and is lost when the interferograms are transformed to spectra.

According to the mark/space ratio of the chopper blade employed, the unit shown in Figure 2 can be used for double-beam differencing to enhance dynamic range (cf. Hall, Vrabec & Dowling 1966) or beam switching, as

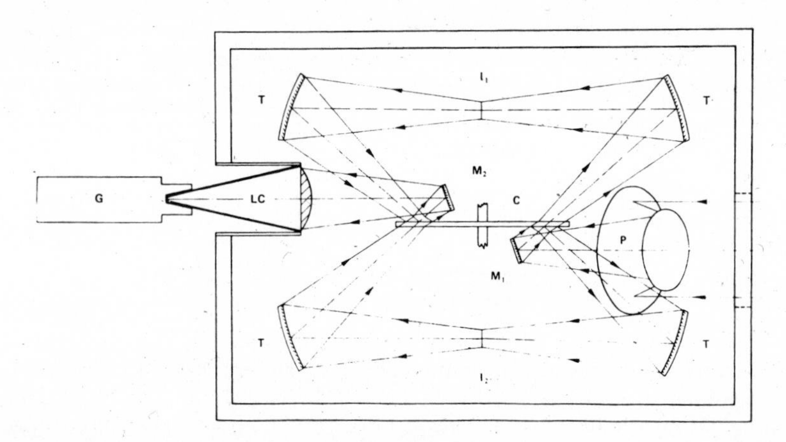

FIG. 2. Beam switching unit designed to replace the beam condensing sample and detector optics of the Michelson interferometer shown in Figure 1. Collimated radiation emerging from the beam splitter module is condensed by an off-axis paraboloid – plane mirror combination *P* to form an image in the plane of the chopper blade *C* after reflection from a small plane mirror M_1. The geometry of the chopper blade is such that the beam may be alternately transmitted or reflected to be re-imaged at positions I_1 and I_2 by means of toroidal mirrors *T* and subsequently recombined at mirror M_2. Both beams then follow identical paths to a Golay detector *G* via a tapered light cone and condensing lens system *LC*. (Adjustable beam attenuators and sample holding arrangements etc. positioned at I_1 and I_2 are not shown.)

required. For the purpose of obtaining two separate interferograms, however, the following mode of operation (cf. Martin 1966) was adopted. It was arranged that signal channels *A* and *B* were allowed to fall on the detector for equal and alternate periods of time with an equal blank period between each, as illustrated in Figure 3. If this period is of duration, say, $1/4f$, then it is intuitively obvious that a measure of the difference $(A - B)$ between the two channels is given by the amplitude of the periodic signal of frequency f. Similarly a measure of the sum of the two channels $(A + B)$ might be obtained from the amplitude of the periodic signal of frequency $2f$. In principle, therefore, it should be possible to recover *A* and *B* separately by summing and differencing the outputs of the two signal channels operating at frequencies f and $2f$, provided that the respective proportionality factors of each channel are known.

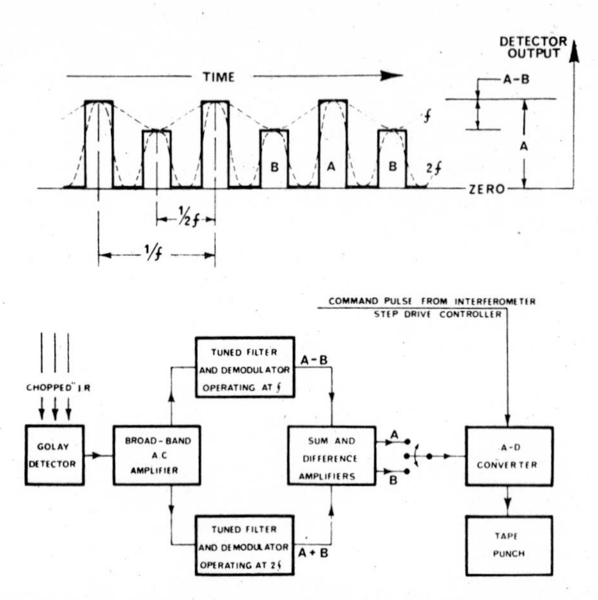

FIG. 3. Schematic representation of the 'sum and difference' method of separating interleaved optical signals as described in the text. In practice, the output waveforms from the Golay detector show considerable deviations from the square-wave pulses shown in the diagram.

In more exact mathematical terms, the square waveforms shown in Figure 3 may be expressed (cf. Hsu 1967) by the complex Fourier series given below:

$$F(t) = B + \sum_{n=0}^{n=\infty} \{A \sin(n\pi/4)/[4(n\pi/4)] - 3B \sin(n\pi 3/4)/4(n\pi 3/4)\} \times \exp(i2\pi nft) \qquad (2)$$

This expression may be simplified to give:

$$F(t) = B + \sum_{n=0}^{n=\infty} (n\pi)^{-1}\{\sin(n\pi/4)[A + (-1)^n B]\}\exp(i2\pi nft) \qquad (3)$$

The amplitude of the fundamental wave and its second harmonic, corresponding to the coefficients of the above Fourier series for $n = 1$ and $n = 2$ respectively, are thus related to the amplitudes A and B of Figure 3 as below:

$$C_1 = (\pi)^{-1}\sin(\pi/4)[A - B] = (\pi\sqrt{2})^{-1}[A - B] \qquad (4)$$

$$C_2 = (2\pi)^{-1}\sin(\pi/2)[A + B] = (2\pi)^{-1}[A + B] \qquad (5)$$

The simultaneous Equations (4) and (5) can be solved to give

$$A = \pi[C_2 + C_1/\sqrt{2}] \tag{6}$$

$$B = \pi[C_2 - C_1/\sqrt{2}] \tag{7}$$

Thus if X and Y are the actual outputs obtained from the signal channels f and $2f$ respectively, for the same frequency bandwidth and for a detector of uniform frequency response, then Equations (6) and (7) may be re-written (omitting a common proportionality factor)

$$A = Y\sqrt{2} + X \tag{8}$$

$$B = Y\sqrt{2} - X \tag{9}$$

The electronic system devised for this purpose is also depicted in Figure 3. The sequence of operations is as follows: the multiplex wave-form from the Golay output is amplified by a broad-band amplifier, and is fed in parallel to tuned filters operating at f and $2f$. The outputs of the two tuned filters are then separately demodulated and smoothed, and fed to a 'sum and difference' dc amplifier circuit, from which two separate dc outputs are obtained which are proportional to the intensity in each optical signal channel. These two outputs are digitized in alternate sequence and transferred to paper tape by a data-logging system which is triggered by 'command' pulses from the unit controlling the inter-ferometer drive unit. The physical appearance of the double-beam interferometer and electronics is shown in Figure 4.

In practice, due to the slow response times of Golay detectors ($\sim$15 msec) it is necessary to use fairly slow chopping speeds: f is typically 6 Hz. In order to obtain high selectivity for each channel it is necessary that the Qfactor is >5, and furthermore the Q of the higher frequency channel should be *twice* that of the low frequency channel in order to obtain an equal bandwidth and time constant for each channel. In practice, Qfactors of 20 and 40 were employed, which were high enough to give reasonable rejection of other frequencies and 'cross-talk' between channels but not too high to make the system critical of the frequency stability of the chopper blade, which is usually limited by variations in the frequency of the mains supply.

In order to take account of the measured fall-off in frequency response of Golay detectors between 6 and 12 Hz, which amounted to $\sim$33 per cent, and the theoretical $\sqrt{2}$ multiplication in amplitude of the higher frequency signal compared to the lower frequency signal imposed by Equations (8) and (9), the gain of the 12 Hz channel should be set at approximately *twice* that of the 6 Hz channel.

Far-infrared interferograms were recorded with the Michelson interferometer, using a stepping drive arrangement for the moving mirror scan. In order to interleave corresponding data points from each interferogram trace on paper tape, it was arranged that the 'sample' interferogram was digitized after an interval corresponding to half the drive pause period, and the 'background' interferogram was digitized at the end of the total pause period. Thus, in order for each interferogram signal to be accurately digitized, it was necessary to set the pause period to be

FIG. 4. Complete double-beam interferometer system, with vacuum lid removed.

approximately five times larger than the smoothing time constant of the electronics. The total scan time for a double-beam run was therefore exactly the same as the total scan time of two separate interferograms recorded in the single-beam mode. The double-beam interferogram tapes recorded were processed by digital computation and/or by a special purpose hybrid Fourier transform computer (Ridyard 1967).

RESULTS

The performance of the double-beam interferometer system may be conveniently discussed with regard to the following criteria. First, the optical and electronic detection system should be linear in response, and should exhibit no interaction (or 'cross-talk') between the two output channels; second, a high degree of reproducibility and photometric accuracy should be obtained when a number of double-beam spectra are recorded of the same sample.

The degree of linearity of the tuned filters and demodulators operating at 6 Hz and 12 Hz was checked by feeding in known ac electrical signals and measuring the dc outputs, and was found to be better than $\pm 0{\cdot}3$ per cent. The linearity of the detector and electronic system was then checked for each optical channel by using calibrated attenuators, and

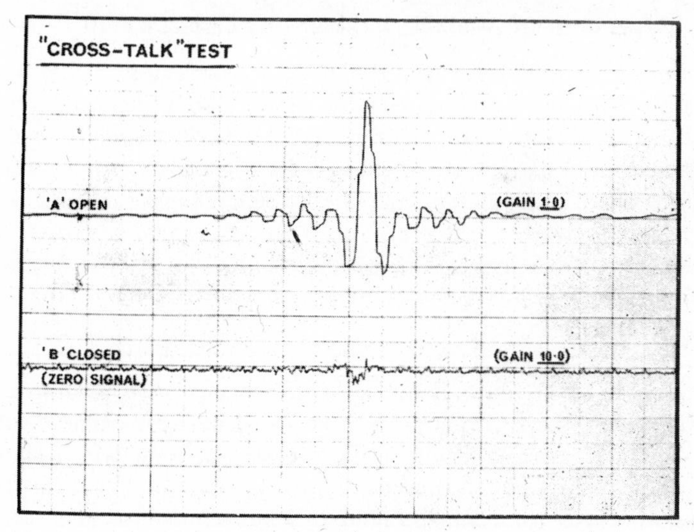

FIG. 5. 'Cross-talk' test showing approximately 1 per cent interaction of channels *A* and *B*.

was found to be good to $\pm 0{\cdot}5$ per cent full scale overall, which was considered satisfactory.

The degree of 'cross-talk' between the two output channels was first investigated by studying the change of signal level in one channel when the optical signal in the other beam was alternatively blanked off or readmitted. By adjusting the gain of the 12 Hz channel to the correct factor (approximately $\times 2$) of the gain of the 6 Hz channel, it was possible to minimize any observed 'cross-talk'. The second test was to observe the residual change in the channel nominally set at zero while an interferogram was scanned using the other beam. The results of such a test are illustrated in Figure 5.

Finite amounts of residual electrical 'cross-talk' were observed, which could vary in magnitude and polarity according to the particular Golay detector employed. 'Cross-talk' levels varying between 1 and 5 per cent of the total signal level were observed during tests carried out with 13 different Golay detectors. In order to investigate the cause of these variations, each detector was tested both for frequency response between 3 and 30 Hz, and for second harmonic content of their electrical outputs. It was found that all these detectors followed the same shape of frequency response curve, but varied widely in second harmonic content of their outputs. Second harmonic levels varying between 2 and 6 per cent of the fundamental were measured, which could be loosely correlated with the 'cross-talk' figures for these detectors which varied between 0·9 and 5 per cent, by an empirical equation of the form:

$$\text{(per cent `cross-talk')} \simeq \text{(per cent 2nd harmonic)}^2/10 \qquad (10)$$

Tests made with another type of far-infrared detector, the room temperature pyro-electric bolometer (cf. Hadni 1967), yielded a 'cross-talk' of ~1 per cent. Rather surprisingly, this detector also had an appreciable second harmonic content (4-5 per cent) which was attributed to distortions in its high-gain preamplifier circuitry. Eight of the 13 Golay detectors tested showed a 'cross-talk' $\leqslant 1{\cdot}3$ per cent, which was considered small enough not to impair the basic accuracy of the double-beam interferometer, where the spectral noise content is usually ~1 per cent full scale in each beam.

In order to test the operational efficiency of the double-beam interferometer, a series of transmission measurements of well-known infrared materials such as crystal quartz or polypropylene were carried out in the 50-400 cm^{-1} spectral region, using a Golay detector which showed <1 per cent 'cross-talk'. A number of consecutive double-beam interferogram traces (see for example Figure 6) were recorded for each sample, and their transformed spectra ratioed, and plotted on the same chart paper. The results of such measurements for polypropylene and crystal quartz are shown in Figures 7 and 8.

Figure 7*A* shows eight consecutively scanned single-beam spectral energy profiles, which served as the 'backgrounds' for the transmission measurements. The variation in their intensities at any one frequency is approximately ± 10 per cent full scale about their mean value, and illustrates the amount of instrumental drift which can occur over a day's operation. However, in the eight double-beam ratioed spectra shown in Figure 7*B*, the spread of values at any one frequency is only ± 1 per cent full scale about the mean for large proportions of the spectrum, and at worst is only ± 3 per cent. Figure 8*A* shows eight successive crystal quartz transmission spectra obtained from ratios against the same background spectrum, which on average show a spread of values up to ± 5 per cent full scale at any one frequency. Figure 8*B* shows the same eight crystal quartz spectra ratioed against their separate double-beam backgrounds, which now only show an average deviation of $\pm 1{\cdot}5$ per cent at any one frequency, in the double-beam traces. Figures 7 and 8 therefore, clearly indicate the improved reliability and accuracy of the simultaneous double-beam method, over the single beam method.

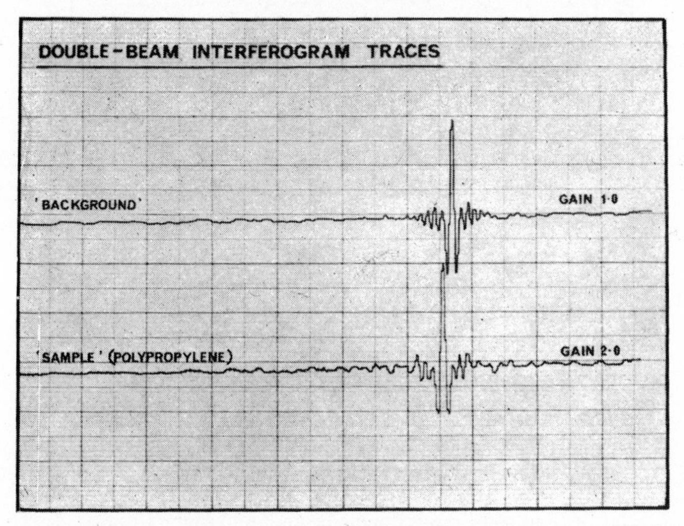

FIG. 6. Double-beam interferogram traces exhibiting identical drift and sampling errors. (The lateral shift between the two interferogram peaks is due to the pen displacement of the double-channel recorder used.)

The average noise in the double-beam spectra (which in these instances can be considered to arise principally from the noise in the sample spectra) may be estimated from the signal/noise value of the recorded sample interferograms S_I, to a good approximation using the relation

$$S_S = S_I\sqrt{N}/M \tag{11}$$

where S_I is the signal/noise of the corresponding spectrum, of M resolved spectral elements, which was computed from a total number of N interferogram points (small effects of 'quantization' in the data-logging electronics can be neglected in this instance).

For the spectra shown in Figures 7 and 8, typical S_I values were $\sim$300 and 100 respectively, and the N values 1024 and 512. For M values of 120 and 60, respectively, the corresponding S_S values are approximately 80 and 40. Thus it would appear that most of the variations shown in Figures 7*B* and 8*B* can be accounted for by the inherent noise level in the spectra themselves, which could amount to ± 1 to 2 per cent full scale.

In order to check the broad range spectral linearity of the system, tests were made with a series of calibrated attenuators in one beam, and the

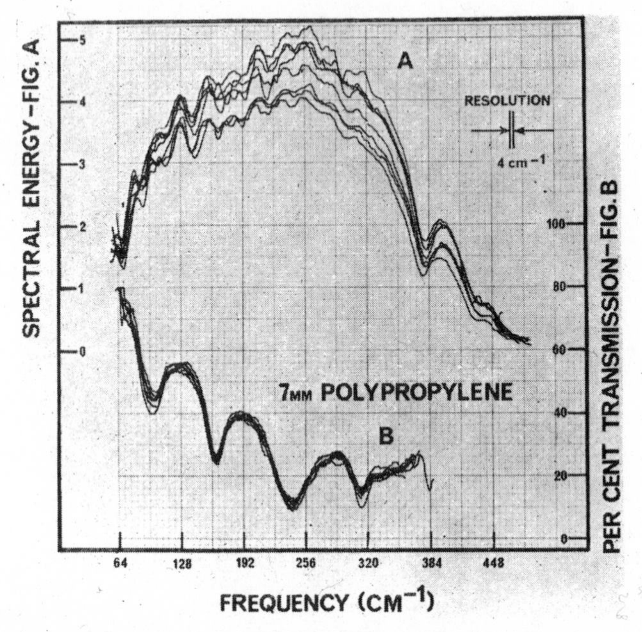

FIG. 7. Results of eight consecutive double-beam interferogram traces: *A*, variation of spectral background intensity; *B*, absolute transmittance of a 7 mm polypropylene sample.

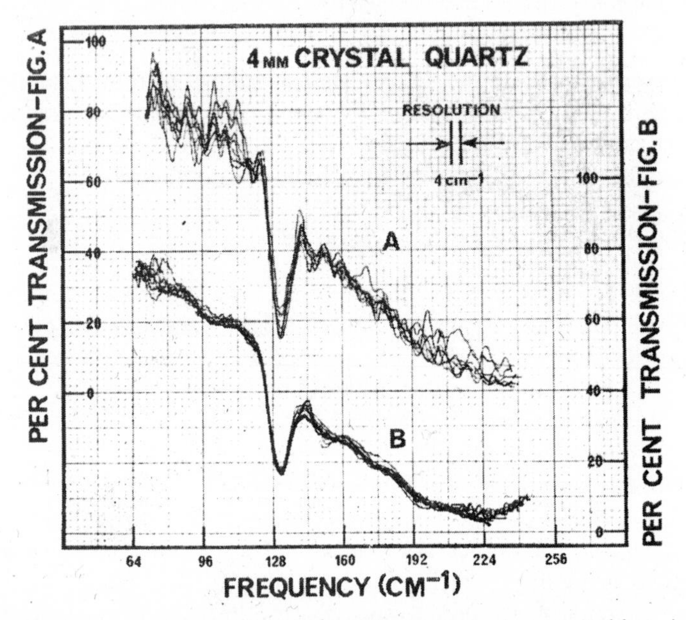

FIG. 8. Absolute transmittance of 4 mm crystal quartz sample measured by: *A*, single-beam method; *B*, double-beam method.

ratios of the plotted spectra compared with theoretical values. The results are shown in Figure 9. It may be seen that the plotted values fall within ±2 per cent full scale of the correct value, and remain relatively flat to ±1 per cent of their mean value, except in the case of 75 per cent transmission where the attenuator setting may have been slightly inaccurate.

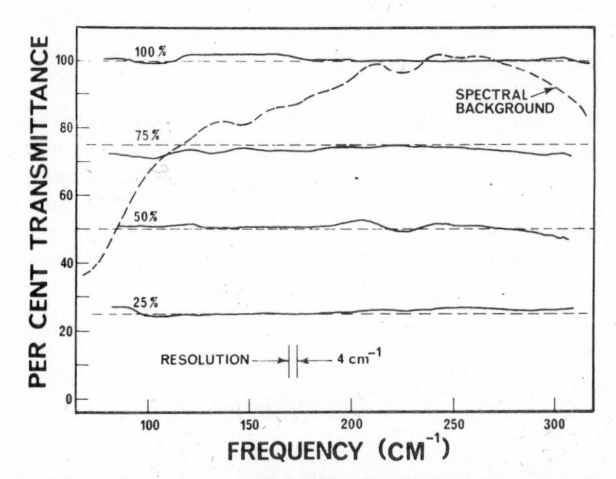

FIG. 9. Spectral linearity tests recorded with the double-beam interferometer. The dotted line shows the spectral background distribution.

CONCLUSIONS

The double-beam interferometer system described here has demonstrated a clear advantage over a single-beam version for far-infrared transmittance measurements. The overall reproducibility and photometric accuracy of ± 1 to 2 per cent full scale are compatible with the noise levels in individual single-beam spectra when Golay detectors are employed.

The method of double-beaming adopted here has the advantage of simplicity, and allows the full working aperture of a single-beam instrument to be retained with only a 50 per cent reduction of the effective duty cycle. The system requires only a modest addition to the electronics of a single-beam interferometer, and can be used with slow detectors. A similar concept of double-beaming a far-infrared Fourier spectrometer has been recently described (Ballantyne 1967). However, the system involves the use of more costly, fast-integrating electronics for the separation of signals, and does not appear to be highly suited for use with very slow detectors such as Golay cells.

The basic accuracy of the 'sum and difference' double-beam method described here depends upon the absence of any higher harmonic content in the output of the detector utilized. The room temperature Golay and pyroelectric detectors tested here showed finite second harmonic content which led to 'cross-talk' figures of ~1 per cent at best. Thus, in order to obtain a maximum accuracy with these particular detectors, it may be advantageous to adopt a system whereby both signal channels are modulated at frequencies well separated and having no exact mathematical ratio.

Acknowledgments

The authors would like to thank their colleagues at Research and Industrial Instruments Company for helpful suggestions concerning this work, particularly Mr. D. Lambert and Mr. G. Fraser for contributions on electronic design, and Dr. D. J. Neale for assistance with Fourier transform computations. The authors also wish to thank Dr. E. H. Putley and Mr. J. H. Ludlow of Royal Radar Establishment, Malvern, for making a pyroelectric infrared detector available for evaluation purposes.

REFERENCES

Ballantyne, J. M., 1967, *Appl. Opt.*, **6,** 587.
Connes, J., 1961, *Rev. Opt.*, **40,** 45, 116, 171, 231.
Fellgett, P., 1951, Thesis, Cambridge.
Fellgett, P., 1958, *J. Physique* (France), **19,** 187.
Golay, M. J. E., 1947a, *Rev. Sci. Inst.*, **18,** 347.
Golay, M J. E., 1947b, *Rev. Sci. Inst.*, **18,** 357.
Golay, M. J. E. 1949, *Rev. Sci. Inst.*, **20,** 816.
Hadni, A., 1967, *Essentials of Modern Physics Applied to the Study of the Infrared*, Oxford, Pergamon Press, pp. 300–334.
Hall, R. T., Vrabec, D. and Dowling, J. M., 1966, *Appl. Opt.*, **5,** 1147.
Hsu, H. P., 1967, *Outlines of Fourier Analysis*, New York, Associated Educational Services Corp., p. 86.
Jacquinot, P., 1954, 17e Congres. GAMS., **25.**
Martin, A. E., 1966, *Infrared Instrumentation and Techniques*, Amsterdam, Elsevier, p. 74.
Milward, R. C., 1968, in *Molecular Spectroscopy*, ed. Peter Hepple, London, Institute of Petroleum.
Richards, P. L., 1964, *J. Opt. Soc. Amer.*, **14,** 1474.
Ridyard, J. N. A., 1967, *Suppl. J. Physique*, **28,** (3/4), C2–62.

InSb Tunable Detector and its Application to a New Far-Infrared Spectrometer

Hiroshi Yoshinaga and Junya Yamamoto

Department of Applied Physics,
Osaka University,
Suita, Osaka, Japan

Abstract—Far-infrared tunable InSb detectors which have very sharp spectral detectivity and fast response in the wavelength region from 30 to 300 μm (excluding the lattice vibration region of InSb) have been made. This followed investigation of the variation of the photoconductivity of n-type InSb with concentration of impurity, temperature and strength of magnetic field, up to 50,000 G. These detectors are much superior to silicon-diode detectors in the microwave region of a few millimetres.

Suitable shortwave cut-off filters must be used to eliminate the higher order spectra of the grating for ordinary far-infrared spectrometers.

A new far-infrared spectrometer using the InSb tunable detector, which is available in the wavelength region from 30 to 300 μm, is under construction. This spectrometer is of very simple construction with only a transmission filter to eliminate near-infrared radiation. Excellent spectra of water vapour in the atmosphere were obtained with preliminary equipment owing to the high efficiency of available energy and the high sensitivity of the detector.

INTRODUCTION

Indium antimonide is an important semiconductor like Ge and Si. n-type InSb has impurity levels or impurity bands which have been studied by transport phenomena (Miyazawa & Ikoma 1967; Sladek 1958) but not yet by optical methods except for some researches by Kaplan (1966). This paper reports on the characteristics of n-type InSb by optical measurements in millimetre and submillimetre-wavelength regions, and the use and development of InSb detectors in these regions obtained as a result of this research. This detector makes it possible to construct very simple, far-infrared spectrometers of high quality.

Figure 1 is a schematic diagram of energy levels and optical transitions among those levels of semiconductors with hydrogen-like impurity atoms in a magnetic field obtained theoretically by Wallis and Bowlden (1958). The transition B shows the cyclotron resonance absorption between Landau levels ($N=0$ and 1) in the conduction band and transition C the cyclotron resonance absorption from the ground state of the impurity levels. The transition A gives absorption in the millimetre-wave region from the ground state to excited impurity levels like (010) or (001). The curve on the right gives the spectral photo-response by transitions B and C in the submillimetre region. The photo-response was studied along with that by transitions A, B and C for different carrier concentrations, in various magnetic fields, electric fields and temperatures.

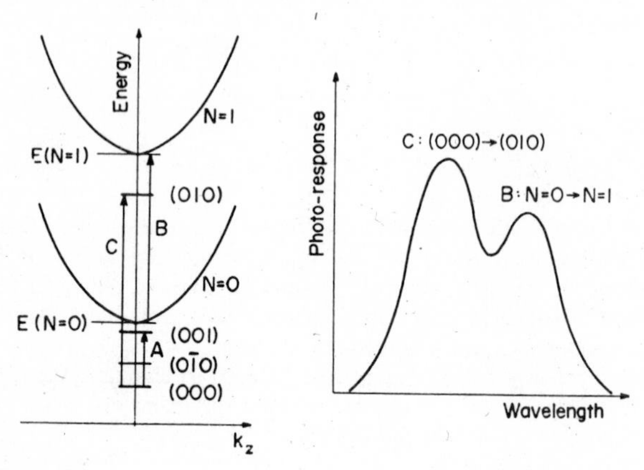

Fig. 1. Schematic diagram of energy levels of InSb, optical transitions (*left*) and spectral photo-response (*right*).

EXPERIMENTAL

Details of typical InSb elements are given in Table 1 and the cryostat used in this research is shown in Figure 2. The maximum field of the superconductive magnet is 50 kOe. LiF/SrF_2 powder filter (under 20 kOe) and a sooted polyethylene filter were used as the cooled filter.

Table 1. InSb Elements

Sample number	Dimension (mm^3)	Carrier concentration (cm^{-3})	Mobility (cm^2/Vs)
K – 22	$6{\cdot}0 \times 3{\cdot}5 \times 0{\cdot}9$	$7{\cdot}0 \times 10^{12}$	$\sim 1 \times 10^4$
TIA – 1	$4{\cdot}7 \times 4{\cdot}5 \times 0{\cdot}2$	$7{\cdot}0 \times 10^{13}$	5×10^5
C – 35S – 1	$4{\cdot}6 \times 4{\cdot}6 \times 0{\cdot}3$	$2{\cdot}4 \sim 2{\cdot}7 \times 10^{14}$	$3 \sim 5 \times 10^5$
C – 35S – 12	$4{\cdot}7 \times 4{\cdot}7 \times 0{\cdot}3$	$1{\cdot}4 \sim 2{\cdot}8 \times 10^{13}$	”

Resonant Photoconductivity in n-type InSb

Figure 3 shows the dependence of the photoconductive response on the magnetic field in the region from 70 to 500 μm. The absorption corresponding to the transition B becomes weaker for stronger magnetic fields. This shows that the energy gap between the donor band and the bottom of the conduction band becomes larger with increasing magnetic field

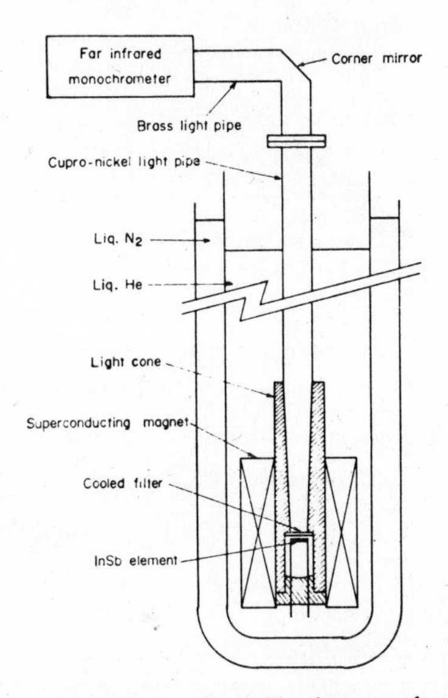

FIG. 2. Cryostat for the InSb photoconductors.

(Yafet, Keyes & Adams 1956). The relation between the resonant frequency and magnetic field is shown in Figure 4. The anomaly near 30 kOe shows the so-called polaron effect by the interaction between electrons and

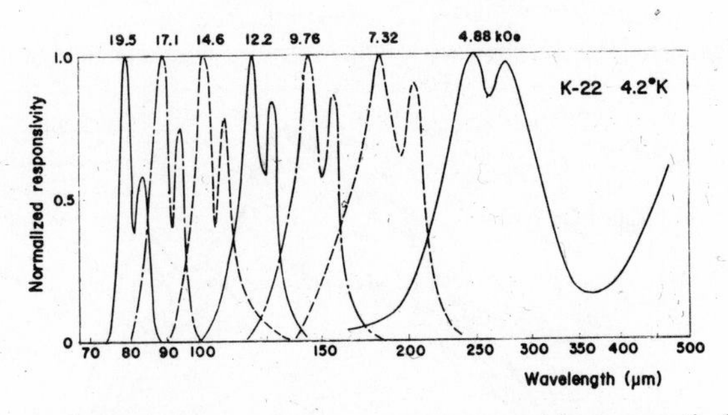

FIG. 3. Spectral photoconductive response at various magnetic fields.

phonons. Figure 5 shows the relation between electric resistance of InSb elements and electric field. The resistance increases at lower temperature in a constant magnetic field, and also increases in higher magnetic field.

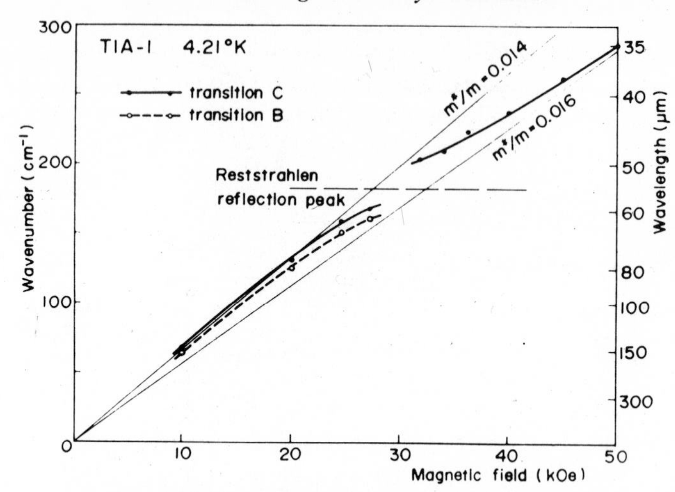

FIG. 4. Resonant frequency versus magnetic field.

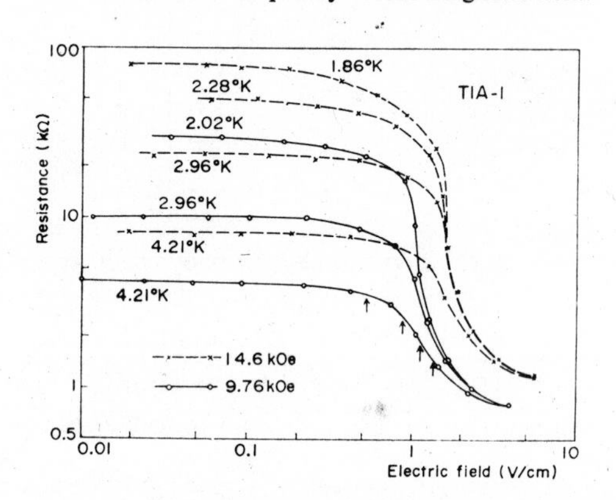

FIG. 5. Electrical resistance versus electric field in various magnetic fields and temperatures

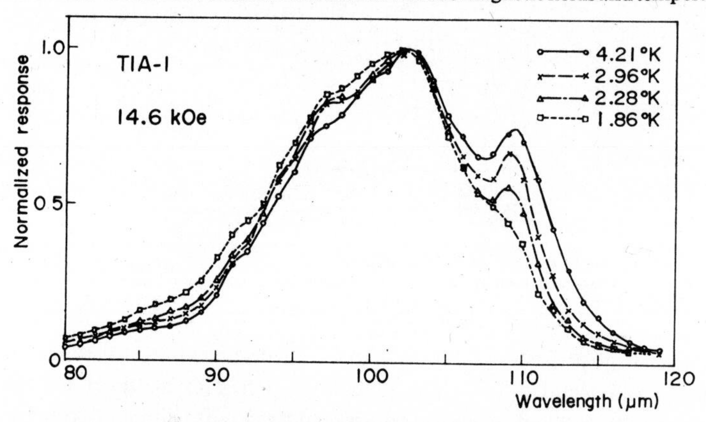

FIG. 6. Spectral photo-response at various temperatures.

The resistance suddenly decreases beyond some strength of electric field owing to electron impact ionization.

The dependence of the photo-response on temperature is shown in Figure 6. It is clearly seen that the response due to the transition *B* decreases at lower temperature. Figure 7 also shows the dependence of the photo-response on electric field. The transition *B* increases with electric field in spite of no change of the transition *C*.

Figure 8 gives the dependence of the electriæ resistance of InSb elements on the electric field for two samples of very different carrier concentration. The higher the carrier concentration, the smaller the electriæ resistance, and the dependence of the electriæ resistance on temperature in a constant

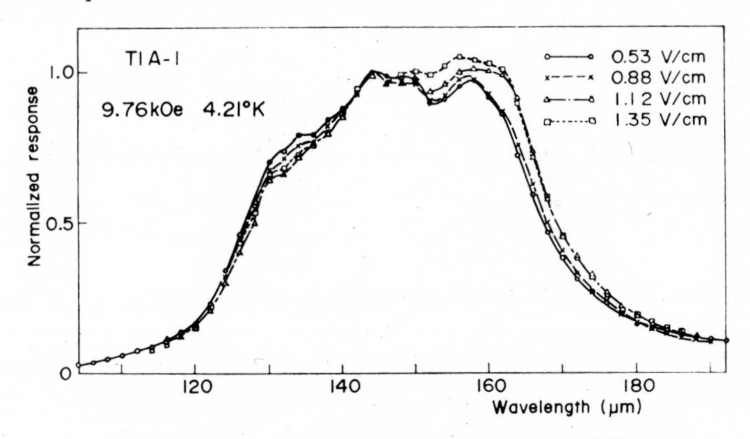

FIG. 7. Spectral photo-response in various electric fields.

magnetic field. The strength of the electric field in which the electrical resistance decreases suddenly is smaller for the element having higher carrier concentration because the ionization energy of the donor band becomes smaller for the element of higher carrier concentration. The spectral photo-response for the same samples used in the above measurements of the electric resistance shown in Figure 9 is easily understandable from the results described above. For the sample of lower carrier concentration, the spectral width of the photo-response is much smaller and the transition *B* decreases markedly at lower temperature. Figure 10 shows the increase of the photo-response under illumination by near-infrared radiation which may increase the carrier concentration and the mobility in the conduction band.

Photoconductivity in n-type InSb in the Millimetre-wave Region

The photoconductive effects due to hot electrons and the transition *A* occur in the millimetre-wave region. A klystron in the 2 millimetre-wave

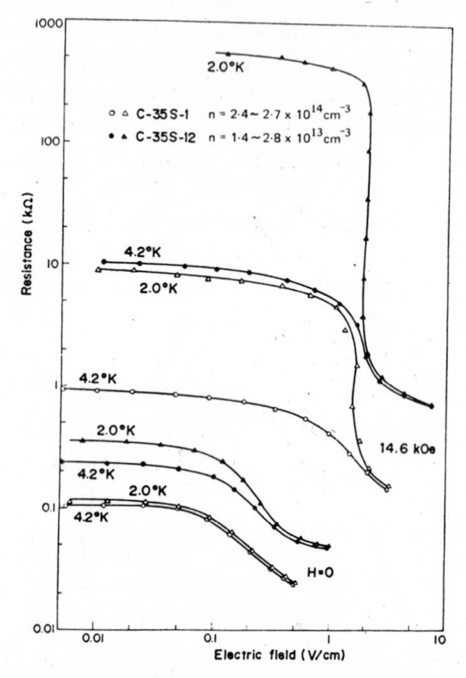

FIG. 8. Electrical resistance versus electric field for samples of two different carrier concentrations at two temperatures, and two magnetic fields.

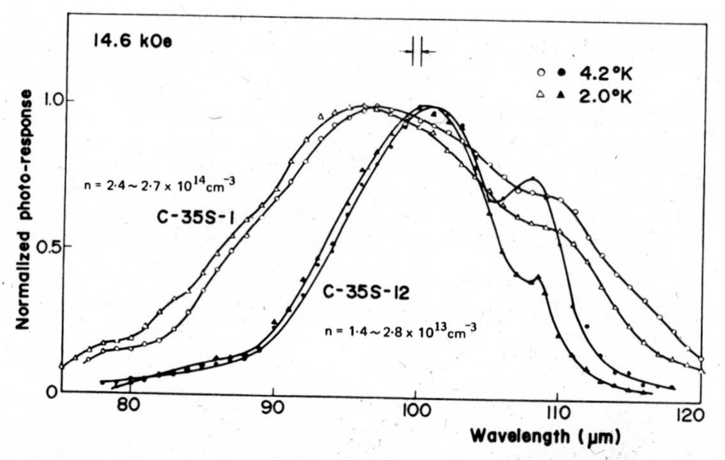

FIG. 9. Spectral photo-responses for samples of two different carrier concentrations at two temperatures.

region was used for the next measurements. Figure 11 shows the dependence of relative change of the electrical resistance on magnetic field for three different wavelengths near 2 mm. The drop of $\Delta R/R$ can be seen at about 2 kOe for 1·98 and 2·10 mm radiations. This shows the decrease of the transition of (000)→(0*mc*). Then the peak which is due to the transition of (000)→(010) appears. Such change of $\Delta R/R$ cannot be seen for 2·40 mm radiation, because the photon energy of 2·40 mm radiation is smaller

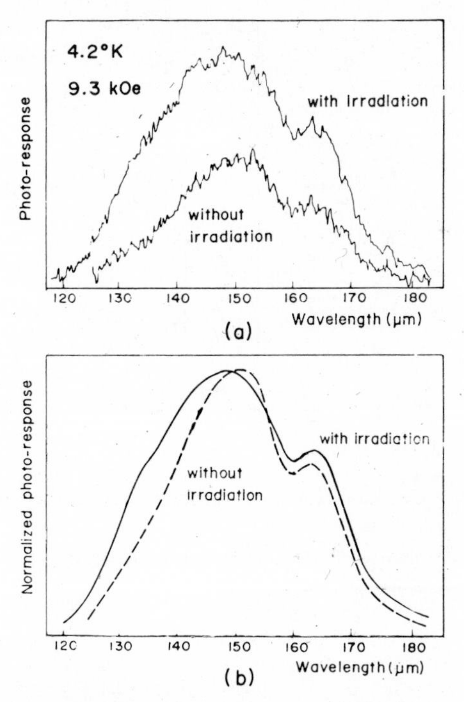

FIG. 10. Effect of near-infrared radiation on photo-response: (*a*) non-normalized; (*b*) normalized.

than the energy gap of those impurity levels and $\Delta R/R$ depends on only the photoconductivity of hot electrons. Figure 12 which illustrates the dependence of $\Delta R/R$ on the magnetic field for the two different carrier concentrations at various temperatures shows that the number of electrons in the donor band increases, and the carrier concentration in the conduction band decreases causing the increase of $\Delta R/R$ at lower temperatures.

From these results the model for n-type InSb shown in Figure 1 has been proved by the optical method. The characteristics of the InSb detector was measured with a Fabry-Pérot interference filter and a mercury lamp compared with a black-body radiation source shown in Figure 13.

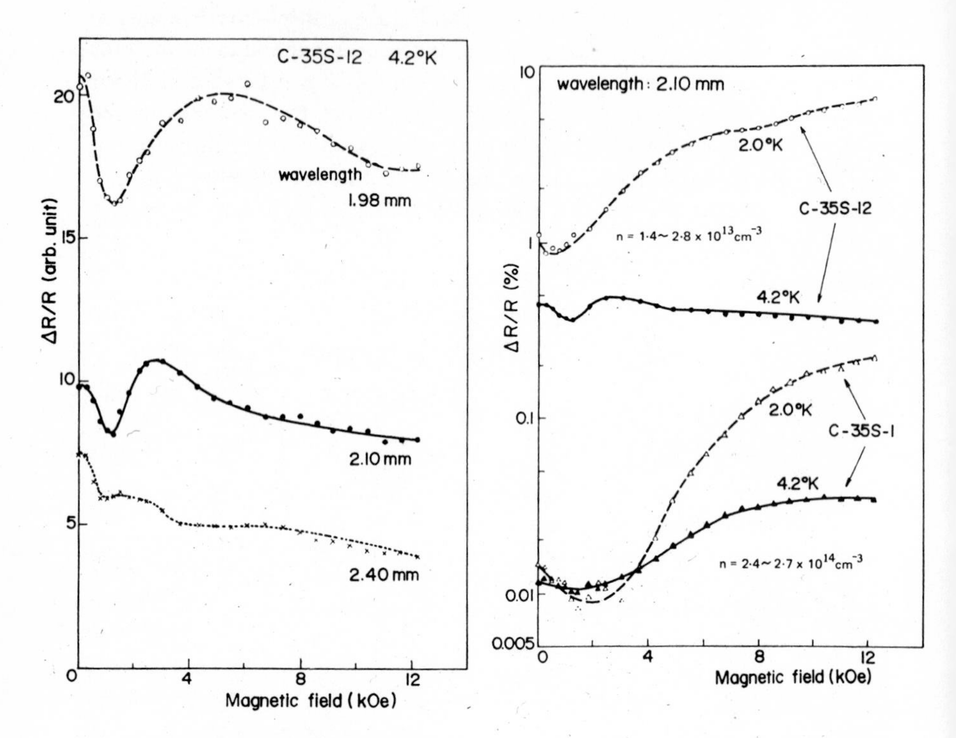

FIG. 11. (*left*) Relative change of electrical resistance versus magnetic field for three different wavelength radiations.

FIG. 12. (*right*) Relative change of electrical resistance versus magnetic field for samples of two different carrier concentrations.

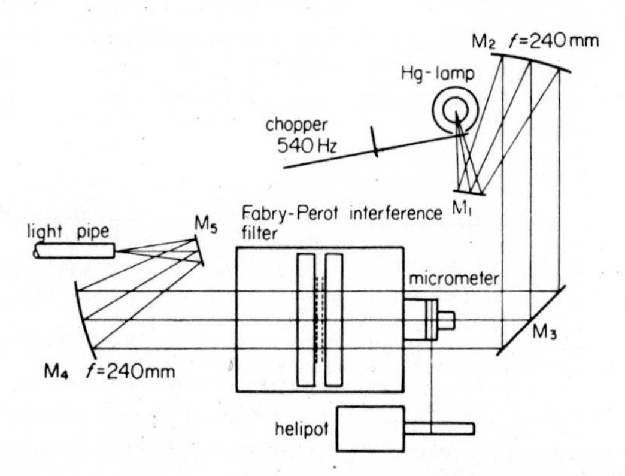

FIG. 13. Optical system for the measurement of detectivity of the detector.

InSb Tunable Detector

Figure 14 shows the spectral responsivity in various magnetic fields. The responsivity at 2·44 kOe does not show cyclotron resonance absorption, but is higher than that in no magnetic field. At 4·48k Oe cyclotron resonance absorption can be clearly seen. The responsivity decreases in the magnetic field above 9·76 kOe due to the *Reststrahlen* reflection loss.

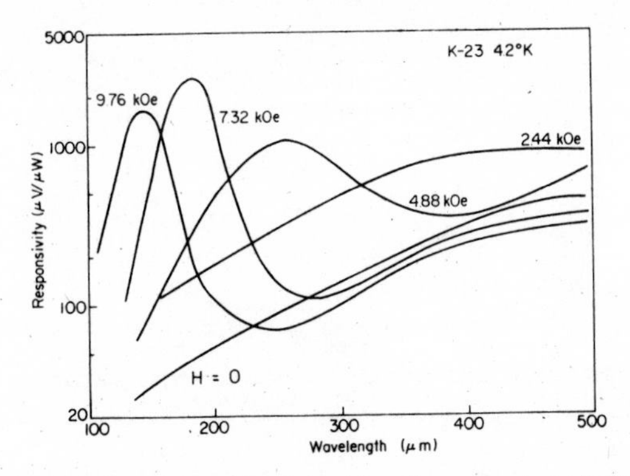

FIG 14. Spectral responsivity in various magnetic fields.

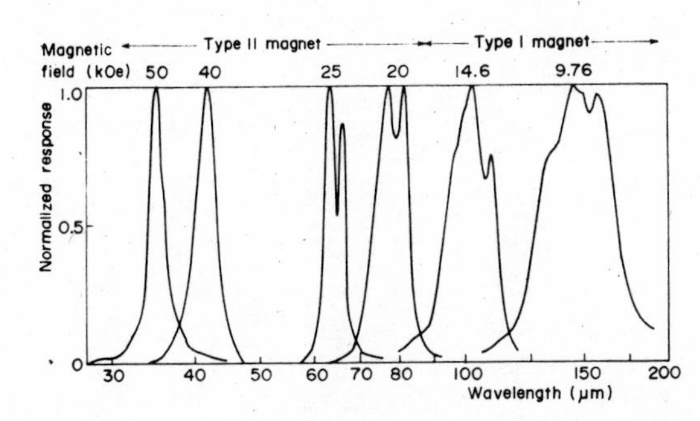

FIG. 15. Normalized spectral response for various magnetic fields.

The maximum responsivity was 2600 μV/μW at 7·32 kOe, that is, for 190 μm radiation. The noise equivalent power was about one tenth of that of the best Golay cell. In Figure 15, the normalized spectral response is illustrated in various magnetic fields. This shows that the InSb detector is a fairly satisfactory tunable detector in the wavelength region from 30 to 200 μm except in the *Reststrahlen* wave region near 54 μm. The response time measured with pulse radiation from HCN laser and a klystron is about 0·4 μs. The response time is somewhat slower, but the detectivity several times higher than that of the silicon-diode. Moreover the InSb detector can be used even for incoherent radiation, but the Si-diode cannot.

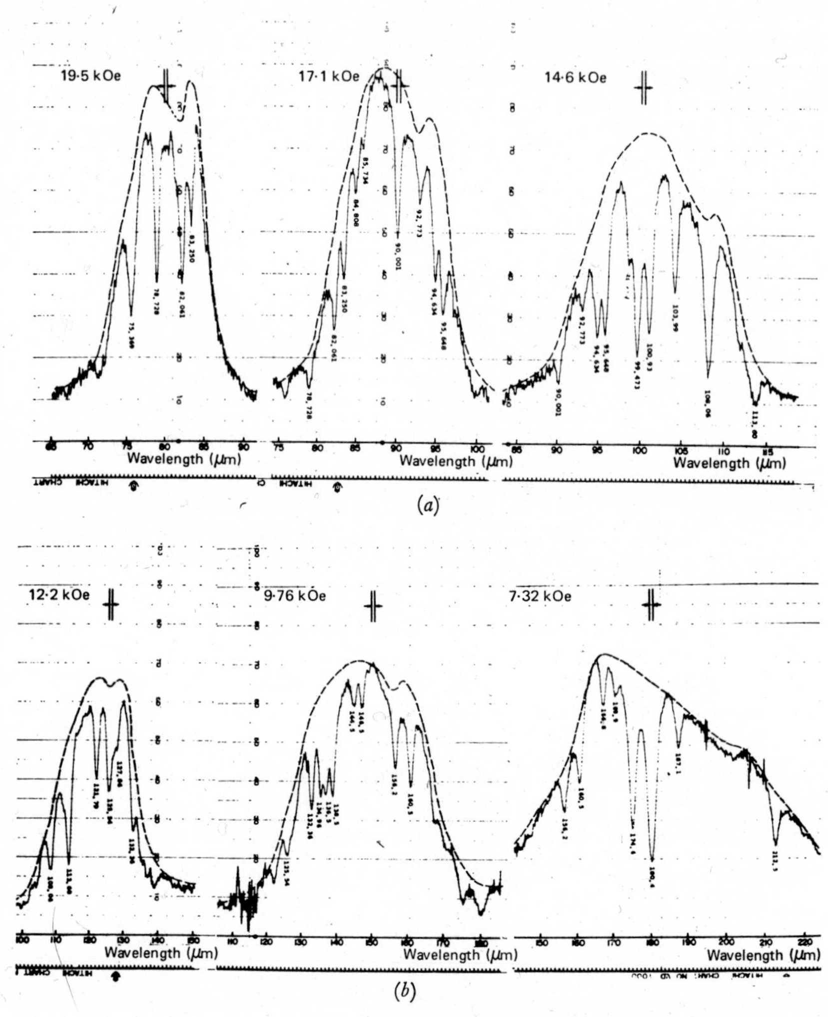

FIG. 16. Spectra of H_2O absorption line in various magnetic fields: (*a*) 19·5 to 14·6 kOe; (*b*) 12·2 to 7·32 kOe.

Figures 16*a* and *b* show the absorption spectra of H_2O obtained with the InSb detector. The radiation source was a mercury lamp and only a cooled filter was used in the measurements. But higher order spectra cannot be seen in these traces. The *S*/*N* ratio in the region of maximum detectivity is much better than that obtained with a Golay cell.

Far-Infrared Spectrometer with InSb Tunable Detector

Many kinds of filters to cut off shortwave radiation, like *Reststrahlen* reflection filters, grating filters and transmission powder filters, have been used in far-infrared grating spectrometers. These filters reduce the available energy at the wavelength to be measured in spite of cutting off the shortwave radiation. Moreover, the optical system of spectrometers becomes more complex and less fast for reflection filters. Even for the best commercial far-infrared spectrometers the efficiency of available

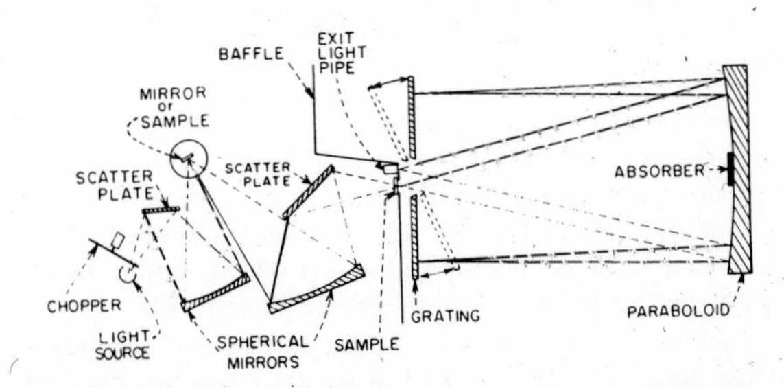

FIG. 17. Optical system of far-infrared spectrometer with InSb detector.

energy is estimated to be at most 40 per cent for the optimum wavelength region and less than 20 per cent for the worst region.

We are now constructing a simple far-infrared spectrometer with the InSb detector shown in Figure 17, which has no filter except a scatter plate and a cooled filter. The magnetic field for the detector can be adjusted to synchronize with the scanning of the spectra. The sample is mounted above the detector element in the light pipe for low temperature measurements. A magnetic field can be applied to the sample with a Helmholtz type superconducting magnet up to 50 kOe. The chopping frequency is 800 Hz. Such a far-infrared spectrometer has not only a high efficiency of available energy, but can scan spectra much faster than ordinary far-infrared spectrometers.

REFERENCES

KAPLAN, R., 1966, *J. Phys. Soc. Japan*, **21**, Supplement, 249.
MIYAZAWA, H. and IKOMA, H., 1967, *J. Phys. Soc. Japan*, **23**, 290.
SLADEK, R. J., 1957, *J. Phys. Chem. Solids*, **5**, 157.
WALLIS, R. F. and BOWLDEN, H. J., 1958, *J. Phys. Chem, Solids*, **7**, 78.
YAFET, Y., KEYES, R. W. and ADAMS, E. N., 1956, *J. Phys. Chem. Solids*, **1**, 137.

Spectromètre à Grilles Double Faisceau

André Girard and Nicole Louisnard

Office National d'Etudes et de Recherches Aérospatiales
92—Châtillon-sous-Bagneux (France)

Abstract—It is known that a spectrometer with selective modulation can be obtained by replacing the entrance and exit slits of a classical spectrometer by gratings whose surface is practically independent of the resolving power. The possibilities of adapting this principle to the case of a double-beam instrument are examined. The experimental results obtained with a prototype instrument are given. The spectral range used was between 5,000 and 350 cm^{-1}; the average limit of resolution is 0·5 cm^{-1}. The increased luminosity obtained with this instrument serves principally to increase the precision of photometric measurement and the speed of scanning.

Le principe du spectromètre à grilles a fait l'objet de publications antérieures (Girard 1963, 1965, 1967).

Il s'agit d'un appareil à modulation sélective, au même titre que le SISAM (Connes 1958) et que le spectromètre multifentes (Golay 1951).

RAPPEL DE PRINCIPE

On sait que le caractère commun aux dispositifs à modulation sélective est le point suivant: l'énergie lumineuse qui parvient au détecteur est relative à de nombreux éléments spectraux parmi lesquels l'énergie d'un seul est modulée. Il suffit de faire suivre le détecteur d'un amplificateur alternatif pour que le signal de sortie constitue une mesure spectrométrique concernant l'élément spectra étudié. Aucun calcul à postériori n'est nécessaire et le balayage du spectre s'effectue par les moyens habituels.

La disposition générale d'un spectromètre à grilles est celle d'un spectromètre à réseau classique dans lequel les fentes d'entrée et de sortie sont remplacées par des grilles, c'est-à-dire un ensemble de zones alternativement transparentes et opaques. Le pouvoir de résolution ne dépend que de la structure des grilles. Si $F(x, y)$ en est la loi de transmission, la résolution est directement liée à l'étendue de la transformée de Fourier de $F(x, y)$. La luminosité ne dépend pas du pouvoir de résolution mais seulement de la surface utile des grilles. Pouvoir de résolution et luminosité sont ainsi découplés. La réalisation d'un spectromètre sans fentes se ramène au problème du choix de fonctions de transmission $F(x, y)$ dont le support est étendu et dont la transformée de Fourier est régulière et de support étendu.

Les grilles habituellement utilisées à l'ONERA sont des grilles à zones hyperboliques réalisées par photogravure sur des matériaux transparents dans l'infra-rouge. Les zones opaques sont métallisées. Une telle grille peut être utilisée soit par transmission soit par réflexion. Les flux transmis

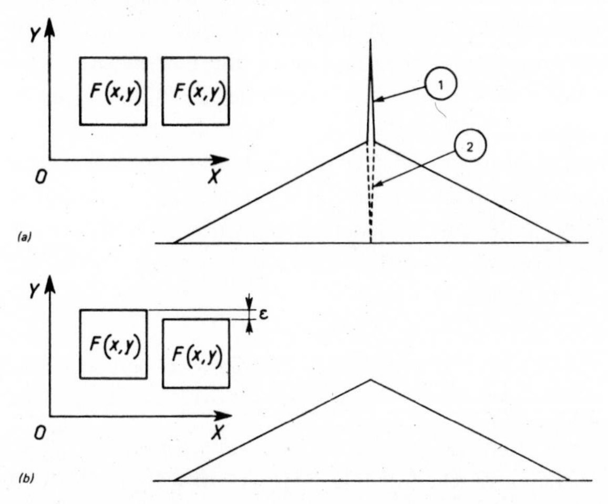

FIG. 1. (*a*) Fonction d'appareil: 1, grilles identiques; 2, grilles complémentaires; FIG. (*b*) Fonction d'appareil obtenue avec deux grilles décalées suivant OY.

et réfléchis sont égaux lorsque la somme des surfaces des zones transparentes est égale à celle des zones réfléchissantes. Cette propriété est très utile pour l'appareil à double faisceau.

Dans un appareil équipé de grilles la fonction instrumentale a la forme indiquée sur la Figure 1*a*. C'est la fonction d'autocorrélation suivant OX de la loi de transmission. Cette fonction se réduit schématiquement à la somme de deux fonctions triangles: le triangle de large base est la fonction qui serait obtenue si les grilles étaient remplacées par des ouvertures définies par leur contour. Il joue le rôle d'un signal parasite. L'autre terme, fonction triangulaire à base étroite, est le terme significatif de la fonction instrumentale. Sa largeur est proportionnelle au pas le plus fin de la grille. Le maximum correspond à la position du réseau pour laquelle l'image de la grille d'entrée est exactement superposée à la grille de sortie. Le flux transmis est alors maximum si les deux grilles sont identiques puisque chaque élément de la grille d'entrée est superposé à son homologue de la grille de sortie. Au contraire le flux est nul si les deux grilles sont complémentaires puisque chaque élément transparent de la grille d'entıée est superposé à son homologue opaque de la grille de sortie.

Si on effectue maintenant le même raisonnement en décalant les deux grilles de la quantité ϵ l'une par rapport à l'autre suivant la direction perpendiculaire à la dispersion, la fonction instrumentale obtenue se réduit au premier terme (triangle à large base, Figure 1*b*): en effet $F(x, y)$ est bidimensionnel et sa fonction d'autocorrélation peut décroître aussi vite suivant OY que suivant OX. On peut utilisei ce fait pour annuler le terme parasite: un mouvement de translation alternatif de l'image de la grille d'entrér suivant OY produit une modulation dont l'amplitude est maximale pour la position centrale de superposition, décroît et s'annule rapidement dès que l'on s'écarte de cette position. Un moyen pratique utilisé pour obtenir ce résultat consiste à imprimer au miroir collimateur un mouvement de rotation alternatif de faible amplitude; l'amplitude est telle que le décalage, suivant OY, de l'image de la grille d'entrée par rapport à la grille de sortie n'excède pas la largeur du pas le plus fin de la grille.

PRINCIPE D'UN SPECTROMETRE À GRILLES À DOUBLE FAISCEAU

Le but d'un spectromètre à double faisceau est d'obtenir directement et automatiquement la courbe de transmission d'un échantillon. Pour cela il est nécessaire de faire le rapport des signaux obtenus avec et sans échantillon. Les solutions actuellement adoptées par les constructeurs comportent toutes l'emploi d'un faisceau de mesure M (avec échantillon) et d'un faisceau de référence R (sans échantillon) issus d'une source unique et mesurés par le même détecteur.

Deux catégories de problèmes peuvent être distingués: le problème de la séparation et de la recombinaison des deux faisceaux; et le problème de la mesure du rapport des intensités.

Separation des Faisceaux

Dans les appareils infrarouge à fentes on utilise habituellement une source unique émettant dans deux directions voisines (l'emploi des lames semi-transparentes n'étant guère possible que dans le visible).

Les deux faisceaux étant ainsi constitués un commutateur optique permet d'envoyer alternativement sur la fente d'entrée du monochromateur le faisceau de mesure et le faisceau de référence.

Ce commutateur doit obligatoirement être situé à l'entrée de l'appareillage afin de filtrer son rayonnement propre et d'éviter des phénomènes parasites importants.

L'adaptation de ce principe au cas d'un spectromètre à modulation sélective est impossible. En effet le récepteur reçoit dans ce cas, non seulement l'energie relative à l'élément spectral étudié, mais également un flux continu très important qui, interrompu périodiquement par le commutateur, provoquerait des phénomènes transitoires très difficiles à amortir. Plusieurs dispositifs permettent de résoudre cette difficulté.

La Figure 2 montre le schéma de principe d'une première solution. Une grille unique est utilisée par réflexion à l'entrée et par transmission à la sortie. Elle est composée de quatre éléments *R*, *M*, *R'*, *M'* (Figure 3).

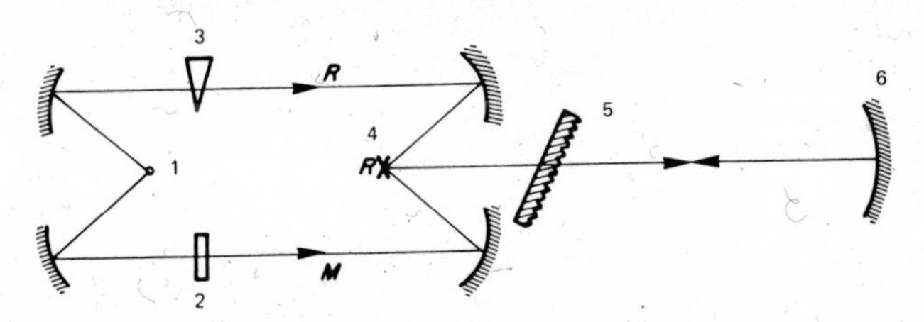

FIG. 2. Schéma optique d'un appareil double faisceau—solution 1. 1, globar; 2, échantillon; 3, atténuateur; 4, grille; 5, réseau; 6, miroir oscillant.

Les éléments *R* et *R'* correspondent au faisceau de référence. Ils sont conjugués l'un de l'autre à travers l'appareil.

Les éléments *M* et *M'* correspondent au faisceau de mesure. Ils sont également conjugués l'un de l'autre à travers l'appareil. *M* et *M'* sont dans le même plan. *R* et *R'* sont dans un autre plan. L'angle formé par ces deux plans est tel que le faisceau de référence et le faisceau de mesure sont réfléchis respectivement par les plans *RR'* et *MM'* dans la direction du miroir collimateur oscillant (Figure 2).

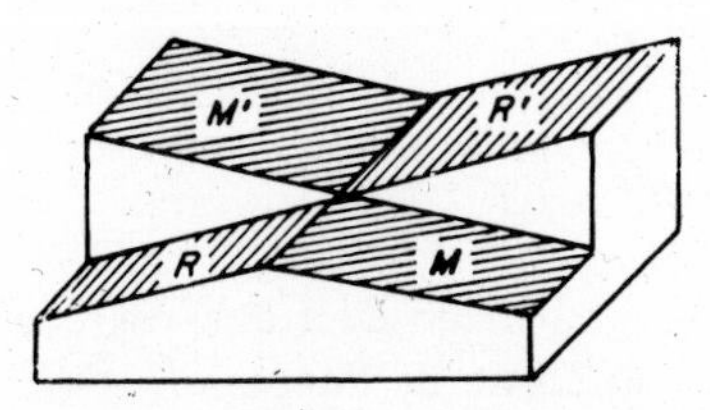

FIG. 3. Grille à quarte éléments.

Les éléments *R* et *R'* sont complémentaires tandis que les éléments *M* et *M'* sont identiques. Par conséquent la teinte plate obtenue est sombre pour le couple *RR'*, brillante pour le couple *MM'*: les signaux résultants des faisceaux *R* et *M* sont donc en opposition de phase. Ce dispositif est parfaitement symétrique pour les deux faisceaux. Il a été abandonné, après expérimentation, en raison des difficultés techniques rencontrées dans la fabrication des grilles.

La Figure 4 montre le schéma de principe de la solution adoptée. Dans cette solution la grille d'entrée et la grille de sortie sont distinctes.

La grille d'entrée, éclairée simultanément et de façon continue par les faisceaux *R* et *M* fonctionne à la fois par réflexion (faisceau *M*) et par transmission (faisceau *R*).

Le récepteur reçoit donc ces deux flux en opposition de phase puisque, si la grille d'entrée à la grille de sortie sont identiques pour l'un des faisceaux elles sont complémentaires pour l'autre ; en l'absence d'échantillon le signal est nul.

Une lame compensatrice identique au support de grille est placée sur la face photogravée de la grille afin de rendre le montage symétrique.

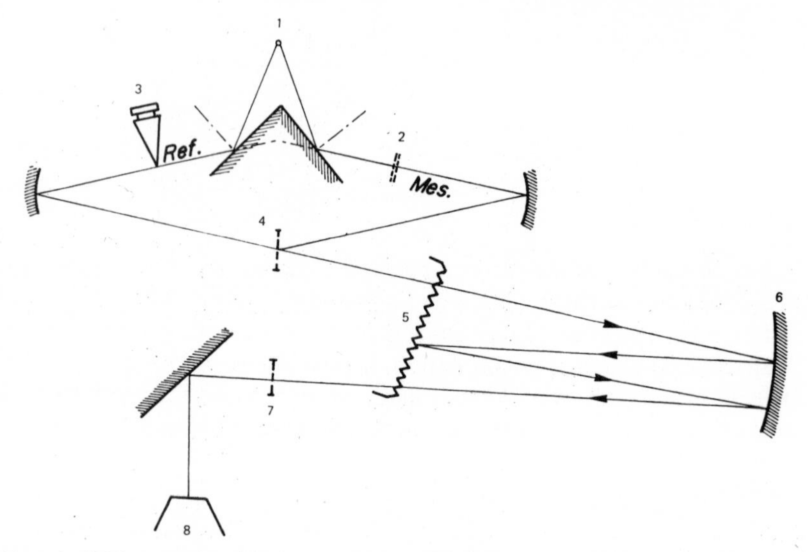

Fig. 4. Schéma optique d'un appareil double faisceau—solution 2. 1, globar; 2, échantillon; 3, atténuateur; 4, grille d'entrée; 5, réseau; 6, miroir oscillant; 7, grille de sortie; 8, détecteur.

On peut envisager également de transposer ce principe de façon à placer l'échantillon à la sortie de l'appareil, le récepteur occupant alors la place de la source et inversement.

Ceci présente l'avantage d'éviter l'échauffement de l'échantillon puisque le flux qui le traverse est déjà filtré partiellement par le monochromateur.

Toutefois deux inconvénients sont à craindre:

(1) Le récepteur recoit les deux flux à comparer sous des angles différents. Il doit donc avoir une ouverture assez grande et une sensibilité indépendante de l'incidence, condition rarement réalisée.

(2) La présence d'un élément mobile rayonnant, l'atténuateur optique, à proximité du récepteur risque de créer des phénemènes parasites.

Mesure du Rapport des Intensités des Deux Faisceaux

On sait que l'on peut distinguer deux catégories d'appareils à double faisceau à fente:

(1) Double faisceau à 'zéro électronique' basé sur la mesure de deux tensions caractéristiques des deux voies de mesure.

(2) Double faisceau à 'zéro optique' basé sur l'emploi d'un atténuateur optique.

Tous ces systèmes comportent sous des formes diverses l'emploi d'interrupteurs optiques des faisceaux. L'adaptation de ces deux méthodes au cas de la modulation sélective se heurte donc à la difficulté déjà rencontrée au paragraphe précédent.

Premier montage ('*zero electronique*'). L'utilisation d'une méthode de zéro électronique suppose que les deux voies de mesure peuvent être distinguées électroniquement par exemple en les modulant à des fréquences différentes. Dans un spectromètre à grilles la modulation de l'intervalle spectral isolé est réalisée par oscillation du miroir collimateur à la fréquence F. Lorsque cette oscillation se fait de part et d'autre de la position moyenne de réglage le signal est modulé à la fréquence $2F$. Par contre si l'oscillation se fait d'un seul côté de la position de réglage le signal est modulé à la fréquence F (Girard 1965).

On peut donc envisager le montage suivant.

La grille est constituée de quatre éléments R, M, R', M' comme il a été décrit au paragraphe précédent (Figures 2 et 3). L'image de R se superpose exactement à R' pour la position moyenne du miroir. Le faisceau de référence est donc modulé à la fréquence $2F$. Par contre les éléments M et M' sont placés de telle façon que l'image de M se superpose exactement à M' pour la position extrême du miroir oscillant. Pour cela il suffit de décaler M' de la quantité $A/2$ dans la direction de la vibration (A est l'amplitude de l'oscillation; voir Figure 5). Le faisceau de mesure est alors modulé à la fréquence F.

Les signaux obtenus sur chaque voie étant des fonctions paires de période $1/F$ et $1/2F$ ils ne contiennent que des harmoniques impairs ($3F$, $5F$ . . . pour l'un, $6F$, $10F$ . . . pour l'autre) et peuvent théoriquement être traités électroniquement de façon indépendante.

Ce procédé a été expérimenté mais a dû être abandonné. Il est en effet très difficile d'obtenir un signal pur à la fréquence $2F$. Un très faible déréglage fait intervenir un signal parasite de fréquence F et les mesures deviennent peu précises.

Deuxième montage ('*zero optique*'). Ce principe est applicable aux deux schémas optiques (Figures 2 et 4) décrits en haut.

Le miroir collimateur oscille à la fréquence F et le récepteur reçoit par conséquent les flux des canaux R et M modulés à la fréquence $2F$, en opposition de phase. Un atténuateur optique est asservi en position de façon à annuler en permanence le signal différentiel, par égalisation des deux faisceaux. Il suffit de lier la plume de l'enregistreur à l'atténuateur préalablement étalonné, pour obtenir directement le coefficient d'absorption de l'échantillon.

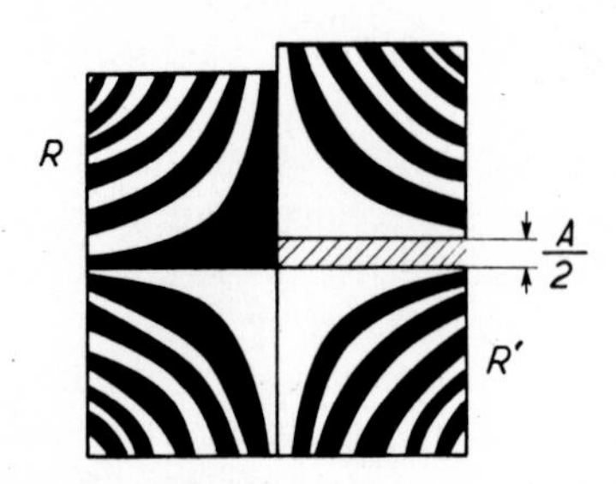

FIG. 5. Solution à zéro électronique. Position des 4 éléments de grille.

Remarques Générales

On sait que la suppression des fentes a pour avantage principal de rendre à peu près indépendantes la résolution et la luminosité des spectromètres.

La modulation sélective a un autre avantage, moins souvent souligné: la lumière parasite diffusée ou émise par les parois et les pièces optiques n'influence pas le signal de sortie puisqu'il n'existe pas de 'chopper' analogue à ceux des appareils à fentes.

Par contre, la souplesse d'utilisation des 'programmes de fentes' utilisés dans les appareils double faisceau classiques est évidemment incompatible avec l'emploi de grilles. L'appareil réalisé et sommairement décrit ci-dessous comporte trois jeux de grilles dont le pas et les dimensions ont été déterminés pour que l'énergie disponible varie dans des limites raisonnables entre 2 μ et 30 μ.

PRÉSENTATION DE L'APPAREIL RÉALISÉ

Etant donnés les besoins de l'utilisateur concerné par ce type d'appareil (essentiellement des chimistes et des physico-chimistes) une résolution moyenne a été adoptée. Le gain important de luminosité dû à l'utilisation des grilles permet d'accroître soit la vitesse d'enregistrement, soit la précision de mesure par rapport à un appareil à fentes de caractéristiques optiques identiques:

L'intervalle spectral couvert s'étend de 2 à 30 μ (330 à 5000 cm^{-1}).

La définition spectrale moyenne est de 0,5 cm^{-1} et reste comprise entre 0,3 et 1 cm^{-1}.

La précision de mesure est meilleure que 0,3 pour cent.

Le schéma général est représenté sur la Figure 6.

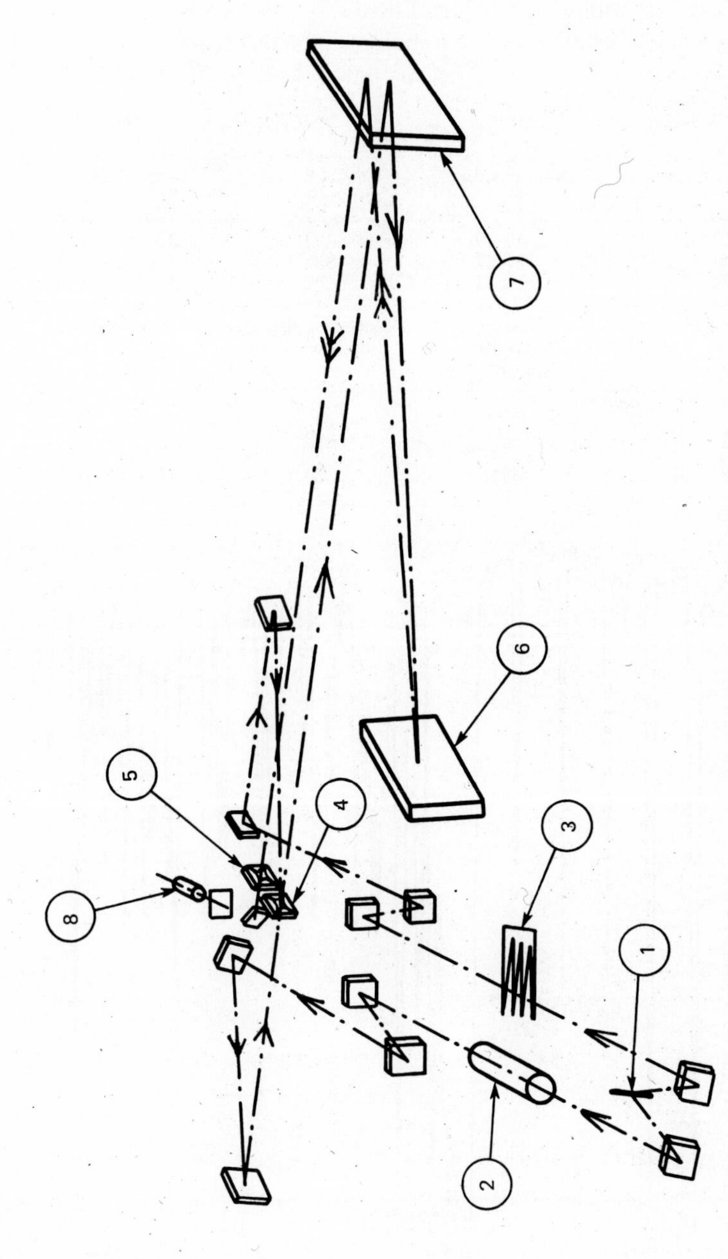

FIG. 6. Schéma général de l'appareil. 1, globar ; 2, échantillon ; 3, atténuateur optique ; 4, grille d'entrée ; 5, grille de sortie ; 6, réseau ; 7, miroir parabolique off-axis F = 1000 ; 8, détecteur pneumatique.

La source est un globar refroidi par circulation d'eau.
Le réseau est utilisé en montage Littrow.
Le collimateur est un miroir parabolique off-axis oscillant à la fréquence de 8 Hz.

TABLE 1

Grilles	Dimensions (mm)	Pas Minimum (mm)	Intervalle Spectral (μ)
1	20 × 20	0·3	2 à 7·5
2	27 × 27	0·6	7·5 à 18
3	27 × 27	2·0	18 à 30
Reseaux	**Nombre de traits par mm**	**Longueur d'onde de blaze (μ)**	**Intervalle Spectral (μ)**
1	300	3	2 à 4·5
2	150	6	4·5 à 9
3	75	12	9 à 14·5
4	45	20	14·5 à 30

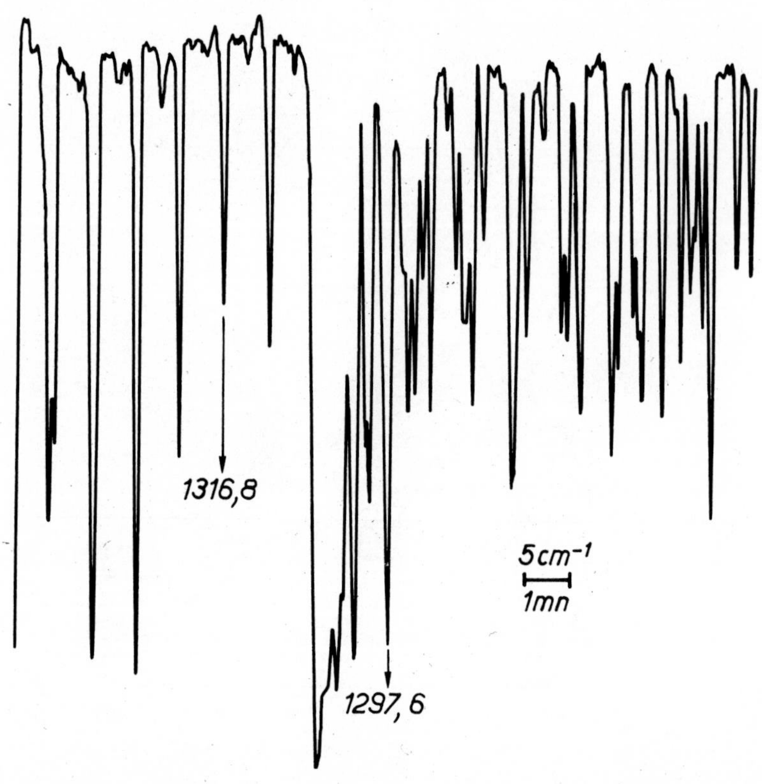

FIG. 7. Spectre du méthane, CH_4, vers 1300 cm^{-1}. Cuve de 100 mm. Pression 50 torr.

L'appareil utilise trois jeux de grilles et quatre réseaux dans le premier ordre permutant automatiquement.

Les caractéristiques des grilles et des réseaux sont rassemblées dans le Tableau 1.

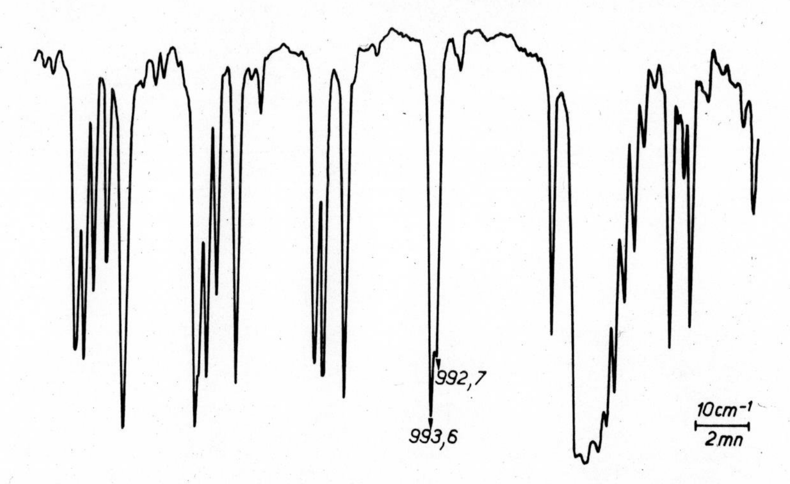

FIG. 8. Spectre de l'ammoniac, NH_3, vers 1000 cm^{-1}. Cuve de 100 mm. Pression 50 torr.

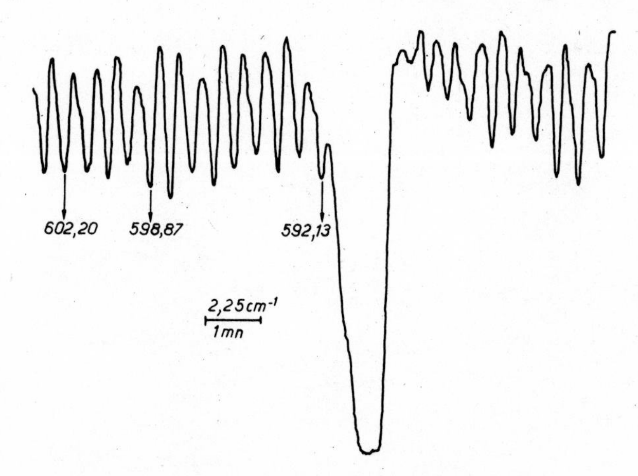

FIG. 9. Spectre de l'oxyde d'azote, N_2O, vers 600 cm^{-1}.

Le récepteur est un récepteur pneumatique ONERA suivi d'un amplificateur à bande étroite centrée sur 16 Hz. Le spectre est enregistré selon une échelle linéaire en nombre d'onde.

Quelques exemples d'enregistrements sont représentés sur les Figures 7, 8 et 9.

REFERENCES

CONNES, P., 1958, *J. Phys.*, **19**, 230.
GIRARD, A., 1963, *Appl. Opt.*, **2**, 79.
GIRARD, A., 1965, *Jap. J. Appl. Phys.*, **4**, suppl. 1, 379.
GIRARD, A., 1967, *Publication O.N.E.R.A.*, No. 117.
GOLAY, J. M., 1951, *J. Opt. Soc. Amer.*, **41**, 468.

Part II

Recent developments in optical production techniques, including the use of new materials

Recent Developments in Optical Production Techniques

Roderic M. Scott

The Perkin-Elmer Corporation,
Norwalk, Connecticut, U.S.A.

Abstract—Space astronomy has developed requirements of considerable severity for stable materials and accurate optical surfaces. Most potential candidates for mirror materials have been examined by precise interferometry on small samples for uniformity of thermal properties and for creep. This application has led also to the development of new methods of surface generation and control as alternates to figuring on pitch. Two such methods are ionic polishing and selective deposition. In addition, a better understanding of the dynamics of polishing on pitch has permitted much improved methods of figuring by the direct control of pressure and velocity of the lap.

All of these activities are made possible by the availability of large, digital computers.

A problem of optical manufacture receiving considerable attention today is that of large optics for space telescopes. These instruments are to be of large aperture and designed to exploit the advantages of the space environment. Thus they should be limited in their resolution only by their aperture and this, hopefully, at ultra-violet wavelengths. The expense of building and launching a large telescope is so great that a long and useful life must be expected. While it is probable that the observatory will be visited from time to time for changes of experiments and equipment, and for service and maintenance, basic elements such as large mirrors should be stable over periods of time approaching ten years. This requirement has led to an extensive study of potential mirror materials.

Experience with satellites so far has taught us that thermal changes and particularly thermal gradients will be a serious environmental problem for any large optical system. Thus the material for a large mirror must either be a very good conductor of heat so that significant thermal gradients cannot be supported, or it must have a very small coefficient of expansion. When one adds the mechanical requirements of low density, high rigidity, long-term stability and suitability for an optical surface the available materials make a rather short list. We have undertaken to study the stability of most of these materials so that some projection of their ability to hold an optical figure for many years can be assessed. It is, of course, not practical to make large optics of accurate figure from each of the materials and store them away or subject them to expected environments over long periods of time. Thus, equipment has been developed to examine smaller samples for instability with great sensitivity.

It would seem appropriate at this conference to describe the interfero-

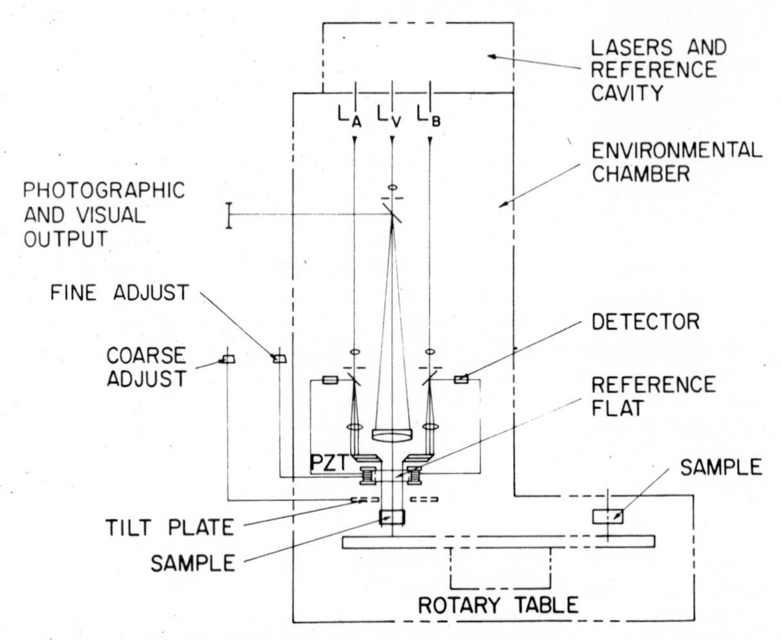

FIG. 1. Schematic of interferometer for testing creep in samples of optical material.

meter developed for the observation of the small creep rates resulting from material instabilities. The device (Pilston & Steinberg 1969) is illustrated in Figure 1. It is a classical Fizeau interferometer whose spacing and wedge angle are servo-controlled. Two He : Ne lasers operating at slightly different frequencies provide signals to control the wedge. A third laser has its frequency shifted in steps to provide the different fringe positions across the face of the sample. Figure 2 illustrates the way in which each of the three lasers are slaved to a single reference cavity whose adjacent modes are the frequencies involved. A typical interferogram appears in Figure 3. The fringes are not separated by a distance corresponding to a wedge separation increase of half a wave because each fringe is at a frequency different from its neighbour by one order in the reference cavity. In the figure the fringe spacing is $\lambda/17$ and the limiting detectable surface error is $\lambda/300$.

The results of about a year of observations on a large number and variety of samples are shown in Tables 1 and 2. Some of the samples were cycled through small temperature ranges—about 10°F—many times and

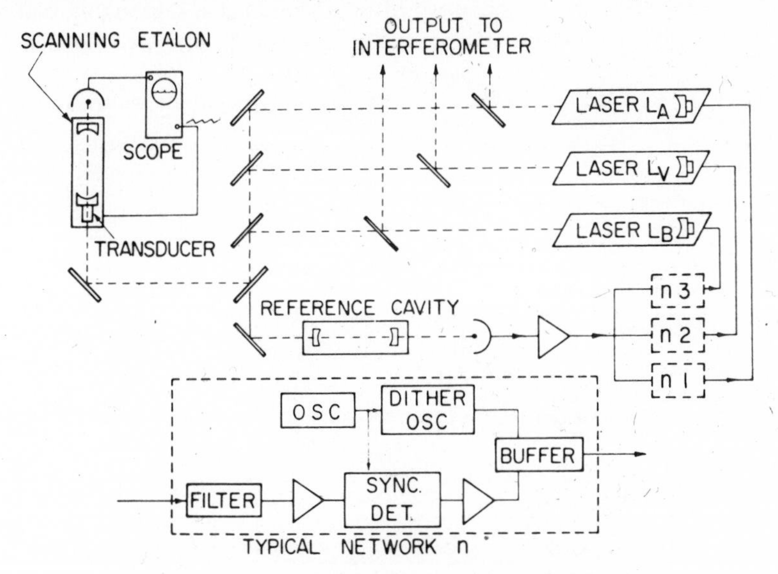

FIG. 2. Frequency control of laser for creep interferometer.

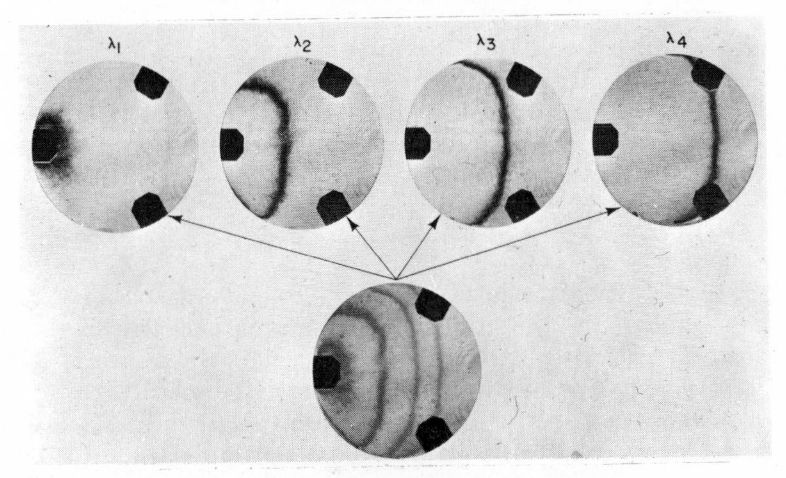

FIG. 3. Fringe system of creep interferometer.

then re–measured at their original temperature. Others were measured at an elevated temperature of 8 or so degrees and compared with their figure at the base temperature.

Some materials are much better than others but all show some creep or non-uniformity of expansion and thus a troublesome problem remains. It is probably safe to extrapolate in time. That is, if λ/50 RMS change of figure occurs in one year, it is likely that λ/5 RMS change of figure will occur in 10 years. However, none of us know what a reasonable scaling with size might be. Should the results of these small samples be scaled up

TABLE 1. Dimensional Stability of Optical Dielectric Mirror Materials

Spec. no.	Material/supplier	Thermal stability (10^{-3}λRMS/°C)	Temporal stability (λRMS/yr.)
	Fused Silica		
70	Fused silica, Corning C–1–0	0·130	0·079
89	Fused silica		0·032
79	Fused quartz, GE–124	<0·15	0·058
80	Fused quartz, GE–125	<0·15	0·057
81	Fused silica, GE–151	0·07	0·039
82	Fused quartz, GE–204	<0·3	0·058
83	Fused quartz, GE–510	<0·15	0·036
84	Fused quartz, GE–124 +510	<0·3	0·038
	ULE		
59	Lightweight Corning	0·088	
74	Corning 7971	0·063	0·051
75	Corning 7971	0·111	0·060
147	Corning 7971	0·081	
	CER-VIT		
61	C–101, Owens, III		0·027
133	C–101, Owens, Ill	0·120	
161	C–100, Owens, Ill	0·038	
	Others		
66	Glass-ceramic, Corning	0·201	0·076
62	Glass-ceramic, Corning		0·065
87	E–6 glass, Ohara		0·057
88	E–6 glass, Ohara		0·058
125	Z–glass, Heraeus	0·052	
149	Solderglass, GE	0·145	
150	Solderglass, GE	0·177	

by the surface area, the volume, the linear dimensions to some power, or taken as they are to predict the effect for large mirrors? There are some qualitative arguments that the scaling law should be a growth of RMS error of figure proportional to the square root of the volume.

Let us now turn to another aspect of optical manufacturing development. There are, today, several methods of achieving high-quality optical surfaces competing with the traditional method of figuring on pitch laps. They may be divided into two types. On the one hand are those which, like the classical methods, remove material differentially at particular points or small areas so that those parts of the surface which are high with respect to the desired surface may be brought down to it. On the other hand, there are those methods which add material to the low areas in a controlled way.

The first technique is exemplified by 'Ionic Figuring'. In this method a beam of energetic ions is swept over the surface to be figured and the material is literally boiled away atom by atom to reduce the surface. After

TABLE 2 Dimensional Stability of Optical Dielectric Mirror Materials

Spec. no.	Material/supplier	Dimensional stability		
		Temporal (λRMS/yr.)	Thermal[a] (λRMS/°C)	Thermal–cycling[b]: 10 cycles[c](λ)
Commercially produced by vacuum hot pressing				
30	GB–2, GAC	0·24	1·01	
90	I–400, BBC, Ni-coated	0·49	—	—
91	S–200–D, BBC, Ni-coated	0·22	—	—
162	HP–50, KBI	—	2·27	—
175	SP, KBI	—	4·65	—
178	I–400, BBC	0·05	1·25	0·003
179	S–200–D, BBC	0·11	1·80	0·009
Hot isostatically pressed by Battelle Memorial Institute				
13	P–40	0·16	0·43	—
15	SP–200–E	0·25	1·31	—
20	P–40	—	0·59	—
43	SP–200–D, Ni-coated	6·46	—	—
60	SP–200–D, lightweight	0·12	—	—
92	SP–200–D, lightweight	0·31	1·15	—
94	SP–200–D, lightweight	0·19	1·39	—
131	P–40, isoforged, lightweight	0·06	1·18	0·016
Pressureless sintered by Stanford Research Institute				
1	FP–17	—	1·50	—
2	FP–17	—	2·94	—
6	FP–17+1 Fe, etched	0·06	0·41	—
7	FP–17+1 Fe, as-machined	0·23	0·52	—
23	FP–17	0·04	4·58	—
40	P–50	0·13	2·91	—
164	P–50+5 Cu	0·05	0·87	0·006
165	FP–17+5 Cu	0·03	0·58	0·006
186	P–50	0·04	—	0·005
189	P–50	—	0·8	—
191	SP	0·06	0·67	0·004

[a] Thermal stability obtained over a temperature interval of 20–80°C.
[b] Thermal cycling: 27–32°C, 90-minute period.
[c] $\lambda = 0{\cdot}6328\ \mu m$.

a long period of experimentation to determine the effects of ion mass, accelerating voltage, beam current, deflection rates and many other factors, successful figuring has now been accomplished and at acceptable rates. Figure 4 shows a flat depressed area about $\lambda/4$ deep excavated by an ionic beam swept in a controlled pattern. The deflection chamber is illustrated in Figure 5. A quite standard ion-beam generator is out of the picture to the right. Perhaps a more meaningful experiment was the parabolization of a sphere. The results appear in Figure 6. The figure of the piece after the first run was $\lambda/35$ RMS deviations from a parabola. The removal of material is

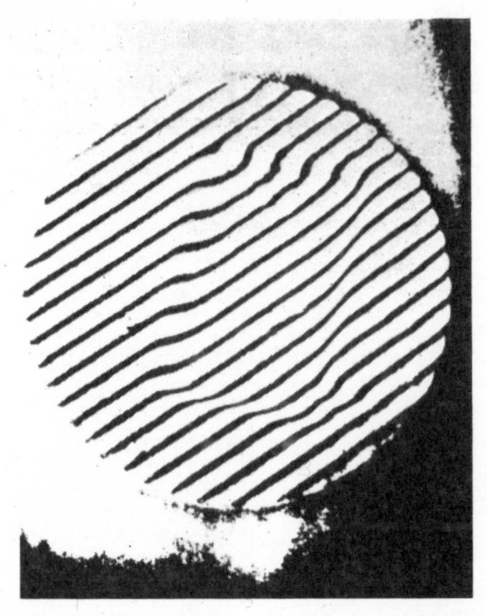

FIG. 4. Square depression produced by ionic figuring.

FIG. 5. Ion-beam figuring tank.

slow, and this feature can be exploited to make it very precise. One great advantage is that the beam of ions may be deflected in any pattern desired so that surfaces of quite non-symmetrical shape are as easily figured as any other.

An example of the method whereby the desired surface is achieved by adding material to the low areas has been described by J. Strong (1938). The real problem has been the generation of masks with sufficient accuracy of detail to be of material advantage in the figuring of large surfaces. This is now done from interferograms by the use of large programs on electronic computers. The real advances in this area have come, first from good

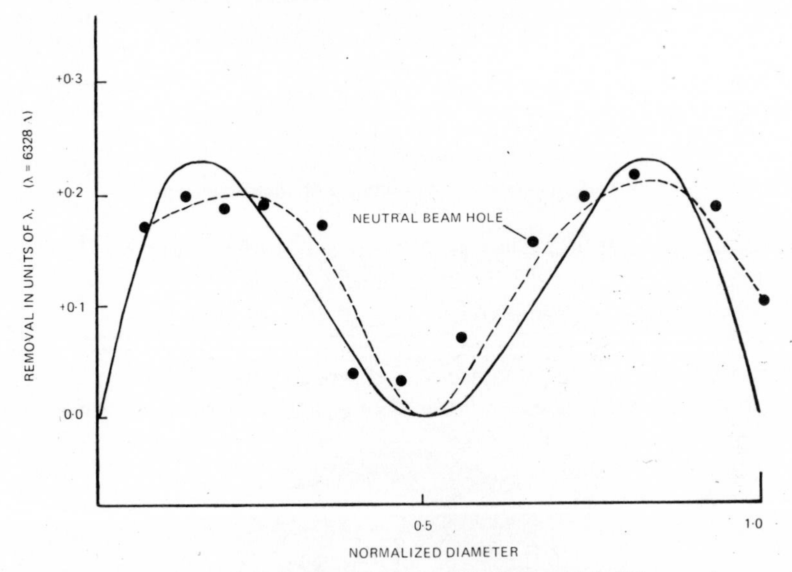

FIG. 6. Parabolization by ion bombardment: expected removal (solid), actual removal (dashed).

interferometers for the measurement of the surfaces and, second, in the analysis of the patterns and subsequent control of the correction method by electronic computation.

At this point let me call your attention to another interferometer which has provided the necessary input data of high accuracy required for the manipulation of the correction devices. Figure 7 is an optical diagram of the interferometer (Polster 1969). The system under test is expected to produce a spherical wavefront centred at the field lens. Multiple-path interference takes place at the right face of the aplanat, which is a sphere centred at the field lens. If the surface or system is aspheric in any sense the correction optics are included on the right of the field lens. A typical interferogram appears in Figure 8. Note the sharpness of the fringes, which makes quite practical automatic fringe-measuring devices which digitize the positions to provide computer input.

I would like now to turn to some developments in the more classical tradition. The shaping and preparation for polishing of optical surfaces by

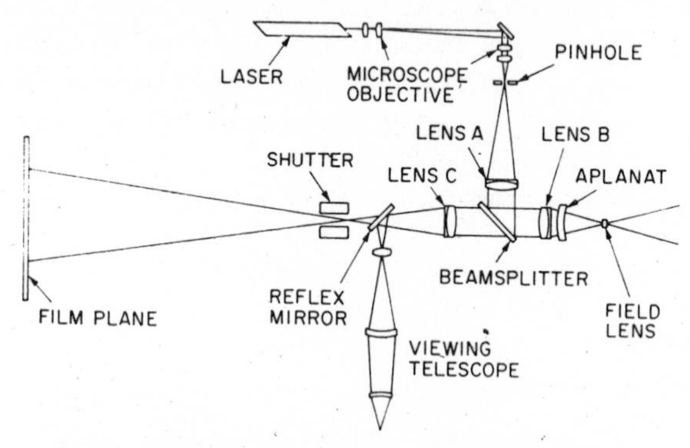

FIG. 7. Spherical wave multiple-path interferometer.

direct diamond grinding has been advanced by the employment of precision, air-bearing spindles. If the spindle carrying the diamond wheel and the one carrying the work run true enough and are stiff enough so that the depth of cut is under good enough control the generated surface will appear quite well polished. Interference fringes, even multiple-beam fringes, may be observed at quite useful contrast from such surfaces.

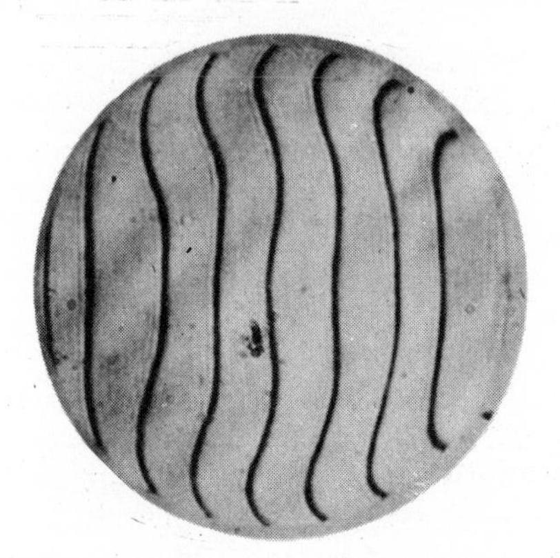

FIG. 8. Multiple-path interference fringes.

Figure 9 is an electron micrograph of a diamond-fined surface. The detail in the surface of the glass has been replicated and this replica shadowed at an angle of 20°. Thus the jagged areas are the shadows of raised ridges on the replica. These come from deep fissures in the glass left by the diamond. Note the absence of long parallel grooves typical of previous diamond grinding. Figure 10 is a fine-ground surface prepared with loose abrasive for comparison.

It has been observed that all fissures are nearly the same depth. A uniform layer of glass may be removed by ion bombardment. Figure 11 shows the results. As the removal continues, some of the fissures get shallower but the deepest ones seem to stay about the same depth for quite some time. This effect results, we think, from the inability of the replicating

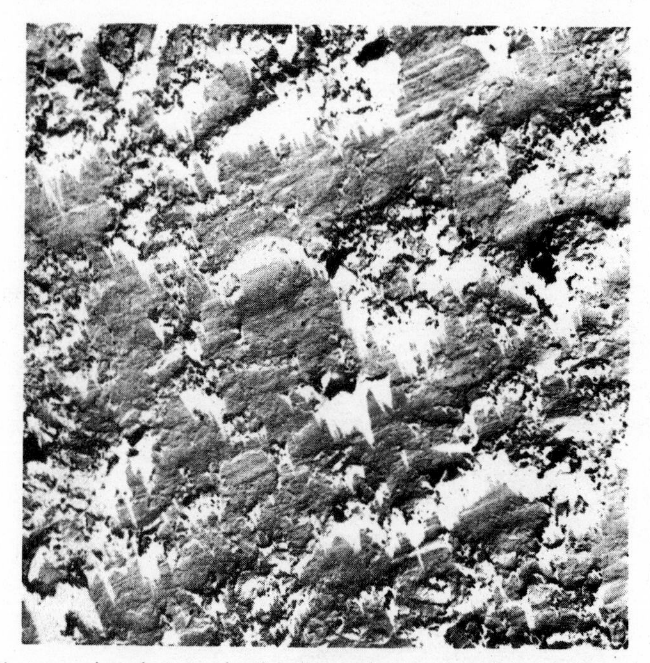

FIG. 9. Electron microphotograph of a diamond ground surface, shadowgraph angle 20°, 16 μ width of field. (|———| equivalent to 1 μ)

material to penetrate beyond a particular depth in the deep, narrow cracks. Finally, however, all the fissures disappear and a good, scatter-free, undamaged surface remains. About two waves must be removed from the glass surface to be sure that all effects of the diamond are gone.

In 1927 F. W. Preston published a paper suggesting relations which we now refer to as the 'polishing equations'. Consider a small area of the surface of a glass piece being polished on a pitch lap. It is postulated that for a useful range of the variables the amount of glass removed from the surface is:

$$\delta_p = k_p P v t \qquad (1)$$

The surface is polished away to a depth δ_p in time t by a pressure between glass and pitch P, when the part of the surface in question moves with a velocity v over the lap. The constant k_p depends on a number of characteristics of the polishing material, of the lap material and of the glass, but is independent of the pressure or velocity. A series of experiments in which carefully weighted small pieces of glass were polished, with first pressure and then velocity held constant, the other being varied, supports the validity of Equation (1). Such a relation arises from the assumption that the coefficient of friction is independent of velocity and the work required

FIG. 10. An electron microphotograph of a fine ground pyrex surface, 5 μ grit, same scale as in figure 9. (|———| equivalent to 1 μ)

to remove a given volume of glass by polishing is a constant, independent of pressure or velocity.

A similar equation can be applied to the flow of the pitch lap. If we consider grooved laps or laps with other forms of relief so that the pitch can flow into the grooves and lower the surface, it will so flow. The amount by which the surface is lowered local area by local area may be given by

$$\delta_L = k_L Pvt \tag{2}$$

The pressure P and velocity v are the same as in Equation (1). The constant k_L bears no relation to the previous constant but rather depends on the mechanical properties of the pitch, the grooving or other relief, and to some extent on the polishing material. No experiments have been performed to establish the validity of this relation directly but certain deductions as to the effect of a relation of this type on the shape and action of laps have been developed and observed. A rationale for this relation arises in this way. If one assumes that the rate of flow is determined by the rise in temperature of the particles of polishing compound at their points of contact with the glass and that these grains are captured and held immobile by the pitch, then those grains which are doing the most work in polishing away the glass will sink fastest into the pitch and the general surface will be lowered. Good pitch must be quite hard at room temperature but its viscosity must decrease very much at a somewhat higher temperature so that this flow can take place. Again the work in each local area must be dependent on the drag force and the velocity there. If the coefficient of friction is constant, the drag force is linearly dependent on the pressure.

Pitch has an elastic deformation with pressure as well, and if the surface of the glass does not match the surface of the pitch the deformation, and thus the pressure, will vary from point to point. Where the pressure is high the polishing will proceed at a high rate and the pitch will flow more rapidly than where the pressure is low. This well-known effect makes possible the very accurate generation of flats and spherical surfaces, and makes so difficult the generation of non-spherical surfaces. Except in the simplest cases, these two relations cannot be applied directly. However, with the aid of electronic computers they may be integrated over long paths which follow selected differential areas on the piece and on the lap. Thus the effect on the piece to change its shape, and on the lap to change it, may be predicted for many useful motions of lap and piece.

One way that this procedure may be applied is to control the path, velocity and pressure of a small lap over a large piece. The lap is small enough so that variations in curvature within the lap area produce only very small variations of pressure over the lap. Programming the path, velocity and pressure then permits controlled polishing and figuring of quite aspheric or assymmetrical surfaces.

A simple example of the application of this development is in the correcting of the angles of corner cubes. One of the packages to be left on the moon is an array of corner cubes to reflect laser light from the earth. Figure 12 is a picture of an astronaut with such an array. A large number of corner cubes of one-second accuracy is required. The method of manufacture involves working them up with mechanical generation accuracy and testing them interferometrically as shown in Figure 13. A fitting with an adjustable weight is placed over the piece so that the pressure varies over the face in contact with the lap by an amount and in a direction to correct that face. The time required for the correction is also computed. A large number of pieces are then run together in a holder on a large ring lap. The

FIG. 11. Electron microphotographs of a diamond-ground surface as more and more material is removed by ion bombardment. (|———| equivalent to 1 μ)

c

d

FIG. 12. Astronaut holding an array of corner cubes which was left on the moon.

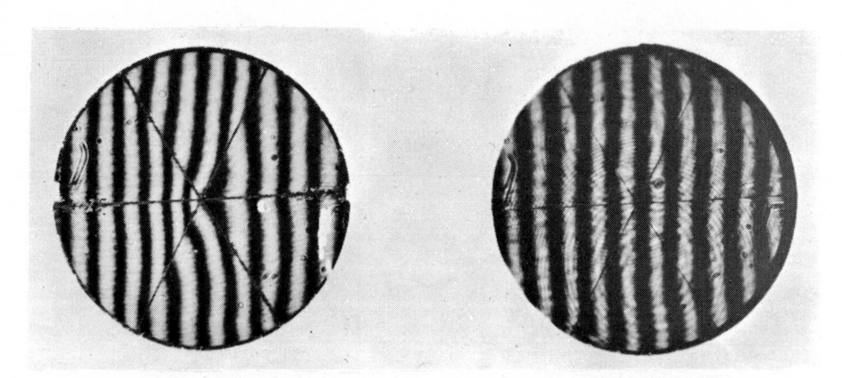

FIG. 13. Interferograms of a corner cube before (*left*) and after correction.

lap is conditioned by a large piece of glass so selected in size, weight and position as to keep the lap flat under the action of the work and the conditioning piece. The corner cubes seldom require a second correction. The greatest difficulty is in keeping track of which piece is which and thus not under- or over-running the time.

In closing I wish again to draw your attention to the increasing rôle of electronic computation in optical processing. We saw it in the reduction of interferograms for the analysis of creep in the material study, as a vital link in the process of controlling the removal of material in ionic polishing or in the design of masks for the selective addition of material. While it was not emphasized with respect to diamond grinding, it should be obvious that such a grinding machine would be computer-controlled and that the advantage of such a method is that it produces a surface good enough for interference testing and thus provides a feedback mechanism. Finally, it should be noted that although the polishing equations are deceptively simple, they require extensive numerical computation before their power can be exploited.

REFERENCES

PILSTON, R. and STEINBERG, G., 1969, *J. Appl. Opt.*, **8**, 553.
POLSTER, H. D., 1969, *J. Appl. Opt.*, **8**, 522.
PRESTON, F. W., 1927, *J. Soc. Glass Tech.*, **11**, 214. (See also K. J. Kumanin, *Generation of Optical Surfaces*, Focal press, 1962).
STRONG, J., 1938, *Procedures in Experimental Physics*, Prentice-Hall, New Jersey.

New Techniques for Optical Polishing

W. H. Steel

CSIRO Division of Physics,
National Standards Laboratory,
Sydney, Australia, 2008

Abstract—New polishers and techniques are reported which enable optical surfaces to be made flat to $\lambda/100$ entirely by machine polishing. The polishers are made on grooved disc of low-expansion glass with a thin coating of pitch or, more recently, Teflon. The flat is placed inside an annular ring and both are polished together on a slow-speed machine.

INTRODUCTION

Over the last decade there has been an increasing demand for flatter surfaces for use in lasers and interferometers, as developments of these instruments make it possible to take advantage of these better surfaces. Typically, surfaces flat to about 5 nm ($\lambda/100$ for 546-nm radiation) are now being demanded. The optical workshop of the National Standards Laboratory has developed several improvements on the usual methods of producing accurate surfaces.

Once surfaces can be made to $\lambda/100$, the problem remains of testing them to this precision against a standard of this accuracy. We have not yet tackled this problem and surfaces described as 'flat to $\lambda/100$' are, strictly, merely matched to some other surface to $\lambda/100$. We test this match by silvering the two surfaces and observing the multiple-beam fringes, a method which is just sufficient for a precision of $\lambda/100$. The absolute flatness of the surface against which the comparison is made is, however, not known to this accuracy.

Pitch Polishers

Smartt (1961) described a method of polishing calcite that later proved successful for polishing glasses, including fused silica. When a soft material like calcite is being polished, a hard pitch is needed for the polishing lap to retain its shape and produce a flat surface, but a soft pitch is required to avoid scratching the surface being polished. Smartt resolved this dilemma by making the polisher of hard pitch and then softening the surface with a suitable solvent. For wood pitch, alcohol is poured on the surface and, after a few seconds, is washed off with water. This leaves a soft layer on the surface which does not scratch calcite but is too thin to deform easily. For hard glasses such as fused silica, we now find that softening of the surface is not necessary.

These polishers are often used for polishing by hand. The polisher is rotated on a machine at a higher speed than is usual for hand polishing

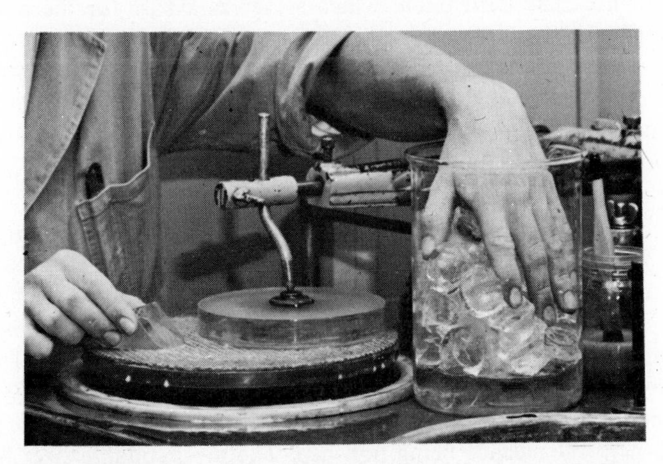

FIG. 1. The pitch lap is kept flat by machine polishing a disc of glass. The smaller component is polished by hand on one side of the lap. The hands are cooled in a beaker of ice.

(~50 rev/min). To keep the polisher flat, the machine is used to polish a larger disc of glass while the smaller flat being made is held on one side of the polisher, as shown in Figure 1. With all hand polishing, it is a problem to prevent warming the component being polished by the hands. Initially, felt boxes were made to surround the component but they made sensitive holding of the component more difficult. Our present technique is to cool the hands in a beaker of ice.

During the manufacture of calcite plates and of other thin components, extensive use has been made of optical contacting. To make a thin plate it is first polished on one side, then this side is contacted to a thicker glass flat to hold the plate rigidly while polishing the second side. Another use of optical contacting has been to join a compensating plate to a coated beam-splitter, so as to match both arms of an interferometer; such beam-splitters have been made with as many as 16 evaporated layers in the coating. These applications have been described by Smartt and Ramsey (1964).

The polishing powder that has been found most successful for polishing soft materials such as calcite is Glassite, a black iron oxide. Recent trials, however, suggest that the anastase form of titanium dioxide is better for finishing calcite surfaces; it polishes more slowly but gives fewer fine scratches.

Pitch-on-Glass Polishers

Otte (1965) has described improvements to both the polisher and its method of use. The hard pitch, which serves as a base for the polisher, can be replaced by a harder material such as a low-expansion glass, Pyrex or Duran 50. The base of the polisher is then a glass disc stuck to a metal lap. The face of the disc is grooved with a diamond saw and then ground flat and polished. The facets left between the grid of grooves are coated individually with a layer of pitch about 0·5 mm thick, a fairly hard pitch being used to polish hard glasses.

With this polisher a technique has been adopted for polishing flats entirely by machine with no hand polishing. The disc of glass used previously to keep the polisher flat is replaced by an annulus inside which the flat is placed. The annulus carries the weight of the polishing arm of the machine, while the flat is separately loaded by weights. The flat is free to turn inside the annulus and, to prevent chipping, plastic tape is applied either to the outer edge of the flat or the inner edge of the annulus, near the surface being polished. The most suitable sizes for the annulus and polisher in relation to that of the flat have been found to give the ratio 3:8:10 for the diameter of the flat, the outer diameter of the annulus, and the diameter of the polisher; the inside diameter of the annulus should be about at least 2 mm greater than that of the flat. Figure 2 illustrates this polisher in use.

The flat being manufactured is first polished to within about $\lambda/2$ on a conventional polisher. It is then transferred to the special polisher which, to prevent heating, is rotated very slowly, about 5–10 rev/min; to obtain these speeds we have either modified commercially available polishing machines or built our own. 'Cerium oxide' polishing powder is added to the polisher once a day only, so that it breaks down to give a fine finish. Once the polisher is flat, it will produce surfaces to $\lambda/100$ with 6–12 hours polishing. The flats have practically no 'turned-down' edge.

Teflon Polishers

Unfortunately, this new pitch polisher, although it retains its shape or 'figure' for longer than a conventional polisher, still changes more rapidly than desirable. Some polishing runs produce good flats but many more have to be interrupted to re-flatten the polisher. In addition, the polisher does not last longer than a few weeks. The hardness of the, pitch is fairly critical, being chosen by experience. During use, however, the pitch becomes loaded with polishing powder and changes its hardness until it is necessary to re-coat the polisher with fresh pitch.

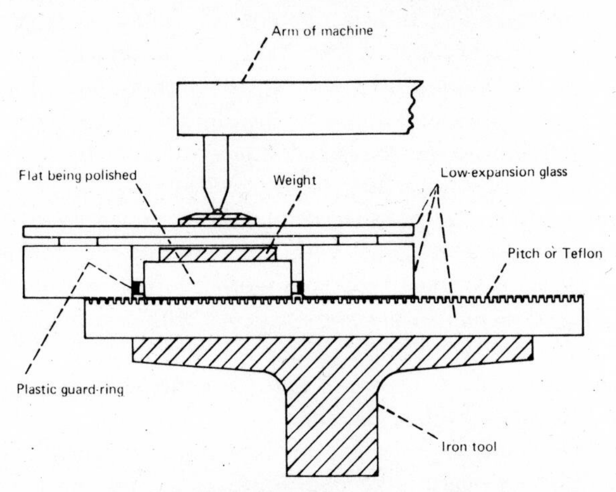

FIG. 2. Diagram showing the method of polishing by machine on glass-based polishers.

A search has been made, therefore, for a substitute for pitch that will retain its figure and its hardness for longer periods. Several materials appear promising and one, Teflon F.E.P. (fluorinated ethylene propylene), has now been used for some years (Otte 1969). The surface of the grooved glass disc is coated with this material. Since it does not flow like pitch, it does not change its figure appreciably over periods of use of several days. Polishing powders do not embed in it, so its hardness does not change and re-coating is not necessary. Once a Teflon polisher has been made flat, it produces flats more reliably than a pitch polisher. The edge is even better than obtained from the special pitch polisher, showing no turned-down edge extending further inwards than 0·5 mm. The microstructure of the polish seems practically as good as that obtained from pitch. With slow polishing, glasses with a high coefficient of expansion can be polished almost as flat as fused silica. The time to polish a flat 75 mm in diameter is about 5 hours.

Making Teflon Polishers

The polisher is again made on a glass disc. Pyrex or Duran 50 have mostly been used, but some recent polishers have been made on Cer-Vit to see if this improves their stability; this work is incomplete. The disc is grooved by a grid of saw cuts, 1 mm wide and 1 mm deep with facets 1·5 mm square left between them. To reduce the effective area towards the centre of the polisher, eight further cuts are made in the surface of the disc along equally-spaced diameters. The disc is then ground flat but not polished, because Teflon will not bond to a smooth glass surface.

The facets of the polisher are coated with Teflon to a thickness of about 0·5 mm by painting on them first one coat of Dupont 856–301 bonding coat, then five coats of Dupont 852–200 F.E.P. to which 30 per cent of graphite powder has been added as a non-scratching filler. Each coat is applied with a soft brush and cured by heating to 350°C. The base of the glass disc is fixed, with double-sided adhesive tape, to a metal plate that screws to the polishing machines.

Since, unlike pitch, the polisher cannot be flattened by use, it is cut flat by rubbing it, on a machine, with a flat disc of ground glass (8 μm grind), This leaves a surface that is smooth enough to show Newton's rings when viewed at almost grazing incidence with an optical flat placed on top of it. The polisher is ready when it is flat to about 1λ; the final figure of the flat polished on it can then be controlled by adjusting the spindle speed of the polishing machine.

This Teflon polisher is used in the same way as the pitch polisher that it replaces. More care is needed, however, in grading the polishing powder. Since this powder does not embed in the Teflon, any coarse particles present produce fine scratches. It has been found necessary to use 'cerium oxide' that has been shaken with water and allowed to settle for about 30 minutes. The material still in suspension is used, well diluted with water.

The flat being polished is placed, as before, inside an annulus and is loaded with an additional weight of about three-quarters its own weight. Figure 3 shows the polisher in use on a double-eccentric polishing machine of our own design. The machine has three separate speed-controlled dc motors and gives, as well as very low speeds, infinitely variable ratios between the speeds of the polishing spindle and each eccentric.

Surface Quality

As stated in the Introduction, the final test used for flatness of the surfaces is the observation of multiple-beam fringes, the surfaces being compared having been coated with evaporated silver. Such fringes from a Teflon-polished flat and a conventionally polished master flat are shown in Figure 4; only a very small region of turned-down edge is seen. It is hoped in future to replace this method of testing by a more precise photoelectric method that does not require silvering the surfaces (for example, that of Roesler & Traub 1966).

Some tests have been made to see if a Teflon polisher gives a surface on the glass with a different microstructure to that obtained from pitch.

FIG. 3. Teflon polisher in use on a special slow-speed polishing machine with complete speed control on the polishing spindle and the two eccentrics. Also shown is a spare annulus, the 'cerium oxide' suspension and ladle, water for keeping the polisher moist, and the ground disc for cutting the Teflon surface flat.

FIG. 4. Multiple-beam fringes between a Teflon-polished flat of 75 mm diameter and a larger standard flat.

Phase-contrast photographs of the surface (Martin & Ramsay 1952) suggest that the Teflon does not give quite as good a surface as the pitch-on-glass polisher. Both surfaces, however, are much better than that obtained by the conventional method of polishing on a fast pitch polisher without 'breaking down' the polishing powder. In addition, electron micrographs of the three types of surface have confirmed the same result, although with less certainty, since each picture covers so small an area.

When used in Fabry-Perot interferometers of 50 mm aperture, the surfaces obtained from either polisher give *finesses* of 30–50. In fact, the present limit to the *finesse* obtainable is no longer the surface flatness but the uniformity of the multilayer reflecting coating that is put on the surfaces. Work is in progress to improve these coatings.

Acknowledgements

The work described here has resulted from the collaborative efforts of the whole staff of the National Standards Laboratory optical workshop, both those who are cited in the references and others.

REFERENCES

MARTIN, L. C. and RAMSAY, J. V., 1952, In *Le contraste de phase et le contraste par interférences*, ed. M. Françon, Paris, *Rev. Opt.*, p. 164.
OTTE, G., 1965, *J. sci. Instr.*, **42**, 911.
OTTE, G., 1969, *J. sci. Instr.* (*J. Phys. E*), ser. 2, **2**, 622.
ROESLER, F. L. and TRAUB, W., 1966, *Appl. Optics*, **5**, 463.
SMARTT, R. N., 1961, *J. sci. Instr.*, **38**, 230.
SMARTT, R. N. and RAMSAY, J. V., 1964, *J. sci. Instr.*, **41**, 514.

Optical Finishing*

Robert E. Hopkins and Charles Munnerlyn†

Institute of Optics,
University of Rochester,
Rochester, N.Y.

Abstract—Today methods of automatic lens design are providing capability far exceeding optical shop techniques. The design potential of lenses is not being achieved in production or even in laboratory research. This paper discusses some possible new approaches to lens making, lens centring, and lens mounting. It presents the view that all optical shop procedures must be systematically challenged in the light of the modern development of lasers, computing machines, tape-controlled machines, diamond-bonded grinding pellets and plastics for polishing.

INTRODUCTION

Within the last five years we have reached a new level of capability in optical design. There are now several sophisticated optical design programmes working on large computing machines. These programmes enable designers to design optical systems which previously could never have been achieved. The best of the modern programs provide the following features:

(1) Active and inexpensive location of local minima of well defined merit functions.
(2) Reasonable cost-effective evaluation of the light distribution in the image.
(3) Meaningful and practical procedures for tolerance analysis.

The state of the art is illustrated in the lens shown in Figure 1. This lens was designed for micro-circuit applications. The theoretical O.T.F. curves for the lens are shown for two off-axis images in Figures 2 and 3. In order to achieve this performance, the designers had to modify and adjust weighting functions dozens of times. They had to watch the behaviour of skew rays at 45° to the meridional and sagittal plane. Probably more computing went into this one design than all lenses

* This work was sponsored by U.S. Army and Air Force Contracts.
† Now located at Tropel, Inc., Fairport, N.Y.

prior to the automatic computer (around 1950). The final adjustment of this lens involved careful appraisal of the O.T.F. in order to aid in the balance of weights on the residual aberrations.

These lenses are so close to perfect imagery that they have to be made almost to perfection. The surfaces must all be spheres and centred. There is next to no chance that lenses with imperfect surfaces or centring can be assembled and tuned to provide the expected performance.

Today the central problem in lens optics appears to lie in methods of manufacture. We are far ahead in design. The potential of most lens designs today is never realized in working lenses. The optical designers and engineers need to pay heed to the following points.

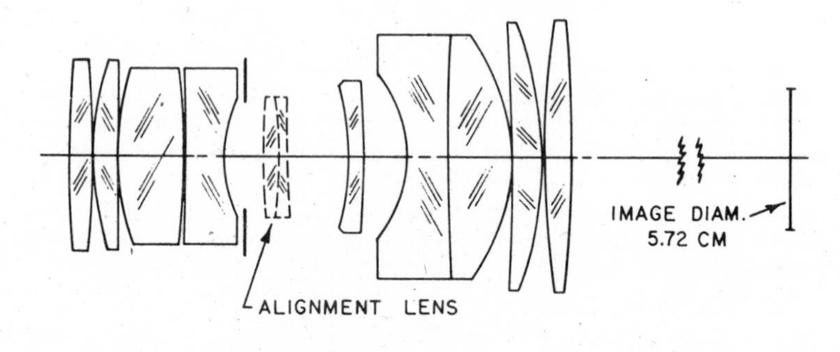

FIG. 1. A one-to-one micro-circuit imaging lens.

(1) Optical design programmes tend to place too much emphasis on image quality. If the image quality is required then the lenses cannot usually be made.
(2) In our emphasis on image quality we have neglected the extremely difficult but equally challenging problem of designing cost-effective designs.
(3) Prototypes are much too expensive. Many new designs or concepts never materialize because of the high cost of prototypes.
(4) In multi-element lenses, centring and surface irregularity cause serious degradation in image quality.
(5) The rule-of-thumb procedures used by most optical shops for tolerancing surface quality and centring are largely based on pre-computer and laser testing experience.

Before we can begin to use computers to help us design more practical optical systems, we need to know more about optical fabrication. We

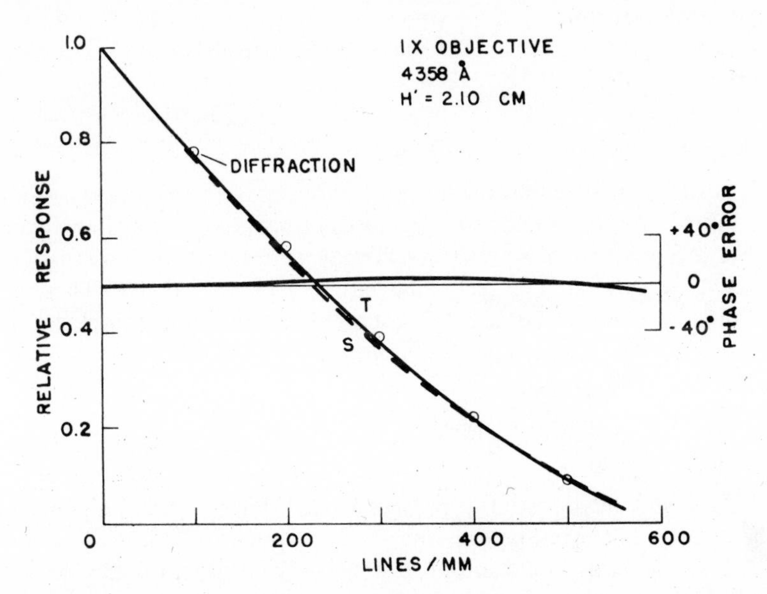

FIG. 2. O.T.F. curves for the lens shown in Figure 1. The O.T.F. refer to an image 2·1 cm from the optical axis.

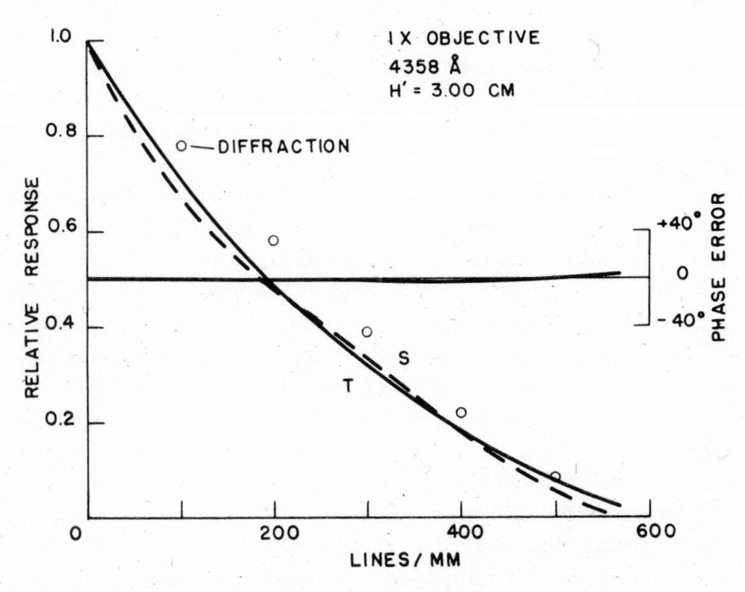

FIG. 3. O.T.F. curves for the lens shown in Figure 1. The O.T.F. refers to an image 3·0 cm from the optical axis.

have attempted this by first studying testing procedures. The purpose of this paper is to describe a few ideas on testing equipment for the optical shop.

Optical Shop Practice

It is well to realize that the finishing of lenses is an old process. Almost everything has been tried, so one must not be discouraged if one cannot invent a completely new technique. Today we have three new capabilities which may turn an old technique into a modern wonder. These new capabilities involve the use of lasers, data-collecting equipment and computers. The suggestions that follow are mostly old ideas clothed in modern technology.

The Grinding Process

It is well known that the process of polishing is aided significantly when care is taken in fine grinding. There appears to be a need for more adequate methods for contour testing of ground surfaces. We have studied the following methods for improved testing of ground surfaces.

Mechanical Probes. These methods are useful if properly and adequately automated. They are expensive and unwieldy for testing large optical surfaces.

The Application of Waxes. We have had surprising success with the use of transparent shoe polish as a wax on ground surfaces. Tests have been made using a laser Fizeau interferometer on fine ground surfaces (aluminum oxide 95). With repeated applications of wax it has been possible to measure surface contours to within one wavelength (Moreau & Hopkins 1969).

Holography. Several methods for measuring surface contours have been described (Hildebrand & Haines 1966; Zelenka & Varner 1968; Shiotake *et al.*, 1968). These methods in effect project interference fringes on the surface. The contours may be determined by viewing from the side. There are also methods for making contour measurements relative to a standard surface (Pastor 1969). If these methods are looked at closely, they are often seen to be cumbersome and of limited usefulness.

The 10·6 μ *Laser Interferometer*. One of us (C.M.) has been able to test opictal surfaces ground with 20 μ grit using a 10·6 μ CO_2 laser interferometer (Munnerlyn & Latta 1968; Munnerlyn 1969). A diagram of the interferometer is shown in Figure 4. The light entering the slit is collimated by the mirror *L*. The grating *G* diffracts the light into the -1 and $+1$ orders. The diffracted -1 order is reflected from the test surface *M* back to the grating. This reflected beam is again diffracted by the grating and the zero-order beam is reflected in the same direction as the +1 order reflection of the original beam from L. The lens L' images M at P'. A detector at P' then measures the interference between the −1 and +1 diffracted wavefronts at the test surface M. Figure 5 shows the contour lines of a ground surface measured on this interferometer. The technique looks promising for testing a surface before polishing. We believe we can test large mirror surfaces. The interferometer can also be used to measure the homogeneity of infrared materials.

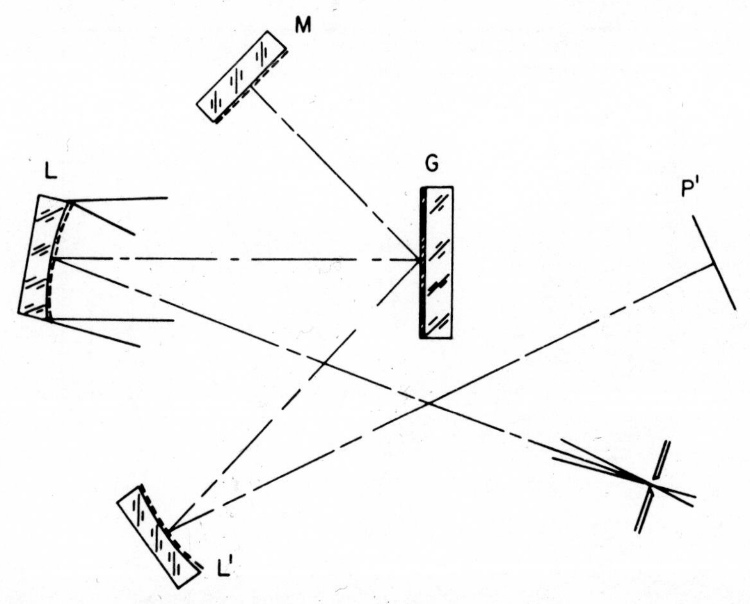

FIG. 4. A 10·5 μ CO_2 laser interferometer. L, collimator mirror; G, diffraction grating; M, test mirror; L', imaging mirror; P, liquid crystal detector.

The making of large aspheric surfaces should be considerably easier using simple floating tools, if these methods for determining surface contours of the ground surface can be worked out. They may also be worthwhile for testing precision spherical surfaces prior to polishing.

Optical Polishing

The standard procedure today is to use test plates to check surface regularity. This ancient procedure is hard to beat but eventually it must be replaced. Figure 6 shows a laser interferometer which can be used to test a wide variety of optical surfaces. The analysis and evaluation of this and many alternative types of interferometer is an extensive subject. For this discussion it is appropriate to list a few of the advantages and disadvantages of this type of interferometer testing as opposed to using test glasses.

Advantages of Laser Interferometer Testing

(1) There is no need for a large collection of test glass plates.
(2) It is possible to separate the test for regularity from the test for radius of curvature.
(3) It removes the danger of scratching a precision polished surface.

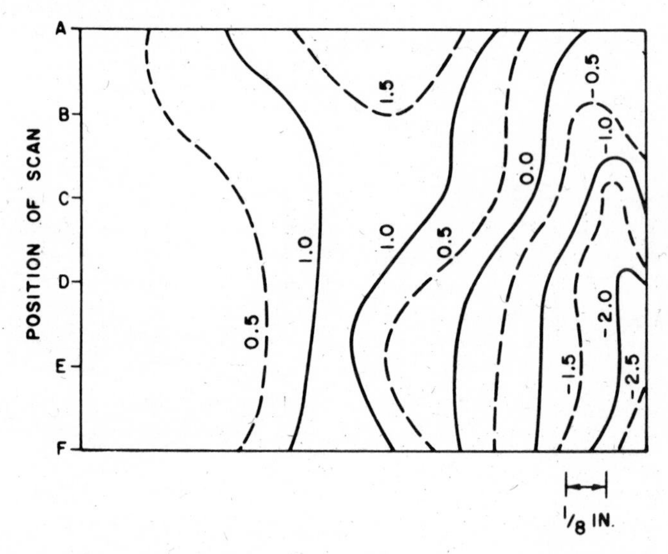

FIG. 5. Contour map of a ground glass surface, measured with a 10·6 μ CO_2 laser interferometer. Deviation of cylindrical concave ground glass surface from cylindrical reference with 5 m radius. (Each increment is 0·5 μ.)

Disadvantages in using a Laser Interferometer

(1) By going to monochromatic laser light one loses the subtle colour changes obtained with white light interferometry.
(2) If one tries to use the advantages of spatial coherence in a laser beam, troubles are encountered with multiple reflections from the auxiliary optics in the interferometer. The multiple reflections cause complex patterns which can be erroneously interpreted.

Most of the disadvantages can be overcome by good design and execution of the interferometer. All the surfaces have to be coated with low-reflection coatings and mounted so that the extraneous fringe patterns are well centred. In order to compensate for the loss of colour differences, it will probably be necessary to use computerized techniques for scanning the fringes and computing the wavefront errors. We now have a program which can take data on fringe locations and prepare a contour map of the surface. It also can compute the rms wavefront deviation.

Besides providing accurate surface regularity measurements, the laser interferometer can be used for accurate radius measurement and can increase the precision of centring.

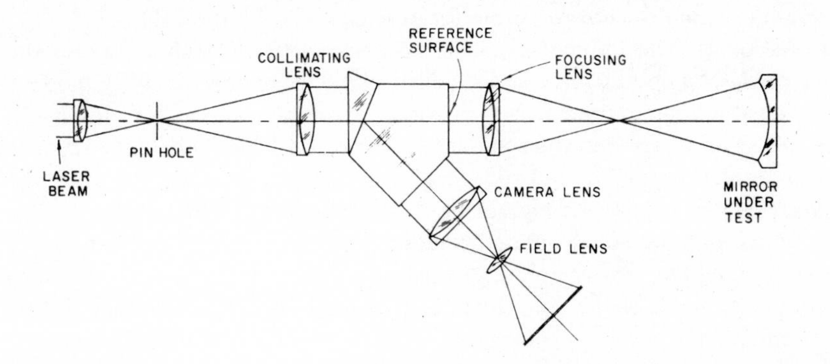

FIG. 6. A Twyman-Green type of laser interferometer for testing a concave surface.

MTF Measurements

The final testing of a lens system is a fundamental part of the manufacturing process. It is of importance to be able to measure the difference between computed performance and actual performance. Only in this way can a lens designer learn to make the proper balance of imagery and cost. How many designers really know how the final performance of their lenses compares with their design data? We suspect most designers have only the vaguest knowledge of their final product. We have come to this conclusion after helping the United States Air Force evaluate the state of the art in O.T.F. measurements. This study is not complete but the results so far show the following:

(1) Few laboratories have adequate O.T.F. or equivalent measuring equipment.

(2) If they do have O.T.F. measuring equipment, they have difficulty in describing where the measurements were made.

The great potential in O.T.F. measurements is apparently not being realized. Many laboratories are using O.T.F. to determine the best focus on axis or in some form of quality control, but these tests could be made using simpler equipment.

The user of O.T.F. data should measure the O.T.F. in several positions of focus in several positions in the field. He should then, with the aid of a computer, locate the position of an image plane to maximize a variety of quality factors. He should then compare this with the design data to determine the quality of the lens. It is only through this procedure that lens designers can make meaningful decisions in design. It appears that no one is doing this today. This is concluded by noting that most benches are designed to turn out O.T.F. curves and do not automatically provide quality factors for the performance of lenses all over the field.

As far as measurement of O.T.F. is concerned, the authors favour the scanning of a point source or a slit with a knife-edge or a slit. With modern data gathering equipment it is possible to compute the O.T.F. for many scans of an image and then to average the results. The knife-edge scanner can be made compact and mounted on a microscope tube. The scanning knife-edge is capable of measuring the O.T.F. of a system.

Measuring the wavefront deformation is also an excellent way to measure the O.T.F. but involves complicated equipment. If, however, one plans properly, much of the equipment can be used also for knife-edge testing or surface testing.

Equipment for the Optical Shop

It is clear that optical shop procedures can be improved by modern testing methods. These methods all depend on electro-optic equipment. The test equipment is expensive, but if the building block approach is used the equipment can be gathered over a period of time and used effectively during the build-up period. The following basic units are recommended for the optical laboratory and shop of the future.

(1) A laser interferometer for testing the regularity of spherical surfaces, aspheric surfaces and wavefront errors in complete lens systems.
(2) A point source of light.
(3) A scanning knife-edge or slit.
(4) A fringe scanner.
(5) A photo-multiplier connected to a digital voltmeter, interfaced with a magnetic tape recorder.
(6) Access to a large digital computer.

REFERENCES

HILDEBRAND, B. P. and HAINES, K. A., 1966, *Appl. Opt.*, **5**, 172.
MOREAU, B. G. and HOPKINS, R. E., 1969, *Appl. Opt.*, **8**, 2150.
MUNNERLYN, C. R., 1969, Thesis, University of Rochester, New York.
MUNNERLYN, C. R., 1969, *Appl. Opt.*, **8**, 829.
MUNNERLYN, C. R. and LATTA, M., 1968, *Appl. Opt.*, **7**, 1858.
PASTOR, J., 1969, *Appl. Opt.*, **8**, 525.
SHIOTAKE, N., TSURUTA, T., IOTH, Y., TSUJIUCHI, J., TAKEYA, N. and MATSUDA, K., 1968, *Japanese J.* of *Appl. Phys.*, **7**, 904.
ZELENKA, J. S. and VARNER, J. R., 1968, *Appl. Opt.*, **7**, 2107.

Initial stages in the Design of a Ruling Engine

J. Dyson, M. J. C. Flude, S. P. Middleton and E. W. Palmer

Division of Optical Metrology,
National Physical Laboratory,
Teddington, Middlesex, England

Abstract—This paper describes the design philosophy and main features of an interferometrically controlled ruling engine under construction in the Optical Metrology Division of the National Physical Laboratory. The engine is designed to provide a ruled area of 25 × 15 cm with a groove positional accuracy of 1/100th fringe, the operating speed being one ruling stroke per second.

Details are given of three independent interferometric control systems which are used to obtain a high overall accuracy from an engine of relatively low mechanical precision. Practical limitations on the frequency response of such control systems mean, however, that mechanical difficulties are not entirely eliminated: the major ones are outlined, together with the mechanical design treatments intended to overcome them.

INTRODUCTION

It was decided some time ago to build at NPL a ruling engine to rule large and very precise diffraction gratings. As it would be foolhardy to embark immediately on a very large engine, it was decided to build first an engine for 25 × 15 cm ruled area to develop the design methods. A secondary objective of this project was to obtain experience of relatively sophisticated electro-optical methods of measurement and control. In parallel with this development a research study was set up at a university to investigate the physics of the ruling process itself in the hope of finding ways of reducing the wear on the ruling tool.

Design Philosophy

It was hoped so to design the engine that its operation would be substantially independent of its workmanship and so could be copied at a reasonable cost. To this end, it was decided not to make a conventional mechanical engine and then to correct its errors, but to seek to avoid them altogether by making the only link between the motions of the blank and of the diamond via the measuring system. Continuous carriage advance is used, and the position of the blank carriage is sensed continuously by an interferometer. The operation can be described as making the moving fringes hard, using them as a rack and driving the diamond carriage from a pinion which is rotated by this rack. The blank carriage is driven by a screw, but the latter plays no part in determining the groove position.

The accuracy of groove position aimed at is 1/100th of a fringe, and

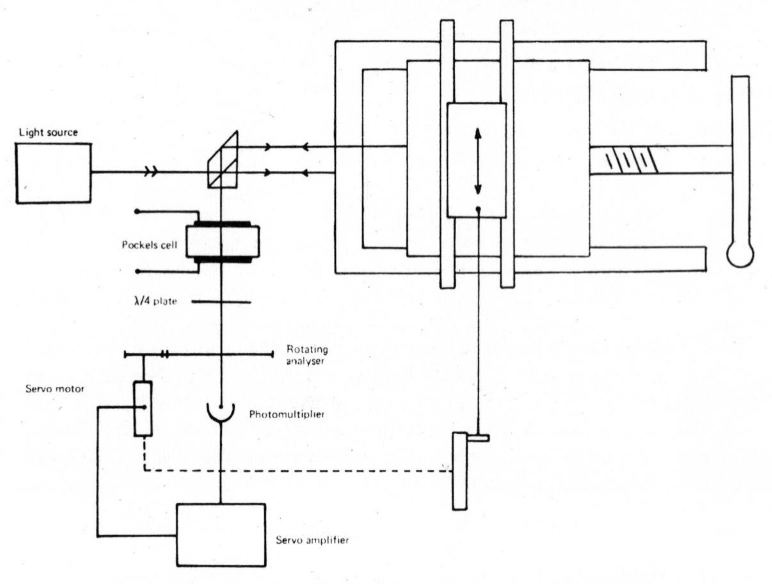

FIG. 1. The general scheme for the main drive of the engine.

to obtain this it is attempted to keep all the identifiable errors in the machine down to 1/1000th of a fringe, peak to peak. It is worth noting that this is about one atomic diameter.

The Main Drive

The scheme is outlined in Figure 1. The motion of the blank carriage is measured by a modified Michelson interferometer fed by a wavelength-stabilized helium-neon laser. One beam strikes a mirror fixed to the carriage, the other falls on a mirror fixed to the frame of the engine. The interferometer is a polarizing one, and emits two beams polarized at right angles to each other and with a path-difference between them varying with the position of the carriage. The two beams pass through a Pockels cell, energized by ac at a frequency of about 2kHz, which adds an additional path-difference varying sinusoidally with time at this frequency. They then traverse a quarter-wave plate, oriented at 45° to the planes of polarization, which converts the two plane-polarized beams to circular polarization of opposite sense for the two beams. These add to form plane-polarized light, of which the plane of polarization rotates at the rate of one rotation for each two wavelengths of additional path-difference.

The light then traverses a rotating analyzer, driven by a servo-motor, and enters a photo-multiplier. The output from this will contain a component of frequency equal to that of the drive voltage on the Pockels cell, which is zero only at the 'parallel' or 'crossed' positions of the analyzer with respect to the plane of polarization, and reverses phase as it passes through these positions. This component is used as the error signal of a servo-system which rotates the analyzer in step with the plane of polarization. The servo-motor is made powerful enough to drive the diamond carriage also.

This system was made up and tested, using simulated fringes and driving a simulated diamond carriage. This test confirmed that the following error did not exceed 1/1000th fringe peak to peak.

Atmospheric Effects

Ruling a grating may take quite a long time and the density of the air will vary during this period. Except when ruling is taking place at the centre of the blank, the lengths in air of the two arms of the interferometer will be unequal, and the path-difference will vary with air density. To avoid the resulting error, the beams are enclosed in vacuum tubes provided with end-windows, and the tube surrounding the beam incident on the carriage is made telescopic and driven by an auxiliary screw to keep the lengths in air of the two beams equal. Under these conditions, any variation of air density is without effect.

The motion of the telescoping tube must be inconveniently precise to avoid errors due to the replacement of a short length of vacuum by air, so use is made of the fact that the end of the tube should track the motion of the carriage to effect a correction. A plane-parallel glass plate is mounted on the tube end and arranged to be tilted by a probe on the carriage. The system can then be designed so that a variation of distance between tube and carriage causes an extra tilt of the plate, which introduces the right amount of path-difference to correct the error.

The Motion of the Diamond

Wear of the ways of the diamond carriage cannot be avoided, so it is necessary to control the distance of the diamond tip with respect to the same datum as that from which the position of the blank carriage is measured. This datum is in the form of a long strip-mirror, placed fairly close to the diamond and parallel with its path.

The diamond tip is inaccessible, of course, when it is ruling, but the rather scanty available knowledge of the lateral forces on the diamond indicates that it can be assumed that its shank is infinitely stiff. Hence, from a knowledge of the lateral positions of two points at different heights on the shank, the position of its tip can be calculated. This could be done by making two separate measurements and feeding them into a computer, but it was felt to be better to use an interferometer to give the desired response directly. This is shown in Figure 2. The interferometer rides on the diamond shank and one beam contacts the strip-mirror twice at a low level, whereas the other contacts it once at twice the height above the blank surface. It is easy to see that the path-difference depends only on the lateral position of the diamond tip, and that a rotation about that point is without effect. The interferometer is a polarizing one, and its output is used as the error signal of a servo which restores the diamond tip to its proper position.

Waltzing of the Blank

This can occur from lack of straightness of the ways and must be corrected to a high degree of accuracy. It is measured by two Michelson interferometers in tandem, giving a signal which indicates its angular position with respect to the strip-mirror referred to above, and is corrected by a servo which rotates a subsidiary platform on which the blank rests.

The Mechanical Design

The mechanical design of the engine was evolved from consideration of two specifications. The first lists major items in the required performance (Figure 3) and it is interesting to note that, although the ruling speed and maximum permissible reciprocating weight have been the dominant factors in the present design, neither appears to have significantly influenced the design of any engine in the past. The second specification gives design features which, though not essential, have been considered highly desirable:

(1) The design should not be irrevocably committed to any one mechanical scheme or arrangement.
(2) Non-standard or difficult workshop processes should be minimized.
(3) The design should remain practicable when scaled up for a considerably larger engine.
(4) The engine should be easy to use and maintain.

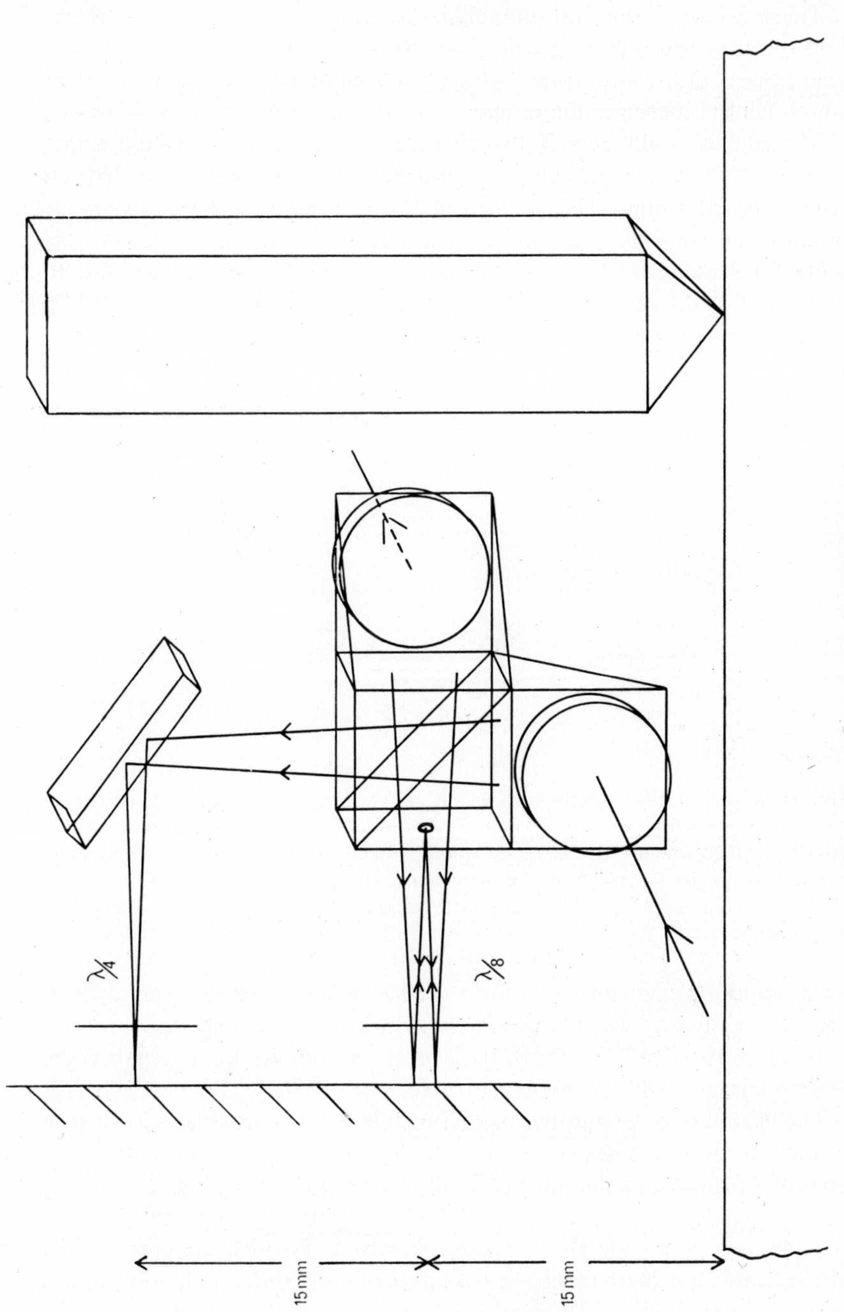

FIG. 2. The diamond path interferometer

These features would presumably be desirable in any engine and therefore hardly constitute a specification; however, they are treated as such here in view of the importance which has been attached to them and their consequent influence in the design.

The engine's mild steel, T-shaped base is constructed on a hollow box principle by screwing and cementing top and bottom plates on a welded and stress-relieved frame. This departure from a cast iron base was made because the steel box section combines high rigidity with considerable scope for alteration. All its surfaces, both internal and external, are smooth and flat so that extensions, mounting pads and stiffening can be added by

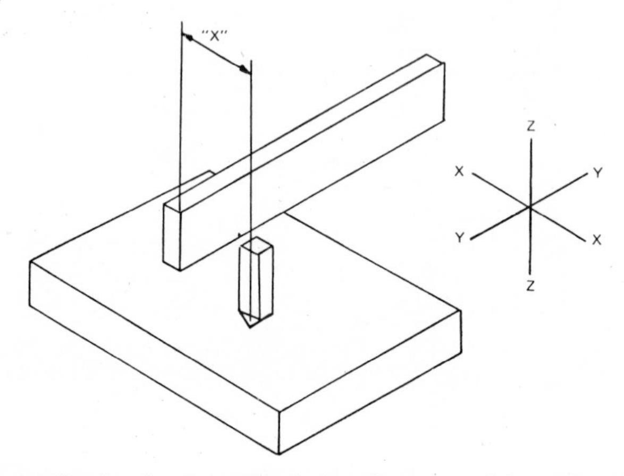

FIG. 3. The specification for the mechanical performance of the engine: 1, maximum blank carriage travel, axis XX, 275 mm; 2, maximum rotation of blank carriage about axis ZZ, $\pm 0{\cdot}06$ seconds; 3, maximum ruling stroke, axis YY, 220 mm; 4, Operating speed, 1 stroke per second; 5, diamond-reference straight-edge distance, X, to remain constant within $\pm 0{\cdot}15$ μm over complete ruling period; 6, maximum weight of reciprocating parts 1 Kg.

screwing and/or cementing without need to machine or seriously disturb the base in any way. Even large extensions with box sections similar to that of the base itself can be added by these means as stiffness, not strength, is the desired characteristic of the final structure.

The flexibility of layout thus made possible was of particular significance in the selection of a diamond carriage system. The diamond carriage is generally regarded as the most difficult component in a ruling engine and, in the present case, the weight limitation and high operating speed considerably aggravate the problems involved. Possible systems can be divided into four main categories in terms of their slideway layouts and a change of category would normally require major modifications to the base. A case can be argued for each system and, in view of the many factors involved, no one system can be considered to have an overwhelming advantage. A system using a slideway at the rear of the blank carriage has, in fact, been chosen as it appears to offer the opportunity of combining adequate performance with excellent ruling point and grating blank accessibility: however, the choice incurs risks which would make it seem dubious unless major changes in layout were not only possible but catered for at all stages of the design.

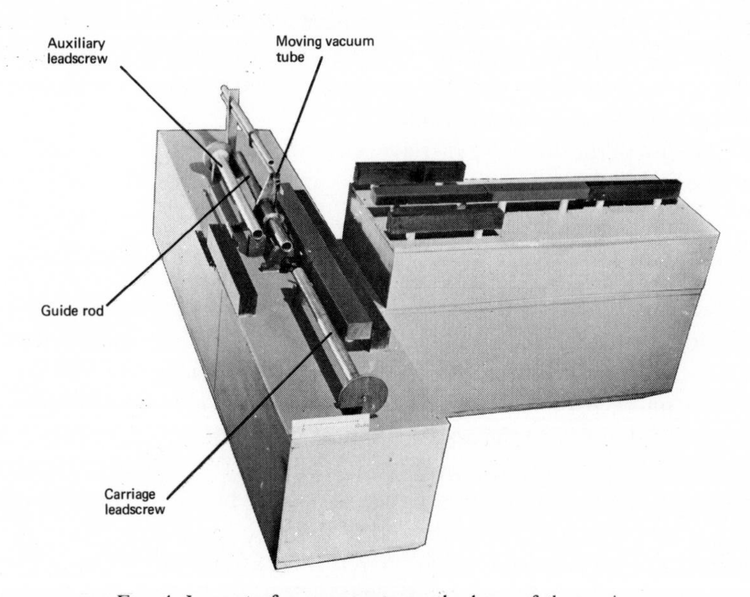

FIG. 4. Layout of components on the base of the engine.

Considering the general question of slideway design, certain kinematic arrangements of carriage constraints permit a slideway construction involving only simple rectangular blocks, each block having one face worked flat to form a guide for a carriage constraint. The advantages of such a construction are:

(1) A flat face on a rigid block can be accurately worked and tested by well-established, standard techniques.
(2) The blocks can be made of virtually any material.
(3) The slideway's position, alignment and geometry can be altered without involving alterations to the main structure.
(4) Misalignment due to long term structural warping of worn surfaces can be corrected with relative ease.

The general layout resulting from this type of construction and the choice of a rear slideway for the diamond carriage can be seen in Figure 4, which is a photograph of a full-scale model. The method of mounting the blocks is not shown, but kinematic block supports and clamping, together with the use of auxiliary adjusting devices temporarily clamped on the base, are expected to make the aligning and fixing procedure tedious rather than difficult.

The geometry of the constraints on the blank carriage is such that if random variations in friction are ignored the line of action of the resultant of the friction forces is independent of the weight of the carriage and blank. The driving thrust from the leadscrew nut is arranged to act along the same line so that, whatever the load, the carriage is in equilibrium under the friction forces and driving thrust alone, i.e. there are no couples tending to make the carriage rotate.

In the case of the diamond carriage, ensuring a pure thrust is complicated by the effects of inertia. Assuming the carriage reciprocates with simple harmonic motion, the peak acceleration rises to nearly 0·5g and the accelerating force required varies during the stroke between zero and a value about twice that needed to overcome friction at the slides. Since the friction forces are nominally constant, the only practicable method of avoiding a fluctuating couple is to make the resultant of the friction forces act through the centre of gravity of the carriage. By this means, the load on each of the five constraints remains constant whatever the friction at the slides or the acceleration of the carriage.

In order to facilitate the possible alteration of the diamond carriage system, what appears to be a raised portion of the base in Figure 4 is in fact a separate box-type frame connected to the base but not integral with it. The complete carriage system can therefore be assembled and tested as an independent unit, and, more important, be removed at any time and replaced with an alternative system without drastically affecting the rest of the engine.

The other items which can be seen on the base in Figure 4 represent twin leadscrew, the vacuum tubes system and the blank carriage counterweight device. Separate leadscrews are used for the blank carriage and vacuum tube systems in order to minimize the load on the carriage drive and to isolate the complete carriage system from the effects of random variations in the friction of the sliding vacuum seal. The extra leadscrew is also a decided asset in the design of a carriage counterweight system and, in view of this, experimental work to produce a seal with sufficiently uniform friction characteristics was abandoned as unjustified. The leadscrew driving the carriage must be a good but not exceptional screw as the only demand made on it is that the carriage motion is smoothly continuous. The vacuum tube leadscrew is nominally identical with it, and, since both screws will be driven by the same types of reduction gearing and synchronous motor the design aim of maintaining the distance between the carriage and vacuum tube support constant within $\pm 5\ \mu m$ should not prove too difficult.

The operating principle of the auxiliary leadscrew components is shown in Figure 5. A sleeve fitted with re-circulating ball bushings is pushed along a guide rod by the leadscrew nut. A flange carries the end of the telescopic vacuum tube an an arm supports the upper end of a strip suspension which

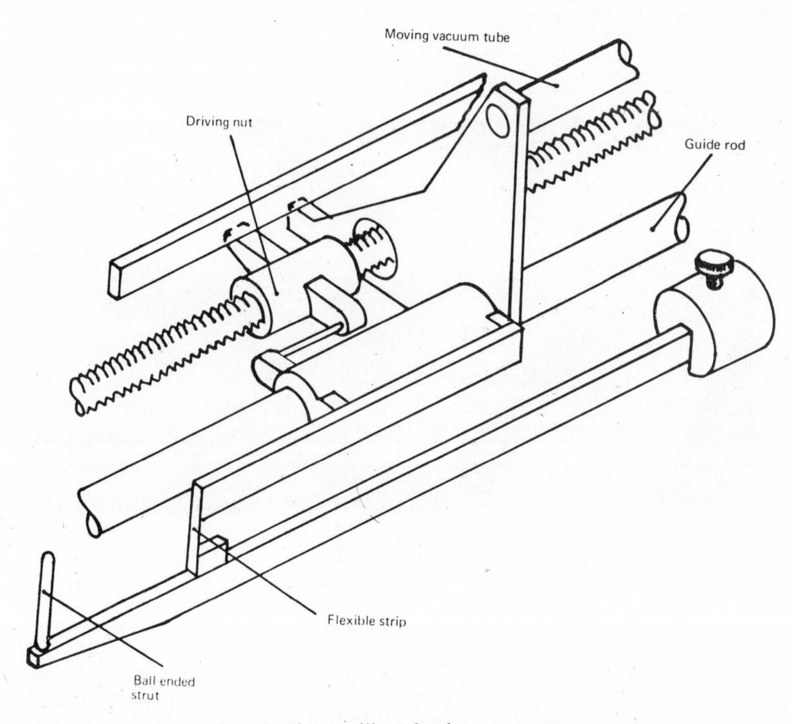

FIG. 5. The auxiliary leadscrew systems.

serves as a fulcrum for the counterweight lever. One end of this lever acts via a ball-ended strut to supply an upward thrust under the centre of gravity of the blank carriage, the other end carries the counterweight which can be moved along the lever to adjust the lift applied to the carriage. The strip suspension and ball-ended strut prevent the system exerting a horizontal force on the carriage and tests on a mock-up indicate that counterbalancing a combined carriage and blank weight of up to 20 Kg should not risk the carriage being subjected to more than 0·01 N.

Figure 6 shows the parts of the engine most closely concerned with the ruling diamond. The features already described as desirable in the diamond carriage and its slideway unfortunately necessitate a carriage size and shape which make the 1 Kg weight restriction a serious problem. Normal construction would indeed be out of the question, but by utilising the high stiffness-to-weight ratio obtainable with thin-walled shells, a carriage of adequate stiffness can be made to appear more practicable. Stiffness tests on full-scale models made of very thin aluminium sheet have been encouraging and carbon fibre reinforced plastic and aluminium honeycomb are factors in reserve; however, the carriage must at present be regarded as the most suspect component in the overall design.

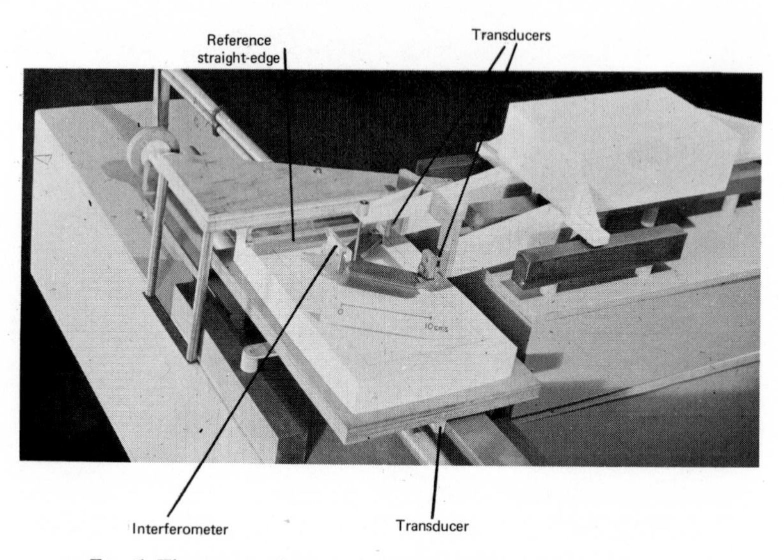

FIG. 6. The two carriages and the diamond path control system.

The proposed form for the carriage, with the exception of a tapering tail section which is relatively unimportant, can be seen in the Figure. All the component parts are hollow, the centre of gravity of the complete moving assembly will lie in, or close to, the horizontal plane containing all the five constraints, and the system is symmetrical about the fore and aft axes. The model shape is not intended to be precise as the optimum distribution of material can only be determined after tests under ruling conditions have shown the relative magnitudes of the sources of strain; however, the final version is not expected to differ widely from the shape shown.

The tracelet, mounted on strip hinges and damped by a dashpot, is fixed to the carriage by two pairs of cylindrical stubs. These are piezo-electric transducers which form part of the diamond path control system. The pairs will work in anti-phase so that the effect of their expansion and contraction is a rotation of the tracelet about a vertical axis mid-way between them and a consequent lateral shift of the diamond and its interferometer.

The reference straight-edge which determines the diamond's path can be seen mounted beneath a frame which bridges one end of the grating blank. This frame, which also carries the fixed reflector of the carriage travel interferometer, is kinematically located on rigid supports built up from the main base. It is removable to facilitate handling the blank, but extreme repeatability is not demanded of its kinematic mounting as the straight-edge can be very precisely rotated in the horizontal plane to optimise its direction relative to the uncorrected path of the diamond.

One other important component, not in fact involved with the diamond, can be seen in Figure 6. The blank carriage slideway arrangement of the engine permits the use of an extremely simple method of applying the corrections dictated by the carriage rotation control system. Rotation of the carriage in the horizontal plane can be obtained by moving either of the two lateral constraint points. Since one of these is almost directly under the reflector of the main carriage interferometer, movement of the other, which can be seen in the Figure, yields a carriage rotation but no displacement of the reflector. Movement of the constraint is obtained by mounting the actual bearing pad on the face of a piezo-electric transducer. The system can therefore be made ideally stiff without adjustment setting up any strain in the carriage structure.

The work described above has been carried out at the National Physical Laboratory.

A New Interferometrically Controlled Ruling Engine for High-Performance Gratings up to 32 x 20 inches

George W. Stroke
State University of New York, Stony Brook, New York 11790 *U.S.A.*

and

Albert E. Johnson
Moore Special Tool Co., P.O. Box 4088,
Bridgeport, Connecticut 06607 *U.S.A.*

Abstract—A brief report is made on the progress of completion and results of initial tests of a new large grating ruling engine which incorporates several new mechanical and interferometric control principles, based on those first put forward by Rowland, Michelson, John Strong, G. R. Harrison and G. W. Stroke, and H. G. and H. W. Babcock, among others (G. W. Stroke 1967). Among the novel features introduced by the Moore Special Tool Company as a part of this work are: a parallelogram diamond-lifting mechanism; a synchronous uniform-velocity diamond reciprocating drive; and a new type of flexure mechanism which permits one to obtain rotation movements without 'lost' motion or unwanted translation components.

A brief discussion is also given relating the new types of interferometrically ruled high-resolution gratings to the various forms of 'holographically' generated gratings, produced by photography, which are now undergoing various stages of development in a number of laboratories, following initial work by Lord Rayleigh (1872), and more recently again by J. M. Burch (1959), N. K. Sheridon (1968), and notably by A. E. Labeyrie (1969).

INTRODUCTION

An exhaustive historical review of grating ruling principles may be found in the 320-page *Handbuch der Physik* article on 'Diffraction Gratings' published by one of us (Stroke 1967). We also stated there:

'There is no simple prescription for building a ruling engine in such a way that it will perform successfully immediately after completion, or, for that matter, during any great length of time after its first successful ruling, without the close attention of the instrument maker, physicist or engineer, who has made it to operate successfully in the first place. This has proved to be true even on the servo-controlled M.I.T. ruling engine, of which the principle of interferometric servo-control was first described by G. R. Harrison and G. W. Stroke in 1955.'

Several ruling engines have been successfully constructed since interferometric servo-control was first successfully demonstrated by Harrison and Stroke at M.I.T. These include engines in France (Jobin and Yvon), Japan (Hitachi and others), several in the United States of America (Bausch and Lomb, Jarrell-Ash and the Babcock engine at the Mount Wilson and Mount Palomar Laboratories, among others).

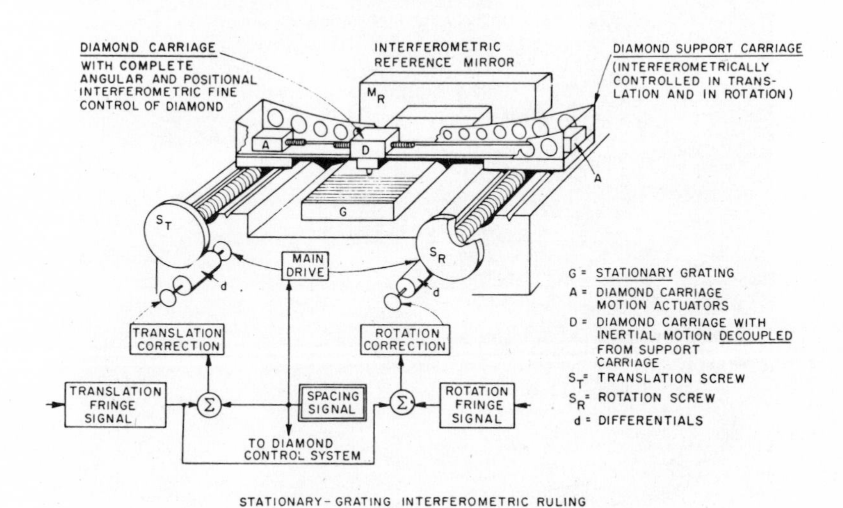

FIG. 1. General design principle.

In view of the preceding quotation, which we still believe to be true, we prefer in this paper to limit ourselves to the description of some novel mechanical features, introduced by the Moore Special Tool Company, and to restrict the general principles of our new engine to that given in describing the illustrative figures.

For simplicity we first give the set of figures, illustrating the general design principles, in a form which we believe to be helpful to readers.

General Engine Design Principles

The general principle of the design of the engine and of the interferometric servo-control system are shown schematically in Figure 1. The distinctive feature of the new engine is that the diamond-carrying platform itself is interferometrically servo-controlled, so as to maintain the diamond properly placed and oriented at all times, with respect to the large interferometric reference mirror M_R, which is firmly maintained against the grating blank itself.

A photograph of the engine in 1965, showing its construction according to Figure 1, is given in Figure 2. Somewhat later views of the engine are shown in Figures 3, 4 and 5 during a period characteristic of the new

FIG. 2. General view (1965) in early development stage.

design approach which we have used. In these figures we show a period during which we were testing various forms of diamond-carriage suspensions, control interferometers and drives. Two of the three features shown in Figures 3, 4 and 5 have been retained:

(1) A three-point (rather than a single-point, as previously used) contact of the diamond-carrying platform (Figure 5) against the diamond 'flat' (Figure 4).

(2) A new type of rotation-insensitive interferometer, used to control each of the three thin teflon 'shoes' with the aid of piezo-electric crystals, so as to maintain the distance and orientation of the diamond, as described. (Only one interferometer is shown in these preliminary tests.) Even though

Fig. 3. Details in later stages of development showing experimental use of linear motor.

Fig. 4. Details of linear motor drive for diamond platform and the diamond 'flat

the 'linear motor' shown in Figures 3 and 4 did in fact provide perfectly stable interference fringes with the three-point contact system used, it was not retained in the final design, but replaced eventually by the mechanical uniform velocity drive of novel design, shown in Figures 6 and 7. Figure 6 shows a general view of the engine, in May 1969, shortly after it had ruled the first test grating with 13·5″ long grooves, for the purpose of testing the mechanical performance of the engine. Figure 7 shows a detailed view of

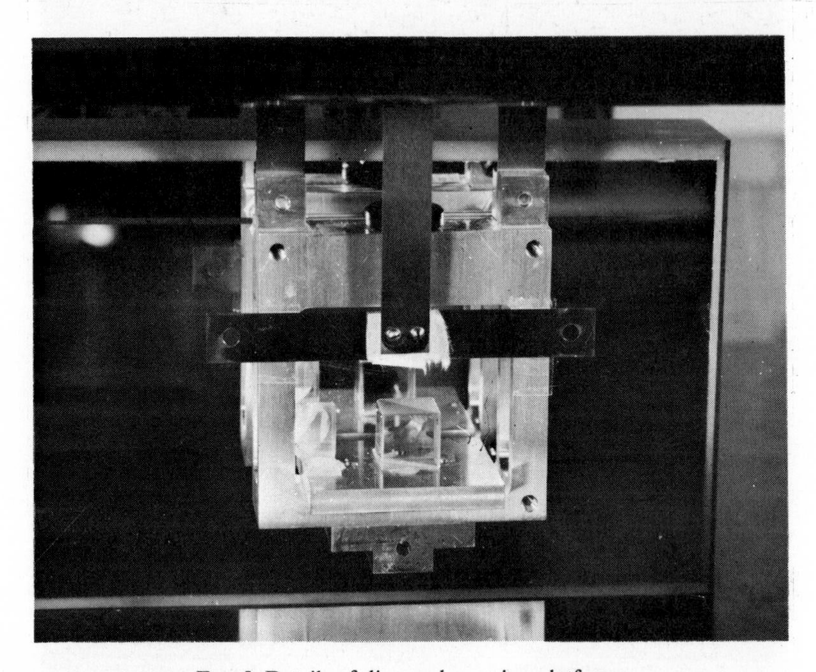

FIG. 5. Details of diamond-carrying platform.

the new uniform velocity drive system further described below. Stroboscopic tests have shown that the uniformity of the linear velocity is approximately 0·1 per cent (this may be compared to the cosinusoidal 'crank mechanism' motions which have been found to be satisfactory in several ruling engines). Interferometric wavefront tests of the grating have demonstrated that the amplitude of the normal periodic error of the uncontrolled engine is just under 0·3 fringes (at 6328Å), and thus ten times smaller than the periodic error of the famous M.I.T. Michelson-Harrison and Stroke engine which, it is well known, was so dramatically reduced with the aid of interferometric servo-control to levels considerably smaller than in uncontrolled mechanical engines, with carefully lapped screws. (The ghost intensity of the new engine, before control, is thus about 100 times smaller than that of the M.I.T. engine before control!) Based on

FIG. 6. General view of engine (May 1969) with mechanical, uniform-velocity drive.

FIG. 7. Uniform-velocity drive system.

extensive previous experience in interferometric servo control, there should be no problem in this case, as in the past, with readily reducing periodic ghost intensities well below acceptable levels. We next give a brief description of some of the mechanical features of the new engine. Several of its early features have already been described by Stroke (1967), including interferometric tests of the drives, as well as the use of a 'long-fringe' interferometer using synthesized laser fringes having a length exceeding at will the half wavelength (Stroke 1967)! In this paper, we limit ourselves to presenting previously unpublished details and results.

SOME MECHANICAL FEATURES OF A NEW CONTROLLED LARGE RULING ENGINE*

The ruling of the larger diffraction gratings required today cannot be achieved by simply extending the carriage travels of the engines which are at present ruling six- or eight-inch gratings. It can be readily shown that, as the size is increased, the volume and hence the weight of the blank is cubed. For example, a 20-inch diameter quartz blank of suitable thickness may weigh 125 pounds. In contrast, the successful engines of the past have done their work with blanks weighing 5 or 6 pounds. Transporting one of these larger blanks over the usual ruling distance would in itself present a difficult problem. However, the problem becomes magnified (perhaps to the fourth power) by having to transport the heavier blank over a longer distance. By using design principles of the past, one would need to fabricate the engine with a material having a modulus of elasticity perhaps a dozen times as high to prevent excess deflection.

With such an impossibility, the only solution rests in the construction of a massive engine. The kinematics to support good geometry are just as necessary as before, but the added essential feature is the strength of the engine components to withstand the heavier carriages, workpieces, and the shifting of mass, while still maintaining reasonable deflection.

Bridge and Flexure Mechanism

It was as a result of these restrictions that the new design approach is to enable this engine to perform all its travels and motions about a stationary grating blank. This was accomplished by means of a bridge straddling the dormant blank. Each end of this bridge rests on a carriage and each carriage is driven by its own lead screw. Advancing these twin carriages provides the line spacing movement.

A most difficult problem arises with the reciprocating motion of the diamond-scribing carriage since it must be mounted on the moving bridge. It seems that what was gained by having a stationary blank was more than

* This section was prepared by A. E. Johnson.

lost by having two motions mounted 'piggy-back'. This feature will perhaps remain a point of controversy, but the elimination of the need to move the heavy blank was an advantage worth striving for.

The 4,500 pound meehanite cast iron base was cast in 1964. The size was approximately 4 feet square by 2 feet thick (Figure 8). The strengthening ribs extend for the full thickness and the average cross-section web thickness is one inch. (It may be of interest to note that consideration was given

FIG. 8. Meehanite cast iron base.

to granite as a material for the base. It was abandoned when we found that its modulus of elasticity may be 1/3 to 1/5 that of a good grade cast iron.) These thick ribs not only provide the necessary rigidity but also provide for a larger thermal mass which is more resistant to small cyclic temperature changes in the environment.

Since the twin carriages supporting the bridge are independently monitored by laser interferometry, the bridge translations as well as its yaw characteristics are under full interferometric control. However, each stanchion of the bridge has to be coupled to each carriage to allow rotation about a vertical axis central with the carriage. An additional freedom must be allowed in one of these two couplings to permit flexure translation between carriages due to temperature changes and/or the non-linearity of the carriage travels (Figure 9).

It is remarkable and fortunate that sensitivity is virtually independent of size This is proved by the fact that this entire bridge and twin carriage system can respond to correction signals equally as well as would be expected of a tiny delicate instrument.

THE DIAMOND-LIFTING MECHANISM

Although not a new idea, the diamond-lifting mechanism (Figure 10) is constructed on the parallelogram flexure principle. It has been only a few years since this type of mechanism was first applied for retracting a scribing

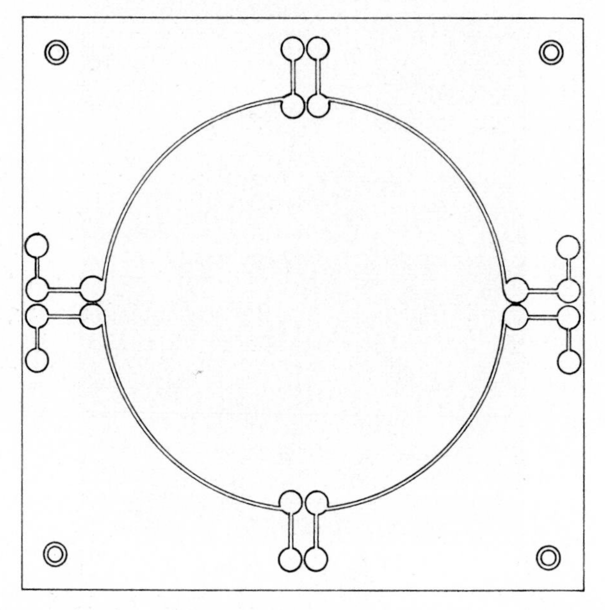

FIG. 9. Diagram showing method used for supporting couplings to allow for flexure

tool. Figure 11 shows a close-up view of the unit which allows adjustment of the diamond about all three axes of rotation. This ability to remove the entire mechanism for adjustment in a convenient optical alignment fixture is novel and eliminates to a great extent the need of the technician burrowing into the heart of the ruling engine at the most critical time. The importance of this can be realized since the engine and the environment must settle to a 0·001°C equilibrium before the start of the ruling process. The device is tremendously rigid in the horizontal plane, and yet offers an extremely sensitive freedom for the raising and lowering requirement. Here again, the importance of this feature cannot be over-emphasized since this diamond lowering must be repetitive over countless cycles to 1/100th of a fringe. An added benefit was realized by having more than one parallelogram device for the alternate use of various settings.

The Reciprocating Drive

If the grooves are to be ruled while the groove spacing is continuously generated, there should be a similarly constant velocity of the reciprocating

FIG. 10. Diamond-lifting mechanism.

FIG. 11. Near view of diamond unit removed from engine for adjustment.

diamond carriage. This quality is essential for the generation of a straight line. However, the two motions must also be synchronous over the entire number of lines ruled in order to eliminate any accumulated error. Hence, there must be a toothed device, e.g. gears, chain drive, or cog belt, for all interconnection between the two motions to maintain synchrony.

Figure 7 shows the unusual timing belt drive which provides the linear velocity for the reciprocation. The drive motor is a finely controlled dc servo-motor. It is resiliently mounted and coupled to the shaft shown. This shaft has two power take-offs—one to the leadscrew drives and one to the reciprocating drive. It is from this shaft onwards that all pulleys must be toothed so as to keep both branches synchronous. The power transmitted to the diamond carriage is first reduced by an unusual ratio of 5 to 17 to the flywheel shaft shown. This allows for a convenient means of simple change gears to provide any multiple of 100 grooves per millimetre. A well-balanced flywheel was found necessary not only to provide inertia to smooth out the reversals, but also to eliminate the many tiny random axial impulses.

From the flywheel shaft there is one pair of timing belts driving a second pair which form an elongated pattern between two small pulleys. These two pairs of belts are separated by a space between them and it is in this space that the connecting rod to the intermediate carriage travels. The connecting rod is rigidly attached to both timing belts so that through the straight portion of the elongated pattern there is a very constant linear velocity. The belts are extremely pliable but are lined with fibreglass thread to be stable and non-stretching. The picture of the drive perhaps explains more than a description.

The proof of the constant velocity is perhaps as unique as the drive itself. Previously, the drives would be of the simple harmonic motion type with modifying cams to compensate for the sinusoidal velocity. The various means of measuring their linearities were hardly any better than one part in fifty. With this new timing belt drive it was found that we could use a 24-inch metallic scale mounted co-linearly with the carriage. It was graduated in fiftieths of an inch. The reciprocation was set at fifteen strokes per minute. A stroboscopic light was synchronized with the scale graduations and in effect was able to freeze them from movement except for a slight drift of perhaps one scale graduation. Over the 18-inch length, this represented a linearity of 1 part in 900, perhaps three times the linearity ever required. This particular feature has been proved entirely satisfactory as shown by a trial grating.

COMPARISON OF APPLICATIONS OF HOLOGRAPHIC AND RULED GRATINGS

For reasons not explicitly indicated in the paper by Labeyrie and Flamand (1969), which reports work carried out in pursuit of that initiated by A. E. Labeyrie in Ann Arbor in 1965, successful holographic gratings have so far been produced only on extremely thin (a few hundred angstroms) suitable 'photo-resist' layers, aluminized to become 'reflection' gratings after exposure (to an argon ion laser) and suitable processing. With thin photo-resist layers, the gratings can only be produced with spacings of about 1,000 grooves per millimetre and finer, however, with practical perfection! For instance, several new types of concave gratings have been produced, all recorded holographically (for a general background see, e.g., Stroke 1966 and 1969) in such a way that one of the two waves originates accurately from the centre of curvature of the very good concave mirror on which the photo-resist layer was coated with extreme (optical) uniformity. Initial experiments (May–June 1969) with one such grating have shown that it achieved 'diffraction' limited performance. Previously (Labeyrie & Flamand 1969) the wavefronts of plane holographic gratings were already found to be essentially plane. Gratings with such fine spacing are of a particular interest for UV and far-UV (possibly even for X-ray) spectroscopy, and would have in fact been very difficult to rule, with large dimensions, even with the best interferometrically controlled ruling engines. Besides, the holographic gratings show, of course, a complete absence of any periodic errors or ghosts. On the other hand, the new types of interferometrically controlled ruling engines are particularly suited for the ruling of the newly needed, large, high-performance (both in terms of resolution and efficiency) *infrared* gratings (e.g. with spacings of say 100 grooves per millimetre and coarser) which now appear to be particularly needed, especially in view of a careful understanding of the advantages of gratings in comparison, for instance, to interferometric infrared spectrometry (Strong 1969). These types of large high-performance infrared gratings with 'coarse' spacings could not, it seems, conceivably be produced with the so far only other successful methods of producing high-performance spectroscopically usable 'holographic' gratings (Labeyrie & Flamand 1969). It thus appears that the holographic gratings (usable for the visible to far-UV, and possibly for X-rays) on the one hand, and, on the other, the interferometrically servo-controlled ruled gratings (usable from the visible to the infrared and far-infrared) have come successfully to solve needs in *complementary* spectroscopic domains, which neither of the two grating types could have solved alone within the current state of the art.

Acknowledgements

In addition to the acknowledgments to colleagues and co-workers, which we gave in the text, one of us (G. W. Stroke) wishes to express his great indebtedness to the National Aeronautics and Space Administration for continued most generous support of this work, most recently under Grant NGR-33-015-068(090). He would also like to use this opportunity to express his very great appreciation to Dr. Henry J. Smith, Dr. Harold Glaser and Dr. Goetz K. Oertel for their continued kind guidance and encouragement, and for most invaluable generously given advice, granted over the course of so many years.

Initial design was carried out jointly with Professor Orren C. Mohler, University of Michigan, and Frank Denton. Tatsuo Harada, on leave from Hitachi Ltd., Tokyo, made significant contributions to the progress of the work and to the design of the diamond-control interferometers. Recently, Vilmars Fimbers of the Moore Special Tool Company has been responsible for a number of very significant innovations in the design and operation of the engine: his participation and continued contribution is acknowledged here.

We also wish to thank Professor John Strong, University of Massachusetts, for a private communication to Professor Stroke.

REFERENCES

BURCH, J. M., 1960, National Physical Laboratory Symposium on Interferometry, Teddington, 1959, H.M.S.O. London, p. 179.

LABEYRIE, A. E. and FLAMAND, J., 1969, *Optics Communications*, **1**, 5. (This paper also contains related references to holographic gratings.)

RAYLEIGH, Lord, 1872, *Proc. Roy. Soc.* (London), XX, 414; *British Assoc. Report*, p. 39, Scientific Papers, Dover Publications, New York, 1964.

SHERIDON, N. K., 1968, *Appl. Phys. Lett.*, **12**, 316.

STROKE, G. W., 1966, *An Inroduction to Coherent Optics and Holography*, Academic Press, New York and London. Second Edition, 1969.

STROKE, G W., 1967, 'Diffraction Gratings' in S Flugge (ed.), *Handbuch der Physik*, Vol. 29, pp. 426–754, Springer Verlag, Berlin and Heidelberg. (This article also contain extensive historical background and complete lists of references to work on diffraction gratings.)

STRONG, John, 1969 (March), private communication with G. W Stroke.

Aberration-Corrected Concave Gratings Made Holographically

Jean Cordelle, Jean Flamand and Guy Pieuchard
Société Jobin et Yvon, 26 *rue Berthollet* 94-*Arcueil-France*

and

Antoine Labeyrie
Observatoire de Paris-Meudon, 92-*Meudon, France*

Abstract—Competitive diffraction gratings are produced by Jobin-Yvon Inc., using a holographic process. In addition to plane gratings, this technique makes it possible to produce concave gratings with improved stigmatic properties. Several particular cases of interest are presented: in one of these, stigmatism is rigorous for 3 wavelengths in the first-order spectrum. Expressions for the low-order aberration terms are given in the general case.

Applications range from precision diffractive imaging to high-aperture spectrographs or monochromators for the visible and ultraviolet, stigmatic X-ray gratings, and also objective gratings for space astronomy.

INTRODUCTION

Diffraction gratings suitable for spectrographic applications have been successfully produced at Jobin-Yvon Inc., by recording interference fringes on photopolymer-coated optical flats, on the basis of unpublished work by Labeyrie.

As described previously (Labeyrie & Flamand 1969), these gratings feature a high diffracting efficiency (blaze up to 90 per cent); they give very pure spectra that are virtually free from ghosts and stray light such as produced by conventional ruled gratings, even when as many as 3000 grooves per millimetre are recorded.

The diffracted wavefront meets the Rayleigh criterion and the resolving power is therefore close to the theoretical value.

Using this technique, we have now produced concave gratings that give, unlike conventional ones, aberration-corrected spectra featuring also rigorous stigmatism in several places.

Grooves as Contour Lines

Concave gratings, often used in ultra-violet instruments, exhibit a strong astigmatism when they are used in the otherwise convenient Rowland-circle arrangement. Rowland himself did show that stigmatic gratings are produced if the groove pattern is distributed in such a way that grooves are contour lines of the optical path, measured from the slit to the diffracted image. In the general case, it has not been possible to generate by mechanical means such complicated groove patterns with the required precision, but this appeared feasible using our photochemical fringe recording technique. Also this technique provides some degree of automatic adjustment of the groove profile which gives a rather uniform blaze over the grating surface (Sheridon 1968).

Stigmatic Properties Deduced From Holographic Theory

Given their fabrication technique, holographic gratings can be considered as a special type of hologram for which both recording waves have a more or less uniform amplitude and 'smooth' phase: not only the reference beam but also the object beam originate from a point source. Also, these holograms work by reflection and the recording surface is curved rather than flat, its shape being preferably spherical for ease of figuring. The basic results of the Gabor theory, as extended to this case, say that the reconstructed beam is identical to the object beam, modified as it is after reflection on the curved substrate. In other words, one reconstructs, instead of the recorded beam itself, its reflected image in the recording surface considered as a mirror.

For spectrographic applications, one convenient way of obtaining a stigmatic focusing grating is thus to choose recording parameters in such a way that beam A' (reflected image of recorded beam A) converges to a real focus. This can be achieved using a plane substrate and beams such as shown in Figure 1 : B diverges from a point source, A converges to a virtual point image. When illuminated with the reference beam B, the grating will diffract A', thus giving, in polychromatic light, a focused spectrum featuring rigorous stigmatism for the recording wavelength.

An alternate way of achieving stigmatic diffraction is depicted in Figure 2: the substrate is spherical, both recording beams diverge, one of them originates from the centre C of the sphere. This case, which is discussed below in more detail, is one for which A' gives a stigmatic focus in C.

Since BC is a revolution axis for the system, the grooves are also the intersection of the substrate with a family of planes perpendicular to this axis. However, these gratings differ from the conventional concave ones since their groove spacing is progressively varying. Before the advent of holographic techniques, Sakayanagi (1967) discovered some of the interesting properties of this type of grating, in a particular case. Though he had proposed to rule it, no attempts have been reported, presumably because of the above-mentioned difficulty of producing non-equispaced rulings, even with a straight diamond motion.

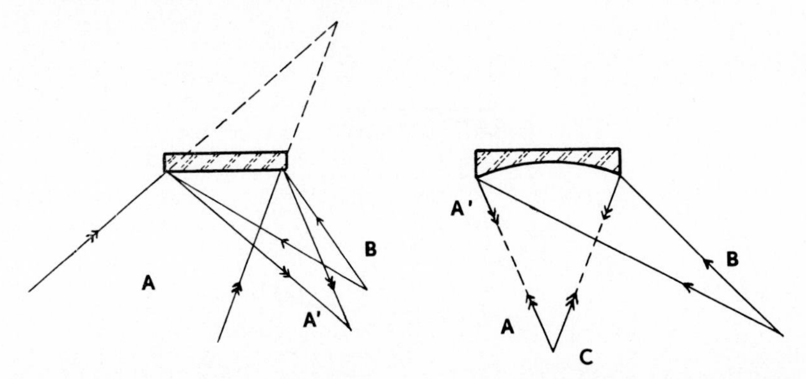

FIG. 1. (*Left*): Recording a focusing grating on a plane substrate: (A and B: recording beams, A': reflected A beam) the hologram reconstructs A' when illuminated with B.

FIG. 2. (*Right*): Stigmatic grating recorded on a spherical substrate: the object point being located at the centre, no extra optics are needed in the recording and/or reconstructing arrangement to obtain a focused reconstructed beam.

Using the above principle, a variety of gratings can be made with different substrate shapes: spherical, parabolic, elliptical or other kinds.

Another example involves a plane, afocal, grating in which the grooves

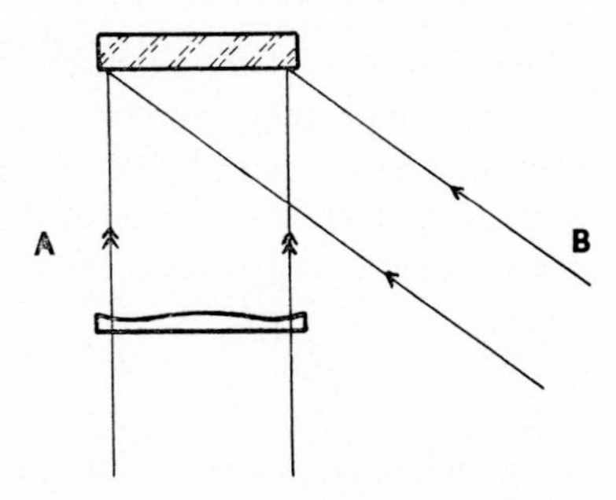

FIG. 3. Plane afocal grating acting as a Schmidt corrector: plane wave A is given spherical aberration opposite to that of the high aperture spherical mirror camera used in the reconstruction. Beam B may be collimated or simply diverging.

are slightly distorted so as to replace the Schmidt corrector plate of a spectrograph's high aperture Schmidt camera. As shown in Figure 3, this can be done by using plane A and B waves, A being pre-distorted (spherical aberration) by transmission through a Schmidt plate. This was actually tested at Jobin-Yvon because of its potential usefulness in both ground— and satellite-based astronomical spectrographs (Figure 4).

Finally the possibility of making crossed dispersion gratings should also be mentioned. In this case, successive exposures are made, the object point being suitably displaced between them (Stroke, Westervelt & Zech 1967). As a result, several spectra are reconstructed giving the same pattern which is obtained in cross-dispersion spectrographs.

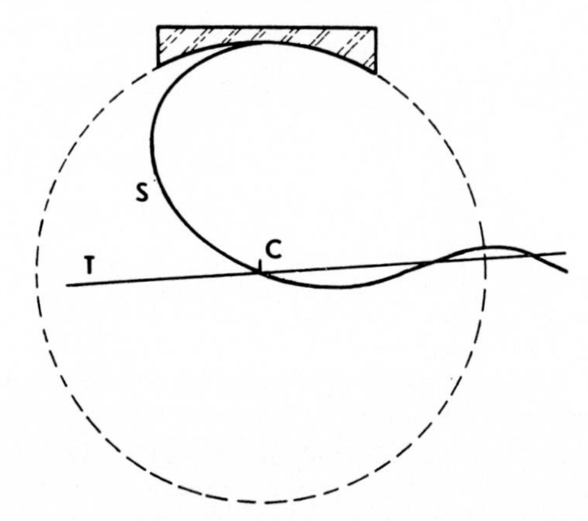

FIG. 4. Astigmatism curves of grating recorded as in Figure 2: the sagittal and tangential focal curves cross in 3 points, which are also the 3 points of rigorous stigmatism.

Spectral Variation of Stigmatism

The results given by the simple theory presented above are valid when the reconstructing wavelength is the same as the recording wavelength (or its harmonics when used in the higher diffracted orders). For most spectrographic applications, more elaborate aberration calculations are needed in order to evaluate the image quality in every part of the spectrum. These are based on the Fermat principle extended to wavefront reconstruction: the optical path CD from a point object C to its stigmatic diffracted image D is constant along every single groove, it increases by $k\lambda$ (k constant integer) from any groove to the next one in such a way that $CD+nk\lambda$ (n, number of the groove) is constant.

Let the recording surface be called Σ, the recording wavelength λ_0, the recording point sources A and B, B being considered as the reference. The reconstructing beam will be supposed to originate from a point C, and the reconstructed point is named D. One point I on the recording surface is taken as the origin for optical paths.

One fringe is such that:

$$MA-MB=N\lambda_0$$
$$IA-IB\ =N_0\lambda_0$$

The number of recorded fringes between I and M is:

$$n = N - N_0 = [(MA - IA) - (MB - IB)]/\lambda_0$$

When reconstructing with wavelength λ, the Fermat condition for stigmatism is:

$$MC + MD = IC + ID + nk\lambda$$

or: $$(MC - IC) + (MD - ID) = k\lambda/\lambda_0[(MB - IB) - (MA - IA)] \quad (1)$$

When the condition is only approximately verified, the transverse aberrations may be computed from the wavefront distortion ϵ:

$$\epsilon = (MC - IC) + (MD - ID) - k\lambda/\lambda_0[(MB - IB) - (MA - IA)] \quad (2)$$

These expressions, found by Pieuchard correspond to those derived by Armstrong (1965) in the case of transmission holograms.

From Equation (1), more properties of interest can be found in the above-mentioned case (Figure 2) of a spherical recording surface for which B and C are superposed, A and D being located at the centre of curvature of the substrate.

As shown by Labeyrie and confirmed by Equation (1), stigmatism is rigorous in the first order spectrum for λ_0 and also $2\lambda_0$, the corresponding points being located respectively in D (centre) and in B (Littrow mounting).

In addition, a third stigmatic point was found by Flamand who remarked that Equation (1) is satisfied if $MD/MC = I - k\lambda/\lambda_0$, which is the case when D is the harmonic conjugate of B with respect to the grating sphere. The existence of this 'Flamand point' is of particular interest for many applications since the corresponding stigmatic wavelength can be pre-determined to fit any particular need, merely by adjusting the position of point B in the recording arrangement. For the same type of grating, it can be remarked that one obtains perfect stigmatism for other couples of wavelengths when a permutation of the source A and the observation point B among the three stigmatic points is made.

The properties of this type of grating, which we have verified experimentally with apertures up to $F/1$ for the diffracted beam, constitute a marked improvement in comparison with those of conventional concave gratings.

Aberrations of Lower Order

Though expression (2) can readily be calculated using computers, it is of interest to expand it in terms of the grating aperture and parameters such as the polar coordinates of points A, B, C and D, using the method introduced in optical computing by Maréchal (1952).

Let the recording sphere be a surface of equation $X^2 + Y^2 + Z^2 - 2RX = 0$

(Figure 4). The distance of point A (x, y, z) to point M (X, Y, Z) of the sphere may be expanded as follows:

$$
\begin{aligned}
MA = {} & l_A \\
& -\frac{1}{l_A}(yY + 2Z) \\
& +\frac{1}{2l_A^2}\left[Y^2\left(1-\frac{x}{R}-\frac{y^2}{l_A^2}\right)+Z^2\left(1-\frac{x}{R}-\frac{z^2}{l_A^2}\right)-2\frac{y^2}{l_A^2}YZ\right] \\
& +\frac{1}{2l_A^3}\left[yY+zZ\right]\left[Y^2\left(1-\frac{x}{R}-\frac{y^2}{l_A^2}\right)+Z^2\left(1-\frac{x}{R}-\frac{z^2}{l_A^2}\right)-2yz\frac{y^2}{l_A^2}\right] \\
& + \text{fourth order terms}
\end{aligned}
$$

Let A be located in the x, y-plane; its coordinates are:

$$
\begin{aligned}
x &= l_A \cos \alpha \\
y &= l_A \sin \alpha \\
z &= 0
\end{aligned}
$$

Thus:

$$
\begin{aligned}
MA = {} & l_A \\
& - Y \sin \alpha \\
& +\frac{Y^2}{2}\left(\frac{\cos^2\alpha}{l_A}-\frac{\cos \alpha}{R}\right)+\frac{Z^2}{2}\left(\frac{1}{l_A}-\frac{\cos \alpha}{R}\right) \\
& +\frac{\sin \alpha}{2l_A}\left[Y^3\left(\frac{\cos^2\alpha}{l_A}-\frac{\cos \alpha}{R}\right)+YZ^2\left(\frac{1}{l_A}-\frac{\cos \alpha}{R}\right)\right]
\end{aligned}
$$

Similar expressions being obtained for points B, C and D, also located in the x, y-plane, expression (2) can be expanded as follows:

$$
\begin{aligned}
\epsilon = {} & -Y\left[\sin \gamma + \sin \delta - k\frac{\lambda}{\lambda_0}(\sin \beta - \sin \alpha)\right] \\
& + Y^2\left[\frac{\cos^2\gamma}{l_C}-\frac{\cos \gamma}{R}+\frac{\cos^2\delta}{l_D}-\frac{\cos \delta}{R}-\frac{\lambda}{\lambda_0}\left(\frac{\cos^2\beta}{l_B}-\frac{\cos \beta}{R}-\frac{\cos^2\alpha}{l_A}+\frac{\cos \alpha}{R}\right)\right] \\
& + Z^2\left[\frac{1}{l_C}-\frac{\cos \gamma}{R}+\frac{1}{l_D}-\frac{\cos \delta}{R}-\frac{\lambda}{\lambda_0}\left(\frac{1}{l_D}-\frac{\cos \beta}{R}-\frac{1}{l_A}+\frac{\cos \alpha}{R}\right)\right] \\
& + Y^3\left[\frac{\sin \gamma}{l_C}\left(\cos^2\gamma-\frac{\cos \gamma}{R}\right)+\frac{\sin \delta}{l_D}\left(\frac{\cos^2\delta}{l_D}-\frac{\cos \delta}{R}\right)\right] \\
& - \frac{\lambda}{\lambda_0}\left[\frac{\sin \beta}{l_B}\left(\frac{\cos^2\beta}{l_C}-\frac{\cos \beta}{R}\right)-\frac{\sin \alpha}{l_A}\left(\frac{\cos^2\alpha}{l_A}-\frac{\cos \alpha}{R}\right)\right]
\end{aligned}
$$

The polar equations of the tangential and sagittal focal curves are obtained by cancelling respectively the terms in Y^2 and Z^2. The corresponding curves are shown in Figure 5 for the case of the grating depicted in Figure 3.

The term in Y^3 is seen to vanish when points A, B, C, D are all four located on the circle of equation ${}_l = R \cos \alpha$, which is the Rowland circle. In this case, the tangential curve is the Rowland circle itself.

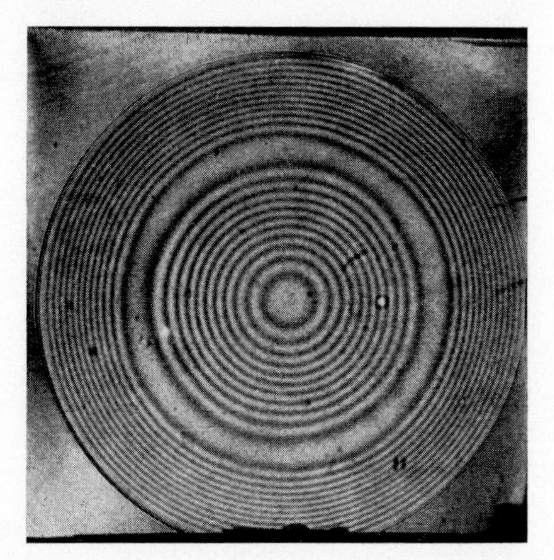

FIG. 5. Michelson-Twyman interferogram of Schmidt grating (see Figure 3): this plane grating exhibits fringes identical to those of the Schmidt corrector plate used in the recording step (size 75×75 mm, 1200 grooves mm^{-1}, first order wave).

These properties and others have been experimentally checked and found to be in good agreement with the above theory. More work is needed in order to optimize operating parameters in every case of interest. Because of the large number of possible combinations (including the possible use of correcting lenses during the recording and/or reconstructing steps), and the many different needs, grating imaging will probably develop as a distinct field in optical computing.

CONCLUSION

Holographic concave gratings have significant advantages regarding size, beam apertures and stigmatic properties. We believe that these new features will be best exploited in ground- and satellite-based astronomical instruments, for which large gratings and high aperture spectrographs are critically needed. Also the application to X-ray dispersive devices seems quite promising since conventional concave gratings are strongly affected by astigmatism when used at grazing incidences.

Finally, it appears that our holographic process can be useful for monochromatic imaging devices and for those applications of holography where 'diffraction limited' imaging quality is required.

Acknowledgment

We wish to thank J. P. Priou for skilful collaboration and we acknowledge support from the Centre National d'Etudes Spatiales and the Délégation Générale à la Recherche Scientifique et Technique.

REFERENCES

ARMSTRONG, J. A., 1965, *IBM Journal*, 171.
LABEYRIE, A. and FLAMAND, J., 1969, *Optics Communications*, **1**, 5.
LOWENTHAL, S., WERTS, A. and REMBAULT, M., 1968, *Comptes Rendus Acad. Sc. Paris*, **t.267**, 120.
MARECHAL, A., 1952, Imagerie Geométrique, Aberrations, *Rev. d'Opt.*, Paris.
SAKAYANAGI, Y., 1967, *Science of Light*, **16**, 2.
SHERIDON, N. K., 1968, *App. Phys. Letters*, **12**, 3.
STROKE, G. W., WESTERVELT, F. H. and ZECH, R. G., 1967, *Proc. I.E.E.E.*, **55**, 109.

Interferometric Methods for the Study of Non-Linear Anisotropy in a Lasing Gas

H. de Lang

Philips Research Laboratories, N.V. Philips' Gloeilampenfabrieken, Eindhoven, The Netherlands

Abstract—It has been found that an amplifying gaseous medium exhibits saturation-induced anisotropy which has its origin in the quantum-mechanical nature of the transition involved. With the aid of experiments making use of a gas laser provided with external anisotropic reflecting systems it has been possible to measure these effects. The principles of this experimental technique are given.

INTRODUCTION

Qualitative Experimental Evidence and Theory

Four years ago we found experimentally that a lasing gaseous medium exhibits saturation-induced anisotropy (de Lang, Bouwhuis & Ferguson 1965). A demonstration of the effect of this anisotropy on the state of polarization of the mode of a single-mode He:Ne laser ($\lambda = 1{\cdot}152\ \mu$) together with a quantitative theory (Polder & van Haeringen 1965) of this effect was presented (de Lang 1967) at the 'Zeeman Centennial Conference'. Saturation-induced anisotropy, which is in fact a non-linear phenomenon involving the vectorial aspects of the interaction of the radiation with the amplifying medium, has its origin in the quantum-mechanical nature of the stimulated transition. Without going into details of the theory we will briefly summarize here the basic aspects of non-linear anisotropy.

Consider a gas laser of, for example, the Fabry-Perot type with a single mode saturating the amplifying medium to such a degree that the condition of steady-state 'gain = loss' is reached. This saturation, however, will not be of a purely scalar nature; in other words the lasing gas will become anisotropic, the type and magnitude per pass of the anisotropy depending on the state of polarization of the mode (i.e. orientation and axes ratio of its polarization ellipse). This anisotropy induced by the elliptically polarized mode will, in its turn, affect the parameters of this polarization ellipse. So, roughly speaking, the polarization ellipse follows the anisotropy created by itself. In establishing the time evolution of the polarization parameters under these conditions use can be made of the fact that in practical cases this evolution is slow enough to allow a 'quasi-stationary' treatment, i.e. the state of saturation adapts itself so quickly that the state of saturation at a given moment is fully determined by the polarization parameters at that time. This condition being fulfilled, the time evolution of the polarization parameters is merely a function of these parameters themselves.

Coming now to the origin of saturation-induced anisotropy, this is found in the degeneracy of the upper and lower level of the stimulated transition. In linear optics this degeneracy can only be raised with the aid of external fields (e.g. Zeeman effect). In the non-linear case, however, the sublevels manifest themselves in the absence of external fields as non-linear anisotropy, the type and magnitude being dependent on the quantum numbers j of total angular momentum of the levels involved. The results of a calculation first carried out by Polder and Van Haeringen (1965; see also de Lang 1967*a*, *b*; Van Haeringen 1967*a*) as formulated for a standing wave laser mode are expressed by:

$$\frac{d\chi}{dt} = (gkc/2L)[\sin 4\chi/(T + \cos 4\chi)] \tag{1}$$

$$\frac{d\theta}{dt} = -(gkc\Gamma/L)[\sin 2\chi/(T + \cos 4\chi)] \tag{2}$$

Where:

$\tan\chi$ = axes ratio of the polarization ellipse of the mode ($-1 \leq \tan\chi \leq +1$);

θ = azimuth of long axis of polarization ellipse;

g = power gain per pass;

k = 'relative pumping excess', i.e. the fraction by which the pumping strength exceeds that needed for threshold operation. Equations (1) and (2) are valid for $k \ll 1$ which in the non-linear theory implies the restriction to third-order field terms;

c = velocity of light;

L = path length per pass;

$\Gamma = -\rho\Delta\nu\{[(\Delta\nu)^2 + 2\rho^2]^{-1} + 4\ln 2/(\Delta_D\nu)^2\}$

$\Delta\nu$ = natural line width of the transition;

$\Delta\ \nu$ = Doppler width;

$\rho = \nu_{mode} - \nu_{line}$;

$T = -5(j+1)^2/[j(j+2)]$ for $j \longleftrightarrow j+1$ transitions;

$T = 5j(j+1)/[(j-1)(j+2)]$ for $j \longleftrightarrow j$ transitions;

j = quantum number of total angular momentum.

The time evolution of χ as given in Equation (1) is due to the absorptive part of the non-linear anisotropy whereas the time evolution of θ as given in Equation (2) is caused by the dispersive part of this anisotropy. Although the dispersive effect is a very interesting one as it plays an important role in the explanation of polarization effects in Zeeman lasers (de Lang 1967*b*; de Lang & Bouwhuis 1967; van Haeringen 1967*b*; de Lang & Van Haeringen 1969), the experiments to be discussed in the present paper will mainly concern the absorptive part.

A simple analysis of Equation (1) shows that $\chi = 0$ and $\chi = \pm\pi/4$ are time-stationary solutions. Which of these solutions is the stable one, however, depends on the value of the parameter T characteristic for the transition. For example, for the neon $1{\cdot}152\mu$ line ($2s_2 \rightarrow 2p_4$; $j=1 \rightarrow j=2$) we have $T = -20/3$. It can easily be verified that stability occurs here at $\chi = 0$ ($\chi = \pm\pi/4$ is stationary but unstable). This explains the experimental fact that a $1{\cdot}152\mu$ He:Ne laser shows a marked preference for linear mode polarization (de Lang *et al.*, 1965). We call such a transition for which $\chi \rightarrow 0$ a 'polarophobe' transition (de Lang 1967*b*) because in the representation on the Poincaré sphere (Poincaré 1892; Born & Wolf 1959) the locus of the polarization ellipse of the amplified wave moves away from the poles of that sphere.

Apart from the confirmation of the experimentally found linear polarization of the neon $1{\cdot}152\mu$ line, the theory had some very intriguing consequences in that it indicated the existence of laser transitions with a preference for circular polarization ('polarophile' transitions) as well as transitions with no preference at all ('indifferent' transitions). 'Polarophily' is indicated for $j \rightarrow j$ transitions with $j > 1$ as for such cases $T > 1$. Experimentally the existence of 'polarophile' transitions was confirmed (de Lang & Bouwhuis 1966). For example the neon $1{\cdot}207\mu$ line ($2s_5 \rightarrow 2p_6$; $j=2 \rightarrow j=2$) showed preference for circular polarization (see p. 61 of de Lang 1967*b*). Investigation of a $1 \rightarrow 0$ transition (Ne $1{\cdot}523\mu$; $2s_2 - 2p_1$; $j=1 \rightarrow j=0$), however, showed, contrary to the theoretical prediction that it should be 'indifferent', a weak 'polarophily'*.

So far we have tacitly assumed that the only anisotropy in the laser was that non-linearly induced in the gas, in other words the interferometer itself was assumed to be isotropic and therefore to have no influence on the polarization ellipse. In practice, however, the laser interferometer will always show some anisotropy (even in absence of Brewster windows) which, even though it is small, also affects the polarization parameters. This residual anisotropy of the interferometer causes, for example, the polarization azimuth θ to attain a fixed value in a non-ideally isotropic

* This anomaly has been explained later by Polder and Van Haeringen (1967) by introducing collision-induced magnetic sub-level relaxation into the theory. This explanation was supported by experiments carried out by Bouwhuis (1968).

laser operating at a 'polarophile' transition (de Lang 1967*b*; de Lang & Bouwhuis 1967; van Haeringen 1967*b*; de Lang & van Haeringen 1969). On the other hand, the anisotropy of the interferometer can be made useful for a quantitative experimental study of the time evolution of χ (and θ) under the influence of non-linear anisotropy. The principle of this method which we originally introduced in order to check the quantitative validity of the theoretical Equation, (1), is as follows. Suppose the interferometer possesses an anisotropy of known type and magnitude. Under the combined influences of the interferometer and the medium the polarization parameters (in particular χ) will attain an equilibrium value which is a measure of the strength of the influence from the non-linear anisotropy. In the next section we will describe how such measurements can be carried out.

Measurement of Non-Linear Anisotropy Making Use of Controlled Interferometer Anisotropy (*de Lang* 1967*b*)

In order to carry out the measurements it is necessary to have a method of adjusting the anisotropy of the interferometer. In principle, it would be possible to achieve this by inserting anisotropic elements between the

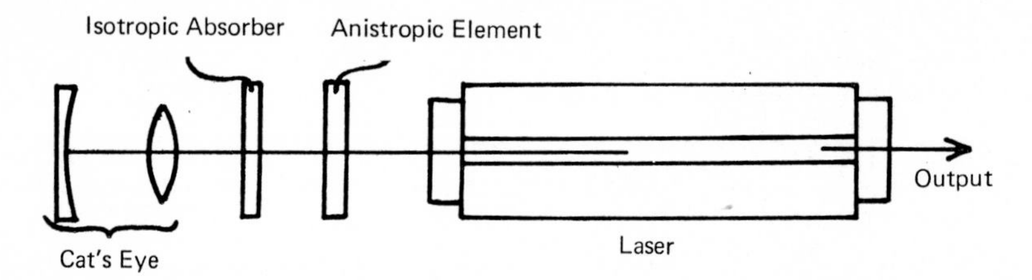

FIG. 1. Planar gas laser with external anisotropic reflecting system.

mirrors of the lasing Fabry-Perot interferometer. Apart from the fact that in this case this auxiliary anisotropy would have to be of a very small magnitude in order to be comparable with the weak non-linear effects to be measured, such a method would require dismantling of the sealed-off laser construction, which is quite impractical. Therefore we have made use of an auxiliary external reflecting system, consisting of a retro-directive lens-mirror system ('cat's eye') and a suitable type of anisotropy as shown in Figure 1. With the aid of such an arrangement the laser interferometer can be given the desired anisotropic properties. As an example we will describe an arrangement for the measurement of the 'polarophoby' of the neon $1{\cdot}152\mu$ line ($2s_2 \rightarrow 2p_4$; $j=1 \rightarrow j=2$). The anisotropy of the external reflecting system consists of a quarter-wave plate and a linear polarizer (see Figure 2). It can be shown that the anisotropy of the interferometer is represented by a Clark-Jones matrix possessing two coinciding circular eigenstates of polarization. The effect of such an anisotropy on

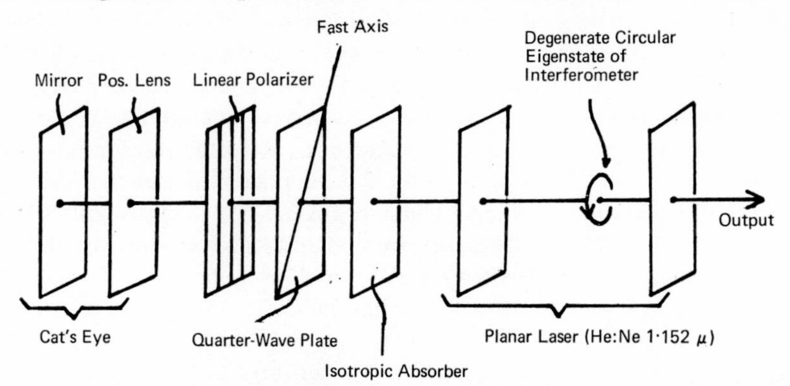

FIG. 2. Laser with external anisotropic system for the measurement of 'polarophoby'. The interferometer possesses a degenerate anisotropy with two coinciding circular eigenstates.

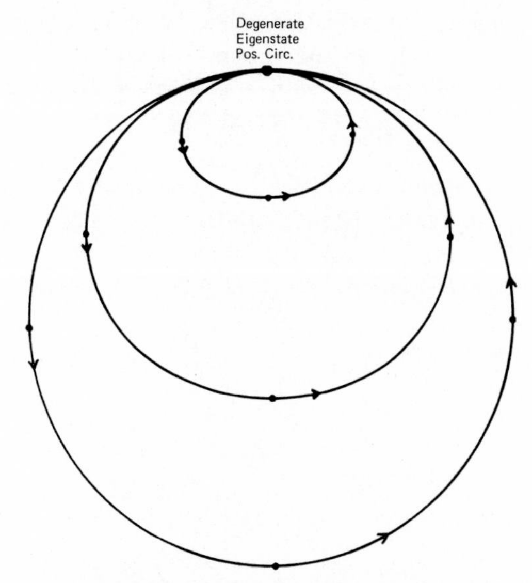

FIG. 3. Flow pattern on the Poincaré sphere of the locus of mode polarization in the interferometer shown in Figure 2, due to interferometer anisotropy (excluding the non-linear anisotropy of the medium).

the polarization of the mode is vizualized in Figure 3 where the flow pattern of the locus of the mode polarization is shown as it appears on the Poincaré sphere. From this flow pattern we see that all flow lines end at the North Pole where the velocity is zero. So, if there were no non-linear anisotropy, mode polarization would be positive circular. However, due to the presence of the non-linear anisotropy of the 'polarophobe' transition there will be a flow pattern away from the North Pole superimposed on the flow pattern shown in Figure 3. Under these combined influences the ellipticity parameter, χ, of the mode will attain an equilibrium value χ_0 which can easily be measured with a photoelectric detection arrangement not to be discussed here. If we have quantitative knowledge of the flow pattern of Figure 3 the value of $d\chi/dt$ at the equilibrium value χ_0 is also known. Thus, apart from the parameters g and k which will not be discussed here, we have a quantitative check of the validity of the theoretical Equation (1). The question remains, however, how to calibrate the magnitude of the auxiliary anisotropy? In principle it would seem possible to derive this from the reflection and transmission of the left laser mirror together with those of the external elements. In practice this is a very unreliable method, and so it is desirable to have a more direct method of calibration. To this end, the axis of the quarter-wave plate is set parallel to the polarizer. It can be shown that then the interferometer anisotropy due to the external system becomes an ordinary, linear birefringence (for suitable axial positions of the 'cat's eye'). Under precautions not to be discussed here, the laser mode will split into two orthogonal, linearly-polarized components with a frequency difference which is a direct measure of the magnitude of the anisotropy due to the external system.

The measurements described yielded a 'polarophoby' for the neon $1{\cdot}152\mu$ line agreeing within a factor of two with the theoretical value expressed by Equation (1). Full details about this measuring technique together with further results concerning the neon $1{\cdot}523\mu$ line have been given by de Lang (1967*b*).

REFERENCES

BORN, M. and WOLF, E., 1959, *Principles of Optics*, Pergamon Press, Oxford, p. 30.
BOUWHUIS, G., 1968, *Physics Letters*, **27A**, 693.
VAN HAERINGEN, W., 1967*a*, *Physics Letters*, **24A**, 65 *b*. *Phys. Rev.*, **158**, 256.
DE LANG, H., 1967, 'Zeeman Centennial Conference, Amsterdam, 1965', *Physica*, **33**, 163 *b*. *Philips Res. Repts. Supplements*, No. 8.
DE LANG, H. and BOUWHUIS, G., 1966, *Physics Letters*, **20**, 383.
DE LANG, H. and BOUWHUIS, G., 1967, *Physics Letters*, **19**, 481.
DE LANG, H., BOUWHUIS, G. and FERGUSON, E. T., 1965, *Physics Letters*, **19**, 482.
DE LANG, H. and VAN HAERINGEN, W., 1969, *Phys. Rev.*, **180**, 624.
POINCARÉ, H., 1892, *Theorie Mathématique de la Lumière*, Vol. 2, Chapter 12, Georges Carré, Paris.
POLDER, D. and VAN HAERINGEN W., 1965, *Physics Letters*, **19**, 380.
POLDER, D. and VAN HAERINGEN, W., 1967, *Physics Letters*, **25A**, 337.

Transverse Deflections of a 45-inch Diameter Lightweight Mirror Blank: Experiment and Theory

W. P. Barnes

Itek Corporation,
Building 10, 3*rd Avenue,*
Burlington, Mass., 01803. *U.S.A.*

Abstract—The central deflection, including both bending and shear effects, of a lightweight circular mirror under a uniformly distributed transverse load equal to its own weight and simply supported at the edge has been calculated.

The calculational approach used is extended to the case of a central load distributed over a small area, with three equally spaced support reactions at the edge. The predicted deflections are compared with the deflections measured using holographic interferometry for a lightweight mirror blank 45 inches in diameter.

INTRODUCTION

In the design of lightweight mirror structures consisting of two thin plates connected by a core structure whose mean density is substantially less than that of a solid core, the influence of shear forces on the total deflection may equal or exceed that of the pure bending behaviour of the mirror blank. Some considerations of the overall bending and shear behaviour of such structures, for circularly symmetric support, is shortly to be published (Barnes 1969). In the discussion below, we consider the case of a centrally loaded mirror supported at three points equally spaced around the edge of the blank. The theoretical calculation for this case is developed and compared with deflections measured using holographic interferometry, on a 45 inch-diameter mirror blank loaned to us by Heraeus-Schott Quarzschmelze GMBH. An exposition of the holographic technique we have used has been published (van Deelen & Nisenson 1969).

EXPERIMENTAL RESULTS

The mirror blank fabricated by Heraeus-Schott consists of a monolithic core and backplate made from their opaque fused silica. Weight has been removed from the core by drilling and milling 2 inch-diameter blind holes

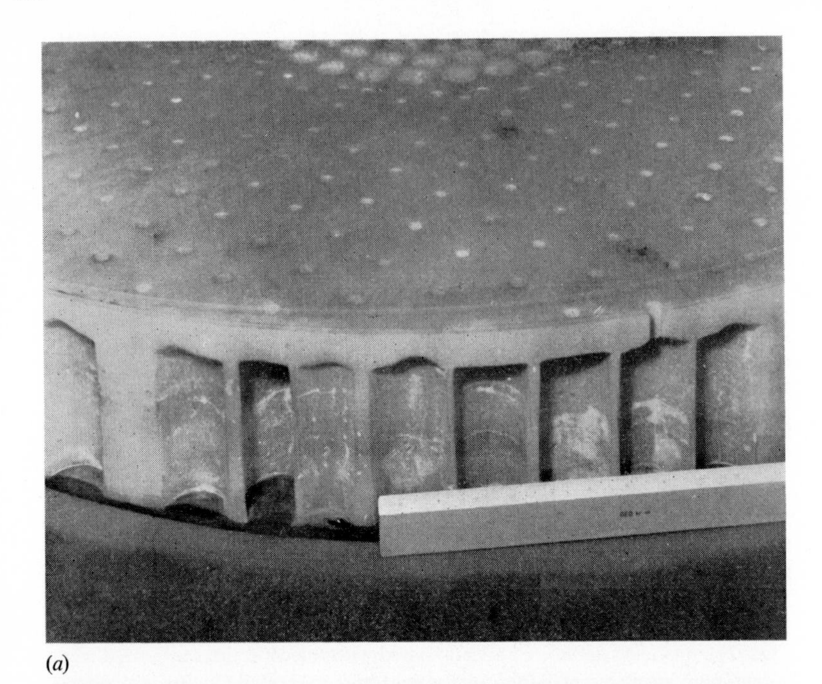

(*a*)

(*b*)

FIG. 1. Heraeus-Schott lightweight mirror blank: (*a*) portion of edge; (*b*) rear view.

to within 1 inch of the rear surface with $\frac{7}{16}$ inch-diameter venting holes carried through the rear plate. To this core and backplate structure is fused a solid disc of clear fused silica. Photographs of the rear and a portion of the edge of the structure are shown in Figure 1, and some dimensional details are considered further in the theoretical treatment. The central area of this blank has been prepared for a Cassegrain perforation.

With appropriate fixturing and force measurement, the holographic interferogram of Figure 2 was obtained with a central load and three-point edge support. The deflection data obtained from this interferogram

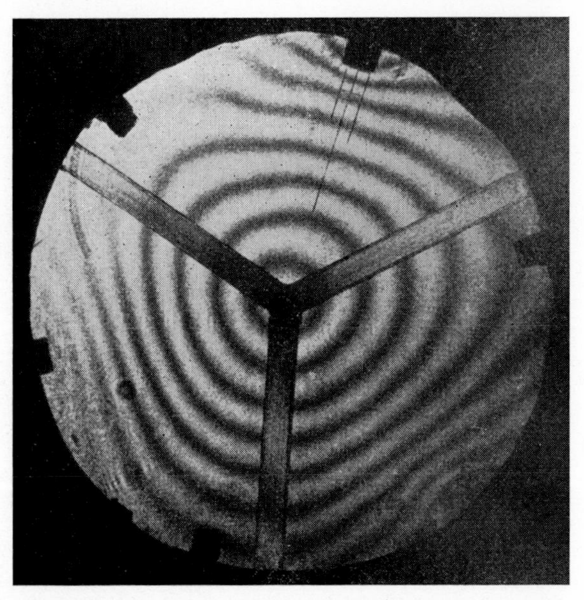

FIG. 2. Holographic interferogram obtained with central load (*c*. 250 pounds) and three-point edge support.

along a diametral line through one support point are shown in Figure 3. A moderate interpolation, at $r=0$, was made to establish the reference $w=0$ at $r=0$, while requiring $\mathrm{d}w/\mathrm{d}r=0$ at $r=0$. The data shown are for a central load of 250 pounds. The total weight of this blank was 430 pounds.

THEORETICAL TREATMENT

Elsewhere (Barnes 1969) I have developed expressions for the flexural rigidity, D, and a shear coefficient, K, from both elementary considerations and the work of Timoshenko and Woinowsky-Kreiger (1959), and Cowper (1966). A representative cross-section of a lightweight structure with uniformly thick ribs is shown in Figure 4, and the expressions obtained are given below.

$$D = \left[\frac{Eh^3}{12(1-\nu^2)}\right][1-(1-\alpha)(1-2t)^3]$$

$$\frac{1}{K} = \frac{2t}{\alpha(1-t)} + 1$$

E = Young's modulus;
ν = Poisson's ratio for the blank material;
h = total blank thickness;
α = ratio of rib thickness to 'rib spacing' (assuming uniform ribs);
t = ratio of plate thickness to total blank thickness (assuming symmetrical construction).

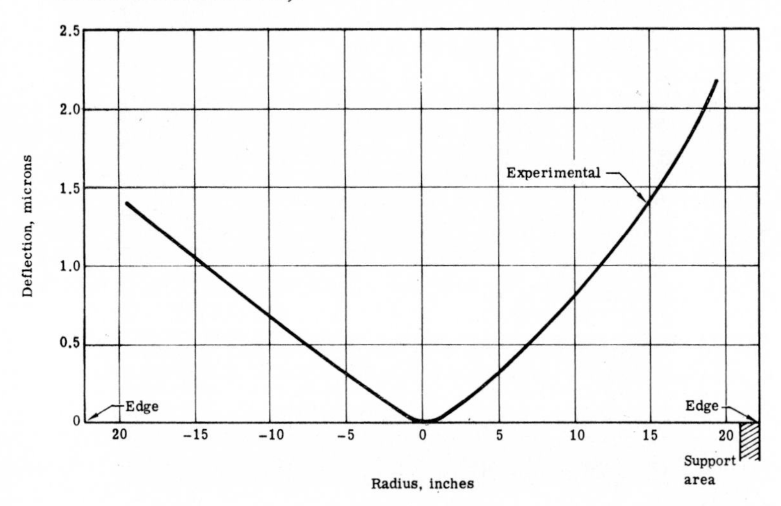

FIG. 3. Deflection versus radius along one diameter through a support.

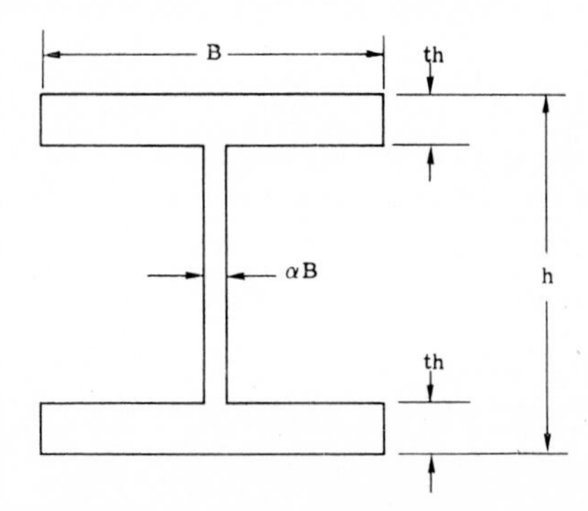

FIG. 4. Typical cross-sectional element (B = cell spacing; t = ratio of plate thickness to total thickness; h = total thickness; α = ratio of rib thickness to cell spacing).

For this blank, the ribs are not of uniform thickness, nor is the front plate, because of local sag during fusing and the subsequent grinding of the front surface to a smooth curve. Measurements of the blank yielded the following:

Front plate thickness $=\frac{3}{4}'' \pm \frac{1}{8}''$.
Backplate thickness $=1'' \pm /\frac{1}{16}''$.
Total thickness $=6\frac{1}{4}''$
Outside diameter $=44\frac{5}{8}''$.

An average value of $t=0{\cdot}14$ was used in the calculations, and the value of α used was calculated by equating the mean core density to that of a core with ribs of uniform thickness, first subtracting the plate weight for the above value of t from the total measured weight of the blank, yielding $\alpha=0{\cdot}21$.

For our case, following Timoshenko and Woinowsky-Kreiger (1959), we may obtain the bending deflection, w_b, in radial coordinates ρ (normalized) and θ.

$$w_b=\frac{Pa^2}{2\pi D}\left\{\left[\frac{3+\nu}{8(1+\nu)}\right]\rho^2-\frac{\rho^2\ln\rho}{4}+\frac{1}{3+\nu}\sum_{m=3,6,9,..}^{\infty}\left[\frac{1}{m(m-1)}+\frac{2(1+v)/(1-v)}{m^2(m-1)}-\frac{\rho^2}{m(m+1)}\right]\rho^m\cos m\theta\right\}$$

$P=$central load;
$a=$outer radius of blank;
$r=$radial coordinate, $\rho=r/a$.

Combining Timoshenko and Woinowsky-Kreiger, and Cowper, we find the derivative with respect to r of the shear deflection, w_s, to be:

$$\partial w_s/\partial r=\partial w_s/\partial a\rho=V_r/GKh$$

where $V_r=$effective vertical reaction, per unit length, on an r face;
$G=$shear modulus of blank material.

Timoshenko and Woinowsky-Kreiger also give V_r as

$$V_r=Q_r-(1/r\partial\theta)(M_{r\theta}),$$

where,

$$Q_r=-D\partial/\partial r(\Delta^2 w_b),$$

$$\Delta^2=\frac{\partial^2}{\partial r^2}+\frac{1}{r}\frac{\partial}{\partial r}+\frac{1}{r^2}\frac{\partial^2}{\partial\theta^2}$$

$$M_{r\theta}=(1-v)D\left[\frac{1}{r}\frac{\partial^2}{\partial r\partial\theta}-\frac{1}{r^2}\frac{\partial}{\partial\theta}\right]w_b$$

Similarly, the derivative with respect to θ of the shear deflection may be written

$$\frac{1}{r}\left(\frac{\partial w_b}{\partial \theta}\right)=\frac{V_\theta}{GKh},$$

$$V=Q_\theta=-\frac{D\partial(\Delta^2 w_b)}{r\partial\theta}$$

If these operations are applied to the expression for w_b, taking the central load to be distributed over a small circle of normalized radius, b, and integrating first for $\theta=0$, then tangentially (constant ρ) we obtain

$$w_s(\text{at } \rho=b,\ \theta=0)=\frac{1}{2}\left(\frac{P}{2\pi GKh}\right)$$

$$w_s(\text{at } \rho\geqslant b,\ \theta=0)=\frac{P}{2\pi GKh}\left(\ln\frac{\rho}{b}+\frac{1}{2}+\frac{1}{3+\nu}\sum_{m=3,6,9,\ldots}^{\infty}\left\{\left[\frac{4}{m}-(1-\nu)\right]\right.\right.$$

$$\left.\left.(\rho^m-b^m)+\left[\frac{4}{m-2}+(1-\nu)\right](\rho^{m-2}-b^{m-2})\right\}\right),$$

and

$$w_s(\text{at } \rho\geqslant b,\ \theta=\pi/3)=w_s(\text{at } \rho=\rho,\ \theta=0)-\frac{P}{2\pi GKh}\left[\frac{8\tanh^{-1}(\rho^3)}{3(3+\nu)\rho^2}\right]$$

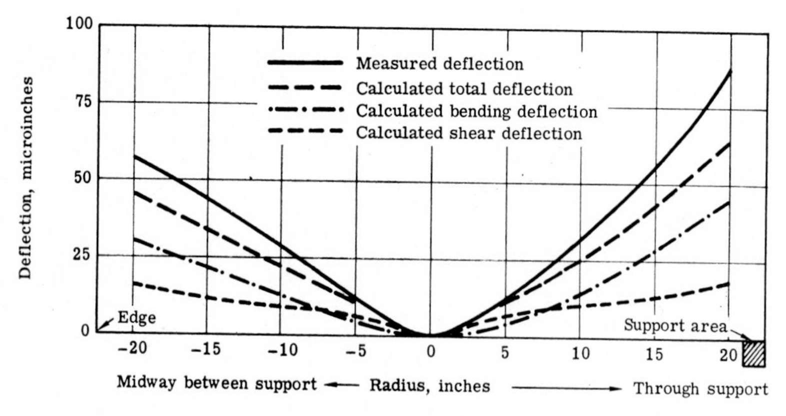

FIG. 5. Calculated deflections and comparison with measured deflections.

The calculated values of w_b, w_s, and w_b+w_s are shown in Figure 5, together with the experimentally measured total deflection of Figure 3.

A linear regression correlation of the measured data (w_m) with $w_b + w_s$, yields:

$w_m = 1{\cdot}384\ (w_b + w_s) - 2{\cdot}4$ (micro-inches) with a correlation coefficient of 0·997.

CONCLUSIONS

It is somewhat premature to draw firm conclusions from the above calculations, but it appears that we may be able to establish a well-founded and fairly elementary theoretical basis for the calculation of the transverse deflections of lightweight mirror structures. At this point in time, our principal need is a more extensive comparison with experiments including a reasonable range of the parameters of α and t.

REFERENCES

BARNES, W. P., 1969, *Appl. Opt.*, **8**, 1191.
COWPER, G. R., 1966, *J. Appl. Mech.*, **33**, 335.
VAN DEELEN, W. and NISENSON, P., 1969, *Appl. Opt.*, **8**, 951.
TIMOSHENKO, S. and WOINOWSKY-KREIGER, S., 1959, *Theory of Plates and Shells*, 2nd ed., **8**, 951. New York, McGraw-Hill.

Optical Imaging by Frequency and Phase Modulation Techniques

A. I. Kartashev and A. N. Korolyev

Committee for Standards,
Measure and Measuring Instruments,
Kwartal Yugo-Zapad 38,
Moscow 189

Abstract—Application of the principles of information theory is considered and the various possibilities of obtaining increased resolution, beyond the diffraction limit, are discussed. The result of frequency modulations or dispersion and of phase modulation or interference are shown to produce the required result, the increase in each case being restricted to one direction. An increase in two directions can be obtained by rotation of the optical system, with a slit pupil, about its optical axis.

INTRODUCTION

It is known that any optical system has a limited resolution which results from the wave nature of light. On the classical theory of Abbe and Rayleigh this limitation is due to diffraction and is defined by the wavelength of the radiation used and the aperture of the optical system.

From the point of view of information theory the total amount of information transmitted by an optical system is determined by the number of independent parameters that are needed for a complete description of the wave field passing through the optical system from the object space to the image space.

The theorem that determines in a generalized form the information capacity of an optical system as a communication channel was formulated by Gabor (1956; 1961) and was called the Invariance Theorem. The total number of degrees of freedom of the wave field for the case of imaging by means of an optical system was given by Lukosz (1966) as the invariant product:

$$N = 2N_t L_x L_y (K'_x K'_y / \pi^2) \tag{1}$$

where 2 is the factor implying the possibility of using the two independent polarization states of light radiation for imaging; N_t is the number of temporal degrees of freedom; L_x and L_y are dimensions of the object field in two mutually perpendicular directions, and K'_x, K'_y are the upper limits of spatial frequencies transmitted by the optical system in these directions. Thus, it appears possible to surpass the upper limit of spatial frequencies of a given optical system by using effectively the degrees of freedom of the wave field of an object, one of the other factors in Equation (1) being decreased accordingly.

It has been shown (Kartashev 1960) that such physical parameters as wavelength or phase of radiation can be used for optical imaging. The use of these parameters provides for a still further increase in the number of degrees of freedom of the wave field and consequently for a higher information capacity of an optical communication channel. In principle, the basis for such an increase is an optical frequency or phase modulation of the initial distribution of illumination in the object plane produced with the aid of dispersion devices (of the interference or ordinary type); other appropriate dispersion devices, located in the image space, play the role of demodulators (Figure 1).

Imaging Processes

The imaging process using the principle mentioned above can be explained in the following way. Let us assume that when an object is illuminated, a certain value of wavelength λ (or phase ϕ) corresponds to each of its elements within the interval (x_1 to x_n). The totality of object elements along the x-axis can be treated as some distribution of illumination $I=f_1(x)$. The application of a dispersion system for illuminating the object introduces an additional condition $\lambda=f_2(x)$. If the dispersion region of the illumination system coincides with the interval ($x_1, \ldots x_n$) in the object plane then a definite value of illumination will be associated with each value of λ; $I=F(\lambda)$. This distribution of illumination is a function of wavelength and is independent of coordinates. If we place another dispersion system for which the relation $\lambda=f_3(x')$ is valid (where x' is a coordinate value of the image space), then, taking into consideration the fact that $I=F(\lambda)$, the distribution of illumination of the image can be obtained in the form: $I=f_4(x')$.

When frequency and phase modulation are used the number of degrees of freedom of the wave field can be increased by N_λ and N_ϕ times, respectively. Hence,

$$N_\lambda=\Delta\lambda/\delta\lambda \qquad (2),$$

where $\Delta\lambda$ is the dispersion region of the system, $\delta\lambda$ is the minimum spectral interval resolved; and,

$$N_\phi=\pi/\delta\phi \qquad (3),$$

$\delta\phi=(1-R)/R$ is the width of the interference maximum expressed in terms of phase.

The advantage of the phase modulation method is the possibility of using several orders of interference. In this case the linear dimensions of the object field may exceed those of the system's region of dispersion and the information from outside the dispersion region can be transmitted in accordance with geometrical parameters of the optical system.

For imaging purposes the number of degrees of freedom of the wave field can be represented, in a generalized form, by the expression:

$$N=2N_tL_xL_y(K_x'K_y'/2)(\Delta\lambda/\delta\lambda)(\pi/\delta\phi) \quad (5)$$

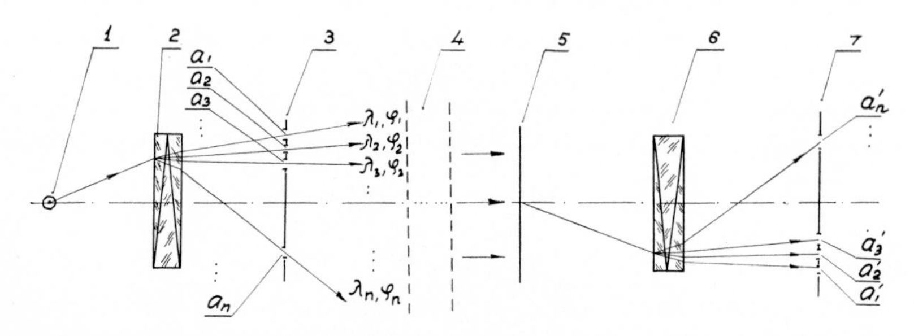

FIG. 1. Principle of optical imaging 1, source of light; 2, first dispersion system (frequency or phase modulator); 3, object; a_1, a_2, a_3–a_n, elements of the object; 4, diffuse or turbulent medium; 5, receiving optical system; 6, second dispersion system (frequency or phase demodulator); 7, image.

Frequency Modulation—Dispersion

Let us consider the imaging process employing dispersion systems which split the light radiation into components having various wavelengths. The wide continuous spectrum of the radiation source obtained with the aid of a dispersion system is projected into the plane of the transparent object. In this case a certain value of the wavelength λ will be associated with each zone of a one-dimensional object in the x-direction that corresponds to the direction of dispersion of the system. The object transforms the distribution of illumination in the spectrum changing the radiation intensity for each wavelength value in conformity with the distribution of optical density in the x-direction. To reproduce the image of the object it is sufficient to obtain the spectrum of the light radiation that passed through the object and to detect it with a non-selective receiver. An optical system whose spatial resolution is zero (e.g. a light guide) can be employed to transmit the light radiation.

It is obvious that this method cannot be treated in an ordinary way. Speaking the language of communication theory, it is possible to say that in this case the information about each element of the object is transmitted through a separate frequency channel and the process of transmitting the image is equivalent to frequency modulation in the object space and frequency demodulation in the image space. In this connection the system has two attractive and advantageous features: firstly, diffraction does not impose restrictions on the resolving power, and secondly, phase distortions of the light wave behind the object do not disturb the process of imaging.

Therefore, a non-uniform or diffusive medium whose diffuse-transmission factor is independent of the wavelength may separate the object and the receiver system without prejudice to the transmission of the image if only the light flux reaching the receiver system is strong enough to be detected.

Phase Modulation—Dispersion

Now, let us consider another method in which phase dispersion systems are employed in the object and image spaces. In this case the object is

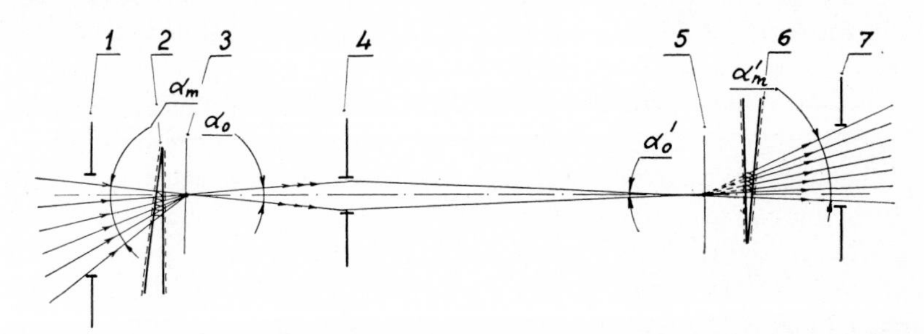

FIG. 2. Principle of optical image transmission by a phase modulation technique. 1, Pupil of the illuminating beam; 2, first multi-beam interference Fabry-Pérot wedge; α_m, aperture angle of coherent beams due to multi-beam interference; 3, object plane; α_0, original aperture angle of the transmitting optical system 4; 5, image plane; 6, second multi-beam interference Fabry-Pérot wedge; 7, α_m, pupil and aperture angle of the detecting optical system.

illuminated in such a way that the object plane coincides with the field of interference in which there is a certain gradient of the path (or phase) difference in the direction coinciding with that along which the optical density of a one-dimensional object decreases. Interference wedges are used as phase dispersion systems; the object is illuminated with a source of white light (Figure 2).

When the direction and the angle of the wedge K_1 are appropriately chosen, each zone of the object located along the x-direction will be illuminated by light rays having a certain path difference Δl which is due to the thickness of the corresponding zone of the interference wedge. The phase difference of the rays given by

$$\Delta\phi = \Delta l/\lambda \tag{6}$$

where λ is the light wavelength, will change periodically in the x-direction and take equal values at those places where the path difference of the rays is equal to a multiple of λ. The points having equal phases will be separated by distances

$$l = \tfrac{1}{2}\lambda V \tan \gamma\delta \tag{7}$$

where γ is the angle of the wedge K_1, and V is the magnification of the system.

Assume that the object thus illuminated is projected with the aid of a lens onto the image plane. To prevent two neighbouring interference orders from falling into the area of the diffraction maximum, it is necessary to make the resolution of this system, expressed in terms of length, equal to an interval l in the object space. Then, the sections of the object, illuminated by light beams having equal phase differences and path differences that differ not less than by λ, will fall into different areas of diffraction maxima.

In order to determine the illumination distribution within the limits of the diffraction spot, i.e. beyond the limits of the system's resolving power, it is sufficient to place an interference wedge K_2 in the image plane. When corresponding sections of the wedges K_1 and K_2 producing equal path differences for the beams are conjugate, interference in white light can be observed. As the images of all separate elements (zones of a one-dimensional object) will then be determined by the width of an interference maximum (but not of the diffraction one), this will result in an increase of the system's resolving power by $l'/\Delta p$ times, where l' is the width of a diffraction maximum, and Δp is the width of an interference maximum. As is known, the width of the latter is determined by the reflection factor of the coatings on the wedge plates. Since the ratio $l'/\Delta p$ may take values in the range 10 to 20 and higher this gives rise to an increase of the resolving power accordingly by 10 to 20 times, as compared with conventional systems.

Increased Resolution in One Direction

Now, let us consider what will ensue from such an increase in the system's resolving power in the x-direction. With this in view, one should analyze how the illumination of an object varies when the interference wedge K_1 is being introduced into the optical path. Let us assume that the interference wedge K_1 is illuminated by a parallel beam of white light. If a plane wavefront falls on an interference wedge, it is known that on its passage through the wedge a 'set' of wavefronts whose angles differ from that of the incident front by a magnitude of $2\gamma k$, where γ is the wedge angle, and k is the number of reflections of the incident light wave from the interference wedge mirrors, is produced as a result of multiple reflections. Light waves produced on reflection are coherent, have a phase shift and decreasing intensity. When the object is illuminated by such coherent light waves under different angles of incidence relative to the object plane, the spatial-frequency band transmitted by the system in the x-direction is increased. Phase differences of the illuminating beams are compensated by the interference wedge K_2.

It should be noted that the 'set' of wavefronts, produced on passage of the radiation through the interference wedge K_1 and illuminating the object, is distributed over a certain interval along the z-axis which is approximately given as

$$\Delta z = 2tN_{eff}, \tag{8}$$

where t is the thickness of the wedge K_1, and N_{eff} is the number of effective beams for the mirrors of the wedge K_1. Owing to the fact that light waves produced by these fronts are coherent, in this case, it is in principle impossible to distinguish between elements of the object lying within the zone Δz along the z-axis.

In this case, the consequence of increasing the resolving power in the x-direction is a decrease of resolution in the z-direction while the number of degrees of freedom of the wave field necessary to transmit three-dimensional information concerning the object remains apparently unchanged.

This conclusion allows us to draw an analogy between the imaging method described and interference microscopy where Ingelstam's uncertainty ratio necessary for the determination of the coordinates along the z- and x- (or y-) axis coordinates of an object element, is known. From this point of view, one may assume that in the method described the object is supplemented with an artificially created phase structure whose law of variation in the z-direction is known and that can be used for the imaging of the object with an increased resolution in the x-direction.

Further progress in frequency and phase modulation techniques designed for imaging purposes has led to the development of a number of original optical systems.

Thus, Koester's method of transmission of two-dimensional images through a single optical channel (fibreglass) was based on a frequency modulation technique (Koester 1968). This procedure implies an application of optical frequency modulation accomplished with the help of a dispersion system for one direction and use of temporal scanning for the other. Synchronously oscillating mirrors, located at the input and output ends of the fibre, scan the direction perpendicular to that of dispersion.

Accuracy of Measurement

The authors have developed an optical system for detecting the position of transparent lines in linear and angular measurements which is based on

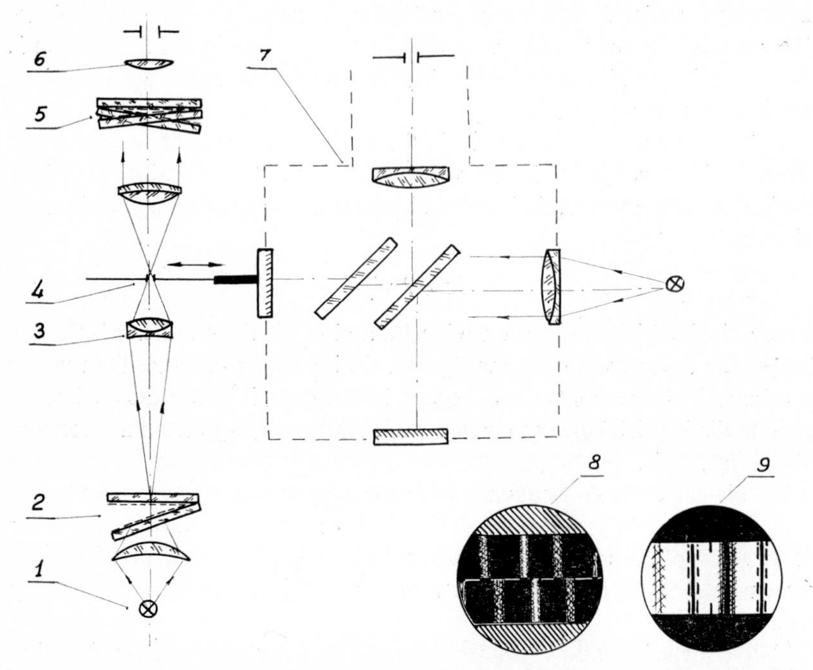

FIG. 3. Optical device for accurate setting on the line using phase modulation of the illuminating beam of light. 1, source of white light; 2, multi-beam interference Fabry-Pérot wedge (phase modulator); 3, lens; 4, transparent line displaced simultaneously with the mirror of a Michelson interferometer; 5, double multi-beam interference Fabry-Pérot wedge; 6, eyepiece; 7, Michelson interferometer; 8, field of vision of the optical system when the interference image of line is displaced; 9, field of vision of the Michelson interferometer when the achromatic fringe is displaced.

a method of phase modulation of the object illumination (Korolyev 1968*a*, *b*). The optical system comprises two multi-beam interference wedges, one of them producing a gradient of phase difference of the object illumination, and the other analyzing the light radiation that passes through the line. The application of phase modulation in such an optical system provides for an accuracy in determining the position of the line of an order of 0·1 μm (Figure 3).

As was stated above, the frequency and phase modulation methods can be used successfully for the imaging of one-dimensional objects or increasing the resolving power of an optical system in one direction.

Since other known methods of increasing the resolving power are also effective for one direction only, the authors have developed and studied a method of the imaging of two-dimensional objects with the aid of optical systems having an increased resolving power in one direction (Kartashev & Korolyev 1967).

Increased Resolution in Two Directions

The task was to make the best use of the bandwidth of such a system for two-dimensional objects. Imaging by rotation of the system about its optical axis was suggested. In this case the direction for which the system has an increased resolving power will change its orientation relative to the

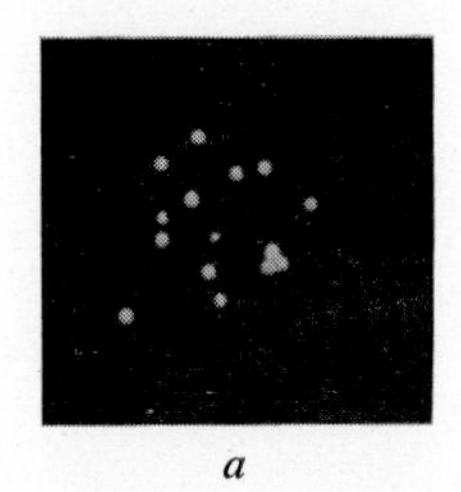

a

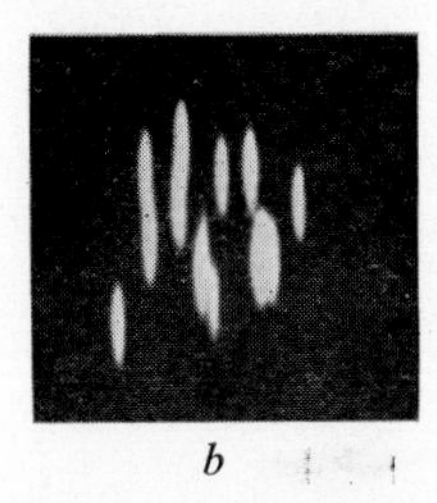

b

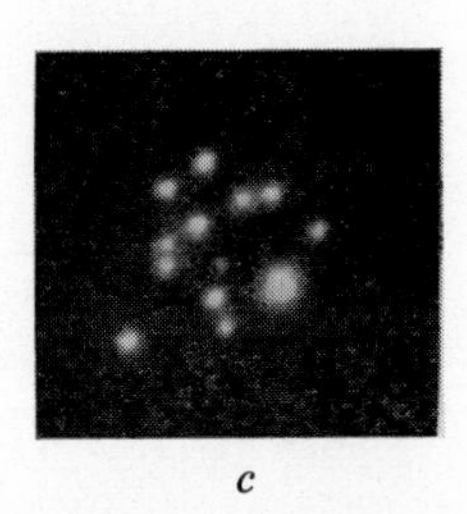

c

FIG. 4. Image of randomly distributed points produced by an objective with a rotatable slit pupil. *a*, original; *b*, image of the object obtained at one of the meridionalpositions. of the slit pupil; *c*, reconstruction of the image by superposition of 36 separate pictures

two-dimensional object and the components of the spectrum of the two-dimensional object will be transmitted into the image space in succession. Then, the system's temporal-frequency band of width Δv will be reduced to a magnitude of $\Delta v = 2\pi\omega$, where ω is the angular velocity of the rotating optical system.

In the process of investigation, the imaging was accomplished by superimposing many times in succession the images corresponding to different orientations of the maximum resolving direction of the optical system. To simplify the investigation, an objective with a slit aperture at the input was used as the optical system.

The illumination distribution in the integrated image of a point obtained by the described method was calculated. This method of imaging was also experimentally investigated and its resolving power for two-dimensional objects was determined (Figure 4).

The present investigation and experiments have shown that any optical system having an increased resolving power in one direction can be used for the imaging of not only one-dimensional objects but objects with an arbitrary two-dimensional distribution of brightness as well.

REFERENCES

GABOR, D., 1956, *Proceedings of the Symposium on Astronomical Optics*, Amsterdam, p. 17.
GABOR, D., 1961, *Progress in Optics*, **1,** 109.
KARTASHEV, A. I., 1960, *Optica i spektroskopia*, **9**, 394.
KARTASHEV, A. I. and KOROLYEV, A. N., 1967, *Optica i spektroskopia*, **23,** 450.
KOESTER, C., ASK, C. and VEER, J., 1968, Optical Society of America Annual Meeting, FD–20.
KOROLYEV, A. N., 1968*a*, *Bulleten isobreteny*, No. 13 Avt. svid. No. 215554. *b Troudy Metrologitsheskich Institutov SSSR*, **101,** (161) (Moska-Leningrad, izd, standartov) p. 129.
LUKOSZ, W., 1966, *J. Opt. Soc. Amer.*, **56,** 1463.

The Use of Dilute Cermets in Infrared Interference Filters

P. W. Baumeister*, G. Borak and L. Stensland

Institute of Optical Research,
100 44 *Stockholm* 70, *Sweden*

Abstract—Cermets, which consist of a dispersion of colloidal metal in a dielectric, are probably useful as thin films in optical interference devices for the infrared part of the spectrum. This paper describes measurements of such films, which were produced by vacuum coevaporation of the two constituents. Gold was used as the metal in these prototypes because of its high electrical conductivity and chemical inertness. Two different types of dielectric host material were chosen, potassium bromide and germanium. A 1·5 μ 10 per cent volume concentration of gold increases the refractive index of KBr from 1·54 to 1·74. Simultaneously the absorption constant k is of the order of 0·01 and decreases to smaller values at longer wavelengths. This absorption is sufficiently low for these dilute cermets to be used as components of optical interference coatings but precludes their use in thicker components, such as lenses. Good agreement is also achieved between the experiments and the Maxwell-Garnett theory for this cermet. The gold-germanium mixtures studied so far are, however, always very absorbing.

INTRODUCTION

A cermet, which consists of a dispersion of a colloidal metal in a dielectric (a ceramic), acts like an artificial dielectric because the metal particles enhance the polarizability of the dielectric. Although the theory of such systems was well established at the beginning of the twentieth century, (Rayleigh 1892; Maxwell-Garnett 1904) they were not technically used until microwave technology needed lenses of high index material. The use of such mixtures in art is much older. The technique of making ruby glass, which consists of gold particles dispersed in glass, was invented in Bohemia in 1670. Thus, although such mixtures are not new optical materials, they have not been studied in thin films with the aim of making films with enhanced refractive index. Previous studies have concentrated on the

* On leave from The Institute of Optics, College of Engineering and Applied Science, University of Rochester, Rochester, New York 14627, U.S.A.

absorption of light in such materials, caused by the metal particles (Weyl 1952; Anders 1959; Hampe 1958 and others). In the infrared wavelength region, a refractive index increase might be achieved without a corresponding increase in the absorption coefficient k. The relatively small value of k can be tolerated in a thin film a few wavelengths thick, while it would be too large to use in a massive component such as a lens. Dilute cermet mixtures seem promising to use in bandpass filters, which exhibit high

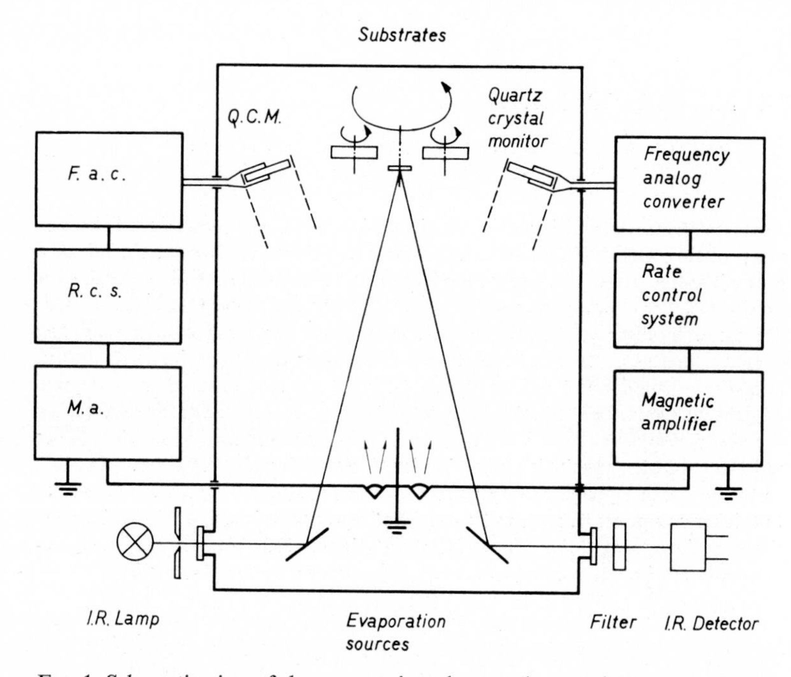

FIG. 1. Schematic view of the evaporation plant used to produce cermet films.

passband transmittance and large offband rejection. Smith (1958) and Seeley (1961) have shown that filters of superior performance are achieved if films of intermediate refractive index are used in the stack.

Production and Physical Properties of the Films

Thin cermet films were produced by co-evaporation of a metal and a dielectric. Separate quartz crystal monitors controlled the deposition rates via feedback loops (Jacobsson & Mårtensson 1965) (Figure 1). Planetary rotary motion was used during the evaporation to improve the thickness uniformity. Typical deposition rates were 400 Å min^{-1} for the dielectric (KBr and Ge) and 40 Å min^{-1} for the metal. The vacuum during evaporation was better than 5×10^{-5} torr. The evaporation sources used were molybdenum boxes for KBr and tungsten boats for Ge and Au. The Au/KBr cermet films were deposited on substrates of polished germanium and glass and the Au/Ge mixtures on glass, germanium, KBr and sapphire. Some very thin cermet films were investigated in an electron microscope. In this case the cermet films were evaporated onto a thin carbon film. The Au/Ge films were in addition evaporated onto NaCl substrates, in which case the film could be floated off the substrates to a water surface. In the KBr matrix gold tends to agglomerate in small spheres, 40 Å to 100 Å in diameter, (see Figure 2). The KBr crystallites are 500 Å to 800 Å in

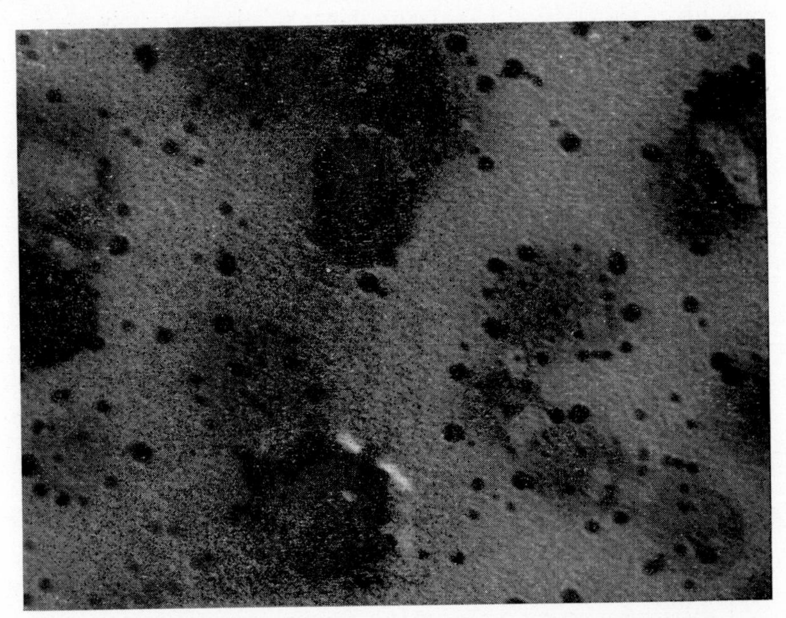

FIG. 2. Electron micrograph of a very thin gold-KBr cermet film.

diameter. This interpretation is confirmed by the electron diffraction pattern from the same film, (see Figure 3). The diffraction pattern shows sharp, spotted rings from the KBr crystallites and somewhat diffuse rings from the smaller gold particles. The width of the diffuse gold rings indicates a particle size of about 70 Å.

The electron micrographs of the Au/Ge mixtures showed that the gold did not agglomerate into spheres. In an effort to increase the gold particle size, thin layers of gold approximately 10 Å in thickness were deposited alternately between thicker films of germanium. These thin gold films are discontinuous and the result is that the gold particles are larger, as shown in Figure 4. The electron diffractograph (Figure 5) shows the presence of both gold particles and almost amorphous germanium.

The optical measurements were made on comparatively thicker films and there is some uncertainty whether the structures of the very thin films used for electron microscopy are representative of the thicker films. As discussed in the next section, the theoretical model does not depend critically on gold particle size and thus the conception of the structure of the thick film obtained from the electron micrographs is sufficient.

Theory of the Optical Constant

The optical dielectric constant of the mixture ϵ_m is computed from a simple model via classical EM theory. Since this is adequately described

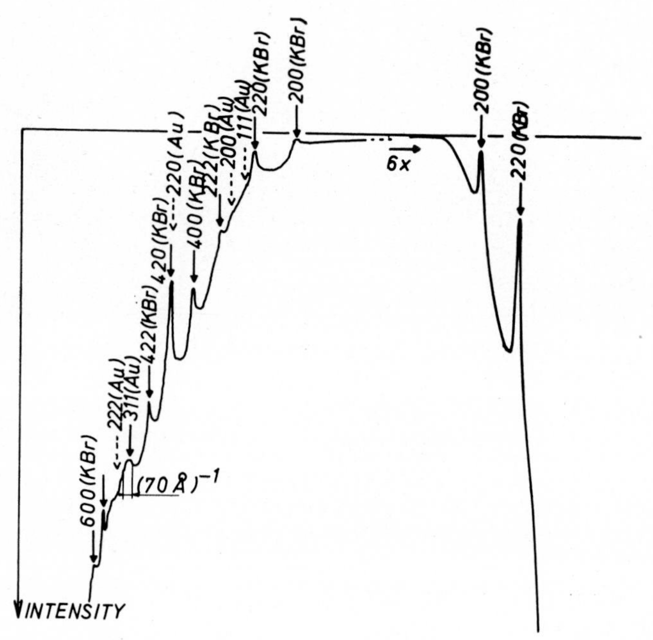

FIG. 3. Densitometer trace of an electron diffractograph of an Au-KBr film.

elsewhere (van Beek 1967), we confine our attention to a discussion of the model and the results.

The model assumes that the metal particles are spheres of diameter d and optical constant:

$$(\epsilon_s)^{\frac{1}{2}} = n_s = n_s - jk_s$$

which are dilutely dispersed in a matrix of dielectric constant ϵ and refractive index $n = (\epsilon)^{\frac{1}{2}}$. Certain restrictions must be placed on d for the model to be valid. If d is comparable to wavelength of the radiant flux, the metal spheres will scatter light according to the Mie theory. This restricts d to

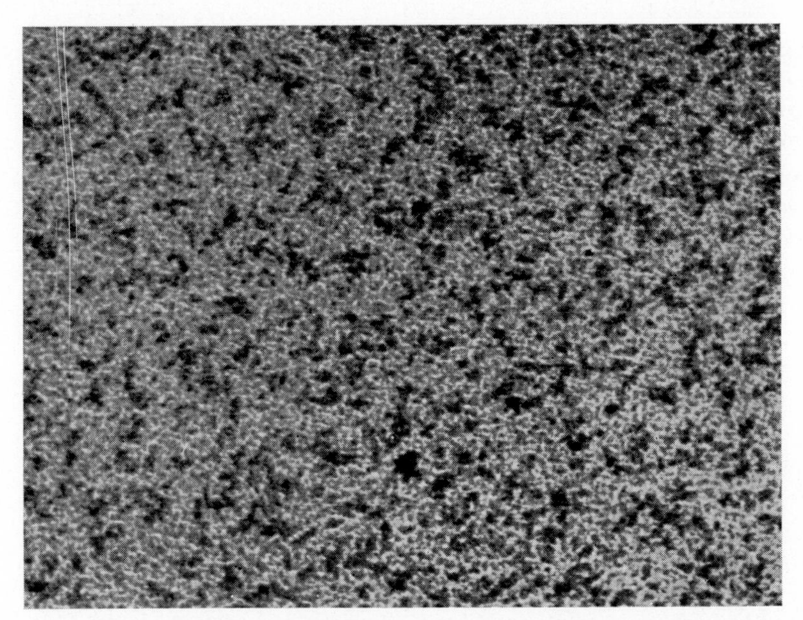

FIG. 4. Electron micrograph of a very thin Au/Ge cermet film.

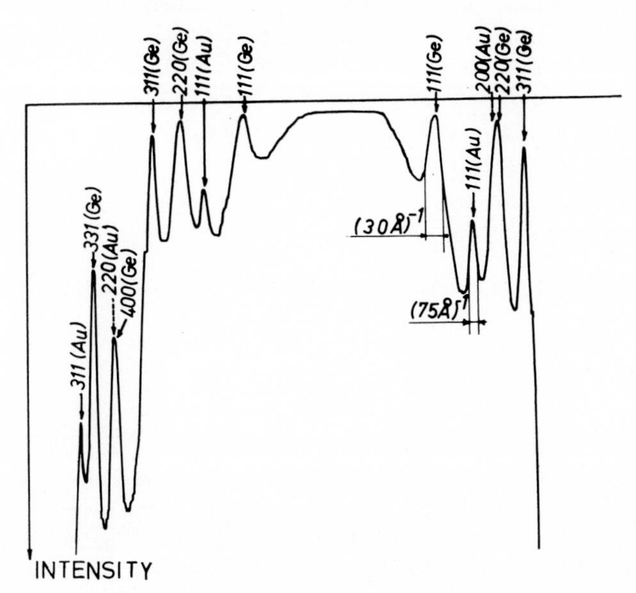

FIG. 5. Densitometer trace of an electron diffractograph of an Au/Ge film.

less than 500 Å. If d is too small, bulk optical constants for the metal will not apply. Although this limit is more difficult to establish, we can obtain an estimate from the fact that according to the Drude free-electron theory, the mean free electron path in gold at room temperature is approximately 70 Å. The bulk infrared optical constants should apply when d is a few hundred Ångstroms. Hampe (1958) attributed some discrepancies in the absorptance of gold particles dispersed in a SiO matrix, to collective electron oscillations.

If such effects are neglected, the induced dipole moment $\mathbf{p}_s$ produced by a sphere in a steady-state oscillating imposed field $\mathbf{E}$ is

$$\mathbf{p}_s = \pi \frac{d^3}{2} \epsilon\epsilon_0 \hat{A}\mathbf{E} = 3V\epsilon\epsilon_0 \hat{A}\mathbf{E} \tag{1}$$

where ϵ_0 has the usual meaning in the MKS units and V is the volume of the sphere. The reduced polarizability $\hat{A}$ is $(\hat{\epsilon}_s - \epsilon)/(\hat{\epsilon}_s + 2\epsilon)$. If f is the volume fraction of the metal, the volume per sphere is V/f and the polarization due to the spheres is

$$\mathbf{P}_s = \mathbf{p}_s/(V/f) = 3f\epsilon\epsilon_0 \hat{A}\mathbf{E} \tag{2}$$

The displacement vector, $\mathbf{D}_m$, in the mixture, can be written in two ways. First, it is the sum of the displacement of the matrix and the polarization of the spheres. It can also be directly expressed in terms of the dielectric constant of the mixture ϵ_m.

$$\mathbf{D}_m = \epsilon\epsilon_0 \mathbf{E} + \mathbf{P}_s = \epsilon\epsilon_0(1 + 3f\hat{A})\mathbf{E} = \hat{\epsilon}_m \epsilon_0 \mathbf{E} \tag{3}$$

Whence,

$$\hat{\epsilon}_m = \epsilon(1 + 3f\hat{A}) \tag{4}$$

The foregoing equation can be corrected for the effect of interactions between the spheres (van Beek 1967), which leads to

$$\hat{\epsilon}_m = \epsilon[1 + 3f\hat{A}/(1 - f\hat{A})] \tag{5}$$

and was originally derived by Maxwell-Garnett (1904). Equation (4) can be written in reduced form, i.e. in terms of the ratios $\hat{\epsilon}_m/\epsilon$ and $\hat{\epsilon}_s/\epsilon$. For a given volume fraction, the reduced optical constant $\hat{n}_m/n$ can be plotted in terms of n_s/n and k_s/n (see Figure 6). For many metals in the infrared $|\hat{\epsilon}_s| \gg \epsilon$, which means that $\hat{A}$ is nearly equal to one and

$$n_m \doteqdot n(1 + 3f)^{\frac{1}{2}} + j \cdot 0 \tag{6}$$

The region around $n_m/n=(1{\cdot}3)^{\frac{1}{2}}=1{\cdot}14$ is plotted on an expanded scale in Figure 7, for f equals 10 per cent.

Optical Measurements

The optical constant n_m-jk_m of the evaporated mixtures was determined from measurements of the spectral reflectance and transmittance at

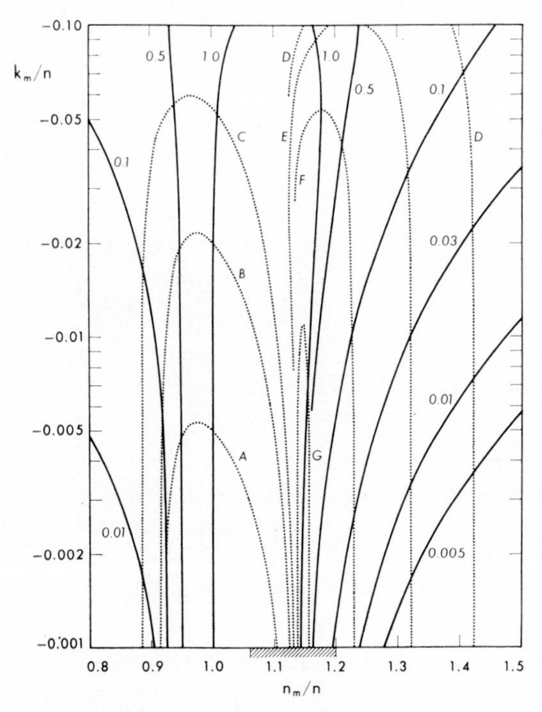

FIG. 6. Contours of reduced n_s (solid curves) and k_s versus the reduced optical constant $(n_m-jk_m)/n$ of the mixture for a volume mixing ratio f of 0·1. The k_s contours (dashed) correspond to: *A*, 0·05; *B*, 0·2; *C*, 0·5; *D*, 1·8; *E*, 2·0; *F*, 2·5; *G*, 5·0.

normal incidence. The optical thickness was computed from the positions of reflectance or transmittance maxima and minima after suitable corrections were made for the dispersion of the refractive index and absorption. The physical thickness was determined by overcoating the film with silver and measuring the displacement of multiple Fizeau fringes in a ditch of the film. If its refractive index is not perfectly homogeneous, the ratio of the optical and physical thickness gives an average value for the index. The refractive index was also computed from the reflectance and transmittance at the quarter-wave points. If both methods of computing n_m give the same result, this indicates that the film is substantially homogeneous. The

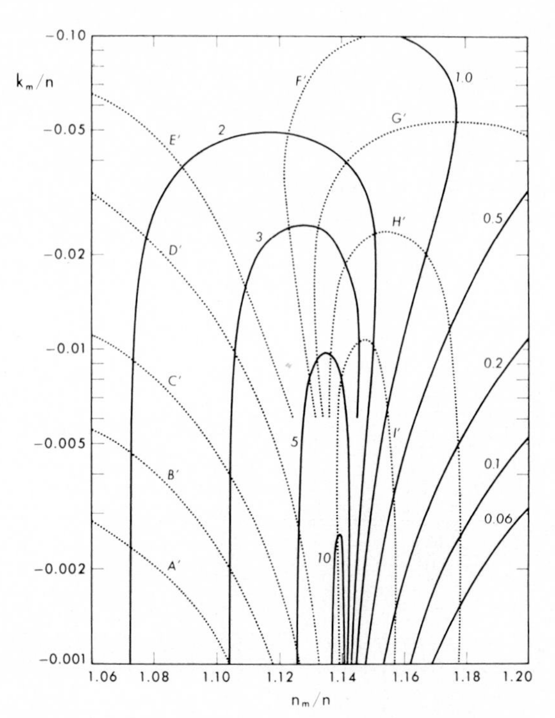

FIG. 7. Expanded plot on the abscissa of the cross-hatched region in Figure 6. The k_s contours have the values: A′, 0·05; B', 0·1; C', 0·2; D', 0·5; E', 0·9; F', 1·8; G', 2·5; H', 3·5; I', 5·0.

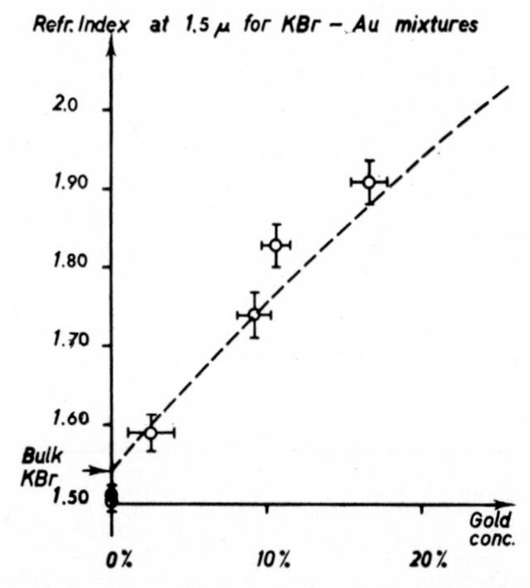

FIG. 8. Variation in refractive index with gold concentration for gold-KBr cermet films at 1·5 μ.

absorption constant k_m was computed from reflectance and transmittance at half-wave points. When a film was highly absorbing, its refractive index was computed from its reflectance at the air-film interface.

EXPERIMENTAL RESULTS

Mixtures of Au and KBr

The variation of the refractive index with gold concentration follows the approximation given in Equation (6), Figure 8. Thin films of pure KBr

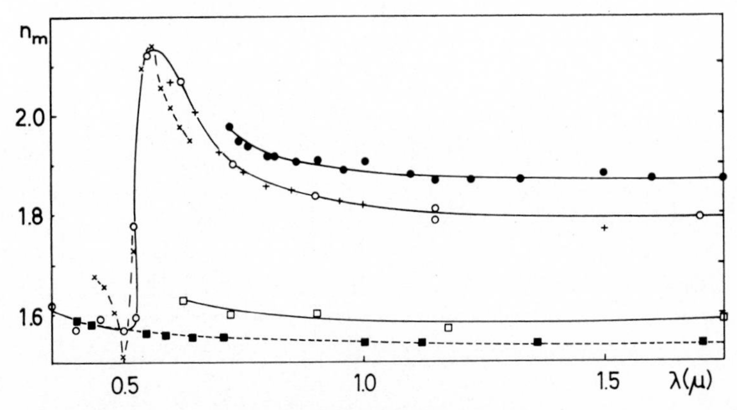

FIG. 9. Dispersion of refractive index for gold-KBr cermet films: –O– Au concentration is 12 per cent of volume; –●– Au concentration is 16 per cent of volume; –□– Au concentration is 2·5 per cent of volume; –■– index of bulk KBr; – × – calculated index of 10 per cent Au concentration (bulk Au constants); – + – calculated index of 10 per cent Au concentration (evaporation Au constants).

have a refractive index which is slightly less than the bulk index. This is a well-known effect by many materials and is due to porosity of the films. In spite of this the theoretical curve is drawn through the bulk value. It seems probable that the mixtures have higher density due to the presence of extra nucleation sites for KBr at gold atoms or particles. The mixtures have in fact a much smoother surface than the pure KBr films. The dispersion of refractive index follows approximately the predictions from the theory, (see Equation (4) and Figure 9). There is a very sharp decrease in the index at 5400 Å, due to the onset of inter-band transitions in gold which abruptly changes the optical constants. Variation in the absorption index is also noticeable (see Figure 10). The agreement with Equation (4) is, however, not so well established here. If a formula for the absorption constant assuming collective electron oscillations (Hampe 1958) is fitted to the maximum value of k_m, the agreement is better.

Mixtures of gold and germanium

None of the Au/Ge films obtained was free from absorption. For a 10 per cent Au/Ge mixture k_m was about 1·0 and only in two cases were

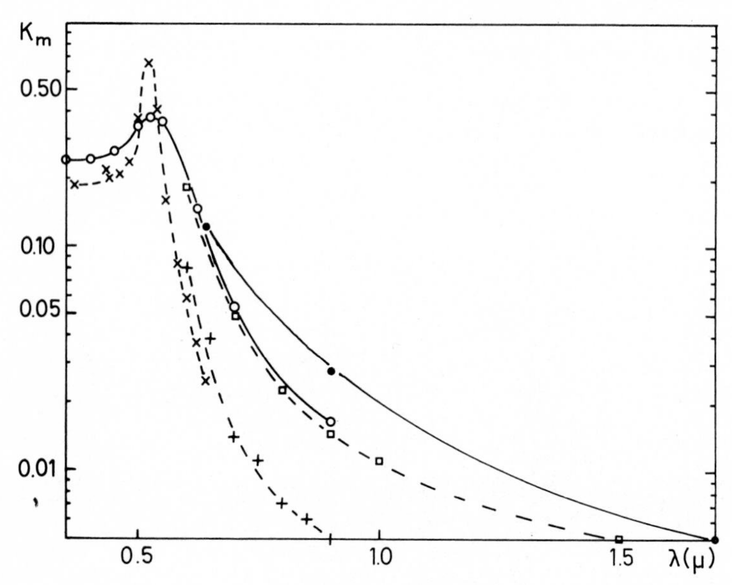

FIG. 10. Absorption constant for gold-KBr cermet films: –○– Au concentration is 12 per cent of volume; –●– Au concentration is 10 per cent of volume; –×– calculated constant for 10 per cent Au concentration (bulk Au constants); –+– calculated constant for 10 per cent Au concentration (evaporated Au constants); –□– curve from collective electron oscillation theory fitted to maximum k_m value.

the films less absorbing with $k_m = 0{\cdot}25$ and 0·5. In these two instances the refractive index agreed well with Equation (6). In the other cases it was considerably higher (see Figure 11). A probable reason for the high absorption is the fact that gold does not agglomerate in spheres when co-evaporated with germanium, as was discussed previously. When cooled down to liquid-air temperature, k_m was reduced typically from 0·95 to 0·80. This change is too small to be due to an absorption change caused by 'freezing out' carriers in doped germanium, and is probably due to the higher conductivity of gold at the lower temperature. As discussed previously the gold particle size was increased by using alternating instead of simultaneous deposition. Still the absorption from such films was very high with $k_m = 1{\cdot}0$ as typical value. Annealing of the films decreased the absorption slightly, the change in k_m being typically from 0·98 to 0·53 after annealing for 4 hours up to 240°C.

CONCLUSION

Previous studies of thin-film mixtures have shown that two dielectric materials can be blended to produce refractive indices intermediate to those of the pure materials (Jacobsson & Mårtensson 1965). In this work it is shown that a metal-dielectric mixture of the gold-alkali halide type can produce a refractive index which is up to 20 per cent higher than the pure dielectric and with comparatively little absorption in the infrared part of the spectrum. Thus these mixtures seem promising for use in this spectral region out to 50μ (where CsI starts to absorb). Not all metal-

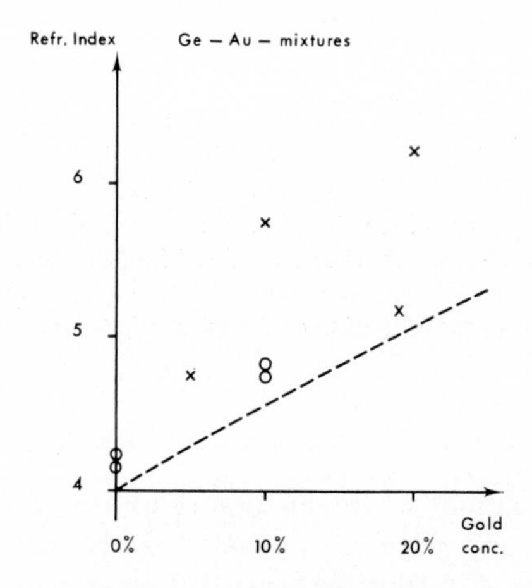

FIG. 11. Variation in real part of refractive index for Au/Ge cermet films: absorbing films at $2{\cdot}5\ \mu$ ($\times$); weakly absorbing films at 5μ($\bigcirc$).

dielectric mixtures will be non-absorbing, as is obvious from the gold-germanium mixtures, however, it may be possible to find a metal, which does not wet germanium to the same extent as gold, and thus forms small spheres more readily.

Acknowledgements

The assistance of A. Keding and T. Widlund in the deposition of the coatings and of Ann-Mari Ekendahl in the preparation of some data is gratefully acknowledged. This work was supported by the Swedish Board for Technical Development.

REFERENCES

ANDERS, H., 1959, Swed, Pat. no. 213,445, applied for in 1959, granted 1967.
VAN BEEK, 1967, 'Dielectric Behaviour of Heterogeneous Systems', in (J. B. Birks, Ed. *Progress in Dielectrics*, Heywood, London, **7**, 69.
HAMPE, W., 1958, *Z. Physics.*, **152**, 476.
JACOBSSON, R. and MÅRTENSSON, J. O., 1965, *Jap. J. Appl. Phys.*, **4**, (suppl.), 333.
MAXWELL-GARNETT, J. C., 1904, *Phil. Trans.*, **203**, 385; 1906, *Ibid.*, **205**, 237.
RAYLEIGH, J. W., 1892, *Phil. Mag.*, **34**, 481.
SEELEY, J. S., 1961, *Proc. Phys. Soc.*, **78**, 998.
SMITH, S. D., 1958, *J. Opt. Soc. Amer.*, **48**, 43.
WEYL, W., 1951, *Coloured Glasses*, Sheffield, The Society of Glass Technology, p. 380.

A Self-Aligning Laser Doppler Velocimeter

M. J. Rudd*

British Aircraft Corporation, Filton, Bristol, and Cavendish Laboratory, Cambridge

Abstract—This paper describes a new type of laser Doppler velocimeter for measuring fluid velocities. Its main property is that it is self-aligning in that the signal and reference beams must always coincide and hence is extremely easy to set up and is not susceptible to vibration. The theory of the operation is given both in terms of the Doppler effect and a 'fringe' model, and the various optical criteria for successful operation are discussed. Some experimental results are given, and the mode of the scattering of the laser light by particles in the fluid is considered.

INTRODUCTION

There are many situations where one desires to measure the velocity of a substance without any physical contact and the Laser Dopplermeter appears to be the ideal instrument for such a requirement. The velocity is determined by observing the Doppler shift in frequency of the laser light scattered from the medium. This scattering is usually produced by small impurities rather than by the medium itself.

Since the speed of the object one wishes to measure is very small compared with the speed of light, the fractional frequency change is very small, too small to measure spectroscopically. However, one can observe the change by beating the scattered light with the original laser light, heterodyning, and detecting the beats with a square-law photodetector. This has an additional advantage of giving noise-free amplification.

The first Dopplermeter was operated in 1964 and later developed by Foreman *et al.* (1966), as shown in Figure 1. This focused the laser beam into the fluid to be measured, and used the transmitted light as a reference beam. The light which was scattered at an arbitrary angle was collected, and recombined with the reference beam by an optical system. The two beams then fell upon a photo-cathode where they beat together. However, in order to obtain a good signal the two beams have to be very accurately aligned over the photo-cathode, to better than a fringe. Goldstein and Kreid (1967) improved this by shining two confocal beams into the

* Now in the Acoustics Department, British Aircraft Corporation, Weybridge, Surrey.

medium (Figure 2). Light was scattered from one beam into the other and was therefore automatically aligned. However, the two beams still had to be made to overlap accurately at their foci. Pike *et al.* (1968) have been developing this system at A.E.R.E., Harwell, and have extended it to a homodyning system, without a reference beam (Bourke *et al.* 1969).

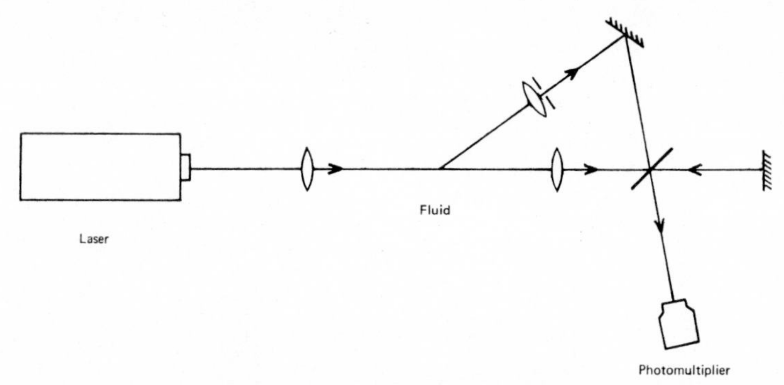

FIG. 1. Foreman's optical system.

Doppler and Fringe Models

The Doppler model described above gives an expression for the frequency shift:

$$\nu = (2v/\lambda)\sin\theta$$

where ν = frequency shift, v = velocity of medium, λ = wavelength of radiation and 2θ = scattering angle.

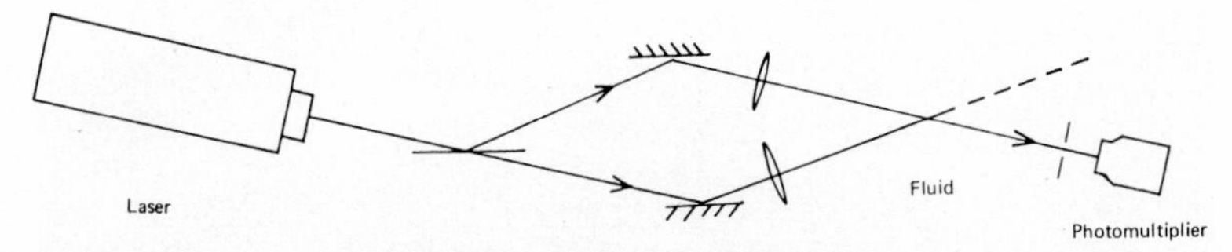

FIG. 2. Goldstein's optical system.

Although this Doppler model is the obvious one, it is frequently easier to consider the operation of the instrument in terms of a fringe model (Rudd 1969*a*). These fringes may be real or virtual. Thus, where the two incident beams of the Goldstein system overlap at their foci, they will interfere and produce a set of fringes. As a scattering centre crosses these fringes, it will scatter more, or obstruct more light from the bright fringes than the dark ones. Thus the total amount of light transmitted will fluctuate and the frequency of this fluctuation is given by the same expression as the Doppler model above.

Simplified Laser Dopplermeter

The author has been employing a simplified Dopplermeter which is easier to align and gives a better signal than earlier designs (Figures 3 and 4). The light from an Elliotts HN2–8 $\frac{1}{2}$mw laser is diverged by the micro-

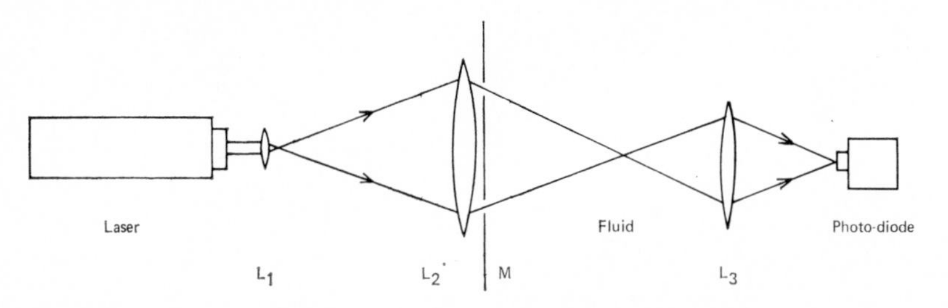

FIG. 3. Author's optical system.

FIG. 4. View of apparatus.

scope objective L, to cover a mask M. This mask contains two slits (width a and separation b) which produces two beams. These two beams are focused into the medium, which in this case is water in a pipe, by lens L_2 (distance from focus l) and they form their fringes at the focus. The light is collected by a third lens L_3 and focused onto a photo-detector. The signal is sufficiently high for a silicon photo-diode to be used and in fact no advantage is gained from a photo-multiplier. The signal is passed through a preamplifier and fed into a Tektronix 1L5 spectrum analyser and finally displayed on a chart recorder via an integrator. The self-aligning property of the system stems from the fact that the two beams have been focused by the same lens L_2. Thus if the lens is free from defects, they must converge on the same point. The system also has quite a high immunity from vibration since the single lens will keep the beams aligned. Other methods of producing two beams can be conceived, but the one described is the simplest.

It is perhaps interesting to consider the shape of the fringe region at the focus. The fringe spacing is $\lambda l/b$ and the width of the region is $\lambda l/a$; thus there are b/a fringes through which the scattering centre can pass. This will give rise to a finite width of the doppler spectrum because one only has a finite number of cycles of the doppler signal. This uncertainty is approximately a/b. This is analogous to the 'Doppler ambiguity' which arises in the Doppler model due to the finite receiving aperture.

If the height of the slits is approximately equal to their separation, then:

Height of scattering volume $= \lambda(l/b)$.
Width ,, ,, ,, $= \lambda(l/a)$.
Length ,, ,, ,, $= \lambda(l/a)(l/b)$.

Typically the scattering volume will be $10\mu m \times 100\mu m \times 1mm$ and so a very good spacial resolution of $10\mu m$ can be obtained in one direction.

Coherence Considerations

Chromatic coherence. The effect of the finite spectral width of any light source is termed chromatic coherence. It has been stated (Goldstein and Kreid 1967) that the line width of the light source of a Dopplermeter must be considerably less than the Doppler shift to be observed. However, this is not the case. In fact the chromatic coherence requirement is that:

$$\Delta\lambda/\lambda \leqslant \Delta v_D/v_D$$

where $\Delta\lambda$ is the spectral width of the radiation employed and Δv_D is the expected spread in the observed Doppler shift. This coherence requirement is not at all stringent and therefore conventional light sources could be used. A highly 'coherent' laser is not essential for a Dopplermeter. This result can be readily understood from the fringe model of a Dopplermeter. The maximum number of fringes which can be produced is $\lambda/\Delta\lambda$ although the actual number may be less than this owing to geometrical considerations. The error in determining the Doppler shift is equal to the reciprocal of the number of cycles of the wave train and this error is $\Delta v_D/v_D$. Thus it can be appreciated that there must be a sufficient number of fringes to give this accuracy.

Spatial coherence. The spatial coherence of light beam is related to the size of the optical source. Again a requirement can be deduced from the fringe model. For the simplified optical system described, it can be seen that the effective size of the source when projected into the operating region must be less than the fringe spacing. Thus the maximum permissible effective source diameter is given by

$$d < \lambda/2\sin\theta$$

This is generally a much more stringent requirement than chromatic coherence, since d is only a few micrometres. This means that if we were to use a conventional light source, the light would have to be condensed upon a small pinhole only a few micrometres across and thus very little light would get through. A laser, however, can be focused into a very small source and thus generally the spatial coherence is quite good enough. However, at large angles, when $\sin\theta$ is large, one may be limited by lens aberrations.

In the optical systems of Goldstein and Foreman the requirement for spatial coherence is different. The projected source size must not be any greater than the diffraction-limited focus size if there is not to be a significant reduction in fringe visibility. This criterion is given by

$$d < (\lambda/2\sin\theta)(\Delta v_D/v_D)$$

since there are $v/\Delta v_D$ fringes of size $\lambda/(2\sin\theta)$. This is much less stringent than the requirement above with a typical source size of about 100μm, and thus it does not rule out the use of a conventional light source.

Temporal coherence. Temporal coherence relates to the path difference involved between two interfering beams from the same source. The coherence length from a typical gas laser is about 30 cm, so that if we are only looking at one particle at a time, in a beam, then there is no problem. However, if several particles are simultaneously present, the wavefronts scattered by them will each differ in phase. On the fringe model this can be interpreted as the particles being in different parts of the fringe pattern, and hence each Doppler wave-train will have a different phase. Thus the net amplitude of the wavefront scattered by all the particles will be proportional to the square root of their number (by analogy with the random walk).

If, however, one uses a conventional light source, which has a coherence length of 1mm or so, then in the systems of Goldstein and Foreman one would have to adjust the path lengths very carefully so that they were very nearly equal. However, in the simplified system the path lengths are automatically made equal and the problem does not arise.

Effect of Inhomogeneities

So far we have assumed that there are no inhomogeneities or irregularities in the optical system. One might well expect that any distortions which do not significantly affect the effective fringe pattern will not have

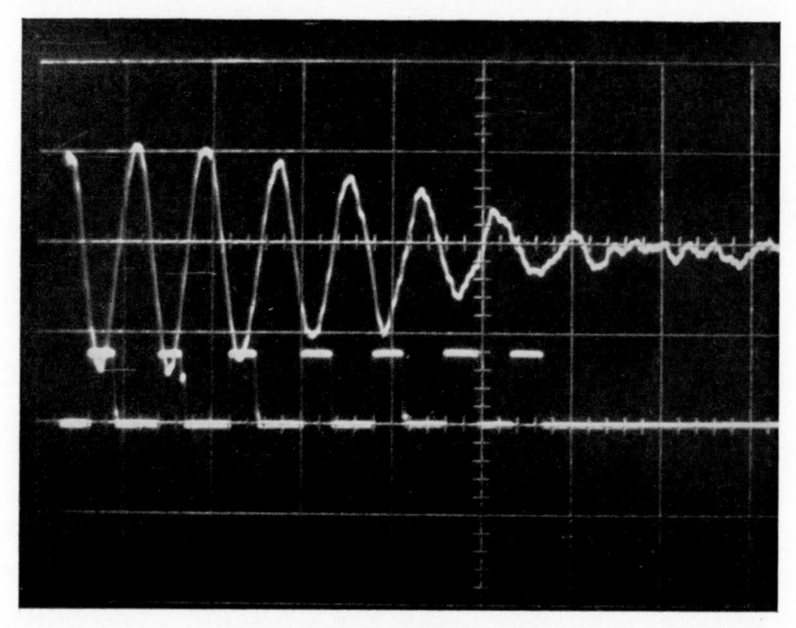

FIG. 5. Typical Doppler signal.

any significant influence on the performance of the system. One requires a total optical path difference of less than about $\lambda/4$ across the width of the beam. This requirement seems to be quite easily met. Ordinary perspex windows have been used on the tank containing the working fluid, and warm water has been added to cold water in the tank, causing easily visible striations without having any observed effect on the performance of the system. The system can also be used inside a flame.

Performance of Dopplermeter in a Fluid

The author has been using the Dopplermeter to examine the properties of drag-reducing polymer additives in turbulent pipe flow (Rudd 1969*b*). A typical signal (Figure 5) and a spectrum (Figure 6) obtained from the flow are shown. The averaging time was, for the spectrum, 1 sec and the whole scan took 50 sec. A signal/noise ratio of 50 dB can be readily obtained and one of 90 dB has been obtained. There is frequently an extra-low-frequency content to the spectrum due to spurious particles crossing the beam.

The accuracy and spatial resolution of the system are demonstrated by the velocity profile (Figure 7). The scatter in the points (about $\frac{1}{2}$ per cent)

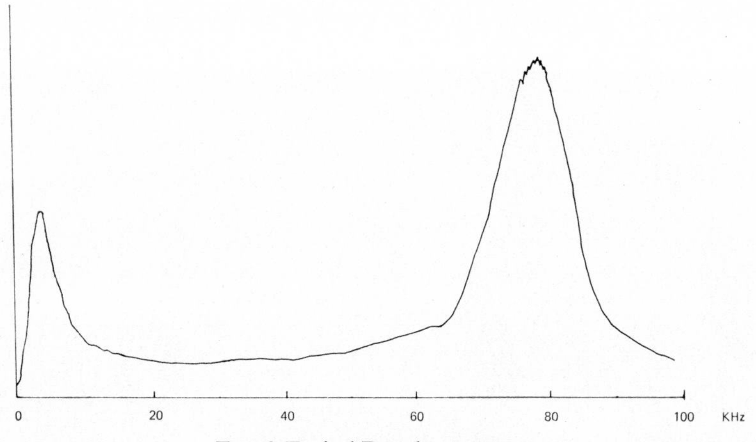

FIG. 6. Typical Doppler spectrum.

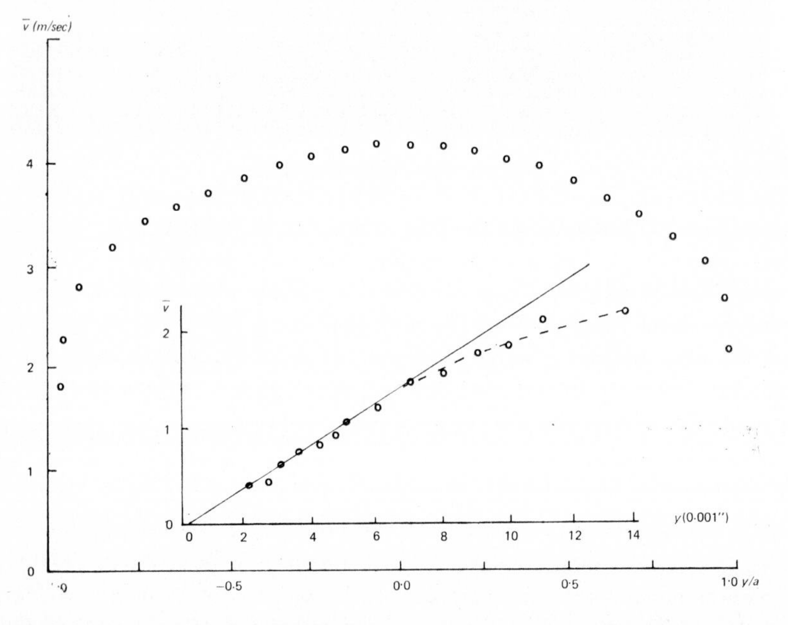

FIG. 7. Velocity profile across a pipe. (Inset shows region very close to the wall.)

does not represent the ultimate accuracy of the instrument, but arises since one is only sampling the turbulence for a finite time. The inset shows the profile close to the wall where there is a rapid change in velocity and the experimental points are separated by only 12·5μm (0·0005″).

Mechanism of Scattering

Frequently there are sufficient scattering centres already in the fluid to give a good signal without any being added. Foreman *et al.* (1966) have shown that the scattering was not due to the fluid itself, but to small contaminant dust particles. However, when working with 0·01 per cent polymer solutions where the molecules are around 1000 Å long, quite a significant amount of scattering is produced by these molecules. Whenever the Dopplermeter has been used with gases, smoke has always had to be added to the flow.

An estimate of the density of scattering centres can be obtained from the output noise. A discreet number of centres gives rise to a 'shot noise' in the output. Typically this density is greater than 10^6 centres/cc. This means that the centres must be less than 1μm in diameter to account for the observed absorption. The large particles (> 10μm) do not appear to be very important scattering centres.

Applications

So far the Laser Dopplermeter has been employed mainly in fluid mechanics research and now it is yielding results which are of use in their own right, rather than using a Dopplermeter for its own sake. However, with slight modification, the instrument can be used to measure the velocity of solid opaque objects such as metals. Indeed some instruments have already been used for this purpose. Some work has also been done on the measurement of remote wind velocities at ranges of a few hundred feet but the work was rather difficult. However, the instrument seems to be mainly a short-range device.

Acknowledgements

The author wishes to gratefully acknowledge the aid and advice of Dr. A. A. Townsend of the Cavendish Laboratory, Cambridge, and Dr. B. M. Watrasciewicz of the British Aircraft Corporation, Bristol.

REFERENCES

Bourke, P. J., Butterworth, J., Drain, L. E., Egelstaff, P. A., Hutchinson, P., Moss, B., Schofield, P., Hughes, A. J., O'Shauchnessey, J. J. D., Pike, E. R., Jakeman, E. and Jackson, D. A., 1969, *Phys. Letters*, **28A**, 692.
Foreman, J. W., George, E. W., Felton, J. L., Lewis, R. D., Thornton, J. R. and Watson, H. J., 1966, *J. Inst. Elect. Electron Eng.*, **QE–2**, 260.
Goldstein, R. J. and Kreid, D. K., 1967, *J. Appl. Mech.*, **34**, 813.
Pike, E. R., Jackson, D. A., Bourke, P. J. and Page, D. I., 1968, *J. Sci. Instr.*, **1**, 727.
Rudd, M. J., 1969a, *J. Sci. Instr.*, **2**, 55.
Rudd, M. J., 1969b, *Nature*, **224**, 587.

On the Use of Optical Interferometry for Studying Refractive Index Distributions in Liquids at Temperatures up to 500°C

Silas E. Gustafsson

Department of Physics,
Chalmers University of Technology,
40220 *Gothenburg* 5, *Sweden*

Abstract—Interferometric measurements with a special experimental technique have been introduced for the study of transport properties of transparent liquids at high temperature. The present investigation includes strongly corrosive melts with melting points from 250 to 420° C. A special thermostat together with a container for the liquid, capable of transmitting an undistorted, plane wave-front had to be designed. The main problem was to avoid the influence on the wave-front of the large temperature gradients in the walls of the thermostat. This has been accomplished by using a special arrangement with vacuum chambers passing through the critical region. The transport properties which have been studied are thermal conductivity and interdiffusion, and in both cases new optical techniques have been introduced. In order to get all the desired information, it turned out that the refractive index of these liquids had to be remeasured. By introducing a simple interferometric method the errors in the measurements could be decreased by approximately two orders of magnitude.

INTRODUCTION

Density or refractive index differences in transparent liquids can be studied conveniently and accurately by using experimental procedures based on the recording of path differences by optical interferometry. Due to the high sensitivity and the large amount of information that can be obtained from an interferometric recording, it is considered to be one of the most reliable methods for studying transport phenomena in transparent liquids resulting from a concentration or temperature gradient.

The present investigation is focused on the problem of applying these methods to the study of density changes in transparent materials at high temperature. From the experimental point of view it is most convenient to keep the interferometer and the auxiliary recording equipment at room temperature, while the material to be studied is placed in a high-temperature thermostat, which can be designed to meet the requirements of the particular experiment. Since the test region has a temperature which deviates from that of the interferometer, the temperature gradients in the walls of the thermostat will be recorded as superimposed upon the process which is of primary interest. In order to avoid the influence of these temperature gradients on the wavefront a special arrangement of vacuum chambers has been designed (Gustafsson, Halling & Kjellander 1968).

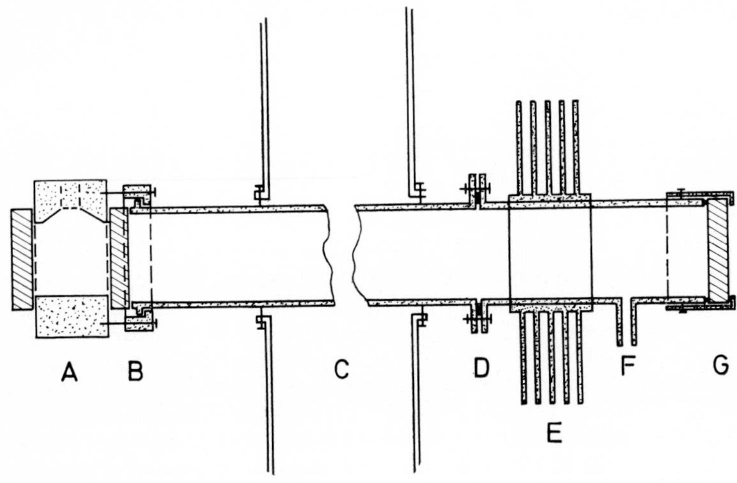

FIG. 1. Vacuum chamber for reducing disturbing optical path differences caused by temperature gradients in the wall of a high temperature thermostat: *A*, test region; *B*, optical flat limiting the test region; *C*, wall of the thermostat; *D*, thermally insulating gasket; *E*, flanges to insure a complete adjustment of the protruding vacuum chamber to ambient temperature; *F*, connection to vacuum pump; and *G*, optical flat with its holder.

The basic principle involves the possibility of changing the optical path difference between two arbitrary rays through the chamber by varying the pressure, and in particular of making it close to zero.

EXPERIMENTAL

The general principle of construction of the vacuum chamber is shown in Figure 1. The main part is a stainless steel tube with an approximate diameter of 3 cm which passes through the wall of the thermostat. Both ends are pressed towards a quartz plate and kept in position by holders made of a material which compensates for the thermal expansion of the screws. The quartz plate inside the thermostat can also serve as a limitation of the test region. Outside the thermostat wall the construction is such as to insure a complete adjustment of the temperature of the quartz plate to that of the surrounding air. This is the reason for using a particular gasket to minimize the thermal conduction in the steel and for using a series of cooling flanges around the tube. These can of course be substituted by a water-cooling arrangement, but the temperature of the coolant must then be adjusted very carefully to that of the room.

THEORETICAL CONSIDERATIONS

It is easy to make a quantitative estimate of the maximum pressure that can be tolerated in the vacuum chamber. Starting with the molar refractivity, A, of a gas, we write:

$$A = (RT/p)(n^2 - 1)/(n^2 + 2)$$

where R is the gas constant, T the temperature, p the pressure, and n the refractive index. If n is close to unity, as in this case, we may write:

$$n = 1 + (3pA/2RT) \quad (1)$$

where A is a constant for each particular gas independent of both temperature and pressure, at least to the extent that may concern us here. The optical path difference (ΔZ) between two arbitrary rays through the evacuated chambers can be given as

$$\Delta Z = \int_0^{Lv} \Delta n dl \quad (2)$$

where L_v is the total length of the regions where the disturbing temperature gradients exist. ΔZ is mostly a very complex function, but since we are only interested in an estimate of the disturbance, we may express it in the following simple form:

$$\Delta Z = L_v(\bar{n}_1 - \bar{n}_2) \quad (3)$$

where $\bar{n}_1$ and $\bar{n}_2$ are mean values of the refractive index at the two positions being considered. According to Equation (1) the discrepancy between $\bar{n}_1$ and $\bar{n}_2$ can be attributed to a difference in the mean temperatures $\bar{T}_1$ and $\bar{T}_2$. It is evident that the path difference ΔZ can be made arbitrarily small provided the pressure can be lowered far enough. In order to estimate the maximum pressure that can be allowed inside the chamber, it is necessary to consider the sensitivity of the recording interferometer. Suppose that $m\lambda$ is the maximum permissible value of ΔZ not to be measurable with the interferometer, then the maximum pressure, p_m, can be given as

$$p_m = (2m\lambda R/3AL_v)(\bar{T}_1\bar{T}_2)/(\bar{T}_2 - \bar{T}_1) \quad (4)$$

where λ is the wavelength. It is interesting to note that the pressure must be lower when working close to room temperature. Choosing typically unfavourable values such as $\bar{T}_1 = 300\ {}^0K$, $\bar{T}_2 = 400\ {}^0K$, and $m = 10^{-2}$, the maximum pressure p_m turns out to be 0·3 mm Hg. The wavelength is taken to be 6328 Å and L_v can be estimated to 30 cm. A value of R/A for air is given by Born & Wolf (1964). With our design of vacuum chambers the pressure can be lowered to 0·1 mm Hg without any difficulty. The possibility of obtaining an extremely low pressure is limited by the fact that one end of the chamber, although hand-lapped, is simply pressed towards a quartz plate, because of the difficulty of finding a proper material which can be used as a sealing gasket at high temperature in the presence of corrosive vapours. On the other hand, it has not been found possible to detect experimentally any influence from the air in the chamber as soon as the pressure fell below 1 or 2 mm Hg. The two interfering rays had a lateral shear in the test region of less than 0·5 cm. It should be noted here that it is only the component of the temperature gradient perpendicular to the optical axis which disturbs the measurements.

Concerning the influence on the test region caused by the evacuated chambers, the only objection that can be made is that the thermal radiation through the light passages may cause an unwanted temperature decrease in the test area. Practical experience from measurements up to 500°C indicates that this effect is completely negligible. However, there is a possibility of reducing the radiation considerably by using two L-shaped vacuum chambers, with a quartz prism in each of the two corners. The prism should then be arranged so that the visible radiation is subject to total reflection while the infrared and thermal radiation is not. This arrangement is now being investigated.

In conclusion, it can be said that this rather unsophisticated experimental arrangement has opened up the field for using optical interferometry for measurements of refractive index distributions at temperatures far from room temperature. Work is now in progress to extend the temperature range from 500°C by another 300 or 400°C. The quality of the interferograms obtained from a plane source diffusion experiment at a temperature of 316°C is shown in Figure 2.

The equipment, which is discussed above, has been used for the investigation of thermal conductivity, inter-diffusion, and refractive index of fused alkali nitrates. These transparent liquids, which are corrosive ionic melts, have melting points in the temperature range between 250°C and 420°C. The particular result of the work is outside the scope of this paper but the optical techniques developed for these measurements will be discussed here in some detail.

The new non-steady-state method for thermal conductivity measurements (Gustafsson 1967) is specially designed to be used in connection with optical interferometry. The conventional arrangement with a continuous line source (hot wire cell) is replaced by a continuous plane heat source consisting of an electrically-heated metal foil. There are several reasons for utilizing a plane source (hot foil) instead of a line source. First of all it is necessary to establish the proper optical path differences in an experiment by choosing a suitable liberation of heat per unit length of the heat source. It is intuitively quite clear that this liberation of heat must be the same whether we use a 'hot wire' or 'hot foil' oriented along the optical axis. In the second case, however, the energy is distributed

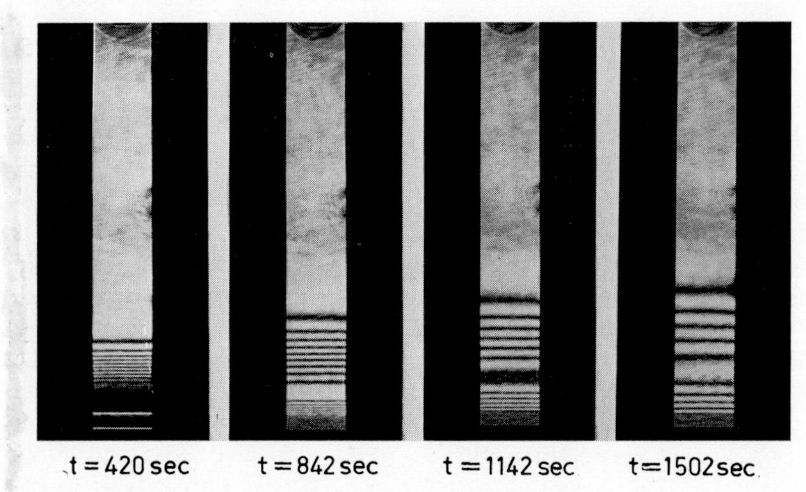

FIG. 2. Interferograms recorded at four different times of an experiment where silver nitrate is diffusing into sodium nitrate at a temperature of 316·1° C. The cell width is 0·514 cm, the vertical shear is 0·07 cm, and the mass of the crystal is 3·34 mg.

over a much larger volume, which means that the temperature increase in the liquid becomes very small. This small temperature increase is important, in the first place because we do not have to consider any temperature correction of the measured coefficients, and in the second because the start of the convective motion in the liquid is greatly delayed. From the experimental point of view it is important to avoid the convection as long as possible in order to extend the measuring time of the experiment, and thus to improve the precision of the measured optical path differences. Another advantage of this new technique, which is important when working with electrically conducting fluids, is that the resistance of a proper metal foil is much less than that of a wire. This means that the voltage across the foil can be kept low enough to prevent current flow and heating in the fluid.

The description of the temperature distribution around a 'hot foil' can readily be done by using the idea of the instantaneous point source, which has proved very useful in the theory of conduction of heat in a homogeneous solid. If it is assumed that the y- and z-axes in an orthogonal coordinate system define the plane which contains the foil within the region $-h \leqslant z \leqslant h$ and $-d \leqslant y \leqslant d$, we get the temperature distribution around the continuous plane heat source by integration as

$$T(x, y, z, t) = T_0 + \frac{Q}{8\rho c(\pi \mathscr{H})^{3/2}} \int_0^{2(\mathscr{H}t)^{\frac{1}{2}}} \exp(-x^2/\sigma^2) \times$$

$$\left[\operatorname{erfc}\left(\frac{y-d}{\sigma}\right) - \operatorname{erfc}\left(\frac{y+d}{\sigma}\right)\right] \left[\operatorname{erfc}\left(\frac{z-h}{\sigma}\right) - \operatorname{erfc}\left(\frac{z+h}{\sigma}\right)\right] \mathrm{d}\sigma \quad (5)$$

where T_0 is the initial temperature of the fluid, t is the time, Q is the constant heat liberation per unit area, ρc is the specific heat per unit volume, $\mathscr{H}$ is the thermal diffusivity, and 'erfc' stands for the complementary error function. This equation seems rather awkward but it is greatly simplified if we choose to study the temperature distribution in the x, y-plane and assume that h is large enough to make the dependence of Equation (5) on z disappear. This is not very difficult to achieve because the time of an experiment is limited to about one minute because of the onset of convection. The optical path, R, is then obtained by using a relation similar to equation (2) and assuming that $(\partial n/\partial T)$ is a constant over a limited temperature range: n is the refractive index and T the temperature. The integration is performed along a path parallel to the foil (x being constant) from end to end of the vessel containing the fluid. With reasonable dimensions of the vessel the optical path difference becomes

$$R(x,t) = (Q^*/\rho c)(\partial n/\partial T)\{(t/\pi\mathscr{H})^{\frac{1}{2}}\exp(-x^2/4\mathscr{H}t) - (x/2\mathscr{H})\operatorname{erfc}[x/(4\mathscr{H}t)^{\frac{1}{2}}]\} \quad (6)$$

where Q^* is the total power supplied to the foil divided by $2h$. It is interesting to note that this expression is identical to the one obtained for a single line source. The 'hot foil' can then be pictured as a number of 'line sources' distributed along the y-axis and parallel to the z-axis. An important consequence of this is that Equation (6) gives the optical path exactly, even if the heat transport along the foil in the y-direction would be considerable, because it is irrelevant where the 'line sources' are situated when calculating the optical path.

When deriving the equations above we assumed that the thickness of the heater was zero and that the radiation from the foil was negligible. It is fairly simple to calculate the maximum thickness of the foil that can be allowed if the equations are to be used in the present form. The radiation becomes more important if the measuring time is extended and the temperature is increased, but it does not interfere as long as the temperature is below 500°C. For very high temperatures the equations are modified to some extent (Gustafsson 1969). In the actual experiments we have used metal foils made of silver or platinum with h ranging from 2 to 10 cm and d from 0·6 to 3 cm. The thickness was slightly less than 0·001 cm.

Since this is a non-steady-state method it is possible to obtain both the thermal diffusivity and thermal conductivity from one single experiment, which is an excellent check of the internal consistency of the measurement. However, in order to obtain the thermal conductivity we must know $(\partial n/\partial T)$ or how the refractive index depend on the temperature. This quantity $(\partial n/\partial T)$ must be known with an accuracy of less than one per cent whilst the differences between recent published data from various laboratories are sometimes of the order of 50 per cent. This situation called for a re-determination of the refractive index of the fused alkali nitrates with a method capable of a much higher accuracy.

When looking for a convenient method for measuring the refractive index, we had to consider that only a narrow field of view was available when working with the high temperature thermostat. The well-known principle of rotating a plane, parallel, glass plate was utilized in such a way that the plate was immersed in the liquid and the optical path difference between a ray passing through the glass plate and another passing only through the liquid was recorded with an interferometer. The equation from which the refractive index n_r can be calculated then takes the form

$$(n_r^2 - \sin^2\alpha)^{\frac{1}{2}} - n_r + 1 - \cos\alpha = m\lambda/dn_v \qquad (7)$$

where $n_r = n/_v \geqslant 1$ (a similar equation is valid if $n_v/n \geqslant 1$), n is the refractive index of the glass plate, n_v the refractive index of the surrounding medium, m the number of fringes recorded when changing the angle of incidence from zero to a certain value α, λ the wavelength, and d the thickness of the glass plate. In any one experiment m may be considered a function of α, in which m varies very slowly as long as α is close to zero. The function is symmetrical about $\alpha = 0$ (glass plate perpendicular to the optical axis), and this is used to determine the angular reading corresponding to normal incidence. Suppose now that we measure the two angles for two different numbers of fringes in one particular experiment. In that case λ, d and n_v are true constants and we may write

$$m_i^{-1}[(n_r^2 - \sin^2\alpha_i)^{\frac{1}{2}} - n_r + 1 - \cos\alpha_i] = m_j^{-1}[(n_r^2 - \sin^2\alpha_j)^{\frac{1}{2}} - n_r + 1 - \cos\alpha_j] \qquad (8)$$

This is an interesting expression in so far as it is possible to measure n_r

without knowing either the wavelength or the thickness of the glass plate. But if we do know λ and d we may calculate explicitly the refractive index of the surrounding medium n_v from Equation (7), thereby making the method an absolute one. During the high temperature work we calculated the refractive index directly from Equation (7), and although only rather primitive angle-measuring equipment was available the accuracy of the measurements were estimated to $\pm 3.10^{-5}$, which compares very favourably with the estimated error of earlier determinations, $\pm 3.10^{-3}$. A more complete report of the work is given by Wendelöv, Wallin and Gustafsson (1967) and Wendelöv, Gustafsson, Halling and Kjellander (1967). The technique is now being further developed in order that the precision may be made comparable with that of the most precise methods. The angle of rotation is now measured with a precision goniometer, giving a much higher accuracy for the refractive index.

Since the general problem of using optical techniques for transport studies at high temperature has been solved, it seems to be a rather straightforward application to measure inter-diffusion in transparent liquids. This is, however, a very difficult technical problem if one tries to use any of the conventional cells in order to create an initial condition which resembles a sharp interface in an infinite medium. A technique, which seems more realistic in high temperature work, is the one described by Ljunggren and Lamm (1957), who injected a concentrated solution at the bottom of a cell and thereby initiated bottom layer diffusion experiments. One of the major problems of arranging a diffusion experiment is the establishment of an initial condition which as closely as possible approximates to the mathematical assumption necessary to solve the diffusion equation. No matter which experimental technique is used there will always be an unwanted disturbance or departure from the mathematical ideal. A detailed consideration of various initial conditions and their importance in plane source diffusion experiments (Wallin and Gustafsson, 1969) shows that it is possible to correct for even a rather serious disturbance at the start of an experiment. Because of this possibility one can simply drop a crystal of the diffusing material to the bottom of a properly designed cell, where it dissolves and starts diffusing into the solvent above. This rather simple procedure requires, of course, that the density of the crystal is so much higher than that of the liquid that the amount of material being dissolved before the crystal reaches the bottom is negligible. This procedure has proved very versatile for high temperature applications (Gustafsson, Wallin and Arvidsson, 1968).

A special investigation was made to see how the dissolved substance was distributed over the bottom of the cell. It turned out that a thin layer was formed almost instantaneously which indicates that the simple plane source solution of the diffusion equation can be applied. However, it is not necessary to make an assumption about a rapid distribution, because with a proper dimensioning of the cell the concentration may be pictured as due to a distribution of line sources over the bottom. If the diffusion vessel extends along the optical axis and is rather narrow perpendicular to the optical axis, the line sources can be assumed to extend perpendicular to a light ray passing the diffusion cell. The correctness of such a picture of the process can be checked experimentally.

Because of the finite extension of the cell along the optical axis (z-axis), it is necessary to use the solution for an instantaneous line source at zero time in a semi-infinite medium, together with the method of images, to get an exact expression for the concentration, c, within the cell, which is limited by the plane and parallel windows at $z=0$ and $z=\mathrm{L}$. This gives

$$c(x, z, t)=\int_0^{t_0}\mathrm{d}t'\int_{z_1}^{z_2} m(z', t')[2\pi D(t-t')]^{-1}\times$$
$$\sum_{n=-\infty}^{n=+\infty}\Big[\exp\{-[x^2+(z-z'-2nL)^2]/4D(t-t')\}+$$
$$\exp\{-[x^2+(z+z'-2nL)^2]/4D(t-t')\}\Big]\mathrm{d}z' \tag{9}$$

where x is the vertical coordinate measured from the bottom of the diffusion vessel, t is the time, and D is the diffusion coefficient. The diffusing material is here supposed to be distributed over the bottom between $z=z_1$ and $z=z_2$. We have also taken care of the fact that the diffusing substance may be supplied at a certain finite rate m (amount per unit area per unit time) from $t=0$ to $t=t_0$. The optical path, R, of a light beam through the cell at an arbitrary height x above the bottom can easily be calculated by introducing the refractive index, n, with the relation $n=n_0+(\partial n/\partial c)c$. If $t \gg t_0$ and a proper zero time correction is applied, we get the well-known expression,

$$R(x, t)=n_0L+(\partial n/\partial c)M(\pi Dt)^{-\frac{1}{2}}\exp(-x^2/4Dt)$$

where M is the total amount supplied to the bottom divided by the cell width measured perpendicular to the optical axis. This expression has the same inherent properties as the solution of the thermal conductivity equation, and so far it has not been possible to experimentally detect any departure from this equation due to improper boundary conditions. Two different cells have been used with the approximate dimensions 50 mm × 5 mm and 40 mm × 3 mm. The development of this extremely simple experimental procedure has been very important in high temperature work.

The experimental procedures, which have been described here, can be used with almost any kind of interferometer for recording the optical path differences. However, it is clear that an interferometer which does not record any linear refractive index gradients is to be preferred, because the test region is necessarily enclosed between several different plane surfaces originating from the quartz plates of the thermostat. The shearing interferometer described by Bryngdahl (1962) has proved to be a very versatile instrument for this kind of application. For thermal conductivity and inter-diffusion measurements the interferometer with one Savart plate has been used, while two Savart plates proved to be the better arrangement for refractive index measurements.

A technique resembling the one discussed in this paper for solving the problem with the thermostat should be applicable not only in high temperature work but also for measurements far below room temperature, even if the maximum permissible pressure in the vacuum chambers must be much lower in order that the sensitivity of the interferometer can be assumed to be the same.

REFERENCES

Born, M. and Wolf, E., 1964, *Principles of Optics*, Oxford, Pergamon Press.
Bryngdahl, O., 1962, *Arkiv Fysik*, **21**, 289.
Gustafsson, S. E., 1967, *Z. Naturforsch.*, **22a**, 1005.
Gustafsson, S. E., 1969, *Abstracts of Gothenburg Dissertations in Science*, **12**.
Gustafsson, S. E., Halling, N.-O. and Kjellander, R. A. E., 1968, *Z. Naturforsch.*, **23a**, 44.
Gustafsson, S. E., Wallin, L.-E. and Arvidsson, T. E. G., 1968, *Z. Naturforsch.*, **23a**, 1261.
Ljunggren, S. and Lamm, O., 1957, *Acta Chem. Scand.*, **11**, 340.
Wallin, L.-E. and Gustafsson, S. E., 1969, *Z. Naturforsch.*, **24a**, 436.
Wendelöv, L. W., Gustafsson, S. E., Halling, N.-O. and Kjellander, R. A. E., 1967, *Z. Naturforsch.*, **22a**, 1363.
Wendelöv, L. W., Wallin, L.-E. and Gustafsson, S. E., 1967, *Z. Naturforsch.*, **22a**, 1180.

Measurement of the Refractive Index of Optical Glass

D. W. Harper

Pilkington Brothers Research and Development Laboratories,
Lathom, Lancs.

and

G. B. Boulton*

Applied Optics Department,
Imperial College,
London

Abstract—The designs of two instruments for the measurement of the refractive index of optical glass are described. It is claimed that a production unit manufacturing optical glass requires two methods of measuring refractive index. These methods are: (1) a rapid and accurate one to give an accuracy of $\pm 1 \times 10^{-5}$ in refractive index, and (2) a more accurate system capable of achieving $\pm 1 \times 10^{-6}$. The first method described is a photoelectric V-block system and the second is a two-beam interferometric system using electronic fringe counting.

INTRODUCTION

There is no doubt that considerable progress has been made by optical designers in the past ten years. Progress continues to be made in the specification and testing of lens constructional parameters (e.g. by S.I.R.A. in the U.K.). About two years ago we decided that, as glass manufacturers, we should carry out laboratory work to improve the measurements of the properties of optical glass. Inevitably the measurement of refractive index was the first problem to be considered. A study of the situation indicated that there were two important methods of measuring refractive index, namely, the precision goniometer and the V-block refractometer (the Hilger-Chance refractometer). The basic principles of measurement with the goniometer were established and recorded many years ago by Guild (1929) and Tilton (1935). Little progress, apart from contributions such as the National Physical Labora-

* On leave from Pilkington Brothers Research and Development Laboratories.

tory (N.P.L.) recording goniometer (Flude, Habell and Jackson 1961), has been made since. The V-block refractometer was conceived in 1917 by F. E. Lamplough (see Holmes 1945), was exhibited in 1937 by Chance Brothers Limited and came into production in its present form (the Hilger-Chance refractometer) about 1947. Various modifications have been proposed (e.g. Hughes 1941; Forrest and Straat 1956) but with no significant improvement in versatility or accuracy.

The two above methods were considered to be the only important methods of measurement in regular use. Critical angle methods were rejected because of their dependence on glass surfaces, and oil-immersion methods were considered to be slow, particularly when measurements were to be made of a large range of refractive indices over a wide range of wavelengths.

It must be emphasized that the goniometer and V-block refractometer should not be considered as alternatives, since the attainable accuracies are different, but rather as instruments satisfying different requirements. It was felt that these different requirements existed in the optical glass industry and were defined as:

(i) a rapid, accurate method capable of reliable measurements to $\pm 1 \times 10^{-5}$ in refractive index and better in dispersion; and

(ii) a high-accuracy method capable of measurements to the sixth decimal place in refractive index and dispersion.

The Hilger-Chance refractometer when used carefully can give results accurate to $\pm 2 \times 10^{-5}$. However, to achieve this accuracy means introducing procedures which make the instrument hardly suitable for rapid routine measurements.

Sixth place accuracy can be attained with a goniometer but the procedure necessary to achieve this is lengthy and complicated. Ideally an instrument was required that was capable of the same (or slightly better) accuracy than the goniometer but with a shorter, less complicated, measurement procedure.

It should be appreciated that to measure refractive index to $\pm 1 \times 10^{-5}$ is not easy. Measurements to this order satisfy the majority of requirements on accuracy for V-values and dispersions. Extending the range of measurement to the sixth place meant introducing considerations which were in conflict with one of the basic requirements, namely, ease of use. It was concluded, both from our own studies and from discussions with the N.P.L., that it would be extremely difficult to develop an all-purpose instrument to satisfy the criteria stated above.

At this point it is appropriate to mention why, for measurements of the highest accuracy, the 'target' in absolute index was limited to $\pm 1 \times 10^{-6}$. It was felt that to attempt to measure to, say, seventh place accuracy in absolute index was unrealistic. Apart from the fact that the dispersions obtained would almost certainly apply only to the piece of glass measured, the homogeneity (either for chemical reasons or due to annealing effects) and the required tolerances on the sample would make the measurement academic rather than practical. The aim was not to demonstrate how accurately refractive index can be measured on one selected piece of glass, but rather to discover what sensible accuracy could be achieved with many samples of production optical glass.

A Photo-Electric V-Block Refractometer

To satisfy the requirements discussed in the Introduction two instruments have been designed and are at present under construction. The first instrument, which is for the rapid measurements with an accuracy of $\pm 1 \times 10^{-5}$, is a photo-electric V-block refractometer. In deciding what method of measurement to use considerable emphasis was placed on: (i) the ease of sample preparation, and (ii) the amount of skill and effort required from the operator. After studying a variety of possible methods of measurement and the samples required, it was decided that the right-angled sample with fine-ground surfaces (as used in the Hilger-Chance refractometer) was the easiest sample to prepare. The usefulness of the method depends to a large extent on the ability to make and test this sample to the necessary accuracy. An analysis of the errors of a V-block refractometer showed that for an error in measurement of $\pm 1 \times 10^{-6}$ from this source alone, the relation

$$\Delta n \Delta A = 0{\cdot}15$$

must be satisfied, where Δn is the difference in refractive indices of the sample and the matching liquid, and ΔA is the deviation, in seconds of arc, of the sample from a right angle. It was decided to apply the principle that, wherever possible, only one liquid should be used for all the spectral lines of interest. For a sample liquid refractive index differential of 0·02 a sample angle of $90° \pm 8$ seconds was required. A method has been described in the literature for testing samples to this accuracy (Harper 1968). The method is an optical one and uses the fact that a specular reflection can be obtained from a fine-ground glass surface at close to grazing incidence.

One of the major sources of error in the Hilger-Chance refractometer is that arising from the combination of (*a*) the setting of the slit image in the eye-piece, (*b*) the averaging of eccentricity effects of the scale by a visual process, and (*c*) the actual reading of the scale. A statistical analysis of a large number of readings by several experienced operators showed that for five readings on any one value of refractive index, the error for an 'average' operator would be between $\pm 3 \times 10^{-6}$ and $\pm 4 \times 10^{-6}$. Obviously a considerable improvement was required. This problem was eased by making the setting photo-electric and using a divided circle graduated to 0·2 seconds of arc (the smallest division in the divided circle in the Hilger-Chance refractometer is 3 seconds of arc). The easiest method of making the setting photo-electrically was by having the V-block in an auto-collimating optical system and using a vibrating slit and photo-detector. The optical system is shown in Figure 1.

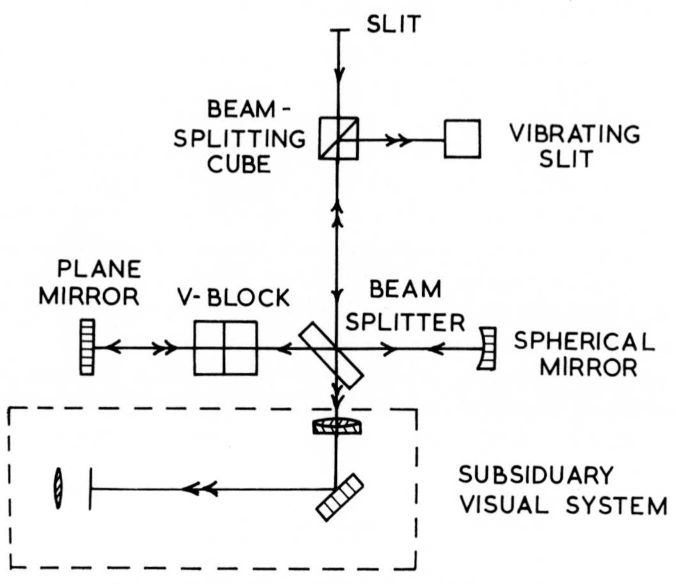

FIG. 1. Optical system of V-block refractometer.

An auto-collimating optical system had certain other advantages: (i) the alignment of the entrance face of the V-block to the beam could be rapidly checked; and (ii) the problem of a 'moving' point of deviation of the beam was eliminated. This second point is illustrated in Figure 2. As the angle of deviation, θ, is increased from zero, the point of deviation moves from A to B. A fixed point of rotation for the telescope means that the beam will illuminate the objective symmetrically at one value of θ only. Replacing the lens in Figure 2 by a plane mirror, with suitable tolerances on its surface, shows that this problem is eliminated.

It was important, in this refractometer, that one V-block should cover a measuring range in refractive index of at least 1·50 to 1·75 to the required accuracy. If this was not the case the necessary changes of V-block for different ranges of refractive index would reduce the usefulness of the instrument. The main limitation to such a feature was slit curvature.

However, using a V-block of refractive index 1·620 and a slit length of 1·5 mm the error introduced was less than $\pm 1 \times 10^{-6}$ over the refractive index range 1·50 to 1·80.

The problems to be solved in the optical design of the system were the tolerances on the various optical elements and their alignment. The approach adopted was to calculate the response of the vibrating slit, firstly to a perfect, diffracted slit image, and then to the image when small aberrations were introduced. The most severe problems arose with aberrations which led to asymmetry in the image. Particularly small

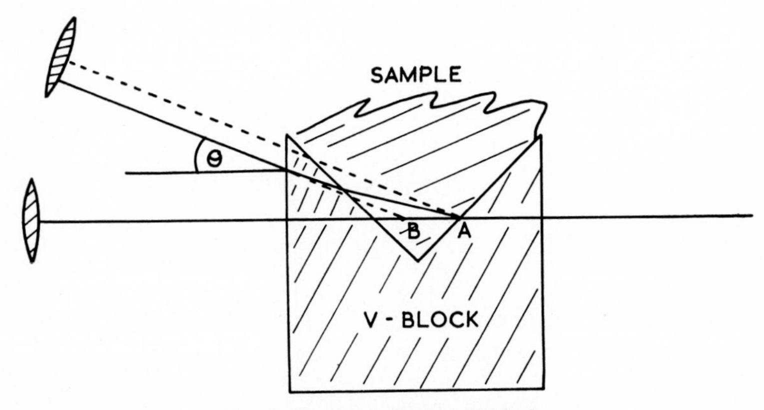

FIG. 2. Ray-path through V-block.

flatness gradient tolerances must be applied to the exit face of the V-block (less than $\frac{1}{150}$ λ per mm) and to the plane-mirror (less than $\frac{1}{500}$ λ per mm).

The response of a vibrating slit detector is quite sharply dependent on the relation between object slit width, vibrating slit width and the vibration amplitude. To ensure an adequate response over the wavelength range 365–1014 nm either one of the slit widths or the vibration amplitude must be varied. It was decided that it was more practical to vary the vibration amplitude. Satisfactory sensitivity has been achieved using a Hewlett-Packard silicon PIN photo-diode as the detector. The electronics basically consist of a synchronous detection system operated at about 42 Hz with a bandwidth of less than 5 Hz.

The most important optical element in the system was obviously the V-block. Serious errors could be introduced due to:

(*a*) a lack of knowledge of the refractive index of the V-block;
(*b*) imperfect construction of the V-block; and
(*c*) imperfect alignment in the optical system

The refractive indices of the glass from which the V-block has been made are known to $\pm 5 \times 10^{-6}$. These values can be checked and improved upon by taking measurements on samples of accurately known refractive index. The samples will initially be prisms, the refractive indices of which have been measured on the N.P.L. goniometric system. Later, samples measured on the interference refractometer will be used.

A new method of construction is being used for the V-block. The methods used at present employ optical contacting. Heat treatment of the V-block is necessary to make the contact permanent. Since the glass used for the V-block must be of the highest quality (i.e. in terms of homogeneity and birefringence) it was felt that this heat treatment was to be avoided. The V-block, shown in Figure 3, has been made in such a way

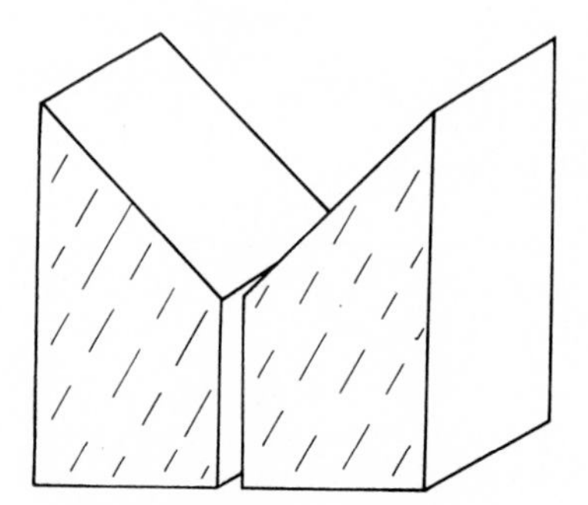

FIG. 3. Construction of V-block.

that it can be assembled and aligned mechanically. It is made from an accurately worked 45°–45°–90° prism. The 90° angle of the prism, which has been removed, is used as a 'zero' sample. This is the best method of establishing and checking the zero for the instrument since the V-block assembly remains undisturbed.

The V-block material was chosen to have a temperature coefficient of refractive index (for the e-line) of about 5×10^{-6} per °C. Since the temperature coefficients of refractive index of most glasses of interest lie in the range up to 1×10^{-5} per °C, this meant that a temperature control of $\pm 0{\cdot}2$°C would give an error of $\pm 1 \times 10^{-6}$ for these glasses. In fact, the instrument will be controlled in temperature to $\pm 0{\cdot}1$°C.

The major sources of error in this refractometer are those arising from the lack of knowledge of the refractive index of the V-block and from the setting on, and reading of, the angular deviation. It is felt that, with suitable standard samples, the first error can be reduced to better than $\pm 3 \times 10^{-6}$ and the second will be less than $\pm 2 \times 10^{-6}$. All other errors have been reduced to $\pm 1 \times 10^{-6}$ or better. Thus it can be seen that the results obtained with this refractometer will be accurate to $\pm 1 \times 10^{-5}$ or better.

An Interference Refractometer

The second instrument, which is for measurements of refractive index and dispersion to the sixth decimal place is an interference refractometer. The main criteria in this case were those relating to accuracy and precision rather than the time scale for the measurement. It was decided that it was important to aim for a system which gave an absolute measurement, i.e. one which did not require calibration with respect to any other refractometer. It was also decided to try to get away from any design which relied on image-forming optics. Inevitably this led to studies of interferometric systems and produced a final design in which the main optical fabrication was essentially the production of a series of optical flats.

Interference refractometers are well known in the study of gases but have received little attention for the measurement of absolute refractive indices of glass. However, measurements have been made using two-beam interferometers in which a parallel plate of known thickness is rotated, in one arm, through a measured angle and the resulting fringe shift is measured, (Hakoila 1952; Shumate 1966). The tolerances on the angle measurement were extremely severe, an error of 0·1 second of arc typically producing an error in refractive index of $\pm 1 \times 10^{-6}$.

The instrument under development introduces a steady change of optical path by moving a prismatic sample across one beam of a two-beam interferometer. Figure 4 shows the optical system of the interferometer. The sample prism moves between two convenient end-points, the base face remaining parallel to its original direction. It is moved along the hypotenuse face of a second, fixed prism with an index-matching liquid providing continuity of the optical path. With this type of system, the precision and accuracy of the refractive index measurements rely solely on linear measurement techniques, the principles of which have been developed to the highest standards in optical metrology.

Two interferometers are provided to obtain the information on optical path change, $(n-1)t$, and thickness change, t of the glass. The upper interferometer measures the optical path change, the beam passing twice through the sample. The measurement of the thickness change is achieved by coating the lower half of the prism base face with a fully-reflecting film and using this as a moving mirror in the second, lower interferometer.

The main experimental problems were the provision of a precise linear motion and a means of placing the prism so that its hypotenuse face lay along the direction of motion. Both of these problems were attacked using a linear slide (Boulton and Cooper 1969). This depended on the use of optical glass flat surfaces as guiding elements for a moving carriage. An experimental system showed that it was possible to move

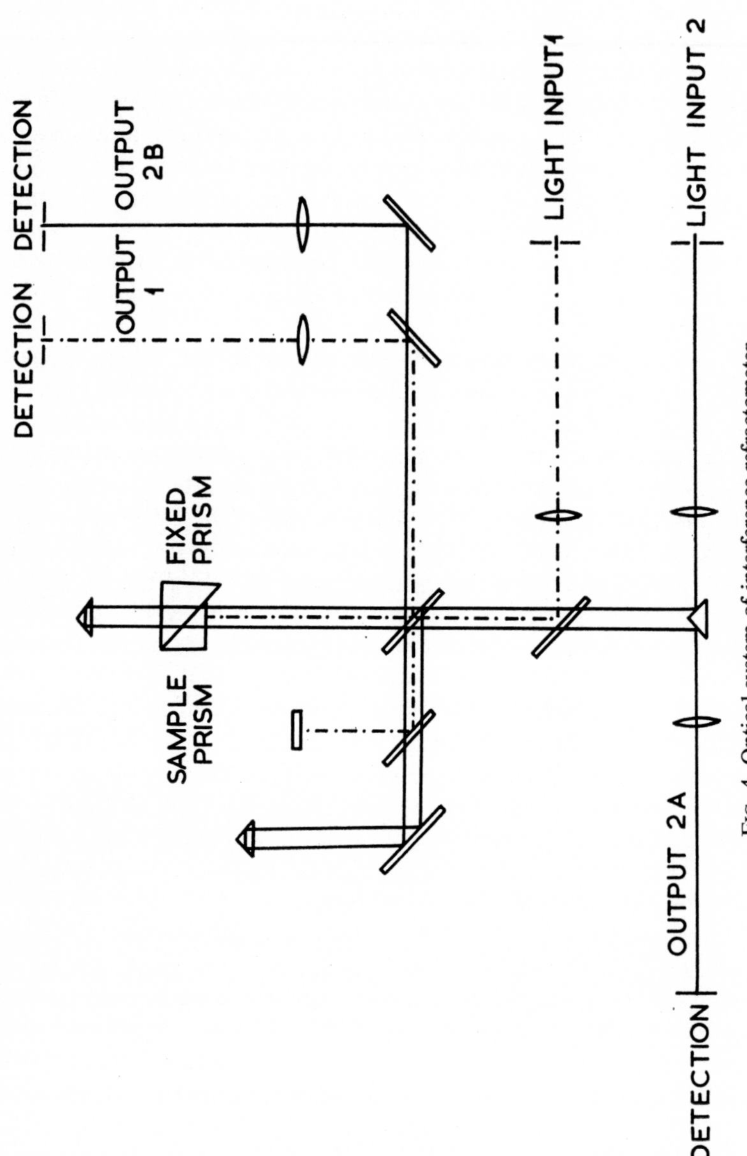

FIG. 4. Optical system of interference refractometer.

two optical flats relative to one another with a very thin air-film between them. This film provided a lubricating effect and also showed a coloured interference fringe system which could be used to obtain correct adjustment of the slide and to gain an idea of the quality of motion.

The instrumental version provides three optical flats arranged as in Figure 5. Flat *A* is laid in a horizontal plane and supports a carriage *B* which rides on three small PTFE pads. A second flat *C* is fixed to the carriage and is brought into close contact with the third guiding surface, *D*, with the formation of the coloured fringe system. The surfaces of *C* and *D*

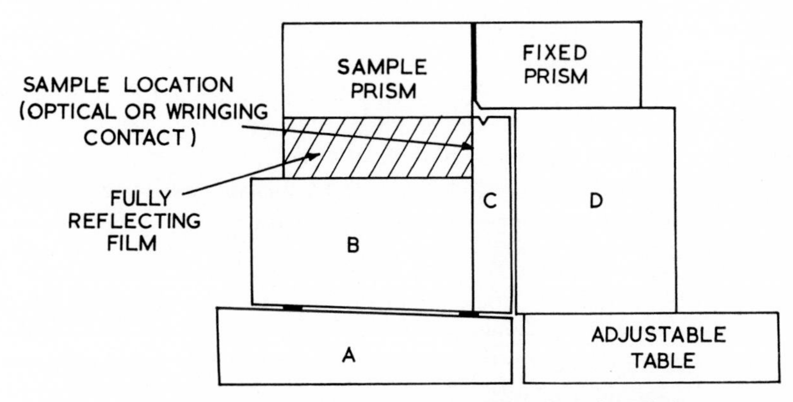

FIG. 5. The instrument slide (view along direction of motion).

are coated with hydrophobic silicone polymer (Midland Silicones, MS 1107 fluid) which prevents the possibility of optical contacting. The motion is controlled by a loop arrangement with towing points at each end of the carriage being connected to a moving nut of a lead screw. Correct adjustment of the slide is achieved when a uniform coloured fringe system is seen to remain unchanged both during forward motion of the carriage and when the direction of travel is reversed at any point during the motion.

An arrangement of the type described above suggested a solution to the problem of precisely locating the sample with respect to the slide. If the flat *C* was accurately polished plane and parallel, it provided a reference surface which was parallel to the direction of motion and to which the sample could be mated. It is planned to investigate both optical and wringing forms of contact to this surface.

It is proposed to look at two methods of measuring the geometrical thickness of glass introduced into the beam. Firstly, a fixed thickness may be used by placing a reflecting étalon in the comparison arm, white-light fringe settings indicating the end-points. The étalon can be calibrated *in situ* by using the lower interferometer as a gauge block measuring system, the calibration being achieved with an end fraction determination at several wavelengths. Such an arrangement might also be used in observing any possible changes in air refractive index during the measurement. The alternative is to use a variable value of t with continuous fringe counting in monochromatic light. This latter method seems of interest from the point of view of automation, although the first could also be automated to some extent.

The technique of electronic fringe counting to be used is standard. Briefly, two output signals with a quadrature relationship are required to sense direction as well as movement of the fringe field. These may be obtained by polarizing the input beam at 45° to the axes of a quarter-wave plate which is situated in one arm of the interferometer. The output signals are each passed through an analyser which is aligned to one or other of the axes of the retardation plate. The signals are then detected by low-noise photomultiplier tubes, S20 cathodes being used since it is required to operate over the region 405–644 nm.

The theoretical considerations relating to this method were basically similar to existing instruments in interference metrology. A full analysis showed that:

(*a*) Errors of obliquity could be made very small by auto-collimation of the entrance apertures of the two interferometers.
(*b*) Corrections due to imperfect collimation could be made very small and allowed for.
(*c*) Suitable spectral lines from ^{198}Hg and ^{114}Cd lamps produced light-wave scales of adequate accuracy for length measurement and for a significant coverage of the visible spectrum for optical design purposes.
(*d*) Corrections due to length measurement in dispersive media were entirely negligible.

The principle parameters affecting the overall accuracy and precision in this system are the atmospheric conditions, the sample, its preparation and location in the instrument, and the slide motion.

It is intended to control the temperature of the instrument to 0·02–0·01°C with sensible equalization of the glass and air paths in respective arms of the separate interferometers. It will also be necessary to monitor atmospheric pressure closely to a fraction of a millimetre of mercury, and to note the CO_2 and water vapour contents.

The tolerances on the sample preparation are reduced to a requirement for two polished faces with an approximate 45° angle between them. The tolerances on this angle and that of pyramidal error are, in fact, removed by interferometer adjustments in the initial setting. Surface flatness needs to be particularly good on the base face, typically to within $\frac{1}{50}\,\lambda$. The hypotenuse face is not so important since irregularities are reduced in proportion to the closeness of the liquid index match. A tolerance of $\frac{1}{20}\lambda$ is set to give a good locating surface. Naturally, the glass is to have the highest standards of homogeneity and birefringence, but the effect of surface refractive index changes are reduced since these are sensibly equal at the two end-points.

The uniformity of the motion could be monitored during each measurement, if necessary, though the conclusion from the initial results was that the slide was consistently performing to within the required limits.

SUMMARY AND CONCLUSIONS

The two instrument designs outlined above are currently under construction for use in Pilkington Research and Development Laboratories. In designing suitable measuring systems the environment and conditions of use had to be carefully examined. Since measurements of refractive index should be a routine procedure it was concluded that each system ought to be capable of electronic recording or setting rather than visual methods, in order to reduce the influence of observer fatigue. Present methods of measurement depend strongly on visual settings (e.g. the N.P.L. recording refractometer is calibrated with readings taken from a visual goniometer). Furthermore, it was thought necessary to control the temperature of each instrument rather than rely solely on the control of a room in which they are placed. This reduces the effects of temperature perturbations due to the proximity of the operator. Finally, it was thought desirable to site the instruments in a room with temperature changes controlled to less than 1°C.

The aim in this paper has been twofold, namely, to show what the requirements are for high-accuracy refractive index measurement in terms of measurement principles, and secondly to demonstrate that a manufacturing concern requires access to two instruments of different but complementary character. Clearly, an optical glass production unit must have available a means of issuing refractive index data which satisfies most of the immediate demands of optical designers. In the development of new glasses there is the problem of accurate measurements on small experimental meltings. The first instrument is ideally suited to providing both these services. There are, however, occasions when sixth-decimal-place determinations are necessary. For example, (i) when glasses for super-quality imaging systems are needed, and (ii) in studies of the effects of fine-annealing schedules on the refractive index and dispersion of both new and existing glass types. The main spectral region of interest here is the visible, 400 nm to 700 nm, and these measurements are to be performed by the interferometer. In the near infrared, tolerance levels on optical materials are reduced and the opinion is held that accurate fifth-place measurements from the V-block instrument are adequate.

Determination of refractive indices to these levels depends on two factors. Firstly, the provision of an instrument which is capable of operation and control to the required limits of performance. The design principles have, as far as possible, attempted to produce systems in which the accuracy is substantiated. Secondly, and equally important, is consideration of the material being measured. Obviously the highest standards of optical uniformity are desired, but also it is of advantage to use a sample with minimal tolerance conditions in preparation. Thus in the first case a modest angle requirement is stated but surface polish is eliminated. The angle can be checked to within the tolerance limits. In the second method extremely good surface flatness is required but no angular or pyramidal tolerance is stipulated.

Information which is given by the instruments above is of course, the starting point of much paper optical design. For a constructional system it is necessary that the designer should be given as accurate and complete data as is possible on the actual material being used. The concern is not only with the absolute refractive index values but also the magnitude and nature of any inhomogeneity. The part of our programme described here deals only with the first point. The second factor, i.e. homogeneity, is to receive thorough consideration in the near future.

Acknowledgments

The authors wish to thank Dr. W. T. Welford of Imperial College for his guidance and criticisms, Mr. A. M. Reid of Pilkington Brothers R & D Laboratories and several members of the staff of the National Physical Laboratory for helpful discussions.

This paper is published by permission of Dr. D. S. Oliver, Group Director of Research and Development, Pilkington Brothers Limited.

REFERENCES

Boulton, G. B. and Cooper, C. V., 1969, *J. Sci. Instr. Ser. 2*, **2**, 532.
Flude, M. J. C., Habell, K. J. and Jackson, A., 1961, *J. Sci. Instr.*, **38**, 445.
Forrest, J. W. and Straat, H. W., 1956, *J. Opt. Soc. Amer.*, **46**, 488.
Guild, J., 1929, *Dictionary of Applied Physics*, Vol. 4, London, Macmillan.
Hakoila, K. J., 1952, Thesis, University of Turku, Finland.
Harper, D. W., 1968, *J. Sci. Instr. Ser. 2*, **1**, 687.
Holmes, J. G., 1945, *J. Sci. Instr.*, **22**, 219.
Hughes, J. V., 1941, *J. Sci. Instr.*, **18**, 234.
Shumate, M. S., 1966, *Appl. Opt.*, **5**, 327.
Tilton, L. W., 1935, *J. Res. N.B.S.*, **14**, 393.

Interferometric Measurement of Temperature Coefficients of Refractive Indices of Optical Glasses

Frances Green

Division of Optical Metrology,
National Physical Laboratory,
Teddington, Middlesex

Abstract—Two polarized light interferometers simultaneously measure the change in length and the change in optical path length of a glass sample as a function of temperature. From these two measurements the temperature coefficients of refractive index may be calculated for the optical glass of which the sample is composed and for the wavelength used in the measurements.

The design of the interferometers is such that the relevant path differences are immune to small movements of the sample and its back plate. The sample, a parallel-sided slab of size 15 mm by 3 mm by 3 mm thick, is positioned in a variable temperature cell which has a temperature range of −195° C to 250° C. A uniform rate of change of temperature results in the output of both interferometers varying in a sinusoidal manner as a function of time. These outputs may be modulated by A.D.P. crystals to produce pulses at each half period of the sinusoidal variation as reference values to aid interpolation.

Present indications are that the cumulative error in the temperature coefficients of refractive index over the complete temperature range, for a given wavelength, will not exceed ±1 in 6th decimal place.

INTRODUCTION

The work to be described followed a request for measurements of the temperature coefficients of refractive index of optical glasses. The request was particularly for low-temperature measurements, as optical systems are being used increasingly in such environments. In the present equipment the temperature control unit has a range of −195°C to 250°C, and a more complicated cryostat is all that would be needed to reach liquid helium temperatures.

The temperature coefficients of many glasses have been measured by Molby (1949) for a few visible wavelengths and six fixed temperatures from steam to liquid nitrogen. He used a Pulfrich-type interferometer and obtained accuracies of about $\pm 1 \times 10^{-5}$. More recently Cleek and Waxler (1969) have made measurements on a few optical glasses at 0·5876 μ over a temperature range of $-195°C$ to $650°C$. The intention in the present work is to extend the range of wavelengths and to show a considerable increase in accuracy.

The method to be described needs no elaborate precautions against vibrations as it uses two extremely stable polarized light interferometers of the type described by Dyson (1968) in a paper entitled 'Optics in a Hostile Environment'. Its hostile environment is in the middle of a large, first-floor laboratory within about 30 feet of several large evaporating plants.

In both interferometers the beam-splitter is a quartz Wollaston prism that splits a circularly-polarized beam from a laser into two beams plane polarized perpendicular to each other. These beams pass through the interferometer where any difference in the two paths introduces a relative phase difference between the two beams so that when they recombine they form, in general, elliptically-polarized light. The recombination is obtained by making the beams pass back through the Wollaston prism with their polarizations reversed by double passage through a quarter-wave plate. The elliptically-polarized beam passes through a quarter-wave plate at 45° and becomes plane polarized, the plane of polarization rotating through 180° for each wavelength of path difference. The light beam then passes through an analyser and the level of intensity of the transmitted beam is dependent on the orientation of the plane of polarization with respect to the axis of the analyzer. As the path difference between the two beams changes the transmitted intensity varies.

The two interferometers simultaneously measure the change in length and the change in optical path length with temperature of a small glass sample. The sample is a parallel-sided slab that is wrung onto a small glass plate with a rectangular hole in it, so that the sample forms a bridge across the hole (Figure 1). The section of the sample and back plate below the hole are aluminised. The lower half of the sample is used to measure its coefficient of expansion and the upper half (in front of the hole in the back plate) the change in optical path length. The temperature coefficients of refractive index can be calculated from these measurements, together with the value of the index of the glass at one fixed temperature (usually 20°C), the thickness of the sample, and the wavelength used.

The Interferometers

When circularly-polarized light is transmitted by a Wollaston prism the light is split into two beams at approximately equal angles to the incident beam. As the angle of the incident beam to the normal to the prism face is changed the splitting angle goes through a minimum (Figure 2). The interferometer for measuring the change in optical path length uses a 45° quartz Wollaston, W_1, at its minimum deviation angle for one of the

FIG. 1. The glass sample and its back plate in the sample holder.

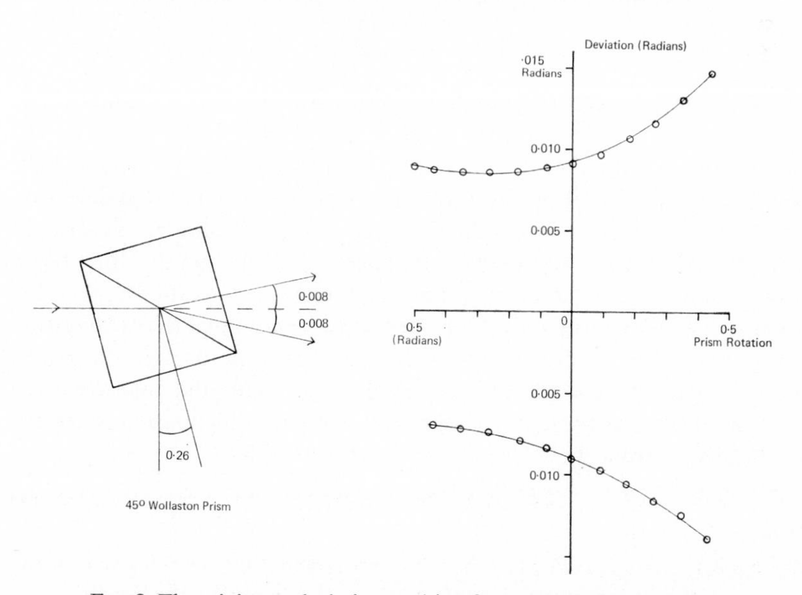

FIG. 2. The minimum deviation position for a 45° Wollaston prism.

beams. When the beams return to the prism the one that left it at minimum deviation retraces its path and the other returns at an angle (equal to the splitting angle) to the minimum deviation beam, but on the opposite side of it. The two beams pass through the prism and emerge at a small angle to the incident beam. In the minimum deviation position the angle between the emerging beams is a minimum and they recombine to form elliptically-polarized light. The minimum deviation position has the additional advantage of stopping the back-reflections from the prism reaching the detector.

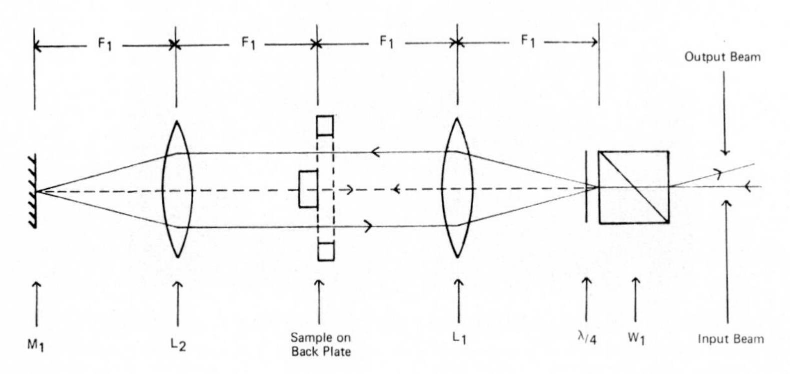

FIG. 3. The interferometer for measuring the change in optical path length with temperature.

In the interferometer (Figure 3) the Wollaston W_1, lens L_1, sample, lens L_2 and mirror M_1 are equally spaced, the spacing being equal to the common focal length F_1 of the lenses. Parallel circularly-polarized light from a laser is incident on the Wollaston prism at its minimum deviation angle. The alignment of the system is such that one beam travels straight through to the mirror M_1 and back through the prism, and the other beam passes round the sample through the holes in the back plate. Between the prism and the first lens there is a quarter-wave plate so that the polarization of the two beams is reversed for their return passage through the prism. The exit beam is deflected into the analyzing system and then onto the detector which is at present a silicon cell. The path difference between the two beams is given by:

$$n\lambda = 2l(\mu - \mu_G) \tag{1}$$

where n is the fringe order number, λ the wavelength, l the thickness of the sample, μ the index of the sample and μ_G the index of the medium. At present the interferometer is operating on the 0·6328 μ He:Ne line. The range of wavelengths will include the 1·15 μ He:Ne line, the blue-green lines from a low-power krypton laser and others within the transmission range of most optical glasses, as suitable lasers become available.

The interferometer for measuring the expansion coefficient uses a 35° Wollaston prism, *W*, which has a small angle prism cemented onto both faces and a quarter-wave plate and plane mirror, *M*, cemented onto one of the prisms (Figure 4). The prisms are such that when a beam is at near-normal incidence to the front face of the system it enters the Wollaston prism at the minimum deviation angle, and one of the exit beams leaves the system normal to the mirror which covers the upper half of the exit face.

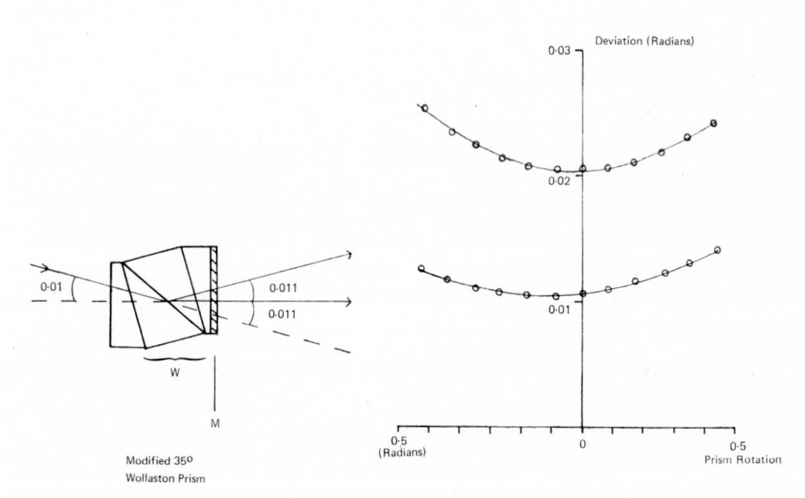

FIG. 4. The minimum deviation position for the modified 35° Wollaston prism.

In the interferometer the prism is at one principal focus of a lens, *L*, and the sample at the other (Figure 5). Parallel circularly-polarized light from the laser is incident on the prism which splits it into two beams that are focused by the lens onto the aluminised section of the sample and back plate. The beam that is normal to the mirror, *M*, in the Wollaston is reflected by the sample back through the lens above the incident beams and onto the mirror. It then retraces its path back to the sample and through the Wollaston prism. The other beam is reflected by one side of the back plate, the mirror, *M*, and the other side of the back plate. The two beams have passed through a quarter-wave plate twice in the interferometer and therefore their polarizations are reversed for the return passage. The path difference is detected in exactly the same way as that of the other interferometer. In this case

$$m\lambda' = 4l\mu_G \qquad (2)$$

where m is the fringe order number and λ' is a constant wavelength (0·6328 μ).

In each interferometer stability is obtained by making both beams traverse the same optics and stay close together. This makes the interferometers immune to small axial translations of the reflecting surfaces. In the interferometer that measures the expansion coefficient, the sample and back plate are also immune to all rotations about an axis through the sample. In the other interferometer the sample is sensitive only to the two rotations that could change the optical thickness of the glass.

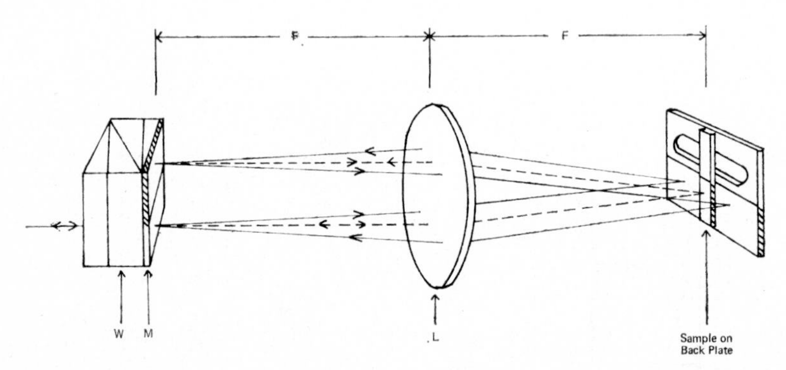

FIG. 5. The interferometer for measuring the coefficient of expansion of a glass sample.

The angles of the Wollaston prisms and the focal lengths, F and F_1, of the lenses, L, L_1 and L_2, have been selected so that the two interferometers can interlock and the measurements can be made simultaneously, i.e. the separation of the two beams in the interferometers must be approximately the same when they reach the sample (Figure 6). There is less than a $\frac{1}{4}''$ clearance between the light paths in places, so the unused parts of two of the lenses (L and L_2) have been removed (Figure 7).

In the path difference Equations, (1) and (2), the index of the medium μ_G can be put equal to unity for reasons that will be explained. The equations then become:

$$n\lambda = 2l(\mu - 1) \tag{1'}$$

$$m\lambda' = 4l \tag{2'}$$

When the temperature of the sample is changed they become, respectively,

$$(n + p)\lambda = 2(l + \Delta l)(\mu + \Delta\mu - 1) \qquad (3), \text{ and}$$

$$(m + q)\lambda' = 4(l + \Delta l) \tag{4}$$

where p and q are the change in fringe order number, Δl is the change in length of the sample and $\Delta\mu$ its change in index.

Subtraction of equation (1′) from (3) gives

$$p\lambda/2 = \Delta l(\mu - 1) + \Delta\mu(l + \Delta l),$$

and (2′) from (4) gives

$$(q\lambda'/4) = \Delta l$$

Hence,

$$\Delta\mu = \frac{2p\lambda - q\lambda'(\mu - 1)}{4l + q\lambda'}$$

If the temperature of the sample is changed continuously at a very slow rate (about 10°C/hr) p and q are sinusoidal-type fringes which can be recorded on a chart together with the temperature. Neither the optical path length nor the length of glasses varies linearly with temperature over such a large range and therefore the fringes are not exactly sinusoidal and cannot be interpolated along the temperature (time) axis.

If the laser has a stable intensity they may be interpolated in intensity, in which case the fringe fractions at 20°C (or the temperature at which the index of the glass is known) are called zero, and the values of p and q at even temperature intervals are read from the chart to within $\frac{1}{500}$th of a fringe. The values of $\Delta\mu$ versus temperature (T) can then be calculated and a curve fitted to the results.

If the source is not sufficiently stable to be able to interpolate at $\frac{1}{500}$th of a fringe the same degree of accuracy may be reached by modulating the output beams. An A.D.P. modulator is introduced before the analyzer in each interferometer and the preceding quarter-wave plate is removed. The induced axes of the A.D.P. are set at 45° and the axis of the analyzer is vertical. This results in a dc level of intensity with a modulation that is zero at every half fringe. The points of zero modulation are detected electronically and recorded as pulses on the chart. The temperature at the half fringe points of p and q are read off and a curve fitted for p versus T and q versus T. The rest of the analysis is the same as for a stable source. This method has the additional advantage in that it would be easily digitized.

The temperature of the sample is controlled by means of a thermocouple connected to the sample holder. There is another thermocouple stuck onto the back plate to record the surface temperature of the glass, and it is this temperature that is recorded with the fringes. The sample in its holder is connected to the base of a stainless steel refrigerant vessel, and the whole assembly is slid inside an outer jacket fitted with fused silica windows (Figure 8). The windows are heated to prevent misting. The space around the sample is evacuated to 0·05 torr. This reduces the heat loss to the outside, but still leaves a residuum of air to transmit changes in temperature of the sample. If air is used at this pressure, its index changes with tem-

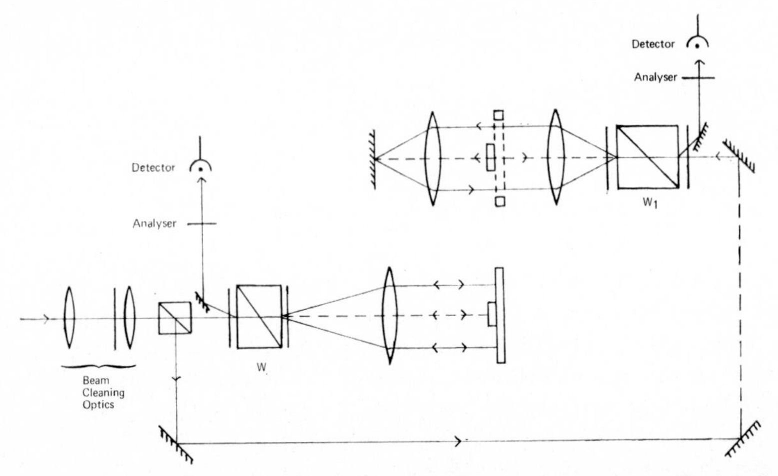

FIG. 6. Plan view showing the relative positions of the two interferometers. The interferometer for measuring the change in optical path length has been displaced for clarity.

perature in the 7th decimal place, whereas helium at the equivalent pressure to give the same thermal conductivity (0·03 torr) changes in the 9th decimal place. These changes in the index of air have no direct effect on

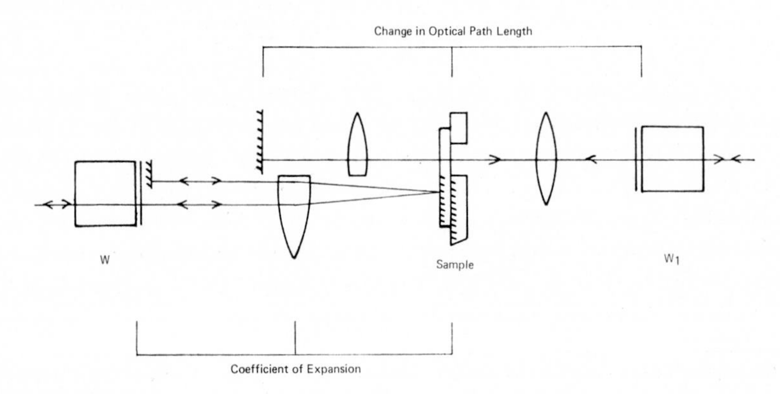

FIG. 7. Vertical section showing the relative positions of the two interferometers.

the accuracy of the coefficients, but to obtain the absolute index at any temperature a small correction for non-vacuum would have to be applied. If helium is used instead this correction is negligible and μ_G can be put equal to unity.

Accuracy

The accuracy expected from current experiments on the absolute index is ± 1 in 6th decimal place. The change in index should therefore be measured to an accuracy that will not affect this value, i.e. it is hoped to measure changes to a few in the 7th decimal place over the temperature range indicated. The temperature coefficient per °C is about 5×10^{-6} and is temperature dependent. The cumulative change from 0→100°C may be 6×10^{-4} and, from 0→ −180°C, 4×10^{-4}. To measure to the 7th decimal

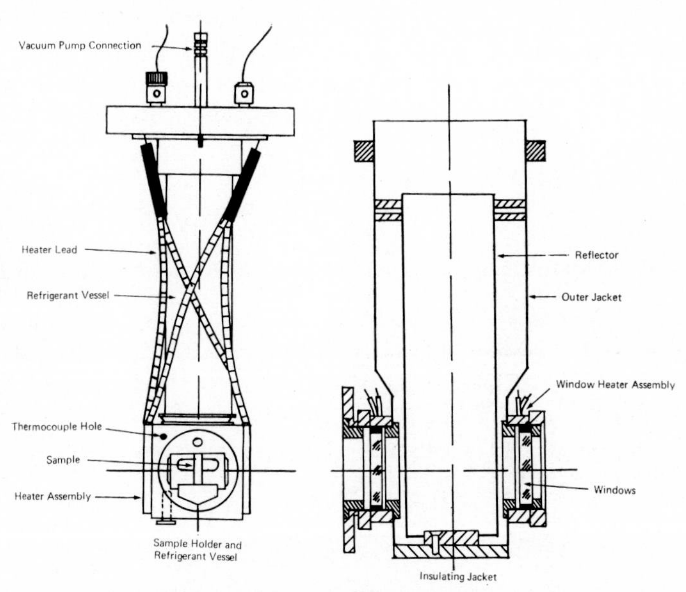

FIG. 8. The variable temperature unit. (Drawings are reproduced by permission of Research and Industrial Instruments Company Limited.)

place we are therefore looking for an accuracy of about four significant figures in all the data needed. This is easily obtained in wavelength and may be reached in source intensity, e.g. by using long-cavity lasers. The many peaks in the Doppler envelope keep the intensity and the mean wavelength constant as the cavity expands.

Because the temperature is being changed continuously the sample size must be such that the difference between its surface temperature and its mean temperature at any instant has no appreciable effect on the accuracy of the results. On the other hand, if the sample is too thin the change in optical path with temperature will be too small to measure accurately. The optimum sample length l is about 3 mm. The specimen then has a mean temperature 0·02°C below that of the surface and in the interferometer there will be an average displacement of one fringe for every 10°C change in temperature. The degree of parallelism and flatness required is that which gives a thickness constant to about one fringe over the sample faces (15×3 mm). The sample length has to be known to 1μ and this can be done with ease on a universal measuring machine.

The whole temperature range is -195 to 250°C, or 20 mV, and measurement is required to about 0·1°C, or 0·005 mV. To obtain this degree of accuracy the temperature recording pen is made to zig-zag across the chart at millivolt intervals.

A detailed analysis of errors shows that the errors in fringe and length measurements have the greatest effect on the accuracy of the results. A decrease in accuracy of fringe measurement has a marked effect on the errors of results within $\pm$ 100°C of the fixed temperature, whereas a decrease in accuracy of length measurement has greatest effect at extremes of temperature.

Acknowledgements

The work described above is being carried out at the National Physical Laboratory, Teddington. I would like to thank Dr. J. Dyson, Dr. K. G. Birch and Dr. R. J. King for much useful discussion and advice, and Mr. M. J. C. Flude for the design and construction of most of the electronics.

REFERENCES

Cleek, G. W. and Waxler, R. M., 1969, private communication.
Dyson, J., 1968, *Applied Optics*, **7**, 569.
Molby, F. A., 1949, *J. Opt. Soc. Amer.*, **39**, 600.

An Interference Phenomenon Associated with Profile Imagery and its Influence on Profile Inspection Measurements*

G. D. Dew
(Presented by K. J. Habell)

Optical Metrology Division,
National Physical Laboratory, Teddington, England

Profile images of cylinders and spheres, produced by conventional techniques of optical inspection, are frequently flanked by systems of interference fringes The mode of formation of these fringes has been studied, and a theoretical treatment is given which enables the intensity modulation in the plane of the profile to be calculated. Quantitative confirmation of this analysis is provided by an experiment in which the fringe pattern is recorded in this plane without recourse to any auxiliary observing system. From the point of view of profile inspection techniques, the results are highly significant. Total destructive interference at the end of the specimen causes the true profile to be completely masked, and the eye sees an artificial edge, the position of which can be predicted from the intensity profile of the interference pattern. The influence of the radius of the specimen on the magnitude of this positional error is examined. Finally, it is shown that defects of focus produce changes in the position of the apparent edge and may, unless special precautions are taken, cause additional errors in 'sizing' measurements.

* A résumé by the author. A full account of this work is to be published in *Optica Acta* (1969) under the title: 'The application of spatial filtering techniques to profile inspection, and an associated interference phenomenon.'

Optical Projection Systems for the Moiré Method of Surface Strain Measurement

A. Luxmoore

School of Engineering,
University of Wales,
Swansea, U.K.

Abstract—The Moiré effect is an important tool in experimental stress analysis. A grid is printed on a body, which is subjected to loads, so that Moiré fringes produced by the grid distortion may be observed by superimposing a reference grid. In certain circumstances it is advantageous to use a lens to project the distorted grid onto the reference grid (or an intermediate photographic plate). This paper describes some possible sources of error in such a system, and outlines the advantages of a telecentric construction for a projection lens. A number of such lenses have been constructed and used by the author, but so far, none has proved ideal.

INTRODUCTION

The use of the Moiré effect for surface strain measurement was first suggested by Weller & Shepard (1948), but their grids were so coarse that they could only detect the displacements produced by slender structural components. Dantu (1958) developed the method to a more practical stage by using finer grids (500 lines per inch), and introducing an interpolation method based on an initial fringe pattern for zero load on the specimen. He showed that the fringes represented contours of displacement, and that strains could be derived from these by graphical differentiation. He also utilized two pairs of line-grids, printed simultaneously, but inclined at 90° to each other, so that the complete two-dimensional displacement field could be determined. The two fringe patterns were separated by using a line-grid as the reference, aligned first with one specimen grid, and then rotated through 90° for alignment with the other grid. This practice has become common among stress analysts, and explains why photographic printing has become the most popular method for depositing the grids. Since that time, a considerable literature has been published on the method, but the number of practical analyses is still small. However, the method has certain advantages over other strain measurement systems, especially for problems involving high temperatures, and it has become a standard tool in experimental stress analysis.

At present, grids of about 1000 lines per inch can be deposited on most engineering surfaces, without undue surface preparation. Finer grids have been deposited (Holister & Luxmoore 1968; Middleton & Stephenson 1968), but most workers have found 1000 lines per inch a useful practical limit. Any Moiré fringes formed with these grids will represent displacements of 10^{-3} in. Most engineering materials reach their elastic limit at strains of 10^{-3}, and displacements of 10^{-5} in. are normally detected if useful measurements in the elastic region are required.

If an initial fringe pattern is present before straining, very small strains can be measured from the changes in the fringe geometry. This is commonly known as the 'mismatch' technique, because the initial pattern is produced by using grids with slightly different pitches. Alternatively, the reference grid can be displaced across the specimen, producing corresponding changes in the fringe locations. If the displacements equal small fractions of the pitch, then this method proves very useful for determining strain gradients, especially when combined with the 'mismatch' process. This can be considered as an alternative to increasing the line density, but the difficulties of producing the minute displacements accurately have discouraged its application to general problems.

These interpolation procedures increase the sensitivity by a factor of 10 or more, and then the main limitation is the accuracy with which the fringe positions can be located.

One of the advantages of the method is the simplicity of observing fringes, by directly superimposing the reference grid on the specimen grid, in incoherent illumination. However, this method is not very convenient under the following circumstances:

(*a*) When the specimen is hot. The reference grid would then expand, probably by an unknown amount and in an uneven manner, thus making analysis impossible.

(*b*) When it is required to change the pitch or position of the reference grid for interpolation purposes.

(*c*) When the fringe contrast can be improved by spatial filtering, or by increasing the sensitivity by diffraction modifications.

To achieve these ends, many stress analysts are working with photographs of the specimen grid, or directly projecting this onto the reference grid. Suitable projection lenses are therefore becoming an essential part of the technique.

Geometrical Problems of Conventional Lenses

Conventional lenses are designed on a perspective basis, so that changes in object distance produce a corresponding change in image size, i.e. a change of magnification. This introduces a location problem, as the strains we are measuring are very small, and small out-of-plane displacements by the specimen produce comparable changes in magnification. In most stress analysis experiments it is difficult to fix the position of the specimen because, as load is applied, the specimen surface can be distorted by bending, twisting, Poisson effects, etc. These out-of-plane displacements are usually much larger than the surface strains, so some compensation must be made for them.

As a first step, we analysed the effect of the surface displacements on the in-plane strains, assuming an initially flat surface. The effect is

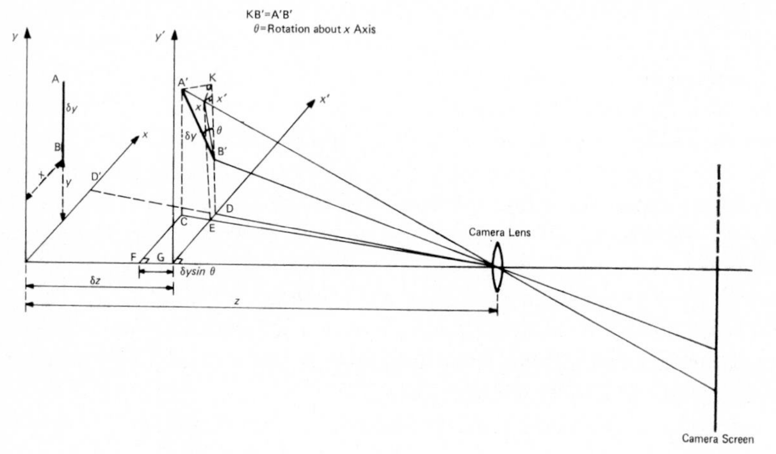

FIG. 1. Perspective effect of line element displaced from its original position.

illustrated in Figure 1, where an elemental line, AB, is displaced and rotated to $A'B'$. This produces an apparent normal strain in the image plane (which is assumed stationary throughout the experiment) of

$$1-\cos\theta[1\pm(\delta z/z)]+y/z\sin\theta$$

and a rotational error of

$$x\theta/z$$

where θ is the angle of rotation about the x-axis, and δz is the displacement along the z- (optic) axis.

In this derivation, the displacements and rotations have been assumed to be small. These formulae can be generalized into two dimensions, by considering a similar elemental line parallel to the x-axis.

There are three possibilities for reducing these errors:

(1) To use a large aperture, with a correspondingly small depth of field, so that any area which deviates too much must be re-focused. This is unlikely to be successful for the non-linear displacements such as bending effects, etc. However, it is a necessary pre-requisite for initial alignment, and has been used as such by the author.

(2) To observe the specimen surface from two separate angles, and hence calculate any out-of-plane distortion, as in photogrammetry.

(3) To make the object distance, z, infinite, by using a telecentric construction. The apparent strain then reduces to a term for simple rotation, viz: $(1 - \cos \theta)$.

Correction of Perspective Images

The determination of surface contours by photogrammetry is a tedious process, but combined with the Moiré effect it can be simplified. In our experiments, it was more convenient to use two off-axis apertures with one lens, instead of two separate cameras.

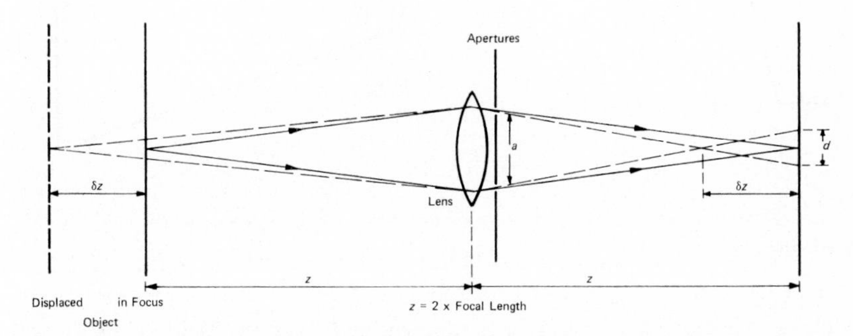

FIG. 2. Determination of surface contours by two off-axis apertures.

The geometry of the system is illustrated in Figure 2. The lens is a 300 mm Apo-Nikkor f/9, with a slot for a Waterhouse plate, and was used at unit magnification. We could observe a 5-in. field with this lens, up to 1000 lines per inch, with negligible distortion (less than 10^{-4} in. over whole field).

When two off-axis apertures are located in the Waterhouse slot, any lines on the specimen will be doubled unless exactly focused. By measuring the separation of these doubled images, the specimen displacement δz can be determined from the relation:

$$\delta z = z(d/a)$$

where d is the separation of the images, and a is the separation of the two apertures.

When using the Moiré effect, the image separation cannot be observed simultaneously, and one aperture must be blanked off. The fringe geometry will be the same in each case, but the fringe locations will be different. This separation of fringes for the two apertures can be used to measure the parameter d, defined as

$$d = (\delta F/F)p$$

where p is the pitch of the grids, δF is the fringe separation due to aperture change and F is the fringe separation in one picture.

With an aperture separation of one inch, we could detect specimen deflections of 0·005 in., by measuring $\delta F/F$ to 0·2, which was fairly easy with a ruled screen.

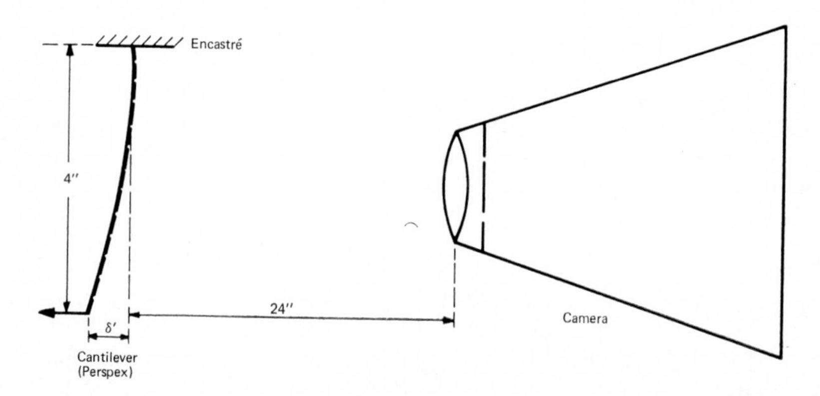

FIG. 3. Determination of surface strain on cantilever, using perspective lens with off-axis apertures.

Once the value of δz has been determined over the field of view, we could determine sin θ and cos θ by graphical differentiation. This is not particularly accurate, but the strain errors are less susceptible to changes in angle than in translation. The strain errors can then be computed, and subtracted from the original calculations.

We have applied this procedure to the problem of a cantilever (see Figure 3) and the comparison between theory and experiment is fair (see Figure 4). Some overall difference between the two is to be expected, as we did not have accurate values for the elastic constants, which were necessary for the theoretical calculations.

Telecentric System

The idea of a telecentric stop is well established, and many commercial lenses are designed on this basis. For our application, it does not solve the problem completely, but it certainly reduces the complications considerably.

Rotation of the specimen can be corrected by the same procedure as before, but more accurately, and with much less computation. Figure 5 illustrates the geometry of the two off-axis pictures, but in this case, we need only consider the actual strains measured from each picture. If we can be sure that the two pictures are taken on either side of the surface normal, then the true strain is given by:

$$\epsilon_T = \frac{\epsilon_1 + \epsilon_2 + 2}{2\cos(\phi/2)\cos[(\epsilon_1 - \epsilon_2)/\phi]} - 1$$

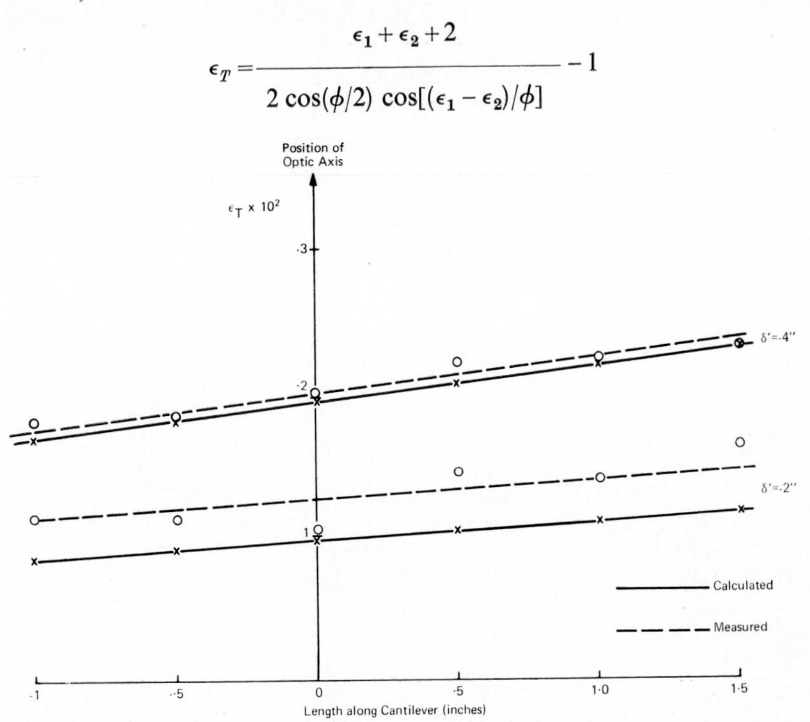

FIG. 4. Comparison between calculated and measured strains on cantilever in Figure 3, for different values of end deflection.

where ϵ_1 and ϵ_2 are the measured strains and ϕ is the angle between the lines of observation.

This process allows us to correct strain measurements down to the limit of sensitivity, which, in our case, is generally around 10^{-4}. However,

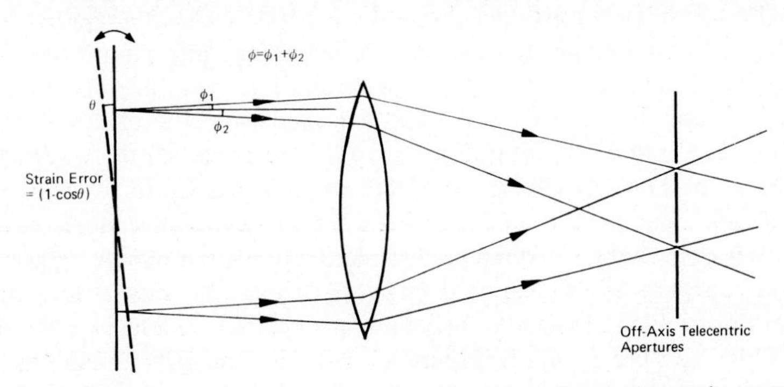

FIG. 5. Correction of strain for rotation of specimen by off-axis telecentric apertures.

in many problems, we find that any rotation of the specimen is often less than 1 degree, which produces an error strain of only 10^{-4}, so that it is not worth making the correction. This is a considerable help in practical cases.

We are hoping to extend the range of our measurements down to strains of 10^{-5}, by using photo-electric detection of the fringes. In this case, correction for rotation will become essential.

Though the telecentric construction overcomes the location problem satisfactorily, it is more difficult to design than a conventional lens of comparable performance. For successful application to the Moiré technique, a lens must be capable of resolving 1000 lines per inch, with negligible distortion (less than 10^{-4} in.) over a 6-in. field of view. This specification can be met by a first-class process lens, but not by the usual telecentric lenses that are available commercially. As the lens diameter

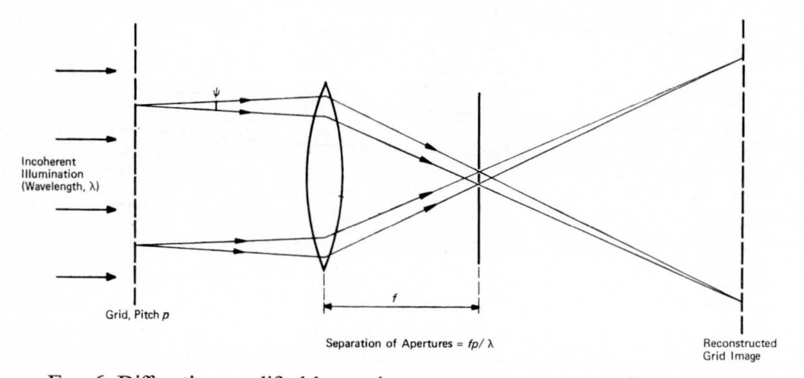

FIG. 6. Diffraction modified lens using two apertures to record regular grid.

must always be slightly larger than the field of view, 6-in. field lenses are very expensive, and successful correction of the aberrations near the edges will be most unlikely. It is the author's experience that these lenses are generally quite successful up to 1-in. diameter, but demonstrate serious distortion, field curvature and chromatic effects outside this region.

When using coherent illumination, the individual diffraction orders can be separated, and any two can be recombined to form an image of the grid. This can be used as a method for increasing the effective fringe density, but, more important, it can alleviate some of the aberration problems, particularly chromatic effects and field curvature.

Unfortunately, most of our specimens are opaque, with a rough surface that diffuses the light, so coherent illumination is not possible. A suggestion by Dr. J. Dyson of the National Physical Laboratory successfully overcomes this problem. Instead of using a single aperture as a telecentric stop, a double aperture is used (Figure 6) with a separation equal to the separation of the diffraction spectra in this position, and aligned in the direction of these spectra, i.e. the line joining the centres is perpendicular to the grid lines. This system selects two beams which, by virtue of the grid diffraction, are relatively coherent, and so form a two-beam interference pattern analogous to the grid lines in the image plane. However, it is rather wasteful of light, and this has proved a problem in some applications.

Practical Telecentric Lenses

As stated previously, our basic specification for a projection lens is 1000 line per inch resolution, negligible distortion and a large field. The first requirement is essential, the second highly desirable and the third depends on the problem in hand. To these should be added the facility for producing an accurately known pitch 'mismatch' and relative displacement between specimen and reference grids, for interpolation.

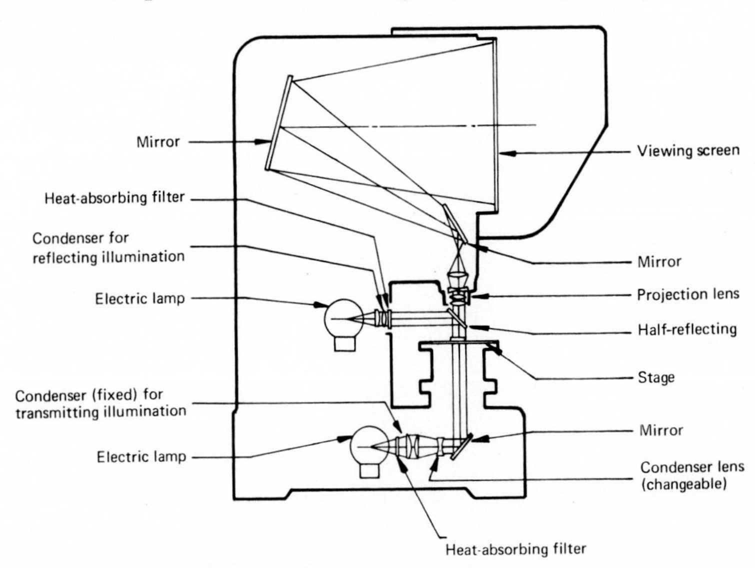

FIG. 7. Optical system of Nikon 6-CT profile projector.

Zero distortion is not essential, as it can always be corrected for by carrying out a preliminary calibration. However, large distortion cannot be tolerated, as it produces a complex fringe pattern, and no useful analyses can be made. We have found that negligible distortion is desirable as it increases the accuracy, and decreases the labour of an analysis.

Profile Projectors

Profile projectors are the most common form of commercial telecentric system, and we have access to two Nikon instruments, the 6-CT and V-16. These instruments are only telecentric on the object side (Figure 7) so we were able to introduce accurate pitch 'mismatch' by manufacturing special screen mountings (Luxmoore & Eder 1967). These altered the magnification by changing the image distance. Magnification changes of ± 2 per cent were possible in both cases.

The lenses, which range from $\times 5$ to $\times 100$, have high resolution and low distortion but this deteriorates outside a 1-in. diameter (this only applies to the $\times 5$ lens, covering a field of $2\frac{1}{2}$ in.). The reference grids have correspondingly low line densities, to match the magnified specimen grids, but these cannot go lower than 50 lines per inch, as the actual grid lines become too distracting. However, the overall magnification increases the accuracy, and the fringe contrast is usually excellent.

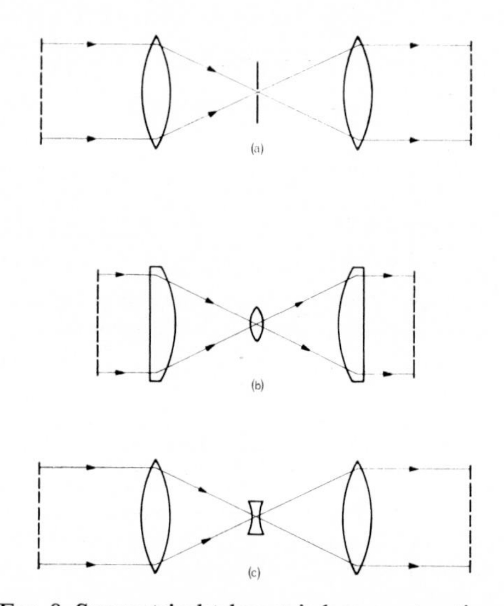

FIG. 8. Symmetrical telecentric lens constructions.

These instruments have proved extremely useful for a number of practical problems, but particularly those involving strain gradients on a microscopic scale.

Symmetrical Telecentric Lenses

The profile lenses do not provide a convenient portable system due to the magnification involved. Symmetrical lenses, such as those illustrated in Figure 8, would overcome this problem, and have the additional advantages of a telecentric construction on both object and image sides, and theoretically, zero distortion. They could also be used in front of other systems, such as the profile projectors, to extend the latter's usefulness.

A system corresponding to Figure 8*a* was constructed for the author by Hilger and Watts Limited using a pair of their $6\frac{1}{2}$-in. f/3 profile lenses. The lenses were adjusted for infinite conjugates, and then screwed into either end of a tube with a central iris. The field curvature was bad, and we could only focus a 1000 line per inch grid over a central half-inch field. Dyson's double aperture increased the depth of field, and we could focus the grid over the full 2-in. field. The distortion was negligible over the central $1\frac{1}{2}$ in., but then deteriorated rapidly. The major disadvantage, apart from the limited field, was the lack of variable magnification, as this proved very valuable with the profile projectors.

The system in Figure 8*b* was suggested by Dr. J. M. Burch for use with a double aperture. Small variations in magnification can be obtained by displacing the central positive lens along the axis. The system was constructed using single element components, and with the object and image located at the centre of curvature of the two field lenses. Fringes cannot

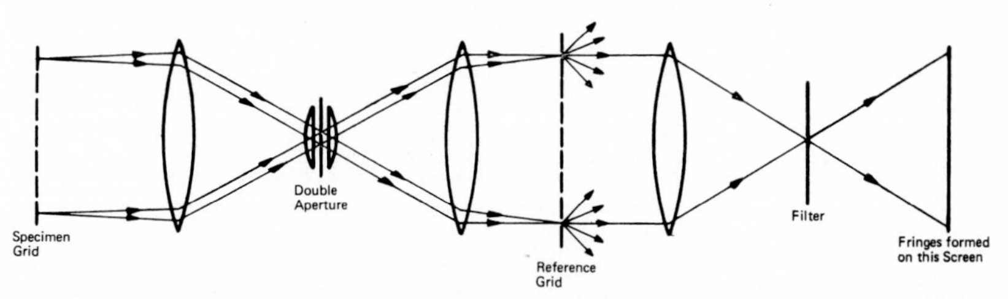

FIG. 9. Symmetrical telecentric lens, with double aperture and auxiliary filter System to separate diffracted orders from reference grid.

be observed directly in the image plane, and a subsidiary lens system is necessary to filter all orders except one (see Figure 9). This system works satisfactorily up to a $1\frac{1}{2}$-in. field, but then chromatism and distortion become significant.

Figure 8*c* illustrates a more usual design for a symmetrical lens where the negative lens is double the power of the outermost, positive components, to give a flat field. We have experimented with this system, using aerial lenses for the positive components and negative spectacle lenses, but the images were always spoilt by chromatic effects.

Lens Hologram Combination

As it seemed most unlikely that conventional lens design would provide a large-field telecentric instrument, we examined the possibilities of a holographic imaging process. A field lens was still essential to provide collimation, but the aberrations it produced would be corrected by making a hologram of the specimen through the lens, and then back-projecting this through the lens.

After some experimentation, a successful arrangement was devised (Figure 10). The hologram was formed in the image plane of a simple lens system, using a collimated reference beam. An identical, but opposite beam, then projected the image back onto the object (or other suitable grid).

This apparatus worked very successfully, resolving 1000 lines per inch over a 6-in. field with negligible distortion. The depth of field was effectively infinite, but the aberrations only cancelled completely in the object position. This apparatus could easily be extended to cover a larger field and finer grids, but it cannot be considered portable. In the author's view, this construction will prove very useful for model studies, where the specimens and loading frames are sufficiently small to be brought to the optical bench, rather than *vice versa*. As it uses standard optical components, the system can be made up for special investigations very quickly.

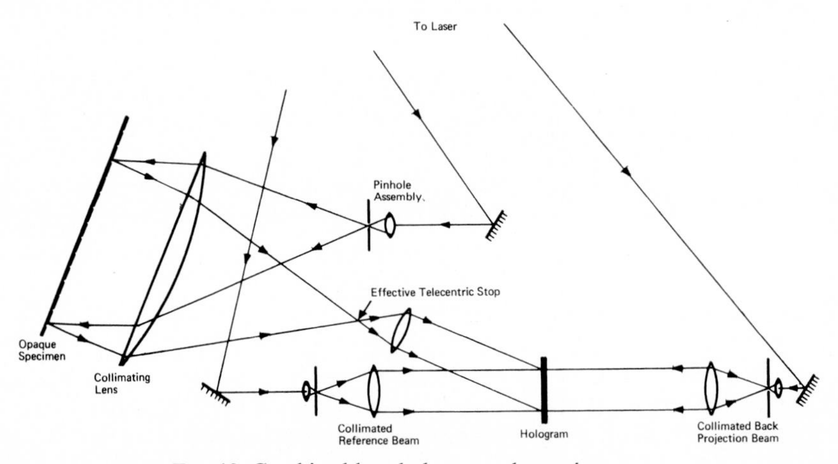

FIG. 10. Combined lens-hologram telecentric system.

Small changes in magnification can be made by adjusting the lens system during back projection, without spoiling the reconstructed image.

Lazer Projection System

Because of the exact collimation of a laser beam, a grid can be projected by illuminating it with laser light, as shown in Figure 11*a*. The divergence of the diffracted orders will cause the grid to be imaged at discrete points in front of the object, where the orders recombine again (Sciammerella & Davis 1968). We have projected photographic grids of 1000 lines per inch over distances of several feet, and by placing the specimen grid at an image position, Moiré fringes can be produced on the specimen surface, where they can be observed by an auxiliary instrument.

Variable pitch control can be introduced by placing the reference grid midway between the pinhole and the collimating lens (Figure 11*b*) so that the line density of the projected grid is halved. Small adjustments of the position of the reference grid will vary the pitch of the projected image in a linear manner.

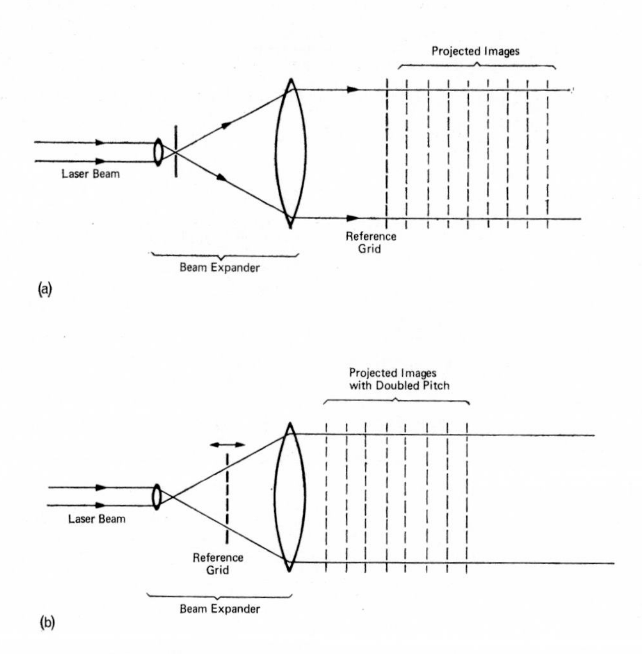

FIG. 11. Laser projection of reference grids: (*a*) by parallel laser beam; (*b*) projection of magnified reference grid with variable pitch control.

The distortion of the projected grid is not so sensitive to aberrations as in the conventional lens design. It is only effected by the beam collimation, which, in turn, is controlled mainly by the spherical aberration of the collimating lens. We have used aerial camera lenses for our work, as these are well corrected for this particular aberration (they can also be obtained with large apertures e.g. a 36-in. f/6·3, covering $5\frac{1}{2}$ in.). The effect of this distortion can be reduced by using the shortest possible distance between lens and specimen, and at 10 in., we can obtain a distortion of less than 10^{-4} in. over a 3-in. field.

It is quite simple to incorporate a double view in order to correct for any specimen rotation. All that is required is a means of displacing the pinhole by a known distance. This can easily be applied to the lens-hologram system as well.

Holographic Interferometry

Our most successful large-field projection systems utilize a laser, so it seemed a sensible step to investigate holographic interferometry as an alternative to the Moiré effect. It would have the added advantages that (*a*) no grid had to be deposited on the specimen; (*b*) it is far more sensitive; and (*c*) it could be applied to surfaces with a complex geometry.

Ennos (1968) has described an arrangement for separating the in-plane displacements, and we are working on a similar system which we hope will be more convenient. So far, the optical problems associated with this method do not make it an attractive alternative.

CONCLUSIONS

Two simple methods have been evolved for correcting dimensional errors due to out-of-plane displacements of a surface. The most accurate and convenient method involves the use of a telecentric lens, but conventional design of these lenses restricts them to small fields. Larger fields may be covered by using a normal, perspective lens, or a telecentric laser-lens combination.

The most useful telecentric lens would be a symmetrical unit relay system which could be used directly with a camera, or in front of a magnifying system. It seems that such a proposition is feasible by conventional design, and could cover as much as a 4-in. field, with negligible distortion. However, to the author's knowledge, no such lens is commercially available, although it is hoped that a suitable item may soon be forthcoming.

Acknowledgements

This work was supported by a grant from the Paul Instrument Fund, to which the author is indebted. He would also like to thank Drs. J. D. Dyson and J. M. Burch for valuable advice on various optical matters, and Mr. C. House and Mr. P. Hollely for assistance with experimental work.

REFERENCES

DANTU, P., 1958, *Ann. Inst. Techn. du Bat et des Trav. Publics*, No. 121, p. 78 (C.E.G.B. translation No. 3089).
ENNOS, A. E., 1968, *J. Sci. Inst.*, **1**, 731.
HOLISTER, G. S. and LUXMOORE, A. R., 1968, *Ex. Mech.*, **8**, 210.
LUXMOORE, A. R. and EDER, W. E., 1967, *J. Sci. Inst.*, **44**, 908.
MIDDLETON, E. and STEPHENSON, L. P., 1968, *Ex. Mech.*, **8**, 19.
SCIAMMARELLA, C. A. and DAVIS, D., 1968, *Ex. Mech.*, **8**, 459.
WELLER, R. and SHEPARD, B. M., 1948, *Proc. S.E.S.A.*, **17**, 35.

Part III

Optical metrology and optical processing of data, including coherent light techniques.

Interferometry with Scattered Light

J. M. Burch

National Physical Laboratory

Abstract—If light is made to pass in succession through two random but equivalent partial scatterers, interference can take place between the scattered-transmitted and the transmitted-scattered fractions of the original input. White-light examples of this interference are Newton's diffusion fringes, Quetelet's rings, and the fringes of equal transit time obtained when an optical system under test is arranged so as to image one scatter-plate onto its fellow.

A more general version of this phenomenon occurs in holographic interferometry when 'live fringes' are being observed on a diffusely reflecting specimen. In that case light scattered by the object is seen by transmission through the hologram, and a comparison beam is provided by light which avoids the object but is scattered by the hologram. Most arrangements require a laser source and a hologram which resembles some complicated transform of the object, but production of this transform is achieved automatically.

The double exposure method of holographic interferometry produces 'frozen fringes' which can be regarded as the result of two juxtaposed scatterers operating simultaneously, rather than sequentially, upon the illumination. Even without a reference beam, it is possible to synthesize many types of interference and diffraction pattern by photographic convolution of a random speckle-pattern with a suitable schedule of slightly displaced overlapping and graded exposures.

Interferometer systems can also be devised which combine a highly irregular wavefront such as a speckle-pattern either with a smooth comparison wave or with a second entirely different speckle-pattern. One can no longer expect to observe recognizable fringe patterns, but if the two disturbances are mutually coherent then any gradual change in their relative phase can be detected quite readily in terms of a pronounced local twinkling of the individual speckles. If on the other hand the changes of phase occur rapidly, or if the two disturbances are mutually incoherent, then it is addition of intensity rather than of amplitude that determines the detailed texture and granularity of the resultant speckle-pattern. The theory of these variations in speckle texture and their possible uses are briefly discussed.

INTRODUCTION

When interference phenomena are described in text-books it is usual to find them classified into two groups: those in which the interfering beams are derived by division of amplitude at a partial reflector and those in which a spatially coherent wavefront is sampled by laterally separated apertures. This paper concentrates on an intermediate possibility whereby division of the light is accomplished by means of processes of diffraction or scattering. It is of course possible to use regularly periodic gratings as beam-dividers or recombiners, but we consider here only the case where the diffracting screens are of an aperiodic diffusing type in which absorbing or phase-changing irregularities are distributed at random across the aperture. When collimated light is sent through such a diffuser, the distribution of phase and amplitude across the emerging wavefront is extremely chaotic, and the pattern formed at infinity is a diffraction halo in which all spatial frequencies are more or less equally represented. Let us consider now some of the ways in which it is possible to use light which has been scattered in this way firstly to produce smooth interference effects and secondly (as the term 'interferometry' implies) to obtain useful information.

Interference produced by Successive Action of Partial Scatterers

The classical examples of this phenomenon are Newton's diffusion fringes and Quetelet's rings, which are produced when a collimated beam of white light is reflected back so as to pass twice through the same weakly diffusing scatter-plate. This is a situation which can arise quite easily by accident, for example when there is an accumulation of dust or a moisture film on the front surface of a back-silvered mirror.

An interesting review of these fringes was published recently by A. J. de Witte (1967). The essential feature is that interference takes place between the scattered-transmitted and the transmitted-scattered fractions of the original illumination. The haloes corresponding to these two fractions are seen superposed and covered with fringes of equal inclination, but to some extent the contrast of these fringes is diluted by a slightly wider halo caused by the doubly scattered portion together with some back-scattered light. One of the most noticeable features of the fringe pattern is the doubly-transmitted portion which remains concentrated into a sharp central spot. It is perhaps worth emphasizing that, unless this spot is very bright, the fringe contrast will be poor; a good compromise between brightness and contrast of the scatter fringes is obtained when between 30 and 50 per cent of the original energy remains in the central spot, undeviated by either traversal of the diffuser.

Instead of using the virtual image of a given diffuser to act as the second scatterer, let us now generate, by mechanical or photographic techniques of replication, two separate scatter-plates S_1 and S_2 whose diffraction spectra are substantially the same in every detail. If we assume that our two scatter-plates, each with a substrate thickness t, are sandwiched together and carefully oriented, we obtain a transmission system in which every individual diffracting irregularity on S_1 is accompanied by a closely corresponding irregularity on S_2 displaced normal to the sandwich by a single plate-thickness t. If a distant white source is observed through the sandwich, scatter fringes of equal inclination are observed in transmission, and these are equivalent to the Quetelet's rings which would be formed in reflection from a back-silvered mirror of thickness $(t/2)$.

Interferometry with Matching Scatter-plates

The above experiment, easily performed, demonstrates merely that scatter fringes can be formed by two separate artificially replicated scatter-plates, provided that they have been derived from the same master. In order to obtain useful information from such a pair of scatter-plates, we now interpose an imaging system of unit magnification so that S_2, instead of being located just behind S_1, is made to coincide as nearly as possible with the optical image of S_1. In that case we obtain 'scatter fringes of equal thickness' (Burch 1953) which delineate interferometrically any

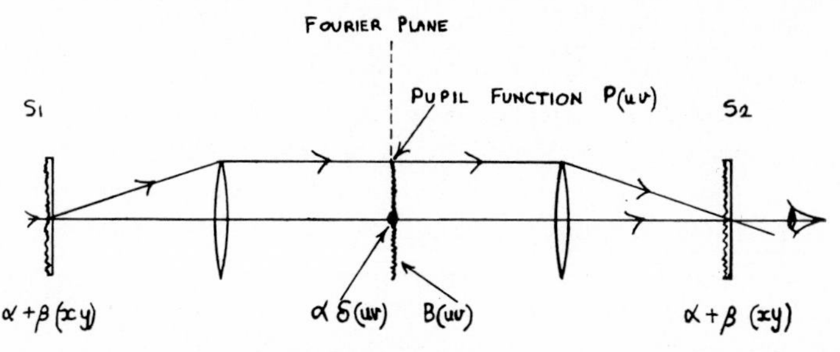

FIG. 1. Use of two matching scatter-plates S_1 and S_2 to test the wavefront error $P(uv) = \exp iW(uv)$ of a through-type imaging system. S_2 is adjusted to coincide in position and orientation with the image of S_1.

variations of optical transit time across the aperture of the imaging system that has been interposed.

Consider for example the simple afocal imaging system shown in Figure 1. Taking x, y-coordinates in the plane of either scatter-plate we can represent their complex amplitude transmittance by the same function $\tau(xy) = \alpha + \beta(xy)$ where α is a complex constant representing the average value of τ and $\beta(xy)$ represents all the local deviations from that average. For a weakly diffusing scatter-plate α will have a modulus close to unity and can without loss of generality be taken as real. For a transparent scatter-plate with small phase-changing irregularities the variations $\beta(xy)$, visualized on an Argand diagram, will be mainly tangential and therefore imaginary, whilst for a plate covered with small absorbing specks the variations will be mainly radial and directed along the real axis.

If we arrange for the first scatter-plate S_1 to be illuminated by a plane coherent wave parallel to the axis, we shall generate in the intermediate Fourier transform u,v-plane an amplitude distribution $T(uv) = \alpha\delta(uv) + B(uv)$ where the δ-function represents undeviated light and $B(uv)$ is the Fourier transform of $\beta(xy)$. We now introduce a complex pupil-function $P(uv)$ which represents firstly the finite aperture of the imaging system and secondly the aberrations which arise within that aperture when a more or less central point on S_1 is being imaged on to the S_2-plane. The disturbance falling on to S_2 is the Fourier transform of $T(uv)$ $P(uv)$ and can be written $\tau(xy) \circledast \pi(xy)$ where π is the Fourier transform of P and the symbol $\circledast$ denotes convolution. What emerges into the observer's eye-pupil after traversal of the second scatter-plate is therefore $\tau(xy) \circledast \pi(xy)\tau(xy)$.

Since however the observer is viewing the intermediate u,v-plane, the fringe pattern that he sees will correspond to an amplitude distribution $A(uv)$ where

$$\begin{aligned} A(uv) &= T(uv)P(uv) \circledast T(uv) \\ &= [\alpha\delta(uv) + B(uv)]P(uv) \circledast [\alpha\delta(uv) + B(uv)] \\ &= \alpha^2 P(oo)\delta(uv) + \alpha P(oo)B(uv) + \alpha B(uv)P(uv) + B(uv)P(uv) \circledast B(uv) \\ &= \alpha^2 P(oo)\delta(uv) + \alpha B(uv)[P(oo) + P(uv)] + B(uv)P(uv) \circledast B(uv). \end{aligned} \tag{1}$$

The first of these terms corresponds to the bright undeviated reference patch. Although it is written here as a delta-function, there must in practice be some blurring, even with a laser source, because of the finite aperture of S_1 which is illuminated.

The second and third terms correspond to the superposed reference halo and error halo, and it is the interference between these two disturbances which generates the wanted fringe pattern.

The fourth term represents doubly-diffracted light which forms a weak background and causes some dilution of fringe contrast. (It is of some interest, however, that if the aberration in the pupil is small, so that $P(uv) \simeq 1$, then the convolution of $B(uv)$ with itself may generate, in addition to the background halo, a sharp central bright spot. With a well-corrected lens system this spot can be observed quite readily if S_2 is adjusted to coincide with the image of S_1 so that the scatter-fringes across the aperture are spread out; a small screwdriver blade or other opaque obstacle is then placed in the centre of the u,v-plane so as to obstruct the direct first term and prevent the second reference term from being generated by S_2. In these circumstances the only light received by S_2 is that which has already been scattered by S_1, and some of this is now 'descattered' into a sharp source-image which has apparently drilled a hole through the screwdriver. The quality of reconstruction depends on what type of scatterer is being employed, but in the case of two identical phase-changing scatter-plates the transform must be such that $B(uv) = -B^*(-u, -v)$. The self-convolution $B(uv) \circledast B(uv)$ is therefore identical to the negative of the auto-correlation function, and this is bound to exhibit a sharp central peak since $B(uv)$ represents a fine-structured random scatterer. The reconstructed image, which can be formed in white light, quickly loses its brightness as soon as the fringe tilt controls are operated to move S_2 slightly out of coincidence with the image of S_1. Weak traces of Zernike fringes can also be seen across the rest of the aperture when this experiment is performed.

Reverting now to equation (1) for the amplitude of the disturbance observed in the u,v-plane, we assume henceforward that the double scattering represented by the fourth term can be neglected in comparison with the second and third terms. Furthermore, since the average value of $\beta(xy)$ is defined as zero, it is known that the halo function $B(uv)$ must vanish in the vicinity of the reference patch $(u=v=0)$. Omitting this reference patch, therefore, we obtain for the observed intensity of the fringe pattern produced by a single source-point at $u=v=0$ the expression

$$H(uv) = \alpha^2 \mid B(uv) \mid^2 \mid P(oo) + P(uv) \mid^2. \quad (2)$$

If we consider only those transmitting regions of the aperture where the pupil function $P(uv)$ can be represented by a pure phase distortion $\exp iW(uv)$, this expression can be rewritten

$$H(uv) = 2\alpha^2 \mid B(uv) \mid^2 \{1 + \cos [W(uv) - W(oo)]\}. \quad (3)$$

In order that the fringe information within the braces shall be displayed as smoothly as possible, it is usually advantageous to replace the single illumination source-point directed towards $u=v=0$ by an incoherent source of finite extent which can be represented by a source function $G(uv)$. The fine structure of the diffraction halo will now appear blurred and smoothed by virtue of the convolution of $G(uv)$ with $\mid B(uv) \mid^2$, and there will be a further blurring in a radial direction if the source is made polychromatic. As far as the fringe pattern is concerned, widening of the source will have no effect on the lateral sharpness with which variations in $W(uv)$ are delineated, but the visibility of the whole fringe pattern will be spoilt if there is any appreciable variation in the phase angle $W(uv)$ within the finite illuminated area over which the reference patch now extends. As the source function $G(uv)$ is changed, the whole fringe pattern will lose visibility and change phase in accordance with the modulus and argument of the complex expression

$$\gamma_S = \frac{\int G(uv) \exp iW(uv) du dv}{\int G(uv) \, du dv}. \quad (4)$$

In a typical scatter-plate interferometer a satisfactory smoothness and brightness of the fringe pattern is achieved by using a directly illuminated reference patch $G(uv)$ whose diameter is between 1 and 2 per cent of the aperture being studied. Loss of fringe visibility may occur if the reference patch is covered by an obstruction or by strong local phase distortion, but it is usually possible to cure this by moving the reference patch off-axis to a smoother region of the fringe pattern. Difficulty may also arise in finding the fringes initially, but with most systems an approximate coincidence of S_2 with the image of S_1 can be accomplished by using the images of bold guide-marks which are engraved for this purpose on the periphery of the master from which the scatter-plates are replicated.

Another difficulty arises because in general the aberration of the imaging system being studied will be a function not only of position within the aperture but also of position in the plane of the scatter-plates. Particularly in the through type system illustrated in Figure 1, the aberration function $W(uvxy)$ may contain not only an arbitrary on-axis aberration term but also several terms representing different kinds of off-axis aberration. Furthermore, until the interferometer has been adjusted there will be tilt and focus terms $c_1u + c_2v + c_3(u^2 + v^2)$ and misalignment terms $c_4(xu + yv) + c_5(xv - yu)$. c_4 vanishes only if the magnification of the imaging system is adjusted exactly to unity, and c_5 represents a residual misorientation of S_2 within its own plane. In order to find fringes initially it is often helpful to illuminate only a restricted region of S_1 and to watch carefully for fringes in the vicinity of the reference patch. Once fringes have been found by searching on the tilt controls it is usually a simple matter firstly to spread the fringes out by the tilt and focus controls and then to reduce c_4 and c_5 to zero by adjusting magnification and orientation about the optical axis. The illuminated area on S_1 is then increased up to a limit determined usually by the effects of off-axis aberration.

One of the most useful applications of scatter-plate interferometry is the testing of large auto-collimating systems. Adjustment in this case is rather simple since errors of magnification cannot arise, and the correct orientation of S_2 with respect to S_1 is ensured once and for all by using a single photographically manufactured '180°-scatter-plate'. In the manufacture of such a plate, two superposed Fresnel images of the ground glass master surface are recorded on the same fine-grained photographic plate, a special jig being used to ensure that the second exposure is made after the master has rotated exactly through 180°. After development the plate is bleached either to give a surface relief image or, if uniform diffraction haloes more than 20° in diameter are required, to give a modulation of refractive index inside the volume of the gelatin.

If a carefully made 180°-scatter-plate is used to test a large concave mirror from its centre of curvature, an extremely faithful contour map of the residual asphericity can be obtained. If the asphericity of the mirror is appreciable, however, there is bound to be lateral aberration of the returning rays, and care must be taken to use a well-corrected viewing system in order to observe or photograph the fringe pattern; the aberrations of this viewing system necessarily include any variations in optical thickness of the scatter-plate itself and will be denoted by $\Omega(xy)$. It can be shown that, because of these imperfections in the viewing system and the off-axis aberrations previously mentioned, the fringe pattern finally observed will lose visibility and change phase in accordance with the modulus and argument of the complex factor γ_A where

$$\gamma_A = \frac{\iint_S \exp i\left[W(uvxy) + f\left(\frac{\delta W}{\delta u}\right)\left(\frac{\delta \Omega}{\delta x}\right) + f\left(\frac{\delta W}{\delta v}\right)\left(\frac{\delta \Omega}{\delta y}\right)\right] \mathrm{d}x\mathrm{d}y}{\iint_S \mathrm{d}x\mathrm{d}y}. \tag{5}$$

Here S denotes the illuminated area of the first scatter-plate and $f = \lambda\rho/2\pi$.

A valuable feature of scatter-plate interferometry is the fact that each small region $\mathrm{d}x\mathrm{d}y$ of the scatter-plate contributes equally to the whole of the fringe pattern observed, and conversely each part of the latter averages the contributions from a scattering area S that may be up to 10 mm in diameter. Even if emulsion shift occurs locally during processing of the scatter-plate, the effect will be a smoothly distributed loss of fringe visibility rather than a systematic distortion of the fringe pattern. If fringes of good contrast are observed, a measurement accuracy of $\lambda/50$ may be feasible provided that atmospheric effects do not supervene. This has been demonstrated and achieved by the careful work of R. M. Scott and his colleagues at Perkin Elmer Corporation (1962) on testing the 36-in. diameter F/4 paraboloid for the Stratoscope II project.

The author's early work on scatter fringes of equal thickness employed scatter-plates that were replicated either by vacuum evaporation through a mask or thermally by moulding small plates of Perspex (methyl methacrylate) against a fine-ground steel master-surface. The test was therefore restricted to optical systems of unit magnification, and an asymmetric system could be tested only by (*a*) cascading it with a similar system (*b*) double-passing it with a (-1) autostigmatic arrangement; or (*c*) double-passing it with a $(+1)$ enantiostigmatic arrangement. The last-mentioned arrangement gives double sensitivity for even aberrations of the system; under test but removes entirely any odd aberrations, the fringe pattern observed being a contour map of $[W(uv) + W(-u, -v) - 2W(00)]$ (Burch 1953; Shoemaker & Murty 1966). In some cases it becomes difficult to interpret the zonal error in a symmetric fringe pattern, and it is worthwhile to use two separate scatter-plates S_1 and S_2 in conjunction with a beam-splitter. By displacing S_2 slightly with respect to the virtual image of S_1 it is possible to introduce a preset fringe tilt (c_1u) or (c_2v) into the above aberration function, without losing the automatic compensation of tilt as well as path length that is characteristic of the enantiostigmatic arrangement.

With photographic methods of manufacture, however, it becomes possible to produce a matching pair of scatter-plates suitable for testing magnifying or demagnifying optical systems, and no doubt this possibility will be exploited eventually. More important, however, is the possibility which we now have of producing holographically a 'matched pair' which will produce smooth fringes from an 'optical system' containing arbitrarily irregular aberrations.

Holographic Interferometry

Let us consider from this point of view the observation of 'real-time' or 'live' fringes on a diffusely reflecting object in a holographic interferometer. In a general way, the object can be regarded as a 'first scatterer' which generates a wavefield of extremely complicated shape in the vicinity of the hologram. In most cases the object is not a partial diffuser and the reference wave used to record the hologram is derived from an arrangement of mirrors that is physically separate from the object. Once the hologram has been recorded, however, it can be regarded as a 'second scatterer' which (*a*) transmits the wavefield generated by the 'first scatterer'; and (*b*) generates by diffraction from the reference beam an exactly similar wavefield. The second scatterer, instead of being identical to the first, resembles some complicated Fresnel or Fourier transform which expresses the way in which a wavefield progressively deforms on its journey from object to hologram. No difficulty arises in generating the required transform; all that is necessary is to use coherent laser light and to use the same geometric locations for both recording and reconstruction.

What we are asserting here is that a scatter-plate interferometer can be considered as a very special case of the more general and now realizable holographic interferometer. Surprisingly enough, however, we have found that there are still advantages to be gained for some purposes by performing holography in a set-up that resembles closely that shown in Figure 1. If for example the second scatter-plate is replaced by a sensitive holographic plate, and the reference source-point is shifted to the edge of the aperture, a hologram can be recorded which functions as a scatter-plate does but corrects in addition for all the residual errors of the imaging system. In the study of shock waves, for example, the arrangement can be used by the double-exposure technique to produce conventional 'frozen fringe' holographic comparisons (Burch *et al.* 1966), or to make interferometric comparisons between separately recorded holograms (Gates 1968*a*), and both of these tasks can be performed with multi-mode pulsed lasers (even, in more static examples, with incoherent light sources!). Several of these possibilities are being pursued at the N.P.L. by J. W. C. Gates and his colleagues (1968*b*).

Related to this work is the discovery by L. H. Tanner (1969) that, if the reference beam in Figure 1 is left central, then a *singly* exposed hologram will produce, upon reconstruction in a separate apparatus, a fringe pattern representing double the symmetric error in $W(uv)$. This occurs because on reconstruction the Fourier reconstruction $B(uv)\ P(uv)$ is seen superposed upon its conjugate reconstruction $B^*(-u, -v)\ P^*(-u, -v)$. Remembering that for a phase-changing scatter-plate $B(uv) = -B^*(-u, -v)$, and putting $W(uv) = W_e(uv) + W_0(uv)$, where W_e is an even function $\frac{1}{2}[W(uv) + W(-u, -v)]$ and W_0 is the corresponding odd function $\frac{1}{2}[W(uv) - W(-u, -v)]$, we obtain for the reconstructed amplitude

$$\begin{aligned} B(uv)P(uv) + B^*(-u, -v)P^*(-u, -v) \\ &= B(uv) \exp iW_0(\exp iW_e - \exp - iW_e) \\ &= B(uv) \exp iW_0\ 2i \sin W_e. \end{aligned} \qquad (6)$$

The brightness of the fringe pattern is therefore proportional to $|B(uv)|^2\ 2(1 - \cos 2W_e)$. With a purely absorbing type of scatterer, we shall expect the interference term to be of the form $(1 + \cos 2W_e)$, but with a 'mixed scatterer' the fringes produced by this single exposure technique will be of poor contrast.

Photographic Convolution by Multiple Exposures

Mention has been made above of holographic 'frozen fringe' methods whereby two or more exposures to the same wavefield are recorded at different epochs on the same photographic plate. When the plate is developed and placed in a reconstructing system, all of these sequentially recorded disturbances are brought to life simultaneously and are made to interfere with each other. When two mutually displaced versions of the same disturbance are recorded holographically with the aid of a reference wave, the displacement may be either lateral or longitudinal, and the result will be well-defined straight or curved fringes at infinity which may be regarded as the Fourier transform of the two-point distribution representing the two exposures.

For some purposes, however, it may not be convenient to provide a reference beam, and it is only the intensity pattern that is recorded at each exposure. Provided that all the displacements are in a direction parallel to the emulsion surface, all of the mutually displaced intensity patterns actually recorded will be the same, and the final exposure energy can be regarded as a convolution of $I(xy)$ the intensity pattern in question with a distribution $\sum_n t_n \delta(x - x_n, y - y_n)$ where $x_n y_n$ is the relative position on the emulsion surface of the nth exposure and t_n is its duration.

When such a plate is developed and viewed in a reconstruction apparatus, the diffraction amplitude at infinity will be the halo amplitude corresponding to a single exposure multiplied by the Fourier transform of the distribution representing the total exposure. By choosing an appropriate schedule of exposures, therefore, the diffraction pattern corresponding to any arbitrary array of pinholes in one or two dimensions can readily be synthesized (Burch & Tokarski 1968; Debrus *et al.* 1969). To synthesize by this means the diffraction halo of continuous functions it is necessary either to change the displacement continuously during the exposure or to use spatially incoherent light in a somewhat different pinhole camera arrangement.

Perhaps the main interest of this type of experiment lies in the simplicity with which it may be set up for instructional purposes. When setting up to manufacture scatter-plates of the 180° type, we have often found it a convenient procedure to displace the master by about 50 μm, rather than rotate it through 180° for the second exposure. The result after bleaching is a beautiful set of white-light Young's fringes, which can be a great help in indicating, firstly what are the optimum conditions for processing these scatter-plates, and secondly over what sort of cone we can expect to produce a satisfactory fringe pattern when testing an auto-collimating system. Nevertheless, since the 180°-scatter-plates operate upon the light twice in succession, rather than once in parallel, their scattering power must be made somewhat less than that which gives best results in the above test with Young's fringes.

Since no reference beam is required for this convolution process, the requirements for coherence in the source are not very stringent, particularly if the photographic plate is kept close to the diffuser. It is possible to use a mercury lamp in most cases.

Interferometry with Speckle-Patterns

In all of the situations considered so far, the wavefronts which interfere have been scattered and are therefore highly irregular, but they match each other closely in every detail so that their phase difference produces recognizable patterns of smooth interference fringes. During the last seven years, however, the availability of continuously-operating visible lasers has brought into prominence the pronounced and rather irritating granularity that results when the phase relationships in a coherent beam are scrambled by an irregular diffuse reflector. Provided that these phase relationships do not change with time, the result that we observe is a fine-grained speckle-pattern, which contains all spatial frequencies up to the highest that our observing system is capable of resolving (Goldfischer 1965). Furthermore, as was shown in Rayleigh's treatment of 'Random Flights in Two Dimensions' (the drunkard's walk), the probability distribution for the resultant amplitude at any point in such a pattern is Gaussian, so that for the resultant intensity the distribution follows a negative exponential law (Rayleigh, Goodman 1965). It follows that the most probable brightness of a point sampled at random from a speckle-pattern is close to zero, and nowhere near its average brightness. To the observer, therefore, there is an obvious difference of texture between a well-developed speckle-pattern and a similar reflector lit by incoherent light, for which the probability distribution of brightness is almost a δ-function centred about the average value. It is this sort of difference in texture or contrast that the observer must learn to appreciate if he is going to extract information visually from a 'speckle interferometer'.

The basic idea of speckle interferometry (Archbold *et al.* 1969) is to combine optically a speckle-pattern either with a second speckle-pattern or with a smooth reference wave of comparable brightness. One can no longer expect to observe recognizable fringe patterns, but if the two superposed disturbances are mutually coherent they will produce what is effectively a single well-developed speckle-pattern. If we think in terms of probability distributions of amplitude upon an Argand diagram, the new distribution is the convolution of the two original distributions. Furthermore, if the two disturbances remain coherent in relation to the integration time of the eye but their mutual phase changes slowly, this will become apparent to the observer in terms of a pronounced local twinkling of the individual speckles. If on the other hand the changes of phase occur too rapidly, or if the two disturbances are mutually incoherent, the amplitude cross-products between the two disturbances give a vanishing time-average, and it is addition of intensity rather than of amplitude which determines the resultant pattern. If we define each disturbance in terms of a probability distribution along a real positive intensity axis, the resultant distribution is again the convolution of the two original distributions.

Let us now consider some of the probability distributions that result from convolution either on the Argand diagram or on the intensity axis. In addition to discussing the limiting case of a very large number of elementary disturbances, Rayleigh gave expressions which represent in terms of infinite integrals of zero-order Bessel functions the probability distributions to be expected from a finite number n of equal or unequal randomly phased disturbances. In particular, for n equal disturbances, the probability of brightness between (x) and $(x+\mathrm{d}x)$ is $W_n(x)\mathrm{d}x$ where

$$W_n(x) = n/2\int_0^\infty zJ_0(z\sqrt{xn})\cdot[J_0(z)]^n\mathrm{d}z. \tag{7}$$

We have used here a relative brightness variable $x = \mathrm{I}/\mathrm{I}_0$ and $\mathrm{W}_n(x)$ is normalized so that

$$\int_0^\infty W_n(x)\mathrm{d}x = 1 \tag{8}$$

and

$$\int_0^\infty xW_n(x)\mathrm{d}x = 1. \tag{9}$$

Equation (9) shows that $x=1$ corresponds to the average brightness of each distribution. In a speckle-pattern produced by n equal disturbances, the *absolute* average brightness will be n times the brightness produced by a single disturbance. If all n disturbances are in phase, as might happen with locally specular polished regions of a single reflecting surface, then the absolute brightness increases to n^2 times that for a single disturbance. It follows that the probability function $W_n(x)$ must vanish for $x>n$.

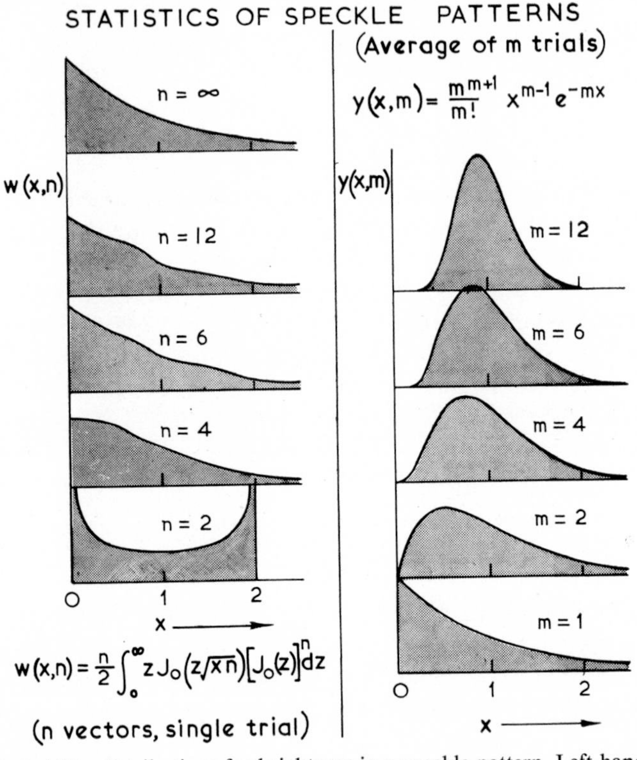

FIG. 2. Probability distributions for brightness in a speckle-pattern. Left-hand column; n equal but randomly phased disturbances. Right-hand column; m incoherently superposed speckle-patterns.

For the trivial case $n=1$ the integral in Equation (7) reduces to Weierstrass' integral and gives a delta-function which vanishes except at $x=1$. For $n=2$ the integral can be evaluated to yield

$$W_2(x)=[\pi\sqrt{x(2-x)}]^{-1}. \tag{10}$$

This distribution is shown at the left of Figure 2 together with approximate curves obtained by numerical integration for the cases $n=4$, 6 and 12. It will be seen that as n increases the distribution changes rapidly from the double-peaked curve for $n=2$ into a monotonically decreasing function similar to the negative exponential $\exp -x$ which corresponds to $n = \infty$. The case $n = 2$ corresponds for example to random sampling of a two-beam interference pattern where peaks and troughs are the turning-points most likely to be encountered. The case $n = 3$ can be evaluated in terms of complete elliptic functions and exhibits a singularity at $x = \frac{1}{3}$.

It is instructive to consider the behaviour of these distributions, as n is increased, in terms of convolution operations upon the Argand diagram. Let $W_n(r)$ represent the probability distribution on this diagram produced by n equal disturbances. If one more such disturbance is added, then the new probability distribution is obtained by convolution as

$$W_{n+1}(r) = W_n(r) \circledast W_1(r) \tag{11}$$

and similarly by induction

$$W_n(r) = W_1(r) \circledast W_1(r) \ldots \circledast W_1(r) \tag{12}$$

in an n-fold convolution.

When only one randomly phased disturbance is used for each trial, the distribution function $W_1(r)$ is a delta-function which produces on the Argand diagram a sharply-defined ring of radius corresponding to the fixed amplitude a of each disturbance; it is an n-fold convolution of this function with itself which determines the final distribution $W_n(r)$.

The convolution theorem states that the resultant of a convolution process has a Fourier transform which is the product of the individual Fourier transforms. Furthermore, since all the distribution functions with which we are concerned are two-dimensional functions which, because all phases are equally probable, have revolution symmetry, each function $W(r)$ corresponds to a zero-order Fourier-Bessel transform

$$\Omega(\rho) = \int_0^\infty r W(r) J_0(r\rho) \mathrm{d}r \tag{13}$$

and reciprocally

$$W(r) = \int_0^\infty \rho \Omega(\rho) J_0(r\rho) \mathrm{d}\rho. \tag{14}$$

Inserting the annular delta function for $W_1(r)$ in (13), we find

$$\Omega_1(\rho) = aJ_0(a\rho). \tag{15}$$

On putting

$$\Omega_n(\rho) = [\Omega_1(\rho)]^n$$

and substituting this in equation (12), we obtain

$$W_n(\rho) = \int_0^\infty \rho J_0(\rho r) a^n [J_0(a\rho)]^n \mathrm{d}\rho \tag{16}$$

in agreement with Rayleigh's result.

Also shown in Figure 2 in the right-hand column are the intensity distributions that result when m mutually incoherent speckle-patterns of the same average strength are seen combined. The formula for this so-called gamma variate is given by Rayleigh, and the implications are discussed, with particular reference to optical radar performance, in Goodman's valuable paper (1965). As far as speckle interferometry is

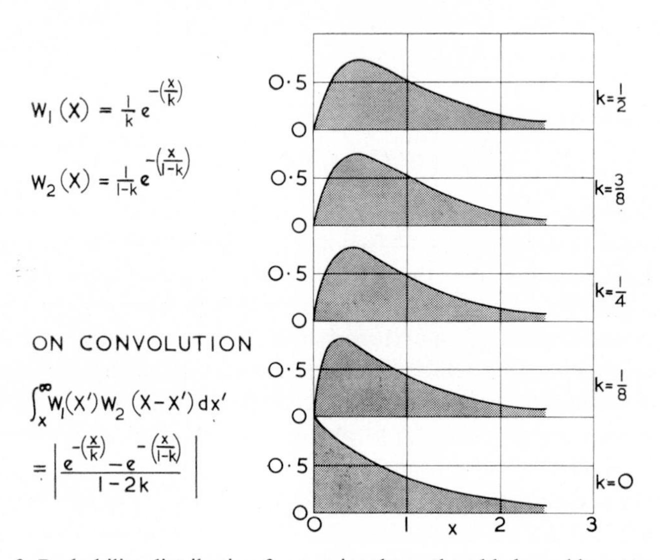

FIG. 3. Probability distribution for two incoherently added speckle-patterns.

concerned, it is principally the case $m=2$ which is of interest. Figure 3 shows in more detail the change in the distribution when one speckle-pattern is progressively 'diluted' by a second mutually incoherent pattern, the relative brightness of the two patterns being in the ratio K to $(1-K)$. The most noticeable feature is in the vicinity of $x=0$; as soon as a second independent pattern is added, it is most unlikely that the truly black regions of the first pattern will coincide with the truly black regions of the second.

With the help of a coarsely ground birefringent plate and a rotatable polarizer set at angle θ to the optic axis, it is not difficult to set up a composite speckle-pattern of this type, in which $K=\sin^2\theta$ and $(1-K)=\cos^2\theta$. We have found that, after a period of practice, many but not all observers are capable of distinguishing, on an absolute memory basis, between a fully developed speckle-pattern for which $K=0$ and a diluted mixture of two patterns for which $K=\frac{1}{8}$. The observation is a demanding one, and it would be difficult to continue making such absolute judgements of speckle texture for more than half an hour at a time.

Our first attempt to extract useful information from such judgements was a mode multiplicity detector constructed by the author's colleague R. W. E. Cook. For detecting multiplicity of transverse modes, e.g. in the output from a powerful argon laser, all that is necessary is to take some of the light and feed it through several closely-stacked ground-glass diffusers, thereby thoroughly redistributing the light laterally but without introducing large path differences. If the resultant pattern is viewed through a polarizer, it should show a fully-developed speckle texture if all of the output is in the central Gaussian mode. If, however, there is more than about 15 per cent energy in the surrounding 'doughnut' mode, an appreciably greyer appearance is observed. For detection of axial mode multiplicity, light from the laser is sampled by means of two incoherent fibre-bundles, one of which is much longer than the other. Light emerging from the juxtaposed ends of the two bundles is repolarized, and one then inspects the resultant speckle-pattern; each bundle on its own generates an irregular blotchy pattern of rather coarse structure, but if there is interference between the left-hand and the right-hand output circles this generates within each blotch a set of several vertical Young's fringes whose visibility can be assessed, approximately but at a glance. By providing several pairs of fibre-bundles with different path lengths, one achieves a compact and portable version of Michelson's interferometer for measuring visibility as a function of path-difference. The fibre-bundles are wound round the outside of the diffuser stack and provide up to $2\frac{1}{2}$ m of path-difference—sufficient to assess a laser of that length.

Another possibility that has been explored at N.P.L. (this volume p. 265) is that of interfering a speckle-pattern against a smooth reference wave (Archbold *et al.* 1969). In this case it is primarily movement or rapid vibration of a diffusely reflecting surface that can be detected. The distributions shown in Figure 4 indicate the effects of coherent and incoherent combination of the two disturbances, and here one obtains a very considerable difference in that the incoherent pattern shows no black regions at all. It is possible that the best contrast is obtained when 70 per cent of the incident energy comes from the reference wave and only 30 per cent from the speckle-pattern, but in the instrument that we have constructed this ratio can be adjusted to suit the observer and the surface being studied. Figure 5 illustrates the way in which the coherent distribution changes into the incoherent distribution as the partial coherence factor γ between the two disturbances is reduced. For a surface vibrating rapidly at a single frequency, the factor γ corresponds to the usual J_0 factor encountered in time-averaged holography.

At this juncture the applications of interferometry with speckle-patterns

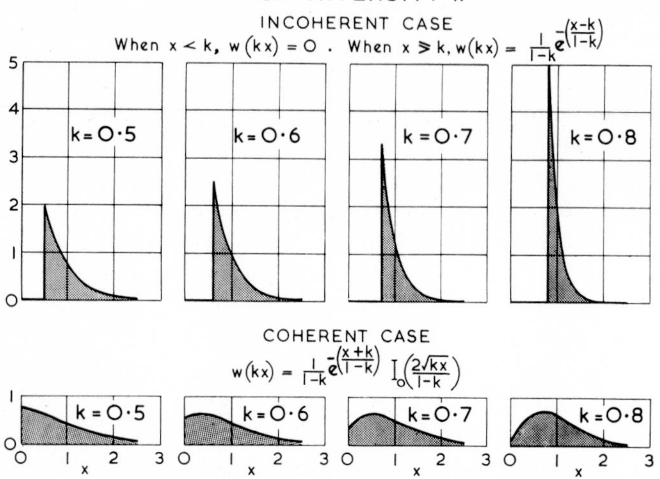

FIG. 4. Probability distributions for coherent and incoherent combination of a speckle-pattern with a uniform reference wave.

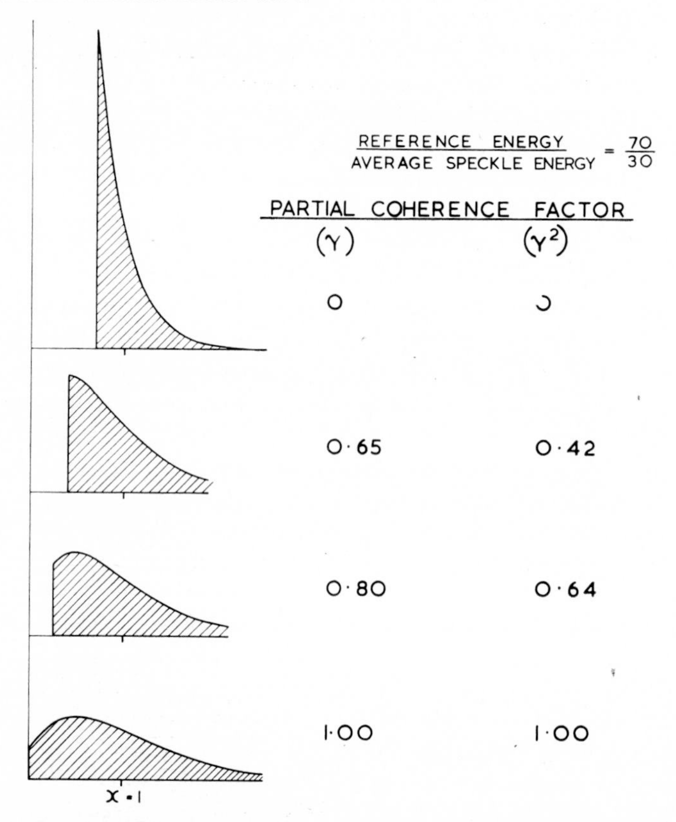

FIG. 5. Effect of partial coherence factor γ on the combination of a speckle-pattern with a uniform reference wave.

are still difficult to assess, but a new application to strain measurement will be described by Leendertz (this volume p. 256) and seems to be a considerable advance. Further work on photographic convolution will be reported by Françon and his colleagues from Institut d'Optique (this volume p. 319). One may therefore hope that interferometry with scattered light, which attracted so much interest in the past, still has much to offer for the future. Let us, in de Witte's words, 'recapture that bygone luster'.

REFERENCES

ARCHBOLD, E., BURCH, J. M., ENNOS, A. E. and TAYLOR, P. A., 1969, *Nature*, **222,** 263.
BURCH, J. M., 1953, *Nature*, **171,** 889.
BURCH, J. M., GATES, J. W., HALL, R. G. N. and TANNER, L. H., 1966, *Nature*, **212,** 1347.
BURCH, J. M. and TOKARSKI, J. M. J., 1968, *Optica Acta*, **15,** 101.
DEBRUS, S., FRANCON, M., MALLICK, S., MAY, M. and ROBIN, M. L., 1969, *App. Opt.* **8,** 1157.
GATES, J. W. C., 1968, *Nature*, **220,** 473.
GATES, J. W. C., 1968, *J. Sci. Inst.*, Series 2, Vol. I, 989.
GOLDFISCHER, L. I., 1965, *J. Opt. Soc. Am.*, **55,** 247.
GOODMAN, J. W., 1965, *Proc. I.E.E.E.*, **53,** 1688.
LIENDERTZ, J. A., 1969, this volume p. 000.
RAYLEIGH, ELDER, LORD, Scientific Papers, C.U.P. VI, 610 and 565.
SCOTT, R. M., 1962, *App. Opt.*, **1,** 396, and 1969, *App. Opt.* **8.** 531.
SHOEMAKER, A. H. and MURTY, M. V. R. K., 1966, *App. Opt.*, **5,** 603.
TANNER, L. H., 1969, *J. Phys. E.*, **2,** 288.
DE WITTE, A. J., 1967, *Am. J. Phys.*, **35,** 301.

Angular Displacement Measuring System

G. Bouwhuis

Philips Research Laboratories,
N. V. Philips' Gloeilampenfabrieken,
Eindhoven, Netherlands

Abstract—An optical system for the incremental measurement of rotation has been designed.

A section of a radial grating is imaged onto a diagonally positioned section with the aid of concave and folding mirrors. Use is made of the two first-order diffracted beams. Rotation of the grating results in mutual phase changes of these beams; when they are combined the intensity becomes a periodic function of the angle of rotation. The period of the photo-signal is a quarter of the grating period. With polarization-optical means a second signal is obtained which is 90° out of phase. This enables the detection of the sense of rotation. Electro-optic modulation has been added to overcome the difficulties at standstill.

With a grating of 64 sec. arc, period digital measuring steps (zero-passages of the signals) of 4 sec. arc are obtained. Further analogue interpolation with a factor of 10 is certainly meaningful. In the prototype the grating diameter was 160 mm with a hole for the axis of 70 mm diameter. The optical elements had a height of 24 mm above the grating.

INTRODUCTION

The object of our investigations on the measurement of rotation is to develop a photo-electric system with digital read-out and with an accuracy better than one second of arc. We started from the linear displacement measuring systems devised in our laboratories (de Lang, Ferguson & Schoenaker 1969) and based on the principles given at the National Physical Laboratory (J. M. Burch 1963) by Guild (1956, 1960). In the linear systems we employed optical gratings and owing to the use of fine diffraction gratings and digital interpolation by optical means we were able to avoid high electronic interpolation factors. Phase gratings were chosen because they yield higher intensities for the signal than absorptive gratings. We use radial reflection gratings for the advantages of manufacture. In the optical system only the two first-order beams of the diffracted light are used. This yields an effective grating period which is a factor of 2 smaller, and, more important, the symmetry precludes defocusing errors and ensures a high degree of modulation in the signal. The principle of eliminating the odd harmonics of the grating errors (to which the centring errors belong) by means of averaging two readings disposed 180° apart, is put into practice by imaging a grating section on to the diametrically positioned area. The imaging system consists of a concave mirror and a correction lens (Dyson 1959) together with auxiliary folding mirrors so that ample space is left for a shaft. In order to obtain directional sensitivity two signals are derived which have a phase difference of 90°. This is accomplished by introducing a shift between the image made in parallelly-polarized light and the image made in perpendicularly-polarized light. The use of discrete polarizations also allows for electro-optical modulation, thus avoiding the difficulties with dc electronics at stand-still.

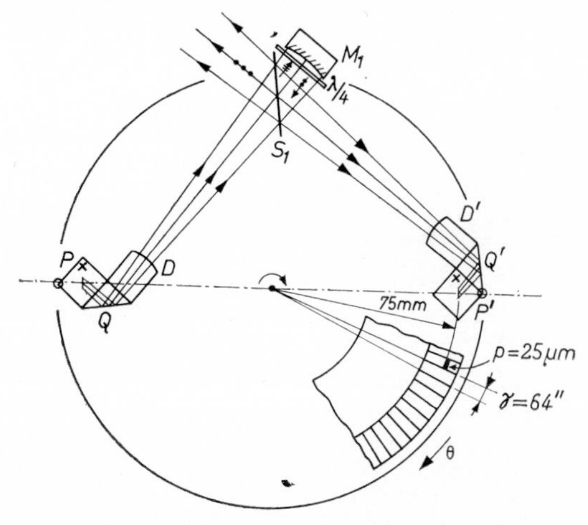

FIG. 1. A grating section is imaged onto the diametrically opposite section. The image-forming element is the concave mirror M_1 used in the neighbourhood of its centre of curvature. With the polarization-separating mirror S_1 the system is partly doubled.

DESCRIPTION OF THE ARRANGEMENT

Diametrical Imaging

The radial grating is 17 cm in diameter; at the peripheral region it carries 20,250 grooves; this corresponds to a period γ of 64 seconds which is in linear measure a pitch of 25 μm. With the reading head a section of about 15 mm is imaged on to the diametrically opposite position (Figure 1). The prism P with the reflecting surface at 45° to the grating plane brings the light-path parallel to the grating and the prism Q directs it to the concave mirror M_1 (the centre of curvature of M_1 lies in the object plane). The polarization-separating mirror S_1 reflects the perpendicularly-polarized component of the light and transmits the parallel component to M_1. After reflection at the concave mirror the rays again arrive at the polarizing mirror S_1, but due to a quarter-wave plate in a diagonal position placed between S_1 and M_1 the polarization has become perpendicular and hence the beam is now reflected by S_1. Via the prism P' and

Q' (identical with P and Q) the image is formed on the grating. With this arrangement the image matches the grating and the movement is in the direction opposite to that of the grating. Of course, the image quality is of great importance. In the neighbourhood of the centre of curvature of a concave mirror the aberrations are small. Dyson (1959) has shown that a much larger field can be used with the aid of one simple correcting lens. In our arrangement this correcting lens is formed by the prisms P and Q and the cemented lens D, of which the outer surface is concentric with the mirror M_1. The aberrations are now sufficiently reduced for our purpose.

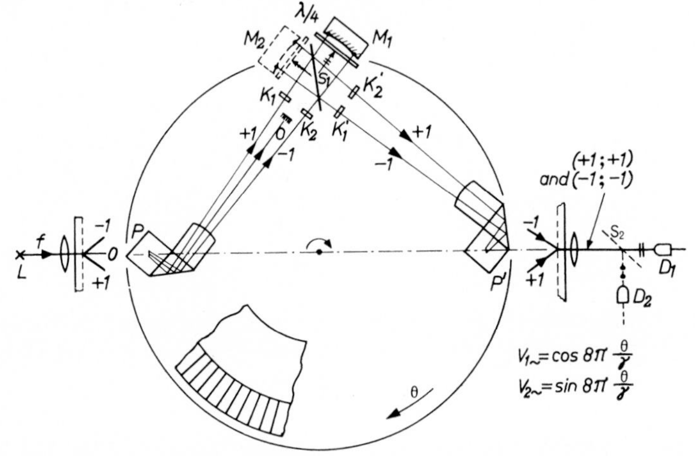

FIG. 2. (a) Only the two first-order beams of the light diffracted at the grating pass through the system.

The parallel polarized part of the incoming light travels via mirror M_1 and the other part via M_2; M_2 is tilted so that two signals with 90° phase difference are obtained.

With the electro-optical crystals K the phase between the $+1$ and -1 order beam is given an auxiliary modulation.

As the system is a telecentric one, the magnification is insensitive to small local changes in the distance between the grating and the optics.

Interferometric Aspects

We considered the optical system as an imaging device, but in some aspects it is more convenient to treat it as a two-beam interferometer. The schematic Figure 2*a* shows the reading-head optics for the case of a transmission phase grating. A beam of nearly collimated light from the lamp L is diffracted at the grating. The zero-order beam is screened off; only the diffracted beams of order $+1$ and -1 are used. They pass through the optical system and are recombined at the diametrically opposite grating section where a second diffraction takes place. Here too, we select the first-order diffracted beams and only those propagated in the axial direction. As these beams have nearly equal optical path length they can interfere; thus their relative phase changes cause intensity fluctuations which are detected by the photosensitive element D_1. If the grating rotates through an angle θ the phase of the -1 order beam decreases and the phase of the $+1$ beam increases by an amount $2\pi\theta/\gamma$ (Figure 2*b*). It can

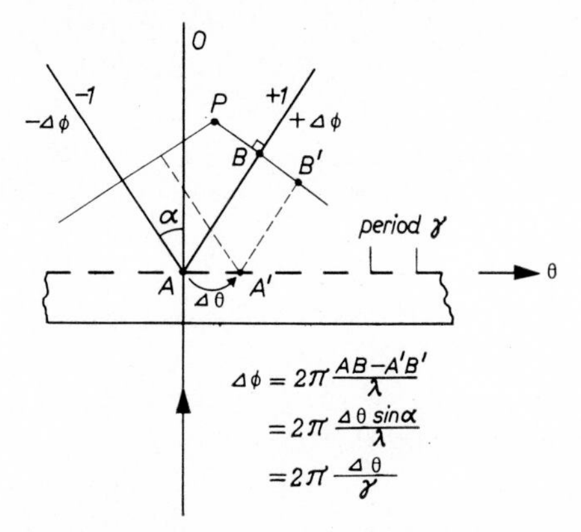

FIG. 2(*b*) At a fixed reference point *P* the phase of a diffracted wave is a linear function of the grating displacement.

be shown (de Lang *et al.* 1969) that the phase difference ϕ between the beams $(+1;\ +1)$ and $(-1;\ -1)$ depends on θ according to the expression:

$$\phi = 8\pi\theta/\gamma \tag{1}$$

As the phase difference is independent of wavelength, white light can be used.

The detected signal V_1 is thus of the form:

$$V_1 = 1 + b \cos 8\pi\theta/\gamma \tag{2a}$$

The beam components $(+1;\ +1)$ and $(-1;\ -1)$ from a grating with a symmetrical groove shape are nearly equal, which makes a large modulation depth possible (b close to 1). The period of the signal is a factor of 4 smaller than the grating period; in our case $\gamma/4 = 16$ sec of arc.

Directional Sensitivity

Only the parallel polarized part of the light has been used. With the perpendicular component a similar ray path with the mirror M_2 has been interlaced; thus the arrangement now contains two distinct information carrying beams. At the detection side the polarization mirror S_2 separates the components and the detector D_2 yields a second signal V_2. The concave mirrors are adjusted so that this signal is in quadrature with respect to the first one:

$$V_2 = 1 + b \sin 8\pi\theta/\gamma \tag{2b}$$

Two signals with a phase difference enable us to discriminate between clockwise and anti-clockwise rotation.

Phase Modulation

In order to avoid troubles with drift in the light source, in the detectors and in dc electronics at stand-still of the grating, the system will be provided with phase modulation. For a reasonable rate of rotation and fast control a high modulation frequency is desirable; this excludes the use of mechanical systems, and therefore magneto-optical or electro-optical means have to be used. The elements K_1 and K_2 (Figure 2*a*) represent electro-optical crystals. The common azimuth of the crystals K_1 is parallel to the principal plane of the mirror S_1 and that of K_2 is perpendicular to that plane. If the same voltage is applied to the crystals, the $+1$ order wave and the -1 wave will undergo an opposite phase change, both for the parallel and the perpendicularly polarized component. This phase shift adds to the phase due to grating rotation, so that the ac part of the signals becomes:

$$V_{1\sim} = \cos(8\pi\theta/\gamma + 4m \sin \Omega t) \tag{3a}$$

$$V_{2\sim} = \sin(8\pi\theta/\gamma - 4m \sin \Omega t) \tag{3b}$$

where Ω is the modulation frequency (typical $\Omega/2\pi = 1$MHz) and m is the amplitude of the phase shift in a crystal (for KD_2PO_2-crystals $4m \approx 1$ radian when the voltage is 250 Volt rms).

In the electronic circuitry the photo-signals are converted into:

$$W_1 = \sin(\Omega t + 8\pi\theta/\gamma) \tag{4a}$$

and

$$W_2 = \sin(\Omega t - 8\pi\theta/\gamma) \tag{4b}$$

that is, the time-harmonic modulation has been translated into a time-linear modulation. As we do not need a high interpolation factor the demands on the circuitry are not extremely severe. The zero-passages of W_1 and W_2 successively coincide when the angle of rotation increases by $\gamma/16$, corresponding to 4 sec of arc.

Interpolation

A method of interpolation to one second of arc is the following: From a 4 MHz generator two signals can be derived; one is a clock pulse of 8 MHz, which is compared with W_1 and W_2 and the other a 1 MHz sinusoidal signal for the electro-optical crystals derived by frequency division. As the clock pulse and the crystal voltage are derived from the same source the phase relation of the two can be made very constant so that reliable interpolation of the zero-passages of W_1 and W_2 is achieved. Because 80 MHz clock pulses are difficult to handle, interpolation to

0·1 sec can better be done with a lower modulation frequency Ω. This lower modulation frequency is not inconvenient because a high accuracy seldom goes together with a high rate of rotation.

A rate of rotation of 1 revolution per second contributes to the frequency of the signal W by 81 kHz. In order not to exceed the bandwidth limitations of the circuitry this calls for a modulation frequency $\Omega/2\pi$ of at least 800 kHz.

The Final Design

Figure 3 shows the arrangement which is under construction. Instead of transmission gratings we use reflection gratings. To allow the light to enter and to leave the arrangement transparent slits are present in the mirrors. There are no further differences in this system with respect to the transmission system described.

The gratings are obtained by photo-etching techniques. The depth of the grooves is chosen such that the intensity of the zero-order beam is practically zero; in this case the first-order beams have maximum intensity. 64 per cent of all diffracted light can reach the detectors.

ERRORS

Sources of error are present in all three parts: in the reading head, in the grating and in the electronic circuitry. Owing to the fact that the electronic interpolation is fairly coarse we may neglect the circuitry errors.

Reading-Head Errors

The diametrical imaging eliminates centring errors and odd harmonics of the grating error. If the two fields are not exactly diametrically opposite a centring error remains; the magnitude of this error can be reduced to very small values with normal precautions.

Due to the use of two diffracted orders which are symmetrical with respect to the grating plane, a change in height of the grating does not introduce systematic errors because the difference in phase between the two beams is not affected, even over the total spectrum. It is easy to realize sufficient symmetry by auto-collimation techniques with the neighbourhood of the grating-track used as a mirror.

We estimated that the errors in the quadrature of the signals, the relative change in intensity between them and the cross-talk can be made negligible with respect to the errors caused by thermo-mechanical distortions in the mounting. The most severe error of the reading head is probably mirror tilt during the measurements. A tilt angle ϵ can cause an error of about 4 ϵ. For a 0·1 sec measuring accuracy the mirror should be stably positioned within 0·025 sec, and this needs a temperature distribution in the mounting base which has to be kept constant within about

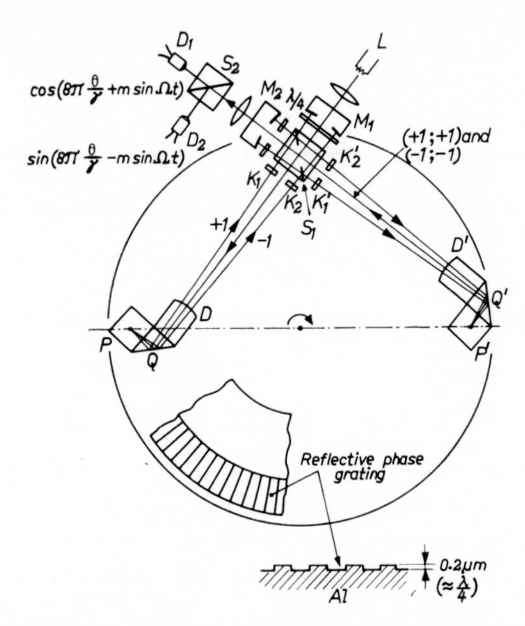

FIG. 3. Scheme of the system which is under construction. Instead of a transmission grating a reflection grating is used here.

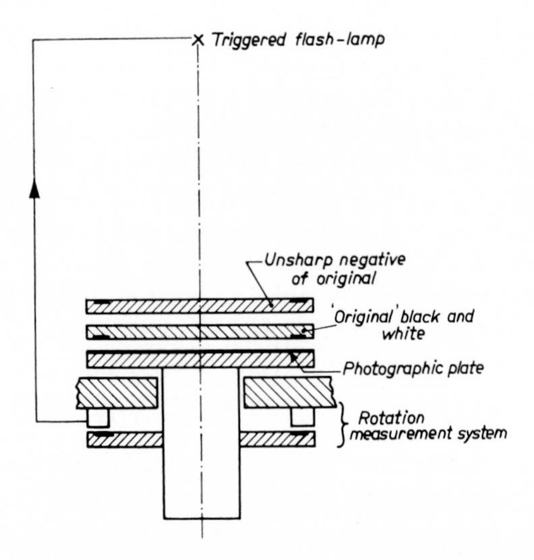

FIG. 4. Breeding of a radial grating with the aid of the rotation measurement system described.

0·02°C if we take a steel base. This requirement demands a sophisticated mechanical design and controlled environment.

Grating Errors

Due to the averaging of the groove position over the field of view and to the diametrical scanning, only low-frequency, even harmonics of the grating error are of interest.

Thermal effects in an all-metal grating are less severe than they were in the reading head. Due to bad heat-conductance, glass gratings need careful shielding from thermal sources.

Bending of the grating does not introduce serious errors: this error is zero at the points of support and also half-way between them. With three points the bending gives rise to an odd harmonic error and thus does not enter into the measurements.

The demands on the periodicity of the grating itself are of course very high, the tolerance being 0·1 sec or arc. McIlraith and Penfold (1967) succeeded in satisfying this demand.

Figure 4 shows a scheme of the arrangement with which we tried out a method of breeding radial gratings. The photographic plate is mounted on the end of a shaft. At a distance of 10–20 μm above this plate an 'original' black and white grating is placed. Above this one an unsharp negative of the original, exposed in the same circumstances, is used in order to increase the homogeneity of the illumination. The flash-lamp is triggered with the signals of the rotation measurement system which contains a phase-reflective grating practically identical with the black and white 'original'. Three or four revolutions are made with 20,000 exposures each. The transverse shifts S of the shaft together with the relative variations ΔI in the amount of light reaching the photographic plate at various angular positions give rise to a linear error of the order of magnitude $S\Delta I$. Shafts can be produced with S smaller than 0·1 μm (Kraakman & de Gast 1969) and ΔI can be reduced to less than 0·1. Hence the maximum value of the grating error will be 0·01 μm, corresponding to angular errors of less than 0·03 sec of arc.

REFERENCES

Burch, J. M., 1963, *Progress in Optics*, Volume II, Amsterdam, North-Holland Publishing Company, p. 73.

Dyson, J., 1959, *J. Opt. Soc. Amer.*, **49,** 713.

Guild, J., 1956, *The Interference Systems of Crossed Diffraction Gratings: Theory of Moiré Fringes*, Oxford, Clarendon Press.

Guild, J., 1960, *Diffraction Gratings as Measuring Scales*, London, Oxford University Press.

Kraakman, H. J. J. and de Gast, J. G. C., 1969, *Philips Technical Review*, **30**, 117

de Lang, H., Ferguson, E. T. and Schoenaker, G. M.r 1969, *Philips Technical Review*, **30**, 149.

McIlraith, A. H., and Penfold, A. B., 1967, *A Radial Grating Division Engine* (presented at M.T.D.R. Conference, Manchester). Pergamon Press, Oxford.

Measurement of Surface Roughness by Interferential Contrast—an Application of Shearing Interferometry to the Study of Phase Objects

C. H. F. Velzel

Philips Research Laboratories,
N.V. Philips' Gloeilampenfabrieken,
Eindhoven, Netherlands

Abstract—Optical methods for the measurement of surface roughness in general have the following characteristics:

(1) they require processing of the optical data;
(2) they investigate the surface point by point.

In this paper a method is described which allows one to obtain a direct measure of the variance and auto-correlation of small surface deviations, averaged over an area which is great compared to the auto-correlation distance of the deviations.

With the aid of a Michelson interferometer two images of the surface to be investigated are superposed. By axial movement and tilt of the interferometer mirrors these images acquire a phase difference and a relative displacement.

A detector in the image plane measures the intensity of the double image; the signal is modulated by varying the phase difference periodically. The modulation depth as a function of the image shift is a measure for the surface roughness; in the case of a phase object (which is approximated by well-polished surfaces) the variance and auto-correlation of the phase follow directly from the modulation depth. The principle of the method is discussed briefly; it will be seen that the coherence of the illumination and the geometry of the interferometer influence the precision to be reached.

A prototype apparatus has been built and preliminary results will be shown.

INTRODUCTION

Weakly modulated phase objects, such as polished glass or metal surfaces, can be studied by numerous optical methods, Among these the phase-contrast method of Zernike (1942) is perhaps the best known. In the book of Françon (1967) several other methods are described. We mention in particular the Linnik interference microscope and multiple-beam interferometry (see also Tolansky 1960) by which a wealth of topographical detail about phase objects can be obtained. Also by recording the far field diffraction pattern of a phase object the geometry and amplitude of the phase modulation can be studied (Maréchal 1958). There are several methods of obtaining profiles of a modulated surface by observing the properties of an image projected on it (Lehmann 1960) or by accurate focusing (Dupuy 1964). When one is interested in certain average properties of a phase object, the data obtained by these methods need processing. Also, the use of electro-optical means of detection is often difficult or leads to unwanted loss of information. For instance, in an interference microscope one could measure the average contrast to gain some idea of the amplitude of surface defects, but the geometrical part of the information is then obscured.

In this paper a method is introduced which gives an objective measure of the auto-correlation and amplitude of phase objects. It consists of the application of shearing interferometry in combination with phase modulation; the closest analogue known to us is a microscope devised by Dyson

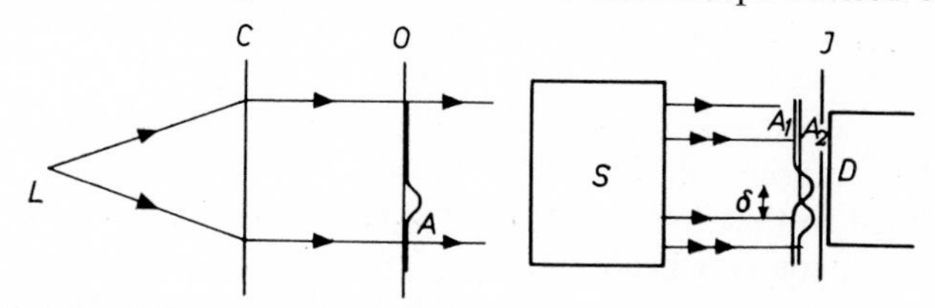

FIG. 1. Principle of the method. With the shearing interferometer S two images of the phase object O are formed in the image plane I. The images have a variable shear δ and a time-dependent phase difference $\phi(t)$. The detector D integrates the intensity in the image plane.

(1963) to measure the height of phase steps. It will be seen that the method is relatively simple and that it can be used for the study of random phase objects to obtain easily some convenient average properties.

Shearing interferometry was applied by Nomarski (1955) to the study of surfaces; in an article by Bryngdahl (1965) many other applications are described. The application we discuss is intended to measure in a simple way some properties of surfaces or thin transparent layers. In this paper the principles and theory of the method are described, together with some experiments performed with a prototype, experimental apparatus.

PRINCIPLES OF THE METHOD: THEORY

The phase object to be examined is illuminated perpendicularly by a plane, monochromatic wave. By a shearing interferometer two images of the object are formed in the same image plane; the two images can be given a variable lateral shear and a phase difference which is a periodic function of time. The intensity in the image plane resulting from the superposition of the two images is measured by a detector which integrates over part of the image plane.

In Figure 1 a diagram is given which illustrates the foregoing description. The phase object, e.g. a glass surface illuminated in transmission or a metal surface illuminated in reflection, is placed in the plane O. A monochromatic point source is placed at L; the wavefront going out from the collimator C is transmitted by the object (we have omitted the case of reflection from the diagram). The transmitted wavefront A is drawn schematically in the diagram. The shearing interferometer forms two images A_1 and A_2 of the wavefront A in the image plane I. The detector D is situated just behind the image plane.

We now will show that the modulation depth of the signal received by the detector in the arrangement described above, measured as a function of the variable shear, is a measure of the auto-correlation (in the direction of the shear) and the amplitude of the deviations from planarity of the phase object.

Calling the space dependent part of the complex amplitude in the object plane $A(x, y)$, the two amplitudes in the image plane can be described, for an ideal instrument, as $A(x, y)$ and $\exp[i\phi(t)]A(x-\delta, y)$, where δ is a shear in the x-direction and $\phi(t)$ is a time-dependent phase term. Deviations from perfect imaging will be treated later in this section. The phase term $\phi(t)$ is chosen so that it varies at least over an amount 2π. The intensity in the image plane now becomes:

$$J = |A(x, y)|^2 + |A(x-\delta, y)|^2 + 2Re[A(x, y)A^*(x-\delta, y)\exp(-i\phi) \quad (1)$$

where the asterisk denotes the complex conjugate of a quantity and Re its real part.

A phase object illuminated by a monochromatic plane wave can be described by:

$$A(x, y) = \exp[i\eta(x, y)] \quad (2)$$

where $\eta(x, y)$ is the phase defect. For instance, with an undulating reflecting surface, we have $\eta(x, y) = 2kz(x, y)$ where k is the wave number of the radiation, and $z(x, y)$ is the deviation of the surface from a plane $z=0$. By inserting this into Equation (1) we obtain:

$$J = 2 + 2\cos[\phi(t) + \eta(x-\delta, y) - \eta(x, y)] \quad (3)$$

The detector integrates the intensity over a surface S in the image plane. The detector signal F can be written:

$$F = F_0[1 + \gamma\cos(\phi(t) + \alpha)] \quad (4)$$

where F_0 is a constant and α a phase term which depends on δ. The modulation depth γ of F is given by:

$$\gamma = \left| 1/S \int_S \exp[i\eta(x, y) - i\eta(x-\delta, y)]\mathrm{d}x\mathrm{d}y \right| \quad (5)$$

For the class of objects we are interested in, weakly modulated random phase objects, we have, to a good approximation:

$$\gamma = 1 - \overline{\eta(x, y)^2} + \overline{\eta(x, y)\eta(x-\delta, y)} \quad (6)$$

where a bar over a quantity denotes its average over the surface S. The last term in Equation (6) is equal to the auto-correlation integral of $\eta(x, y)$ to a good approximation when the dimensions of S are great compared to the correlation length of $\eta(x, y)$. For values of δ greater than this correlation length, this term tends to zero and the deviation of γ from unity gives the mean square phase defect. It can be concluded that for this class of objects useful information about the structure can be obtained.

For random phase objects with a greater modulation depth the statistics should be known before more can be said about the form of the modulation depth curve. When we consider $\eta(x, y)$ and $\eta(x-\delta, y)$ as correlated random processes with a two-dimensional probability density, we have, assuming that the statistics of η are uniform over the surface S, that γ as given by Equation (5) is equal to $|\chi(1, -1)|$ where $\chi(\sigma, \tau)$ is the characteristic function of the two-dimensional random process $\eta(x, y)$, $\eta(x-\delta, y)$ (Beckman & Spizzichino 1963). For random phase defects with Gaussian statistics we have then:

$$\gamma = \exp\left[-\overline{\eta^2(x, y)} + \overline{\eta(x, y)\eta(x-\delta, y)}\right] \tag{7}$$

which resembles Equation (6) quite closely. The form of $\gamma(\delta)$ as given by Equation (7) depends on the auto-correlation integral of $\eta(x, y)$; for high values of δ again the mean square phase deviation is obtained.

We also consider briefly the case of systematic phase objects. Clearly the concepts of correlation integral and mean square deviation do not make sense, in this case; the computation of the form of η from a measurement of $\gamma(\delta)$ is not possible in general so that the method loses much of its attractivity. When we have a sinusoidal phase deviation: $\eta = ka \sin \omega x$ (for instance, a reflecting phase grating with a depth of modulation equal to a) we have:

$$\gamma = |J_0[ka \sin(2\omega\delta)]| \tag{8}$$

where $J_0(z)$ is the zero order Bessel function of argument z. In Figure 2 some $\gamma(\delta)$ curves are shown for a few values of ka. The examples given above show that for weakly modulated random phase objects the interpretation of $\gamma(\delta)$ curves is simple and straightforward. For random phase objects where $\eta > \frac{1}{2}$, say, the situation becomes more complicated and for systematic phase objects it is in general not possible to obtain the structure from the curve. However, in these cases useful information can be obtained about the influence of a surface or a thin layer on the contrast transfer curve of an optical system.

So far we have considered perpendicular, plane-wave, monochromatic illumination, perfect imaging, pure shearing and a noiseless linear detector and source operation. In practice these conditions will not be met, and we must see how the diverse sources of error influence our measurements.

It can be shown from Equation (5) that for a random object illuminated by a nonplanar wavefront, so that $\eta(x, y)$ can be written as the sum of a systematic and a random deviation, γ is given by the product of a stochastic part and an instrumental factor. This is also true when the illuminating wavefront has random deviations from planarity, as long as these do not

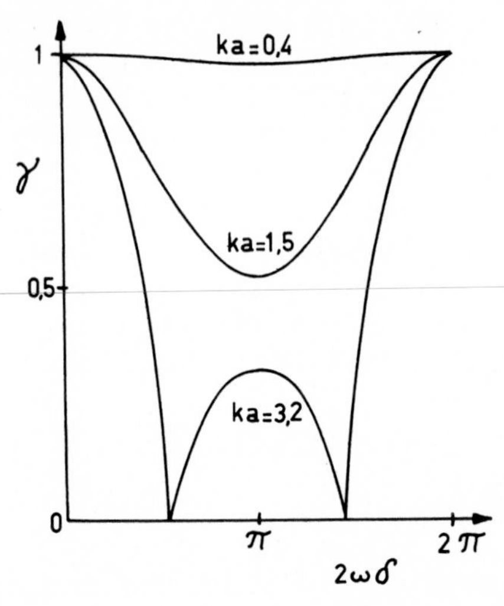

FIG. 2. Graphs of the function $\gamma = | J_0(ka \sin(2\omega\delta)) |$ for three values of ka.

correlate with the random deviations of the object. When both the illuminating wave and the object have systematic phase deviations γ is not separable.

When the illuminating light is not monochromatic, the intensity of Equation (3) should be integrated over the spectrum also. When $\phi(t)$ and $\eta(x, y)$ have a linear dispersion, it can be shown that the influence of a finite bandwidth is negligible as long as the product of the relative bandwidth $\Delta k/k$ with the argument of the cosine in Equation (3) is small compared to 1. When the imaging of the instrument is not perfect (it is at least limited by diffraction) the object amplitude will be convoluted in the image plane by a point-spread function. Because convolutions obey the associative property and because γ has the form of a convolution integral this error leads to the convolution of the integral in Equation (5) by an instrumental function $C(x, y)$. When the deviations from planarity of the illuminating wavefront are slow compared to the point-spread function C, multiplicative and convolutive errors are separable.

The presence of noise impairs directly the measurement of the modulation depth of the detector signal. However, the source noise can in principle be subtracted out by generating a monitor signal of equal modulation depth as the signal to be measured. The detector noise forms an essential limitation to the measurement of $\gamma(\delta)$.

Of the errors we have discussed so far the multiplicative errors can be corrected for by calibrating the instrument. The convolutive errors lead to loss of geometrical detail but do not hinder the measurement of the mean square phase defect when the $\gamma(\delta)$ curve is smooth enough. The bandwidth of the radiation used is not severely restricted. By suitable design of the instrument the signal-to-noise ratio can be made high enough to measure modulation depth differences of the order of 10^{-4} (see Dyson 1963).

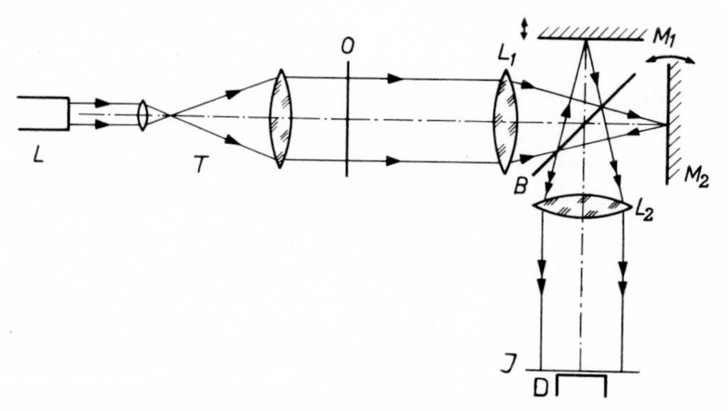

FIG. 3. Diagram of the apparatus. The shearing interferometer is of the Michelson type

Apart from the difficulties of interpretation we mentioned above, there are other theoretical limits to the method we describe here. The concept of a phase object is an idealization which comes near the truth only for surfaces with small deviations of moderate spatial frequencies, or for very thin layers. When the gradients of the phase modulation become greater than 0·1, say, or when the radii of curvature of the wavefront transmitted by the phase object are not great compared to λ, the theory given above ceases to be applicable. Naturally isolated discontinuities in a surface can be tolerated, but there should not be too many of them. In conclusion, the method described in this paper can be expected to be useful for the measurement of the roughness of essentially 'specular' objects.

EXPERIMENTS

Some preliminary experiments were done with an apparatus in which a Michelson interferometer was used for generating the variable shear and phase modulation. The Michelson, though it is somewhat clumsy and sensitive to vibrations, has the advantage that most of its parameters can be influenced independently. Moreover, it is clear that the modulation depth of the detector signal will be much less sensitive to vibrations than the signal itself. A diagram of the apparatus is shown in Figure 3.

The light source L was a He:Ne laser with an output of about 5 mW at 633 nm. The laser beam was expanded by the telescope T and illuminated the object O (again for simplicity the transmission set-up is shown; for illumination in reflection a semi-transparent mirror was inserted). The object was imaged at infinity by the lens L_1. The focal length of the lens was 53 mm and the diameter of the aperture 6 mm. The interferometer consisted of the beam-splitter B and the mirrors M_1 and M_2. The mirror M_1 could vibrate in the direction of the optical axis over a distance of a few wavelengths with a frequency of 125 Hz. The mirror M_2 could be tilted about an axis in its plane. A one-to-one image of the object was formed in the plane I by the lens L_2, identical to L_1. The back focal planes of both lenses lie in the plane of the mirrors M_1 and M_2, situated at equal distances from the beam-splitter. Thus a plane wave in the object space remains plane in the image space. By situating the mirrors in the back focal plane of L_2 pure phase differences and pure shearing between the images in I were generated. The detector D was situated in the image plane I, it covered a circular surface of 2·5 mm diameter in this plane. The modulation depth of the detector signal was measured with the aid of an oscilloscope. The chief source of noise in these measurements was the laser used as coherent point source. It had a noise power of about 0·01 of the average power so that the accuracy of the measurement of γ was limited to 1 per cent.

With the instrument described above we performed some measurements on transparent objects. The instrument curve for transmission is given in Figure 4, curve a. The maximum of this curve is at about 98 per cent due to false light.

We made a random phase object by covering a layer of photoresist on glass with small particles of polishing red, illuminating the layer with UV light and washing away the unilluminated parts of the layer. In this way local variations in thickness of about 2 μm were obtained. It can be shown that for deep, local phase defects the asymptotic level of the curve lies at a value $(1-p)^2$, where p is the part of the surface occupied by the phase defects.

The modulation depth curve is shown in Figure 4, curve b, with the object in focus. By a defocusing of 5 mm the curve becomes shallower, as shown in curve c. For comparison we show the modulation depth curve of a well-polished glass surface, in curve d or Figure 4.

The modulation depth curve for a transparent phase grating is shown in Figure 5, curve c. The period of the phase grating was about 35 μm; the form of the grating was square-wave with a depth of 0·23 μm.

From the theory of given above one expects a triangular form of this curve, with a minimum $y=\cos \alpha$, where α is the phase step of the grating. The theoretical curve has been drawn in Figure 4, curve d. From the comparison of curves c and d the effect of the instrument point spread function C is clearly visible.

For reflection measurements the use of a semi-transparent mirror for the illumination is needed. Moreover, the calibration gives some difficulties

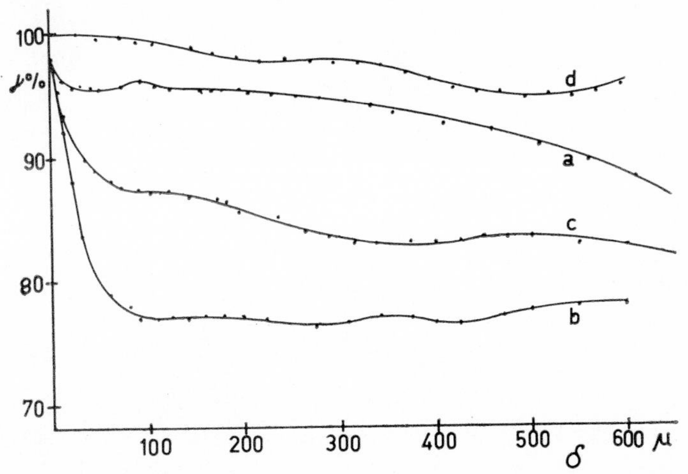

FIG. 4. Measurements on transparent objects: curve a gives the modulation depth for the empty instrument; b, for a random object in focus; c, the same, defocused; d, for a polished glass surface.

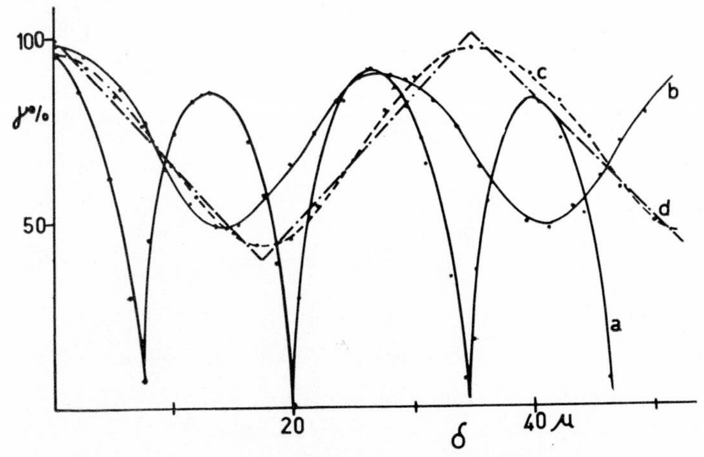

FIG. 5. Modulation depth curves for some phase gratings of square wave form: curves a, b, reflection gratings of different depth; curve c, transmission phase grating; curve d, the theoretical curve belonging to c.

since an ideal reflecting surface cannot be found. The difficulty can be solved by placing a very well polished mirror at a good distance out of focus. The resulting calibration set up is shown in Figure 6. The mirror is situated in the plane M, the object can be inserted in O, the semi-transparent mirror H receives light from the collimator C, sends it to M or O and lets through part of the reflected light to the first objective L_1 of the interferometer.

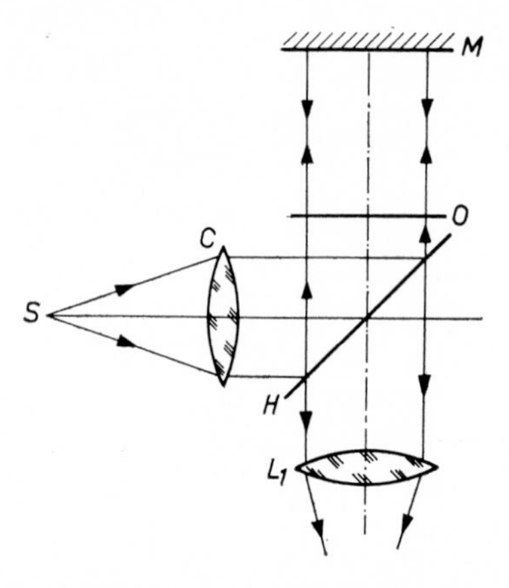

FIG. 6. The reflection set up with the semi-transparent mirror H and the calibration mirror M.

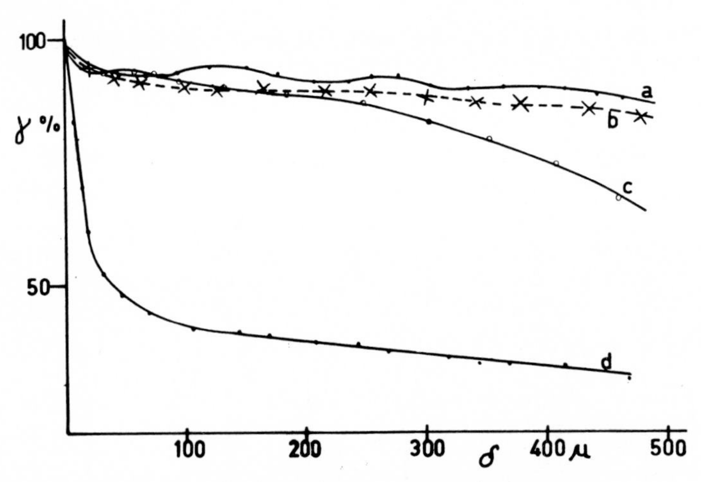

FIG. 7. Measurements on reflecting objects: curve a, the instrument curve; curve b, laser mirror; curve c, aluminium mirror; curve d, random object.

The instrument curve of the reflection set up is shown in Figure 7, curve a. In curve b we show the modulation depth for a well-polished laser mirror, comparable in quality with the optical flat examined in transmission. In curve c the modulation depth is shown for an aluminium polished flat; for higher values of δ the curve keeps falling due to a conoidform defect of the object. Curve d shows the modulation depth for an object with phase steps of equal depth but random form, obtained by a photo-etching technique in aluminium, analogous to the method described before in connection with transparent objects. It is seen that the root mean square depth of the phase defects is equal to 0·06 μm, and the correlation length amounts to 50 μm. In Figure 5, curves *a* and *b*, we show the modulation depth curves for two reflecting phase gratings of different step height.

DISCUSSION

With the existing apparatus we can measure the amplitude of random surface defects in a range from 5 to about 75 nm (for reflecting objects, and about 20 to 300 nm for transparent objects). The spatial resolution is of the order of 5 μm, the greatest correlation lengths we can handle are about 500 μm.

These results can still be improved considerably. The precision of the measurements, and also their sensitivity, can be increased by using noise-subtracting detection methods. In this way one can expect a sensitivity well under 1 nm. It is also possible to decrease the noise power of the laser; a classical light source can be used, because the demands on monochromaticity are not very high. It can be shown that with a detector field of 1 mm^2 and a compact mercury arc source giving 1 watt in the green line, radiating into about 4 steradians, the detector signal will be a factor 100 smaller than with a 10 mW gas-laser. So the use of conventional light sources demands a much higher detection sensitivity. By using radiation with a greater wavelength, surfaces with a roughness greater than 75 nm could be examined; it is also possible to use the method of replicas (Françon 1967, p. 279) to obtain manageable phase modulations.

The numerical aperture of the objectives used in the apparatus was about 0·1. By taking a higher numerical aperture we could obtain a better spatial resolution.

Further improvement could be made by taking a different type of shearing interferometer and by streamlining the modulation depth measurements.

We intend to apply the method to the study of polished surfaces (metal, glass) and of thin, transparent layers (optical coatings, photographic emulsions).

Acknowledgements

The author is grateful to Mr. C. Beye who performed the measurements, to Dr. P. Kramer for his encouragement and to Dr. H. de Lang for his penetrating criticism. The idea for this study arose out of a discussion with Dr. E. T. Ferguson, to whom special thanks are due

REFERENCES

Beckmann, P. and Spizzichino, A., 1963, *The Scattering of Electromagnetic Waves from Rough Surfaces*, New York, Appendix E of Part I.
Bryngdahl, O., 1965, in E. Wolf (ed.), *Progress in Optics*, Volume IV, Amsterdam, p. 37.
Dupuy, Mlle O., 1964, *Rev. Opt.*, **43**, 217, 282.
Dyson, J., 1963, *Nature*, **197**, 1193.
Françon, M., 1967, *Optical Interferometry*, New York.
Lehmann, R., 1960, *Leitfaden der Längenmesstechnik*, Berlin.
Maréchal, A., 1958, *Optica Acta*, **5**, 70.
Nomarski, G. and Weill, Mme A. R., 1955, *Rev. Metall.*, LII, 121.
Tolansky, S., 1960, *Surface Microphotography*, London.
Zernike, F., 1942, *Physica*, **9**, 686.

Some New Modifications of the Method of Three Slits in the Measuring Applications

M. Miler

Institute of Radio Engineering and Electronics,
Czechoslovak Academy of Sciences,
Prague 8, *Czechoslovakia*

Abstract—The method of three slits uses an optical system to project the first slit onto the third, and a second slit in an intermediate position where a phase object is also situated. The first two slits are mutually perpendicular and the third at some determined angle. An anamorphotic system then images both the second and the third slits onto a screen. Variations in the refractive index of the test object then appear as curves. The method has been used to show the variations of other physical quantities, e.g. temperature in a liquid, which can affect the refractive index. Modifications have been introduced in which direct-vision prisms are used to increase the sensitivity and measure the actual refractive index.

In the experimental study of non-homogeneous materials the problem arises of how to determine the change of some physical quantity along a certain section. If the material is transparent, and the required physical quantity is dependent on the refractive index of the material, then it is possible to carry out the measurement by optical methods. The advantage of these methods lies in the fact that the light passing through the material provides the information on the refractive index without any effect on the material.

The method of three slits (Schrägspaltmethode, Philpot-Svensson Method) is a suitable method to use for this purpose. It is a geometric-optical method providing a direct record of local deviations in the optical path determined by a given slit. The optical system imaging the first slit on the third one (Figure 1) is the basis of the method. The object under examination which causes the deviation in the direction of the light beam is affected by the refractive index changes and is placed near to the middle or second slit. The first two slits are mutually perpendicular, the third one is inclined. The refractive index gradient in the slit direction causes a shift of the first slit image in the third slit plane perpendicularly to its image—the vertical direction in our case. A narrow light beam, which can pass through the inclined slit, participates in the further imaging. This beam leaves the slit with a deviation perpendicular to the movement of the first slit image. An anamorphotic optical system and the screen complete the optical system. By the anamorphotic system the third slit is imaged on the screen, and in the plane perpendicular to this image the second slit is imaged on the same screen. Thus the deviations along the second slit are changed into perpendicular deviations.

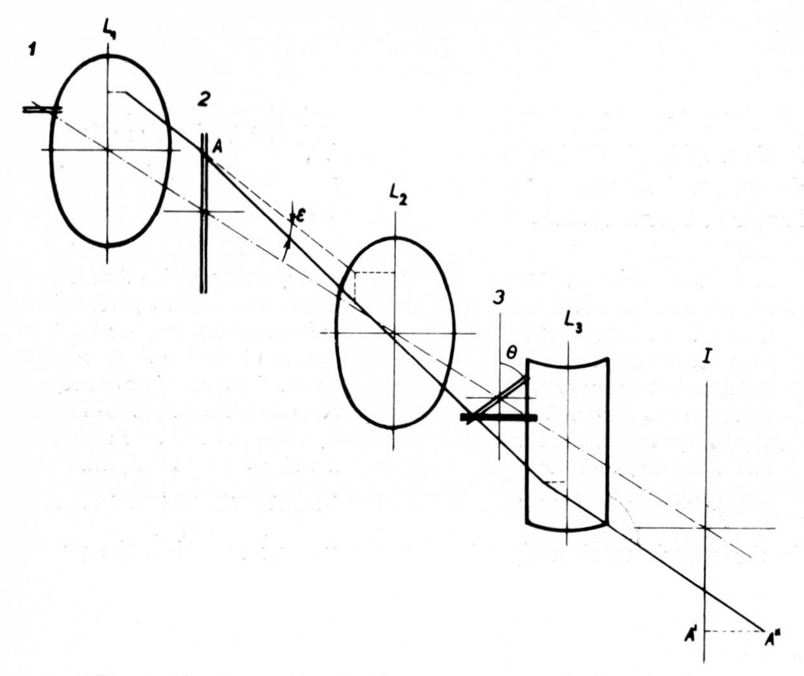

FIG. 1. Diagram showing the theory of the method of three slits.

Any deviations caused by the refractive index gradient perpendicular to the object section cannot appear in the resulting image, because the first slit image in the third slit plane can be longitudinally shifted, without affecting the resulting image on the screen.

For a quantitative description a rough approach will be sufficient. The deviation ϵ of the light beam in the gradient field of the refractive index $dn/d\xi$ is given by the relation

$$\epsilon = l\frac{dn}{d\xi}, \tag{1}$$

where l is the length of the light path in this field. If Γ and Γ' are the magnifications of the optical system in the vertical and horizontal plane, f is the focal length of the lens L_2 and θ is the angle of the inclined slit from the vertical direction, then the distribution of the refractive index in the slit is given by the relation

$$n(x) = n_0 + \frac{1}{\Gamma\Gamma' lf \tan\theta} \int_{x_0}^{x} y(x)\mathrm{d}x \tag{2}$$

where $y(x)$ defines the curve on the screen. From the determined variation of refractive index the variation of the dependent physical quantity may then be obtained.

The commonest application of this method is in the study of liquid materials, particularly with diffusion, sedimentation, electrophoresis and

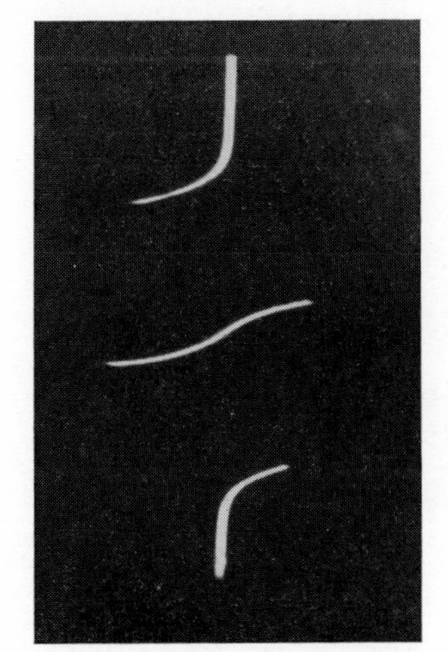

FIG. 2

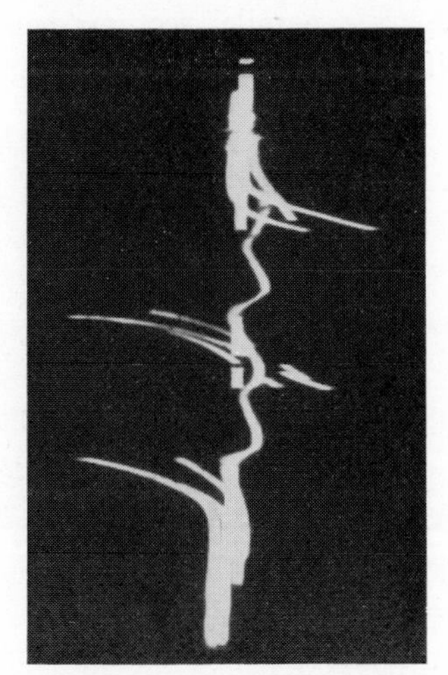

FIG. 3

FIG. 2. Curves obtained on a screen showing variations of refractive index between two heated wafers.

FIG. 3. Superimposed curves for five sections of the space between the two heated wafers of Figure 2.

other processes. These investigations deal with the refractive index gradient in one direction only without the perpendicular component. However, realizing that the gradient of the component perpendicular to the section cannot be found with a given optical arrangement, the method in a more complicated form may be used for recording any given component. In our laboratory the method was used in the study of the temperature processes taking place in cooling liquids of electric machines.

As an example, I will describe the experimentally investigated variation of the refractive index gradient in a section perpendicular to two electrically heated horizontal wafers placed one over the other (Figure 2). The picture was taken after stabilization of the dynamic equilibrium. From the gradient curve of the refractive index we may easily determine the temperature variation, if the temperature coefficient of the refractive index is known.

By moving the object slit into other positions, the whole field can be successively mapped. However, this successive mapping can only be performed at one stabilized equilibrium. With transient processes the measurement should always be performed simultaneously for all sections. In the described classical method of three slits, parallel sections of the object will be imaged at one place on the screen, as a result of the anamorphotic optical system. Thus the gradient curves in separate sections of the object coincide, and they are mutually distinguished only by the different shapes of the curves. In this way the comparison of the gradient curves in separate parallel sections can be carried out successfully. However, a difficulty arises in determining the curve corresponding to a given section

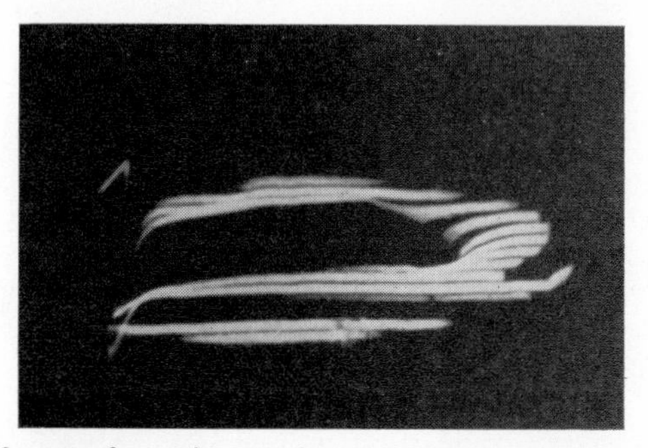

FIG. 4. A series of curves for various sections when a model of a stator winding of an electric machine is heated.

of the object. This problem may be solved by colouring the individual sections or by a slight movement of the screen from the imaging plane along the optical axis of the system. Thus the individual curves will be slightly separated each from the other in the same succession as in the object. Figure 3 showing the gradient curves in five horizontal sections of two vertically placed heated wafers may serve as an example. Figure 4 shows also a plot of the gradient curves in several sections when the model of the end of the stator winding of a rotary electric machine is heated. In order to distinguish the curves from each other, it was necessary to shift the screen far more than in the preceding case. Thus the anamorphotic imaging of the object was obtained which simplified the identification of the curves.

However, the anamorphotic imaging distorts the resulting image, and therefore an attempt was made to remove this distortion by eliminating the anamorphotic system. So long as the screen remains in the plane of the imaging, the distortions caused by the refractive index gradient cannot appear. But when the screen is shifted from the imaging plane, as for the separation of individual gradient curves, the imaging of the gradient curves will remain to some extent unchanged. However, in this case Equation (2) can no longer be used for calculations.

The optical arrangement for this method is evident from Figure 5, which shows the object section which does not pass through the axis of the system. The plane of sharp imaging is denoted by I and that of the screen location by C. The screen shift is Δz. Denoting the magnification of the system by $\Gamma = a'/a$ and $\Gamma^* = (a' - f' + \Delta z)/f'$ then the refractive index will be

$$n(x) = n_0 + \frac{\Gamma \cos \theta}{\Gamma^* l \Delta z} \int_{x_0}^{x} y(x) \mathrm{d}x \tag{3}$$

and the coordinate system is oblique, with an angle θ.

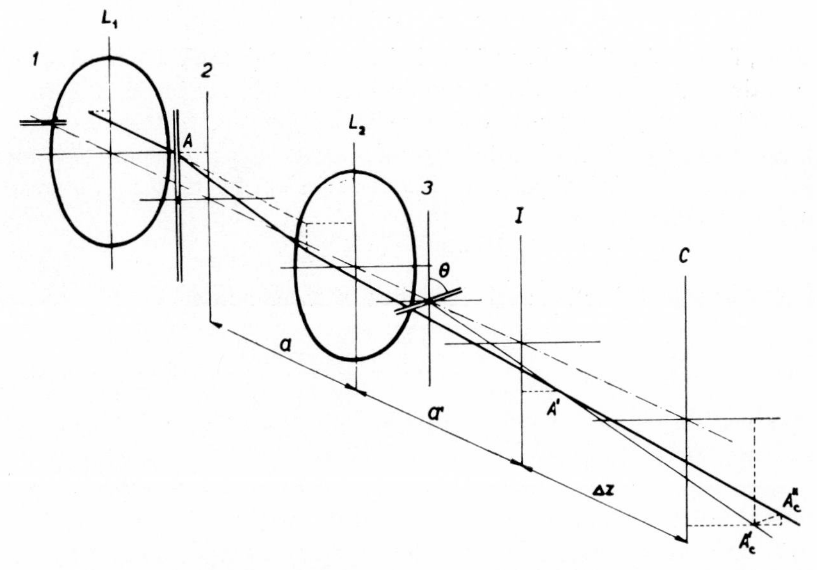

FIG. 5. Diagram of a modified optical system for off-axis objects without the anamorphotic system of Figure 1.

Because the screen movement causes a defocusing of the image, it cannot be arbitrarily large. On the whole, the sensitivity of this method is a little smaller than the classical one.

In the modifications of the three-slit method so far described, the refractive index gradient is recorded on the screen, and it was only possible to determine the refractive index changes by additional calculation. A direct record of this dependence for a given section is performed by prism methods. The basic element of the method is the measuring prism, in which the refractive index is changed in the direction parallel to the refracting edge of the prism. This is also the direction of the section in which the measurement is performed. Owing to the changes of the refractive index, the beams deviate in a direction vertical to the section plane, and these deviations are imaged on the screen after the passage of the light beam through a suitable optical system.

For a small deviation of the beam from the symmetrical path the relation

$$\Delta\epsilon = \frac{2 \sin \psi/2}{\sqrt{1 - n^2 \sin^2 \psi/2}} . \Delta n \tag{4}$$

may be written for the deviation angle $\Delta\epsilon$ of the beam, where ψ is the refracting angle of the prism. Owing to total reflection the magnitude of the refracting angle is limited by the condition

$$\psi < \cos^{-1}(1 - 1/n^2). \tag{5}$$

Thus, for a refractive index of 1·5, it follows that $\psi < 84°$. To increase the sensitivity, and in order that the optical axis should not be deviated, it is advantageous to use the so-called differential prism. In this prism the measuring prism is flanked by two prisms producing, in effect, a plane parallel plate having dispersion without deviation. In this case, for small deviations, the relation

$$\Delta\epsilon = 2\Delta n \tan \psi/2. \tag{6}$$

holds, and the refracting angle is limited by the condition

$$\psi < 2 \sin^{-1}\left(1 - \frac{\Delta n}{n}\right) < \pi, \tag{7}$$

thus resulting in a doubling of the refracting angle compared with the simple prism.

Even with the prism method, however, it is an advantage to use the optical system with three slits for recording changes in the refractive index in a single section of the object, but some changes are necessary (Figure 6). Because the deviations of rays perpendicular to the section are obtained in this case, it is necessary to turn the first slit through 90°, to make it parallel to the object section. It is then not necessary for the third slit to be inclined, and it is oriented perpendicular to both the preceding slits. This slit also performs the function of compensating for the longitudinal gradient in the direction of the object slit.

The arrangement of this method for the simultaneous study of refractive index in several parallel sections may be performed analogously to the previous method for the study of gradients. The anamorphotic lens is removed, and the screen is shifted from the plane of sharp imaging. It is also advantageous to replace the simple differential prism by a 'Fresnel' prism, in which both the refracting faces have a sawtooth shape.

With this last arrangement the problem lies in the refractive index gradient perpendicular to the object sections. The measurement is important only if the beam deviation caused by the refractive index gradient is negligible compared with that caused by the refraction on the prism wall, i.e. if the inequality

$$l \frac{dn}{d\xi} \ll 2\Delta n \tan \psi/2 \tag{8}$$

holds. Thus a very small change in the refractive index when passing from one object section to another is assumed.

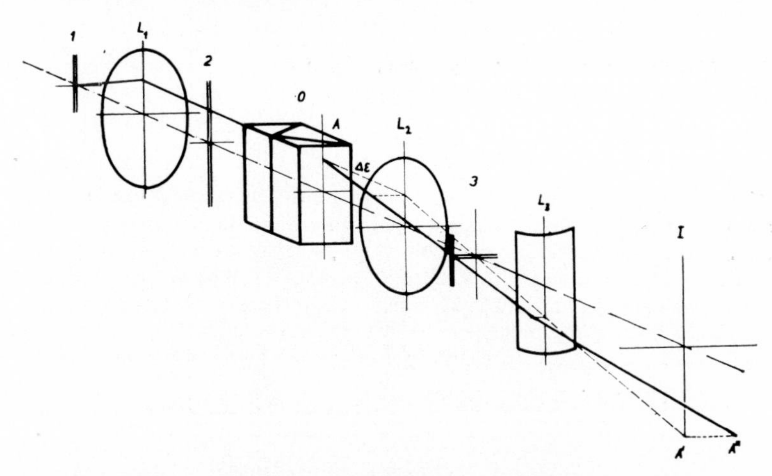

FIG. 7. Results obtained on the screen when using the prism method.

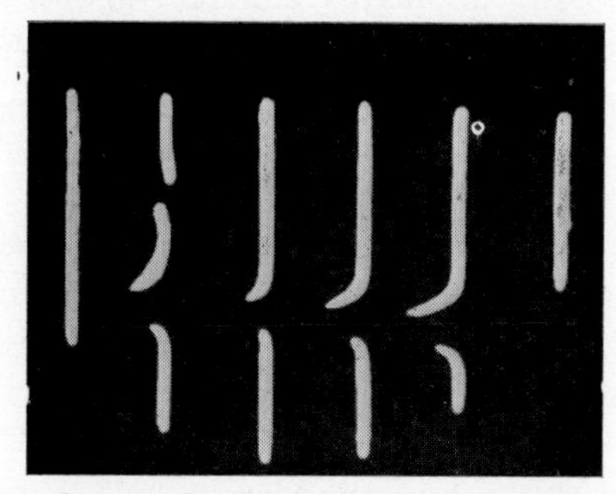

FIG. 6. Showing the use of a direct-vision prism to obtain dispersion.

As an example of the application of the last method the temperature field of the refractive index of the cooling liquid in the neighbourhood of the heated wafer can be used (Figure 7).

The modifications of the method of three slits as described are very useful for many purposes. They extend the application of the classical method to cases of two-dimensional fields of the refractive index, and to quantities depending on it. They are particularly useful for measuring relatively large changes of the refractive index in liquids and solids.

Measurement of Surface Displacement by Interference of Speckle-Patterns

J. A. Leendertz

Department of Mechanical Engineering,
Loughborough University of Technology,
Loughborough, England

Abstract—When a scattering surface is imaged in highly coherent light it has a speckled appearance due to random interference within the resolution limit of the imaging system. The superposition of two coherent speckle-patterns results in a third speckle-pattern with a different intensity distribution which changes as the relative phases of the two interfering patterns changes. Thus measuring the correlation between the initial and final patterns gives a measure of any relative phase change. This permits sensitive measurement of both normal and in-plane displacement on scattering surfaces.

INTRODUCTION

The 'speckle effect' is the name given to the grainy appearance of scattering surfaces when illuminated by highly coherent light such as He:Ne laser light. This effect occurs when the viewing system cannot resolve the fine structure of the surface so that any point in the image plane receives light from an area in the surface containing a number of fluctuations in surface height. The scattered waves from within this area will have random relative phases due to the random structure of the surface so that the resultant intensity in the image plane will be a random function of position. The distribution of intensities has well defined statistical properties and has been investigated by several workers (Allen & Jones 1963; Goldfischer 1965).

The present paper shows how the idea of optical interference between two such speckle-patterns can be used to give a high sensitivity method of measuring surface displacements. Normal or in-plane displacement can be displayed in a continuous manner over the surface of interest.

Interference of Speckle Patterns

In conventional two-beam interference experiments the interfering waves generally vary smoothly in amplitude and phase across the region of interference. This means that well-defined interference fringes can be observed and, by counting fringes or by watching fringe movement, it is possible to measure optical path changes occurring in either beam. However, the amplitude and phase in a speckle-pattern vary randomly from point to point so that superimposing two coherent speckle-patterns does not result in fringes, but gives a new pattern with the same statistical properties but a different detailed amplitude and phase distribution. Consider the intensity at any point in the combined pattern: this will be obtained by vector addition of the disturbances due to each of the component patterns. If the relative phases of the disturbances change by the same amount, δ, at every point then the detailed intensity distribution will change, in a random manner, until δ approaches 2π when the distribution

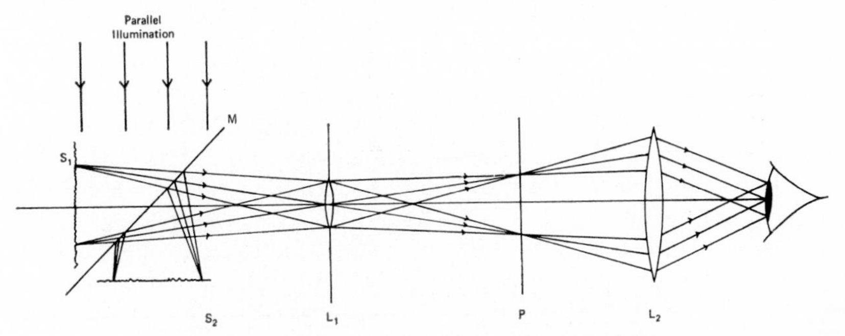

FIG. 1. Arrangement for measuring normal displacement.

will be the same as initially. Thus, by thinking in terms of the correlation between initial and final patterns we can obtain a measure of any change in relative phase which may have occurred between the two component patterns.

Changes in δ are accompanied by a rearrangement of the detailed intensity distribution, but the average intensity over a large number of speckles will remain constant. Thus the unaided eye will be unable to perceive such phase changes directly. However, if a photographic plate is exposed to the combined pattern and is replaced exactly after processing, it will act as a shadow mask, the brightest regions of the pattern falling on the most absorbing regions of the plate. If any rearrangement of the pattern takes place the brightest regions will no longer fall on the most dense parts of the plate, and so the average transmitted intensity will be greater than for $\delta = 0$. Thus if δ varies slowly over the region of interference a system of bright and dark fringes will be seen, each dark fringe being a locus of δ equal to multiples of 2π.

Measurement of Normal Displacement

Figure 1 shows a suitable experimental arrangement for measuring displacements of a scattering surface normal to itself. S_1 and S_2 are two scattering surfaces illuminated and viewed in much the same way as the mirrors in a Michelson interferometer. The lens L images the two surfaces on to the photographic plate, P, so that the speckle-pattern on any small region of P is due to light scattered from a corresponding region on S_1 or S_2. Thus values of δ occurring at the plate will be directly related to the local displacements occurring at S_1 or S_2. The proceedure is to expose the plate to the superimposed images of S_1 and S_2. After processing, the plate is returned to its original position and the image is viewed via L_2 which serves to converge the light diverging from the aperture of L_1 and permits viewing of the whole image at once. Any deformation of either S_1 or S_2 will then show as bright and dark fringes, each fringe being a locus of constant normal displacement, adjacent fringes corresponding to $\lambda/2$ difference in normal displacement for this geometrical arrangement. In practice it is found that the highest fringe contrast is obtained by using a contrast developer when developing the shadow plate, although this causes considerable reduction in image intensity.

It is also possible to obtain the fringe pattern from two separate photographs of the combined speckle-patterns, one corresponding to the undeformed surface and the other to the deformed surface. This may be required if the real-time analysis proves inconvenient. The intensity distribution of one of the images is reversed by making a contact print from it. When this contact print is aligned against the other image, those regions where the two patterns are similar will have a uniform transmission, being a superimposed positive and negative of the same pattern. Similarly, those regions where the pattern has changed will appear highly speckled. These regions correspond to regions of low and high spatial frequency content, respectively, and can be converted to intensity variations using a Fourier filtering arrangement such as that shown in Figure 2. Figures 3*a* and *b* show the resultant speckle-patterns for a pair of surfaces before and after deformation of one of them. Figure 3*c* shows the fringe pattern obtained by the above procedure. In this instance the imaging aperture was made very small in order to produce a coarse speckle-pattern for the purposes of illustration.

Measurement of In-Plane Displacement

The two interfering patterns considered so far have been derived from separate surfaces using a partially transmitting mirror. It is also possible to derive the two patterns from one surface by illuminating it with two separate, coherent beams. This gives rise to two coherent, independent speckle-patterns in the image plane—one due to each illuminating beam—and any change in the relative phases of the patterns can give rise to intensity variations as described above.

A case of particular interest occurs when the two illuminating beams are equally inclined to the surface but on opposite sides of the normal. Figure 4 shows a surface illuminated at angles θ and $-\theta$ to the z-direction, the beams being perpendicular to the y-direction. It will be seen that movement of the surface in either the z- or the y-directions will result in no relative phase change between the two patterns. Displacement in the x-direction, however, will result in a phase change of $2\pi d(\sin\theta)/\lambda$ due to

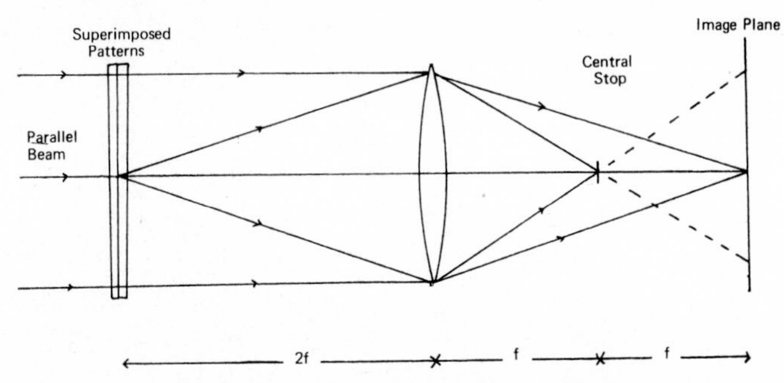

FIG. 2. Fourier filtering arrangement.

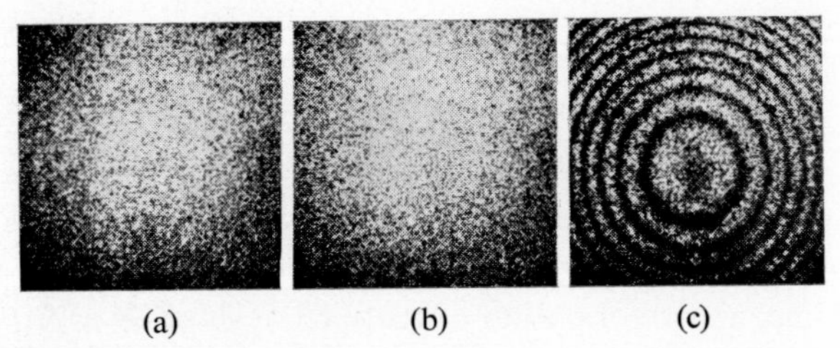

FIG. 3. Speckle pattern (*a*) before deformation; (*b*) after deformation. (*c*) Resultant fringe system showing displacement.

one beam and of $-2\pi d(\sin\theta)/\lambda$ due to the other beam, giving a total relative phase change between the two patterns of $4\pi d(\sin\theta)/\lambda$, where d is the displacement in the x-direction. Thus, if we have a surface in which in-plane strain is occurring, fringes, which are loci of constant displacement in the x-direction, will appear on the image independently of simultaneous displacements in the z- and y-directions.

A convenient experimental verification of this effect is obtained by rotating the surface about the z-axis. The x-component of this in-plane rotation gives straight, equi-spaced fringes parallel to the x-direction, which may be counted and compared with the expected number calculated from the angle of illumination and the measured rotation. In addition, the surface may be rotated about the x- or y-axes to verify that normal movement does not change the fringe pattern. In practice it is found that a rotation corresponding to a normal movement of about 90 wavelengths per inch does not introduce any extra fringes. At larger levels of normal movement the fringe contrast begins to fall. This is due to the fact that tilting the surface produces a change in the phase relationships between the individual scattering points contributing to the intensity of any image point. Thus each of the two patterns changes in a random and unrelated manner causing loss of fringe contrast. This loss of correlation due to large surface movement occurs whenever the scattered light from a diffusing surface is used to record surface position. For example, in

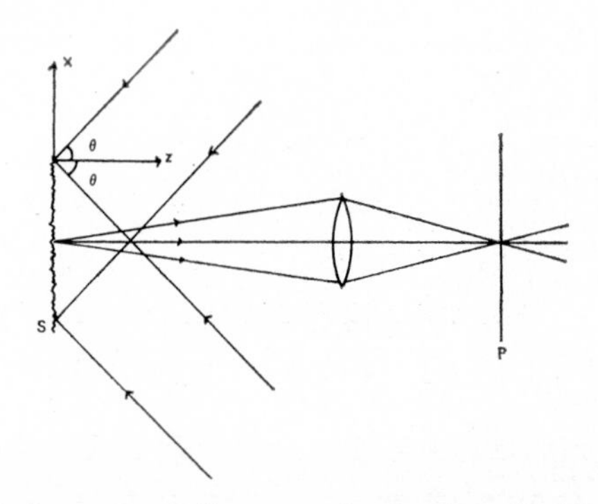

FIG. 4. Arrangement for measuring in-plane displacement.

holographic interferometry the light scattered from a surface is faithfully recorded and reconstructed to interfere with the light from the surface at a later time, any deformation of the surface showing as fringes. In this case the loss of correlation due to large surface movement coincides with the fringes becoming too close to resolve. This limits the amount of surface rotation or bending which can be tolerated in the holographic method of Ennos (1968) for measuring in-plane displacement.

Groh (1968) has in fact used this effect of loss of correlation to provide a medium sensitivity method of detecting rigid body displacements by measuring the change in the intensity distribution of laser light scattered directly from the surface. Thus there is a basic limitation to the amount of normal movement which can be tolerated; but in practice it is found that fringe contrast does not drop appreciably until the normal displacement exceeds the in-plane displacement by about two orders of magnitude.

Figure 5 shows the in-plane displacement, in real time, on the surface of a thin bar with a hole in it. The bar is being stretched in the direction parallel to the edges and the fringes are loci of constant displacement in this direction.

Statistical Calculation of Fringe Form

In this section a calculation will be made of the average light intensity transmitted through the shadow plate as a function of the change, δ, in the relative phases of the two speckle-patterns. This calculation is intended to show the mechanism of fringe formation for real time measurement.

FIG. 5. Real time fringe system showing in-plane surface displacement on a stretched bar.

Consider an area, A, in the xy-plane where the photographic plate intercepts the two superimposed speckle-patterns. Let the total energy fluxes per second through A be E_1 and E_2 due to the two patterns. Let the disturbances due to the patterns at any point $P(x, y)$ be $\bar{F}_1(x, y)$ and $\bar{F}_2(x, y)$. If the phase angle between $\bar{F}_1$ and $\bar{F}_2$ is θ initially, and the resultant intensity is I_3 then

$$I_3 = I_1 + I_2 + 2\cos\theta\sqrt{I_1 I_2}$$

where

$$I_1 = |\bar{F}_1(x, y)|^2, \quad \text{and} \quad I_2 = |\bar{F}_2(x, y)|^2.$$

The resultant intensity when $\bar{F}_1$ is phase shifted by an amount, δ, is given by:

$$I_3(\delta) = I_1 + I_2 + \cos(\theta + \delta) 2\sqrt{I_1 I_2}$$

The photographic plate is exposed to the initial pattern, denoted by $I_3(0)$, and replaced so that the transmitted intensity for a given δ is

$$I_T(\delta) = I_3(\delta) T[I_3(0)]$$

where T is the transmittance of the plate when exposed to light of intensity $I_3(0)$. It is now necessary to assume the form of the photographic response curve. For a small exposure most emulsions show a linearly decreasing transmittance with increasing exposure. Thus we may assume:

$$T(I) = 1 - \alpha It \tag{1}$$

where α is a constant, I is the incident intensity and t is the exposure time. Thus

$$I\ \ (\delta) = I_3(\delta)(1 - \alpha t I_3(0)).$$

The energy absorbed per second in a small area dA over which I_1 and I_2 remain sensibly constant is

$$\begin{aligned} \mathrm{d}E &= \alpha t \mathrm{d}A I_3(\delta) I_3(0) \\ &= \alpha t \mathrm{d}A[I_1 + I_2 + 2\cos\theta\sqrt{I_1 I_2}][I_1 + I_2 + 2\cos(\theta+\delta)\sqrt{I_1 I_2}]. \end{aligned}$$

In order to perform a statistical averaging we must assume that any particular values of I_1 and I_2 occur in a large numbei of elementary areas dA with random phase angles θ, any value of θ being equally probable. Thus

$$\begin{aligned} \mathrm{d}E_{ave} &= \frac{\alpha t \mathrm{d}A}{2\pi}\int_0^{2\pi}[I_1 + I_2 + 2\cos\theta\sqrt{I_1 I_2}][I_1 + I_2 + 2\cos(\theta+\delta)\sqrt{I_1 I_2}]\mathrm{d}\theta \\ &= \alpha t \mathrm{d}A[I_1^2 + I_2^2 + 2I_1 I_2(1 + \cos\delta)]. \end{aligned} \tag{2}$$

Thus we see that for any pair of values of I_1 and I_2, the average absorption is a maximum for $\delta = 0$, and a minimum for $\delta = \pi$. It is now necessary to calculate the probability of any pair of values I_1 and I_2 occurring, and to average the absorption over all possible pairs.

Assuming the scattering surface to consist of a large number of sources of equal amplitude with random phases it may be shown (Rayleigh 1880) that the probability of finding the intensity between I and $I + \mathrm{d}I$ is proportional to exp $(-kI)\mathrm{d}I$ where k is a positive constant. This leads to the result that

$$\mathrm{d}A(I_1, \mathrm{d}I_1)/A = A/E_1 \exp(-I_1 A/E_1)\mathrm{d}I_1$$

where the left-hand side is that fraction of the total area occupied by intensities in the range I_1 to $I_1 + \mathrm{d}I_1$. Thus in equation (2),

$$\mathrm{d}A = A(A/E_1)\exp(-I_1 A/E_1)(A/E_2)\exp(-I_2 A/E_2)\mathrm{d}I_1 \mathrm{d}I_2.$$

Thus the total energy absorbed, E, is given by

$$E = \frac{\alpha t A^3}{E_1 E_2}\int_0^\infty\int_0^\infty[I_1^2 + I_2^2 + 2I_1 I_2(1 + \cos\delta)][\exp(-I_1 A/E_1)\exp(-I_2 A/E_2)]\mathrm{d}I_1 \mathrm{d}I_2.$$

Substituting $u_1 = I_1 A/E_1$ and $u_2 = I_2 A/E_2$ this becomes:

$$\begin{aligned} E &= \frac{\alpha t}{A}\int_0^\infty\int_0^\infty[E_1^2 u_1^2 + E_2^2 u_2^2 + 2E_1 E_2 u_1 u_2(1 + \cos\delta)][\exp(-u_1)\exp(-u_2)]\,\mathrm{d}u_1 \mathrm{d}u_2 \\ &= \alpha t/A[2E2_1^2 + E_2^2 + 2E_1 E_2(1 + \cos\delta)]. \end{aligned} \tag{3}$$

The average incedent energy per second is E_1+E_2, and thus the ratio of maximum to minimum transmitted energy is given by

$$\frac{I_{max}}{I_{min}}=\frac{E_1+E_2-E_{min}}{E_1+E_2-E_{max}}$$

where E_{min} and E_{max} are the values of E in Equation (3) for $\delta=\pi$ and $\delta=0$, respectively.

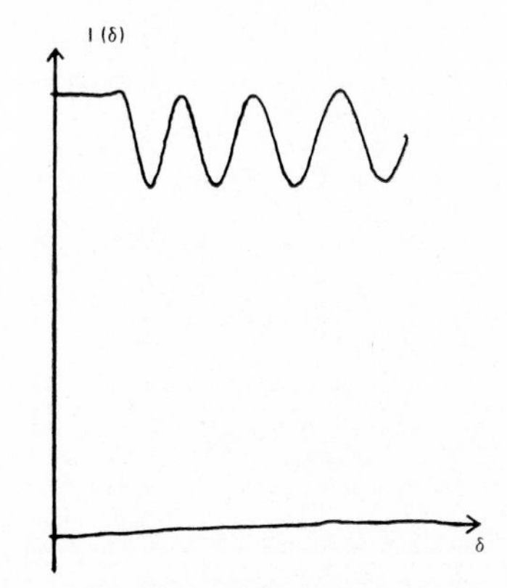

FIG. 6. Experimental trace of transmitted intensity as a function of relative phase angle.

If, for example, we chose an exposure time such that complete darkening of the plate occurs for an intensity of 5 times the average intensity, then from Equation (1):

$$0=1-[\alpha t\times 5(E_1+E_2)]/A$$

i.e. $\alpha t=A/[5(E_1+E_2)]$ and, for $E_1=E_2$ we have:

$$\frac{I_{max}}{I_{min}}=\frac{4}{3}$$

The above analysis is only valid for small values of exposure otherwise Equation (1) indicates a negative value of transmission for appreciable areas of the plate. It will be seen that a useful contrast can be obtained in the fringes, and Figure 6 shows a plot of the measured transmitted intensity as a function of relative phase angle.

CONCLUSION

A coherent light technique has been presented for showing displacements of a scattering surface in a continuous manner over the whole of the surface. High sensitivity measurements of either normal or in-plane displacement can be made in real time. Measurements of in-plane displacement in real time by holographic interferometry have been prevented by the fact that simultaneous normal movement always causes additional fringes to form. In the method presented here, simultaneous normal movement has been eliminated in a basic manner.

The theory of fringe formation has been developed and some of the basic limitations of coherent light techniques for measuring surface displacement have been briefly discussed.

REFERENCES

ALLEN, L. and JONES, D. G. C., 1963, *Phys. Letters*, **7**, 321.
ENNOS, A. E., 1968, *J. Sci. Instr.*, Series 2, **1**, 731.
GOLDFISCHER, L. I., 1965, *J. Opt. Soc. Amer.*, **55**, 247.
GROH, G., 1968, Symposium on 'The Engineering Uses of Holography', University of Strathclyde.
RAYLEIGH, LORD, *Scientific Papers*, Vol. 1, Dover Press, p. 491.

A Laser Speckle Interferometer for the Detection of Surface Movements and Vibration

E. Archbold, A. E. Ennos and Pauline A. Taylor

Division of Optical Metrology,
National Physical Laboratory,
Teddington, Middx.

Abstract—A laser interferometer is described in which a speckle-pattern field, derived from the scattering of coherent light off a rough surface, is superimposed upon a uniformly illuminated field obtained from the same laser source. Individual speckles in the modified pattern, so obtained, undergo cyclic variations of brightness when the surface moves towards or away from the source, resulting in a 'twinkling' effect. If the surface vibrates, the speckle-pattern becomes blurred. The instrument can thus be used to detect nodal regions of a vibrating surface, where the speckles maintain a high contrast. Use of the instrument for vibration analysis is compared with holographic techniques.

INTRODUCTION

When a rough surface is illuminated with laser light, it exhibits a speckled appearance. The same is true for a diffusing material which is trans-illuminated. The speckle-pattern arises from interference between the many individual wavelets scattered from different points on the surface, each of which has random phase with respect to its neighbours. Although there is a continual range of brightness of speckles in any one pattern, the light coming from each speckle has a well-defined amplitude and phase, so that optical interference experiments may be readily carried out with a speckle-pattern field, just as with a coherent beam of uniform intensity. The interference effect will not, however, be in the form of a recognisable fringe pattern, but will produce alterations in the detailed texture and contrast of the speckles. In this paper we describe a laser speckle interferometer in which the light scattered from a coherently lit rough surface is combined with a beam which has not suffered scattering. It will be shown that the visual appearance of the resultant field is extremely sensitive to small surface movements and vibration. In particular, such an interferometer can be used for direct observation of the nodal (i.e. stationary) regions of a vibrating surface.

THEORY

The distribution of light in a speckle-pattern results from the statistical summation of a large number of vectors of equal amplitude but random phase. Rayleigh (1920) treated this under the title of 'The problem of Random Flights in Two Dimensions', and showed that the brightness distribution followed an exponential function, the most probable brightness being zero. Thus

$$W(x) = e^{x} \tag{1}$$

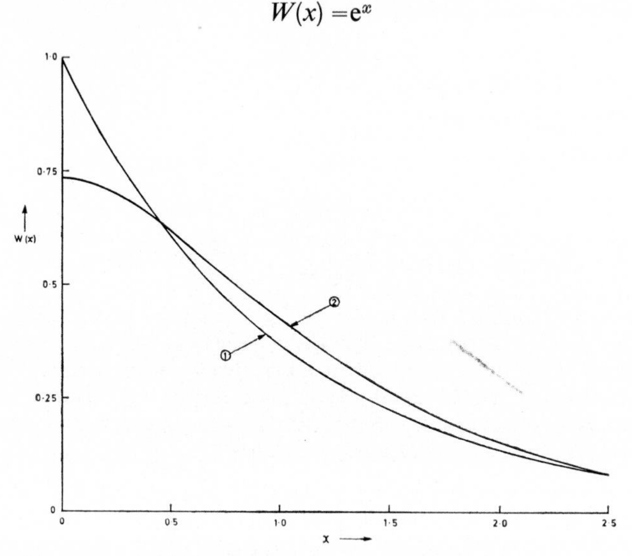

FIG. 1. Speckle-pattern brightness, distribution I: (1) Speckle-pattern alone. (2) Combined with uniform coherent background (of similar brightness to average speckle brightness).

where $W(x)$ is the probability of a sample point having brightness x. The average brightness of a speckle in this case will be given by $x = 1$.

If the speckle-pattern is mixed with a uniformly bright field of coherent light whose brightness is similar to that of the mean speckle brightness, the two fields interfere and produce a speckle-pattern of modified brightness distribution. It can be shown that the new distribution takes the form

$$W(x) = 2 \exp\left[-(2x+1)\right] I_0(2\sqrt{2x}) \tag{2}$$

where I_0 is the Bessel function of zero order with imaginary argument (i.e. $I_0(z) = J_0(iz)$).

Equations (1) and (2) are plotted in Figure 1 as normalized probability distribution curves. It is seen that the modified speckle-pattern has a brightness probability distribution not very different from that corresponding to the speckle-pattern alone; the proportion of completely dark speckles is only somewhat lower.

As outlined in the Introduction, a movement of the surface from which the speckle originates will give rise to phase changes between the two interfering waves, and each individual speckle will undergo a cyclic change in brightness. The change in brightness from maximum to minimum will correspond to a quarter of a wavelength of motion in the line of sight. For a slow speed movement the speckle-pattern will appear to 'twinkle'. For rapid motion the twinkling rate will be too fast for the eye to follow,

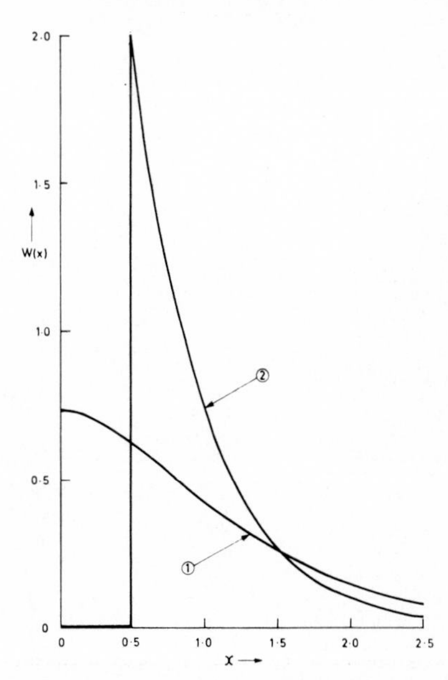

FIG. 2. Speckle-pattern brightness, distribution II: (1) Speckle combined with uniform coherent background of similar brightness. (2) Combined with uniform incoherent background of similar brightness.

and the speckles will become blurred. From the theoretical point of view, the effect will be the same as though the speckle field and uniform field were being added incoherently, since on average their relative phases can be thought of as having a random value. The new probability distribution of brightness which results is now

$$\left.\begin{aligned} W(x) &= 0 \text{ for } x < \tfrac{1}{2} \\ W(x) &= 2 \exp[-(2x-1)] \text{ for } x \geqslant \tfrac{1}{2} \end{aligned}\right\} \quad (3)$$

There will now be no completely dark speckles, the lowest brightness value in fact being half the average brightness. In Figure 2 the distributions given by Equations (2) and (3) are plotted together for comparison. It is seen from this that the 'blurred' field has the majority of its speckles within a narrow band of brightness.

It is instructive to compare the effects of interfering a speckle-pattern and a uniform field with the effect of interfering two speckle-patterns. In the latter case, if the interference takes place coherently, a third speckle-pattern will be produced, differing in detail, but having the same probability distribution of brightness of speckle. On the other hand, if the two speckle-patterns are superimposed incoherently, a new distribution results, which can be shown to be of the form

$$W(x) = 4xe^{-2x}. \tag{4}$$

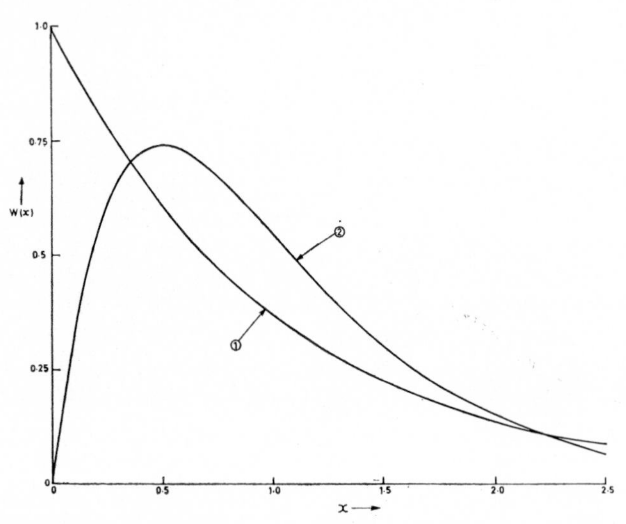

FIG. 3. Speckle-pattern brightness, distribution III: combination of two speckle-patterns of equal brightness, (1) coherently, (2) incoherently.

This probability curve is plotted in Figure 3, together with the distribution curve for two speckles combined coherently. It can be seen that, in the incoherent case, although there are no completely dark speckles, the brightness spectrum is otherwise filled. This implies that it may be a matter of some difficulty to distinguish between two fields of view in which the combination takes place coherently in one, and incoherently in the other.

The size of the speckles in an individual pattern is related to the aperture with which they are viewed. Theory indicates that there is a continuous distribution of speckle sizes, down to a minimum size which has an angular subtense of the order of λ/a, where a is the aperture diameter, λ the wavelength. Thus by reducing the viewing aperture, e.g. by provision of an iris diaphragm in front of the eye, the speckle size may be increased to a value where it may be clearly seen. The apparent brightness of the surface will of course be reduced accordingly.

THE SPECKLE-PATTERN INTERFEROMETER

In order to detect surface movements and vibration we have designed an instrument that is essentially an unequal path interferometer, in which two superimposed fields are presented to the observer, one being the laser light scattered from an object, and the other a uniformly bright field of comparable intensity. Figure 4 shows the optical system employed. Light from a helium-neon laser, polarized in the vertical plane, is diverged by means of a short focus lens on to the object under examination, which is placed some distance away. The divergent beam first passes through a

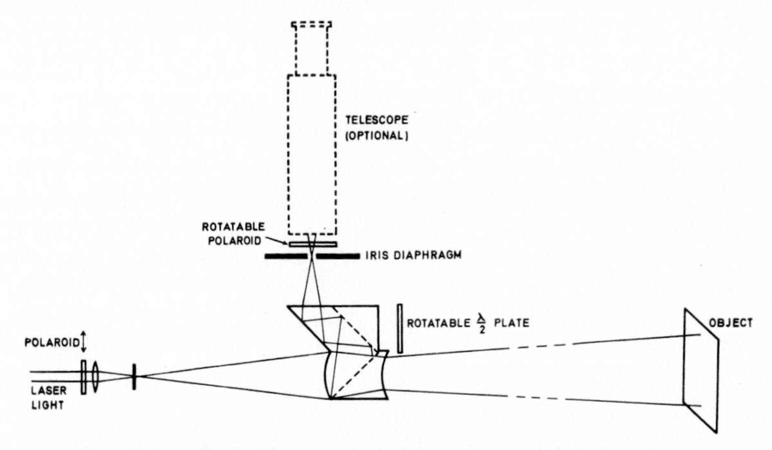

FIG. 4. Optical layout of speckle-pattern interferometer.

beam-splitting prism assembly, entering at a convex spherical surface and leaving by a concave surface of similar curvature. The degree of divergence of the object illuminating beam is thus not appreciably altered. The light scattered by the surface is received at the eye by looking through the top half of the prism assembly, which incorporates a 45° total internal reflection prism. Viewing can be carried out either directly or with the aid of a telescope. An iris diaphragm is positioned in front of the eye or at the entrance pupil of the telescope objective so that the size of the speckles can be varied. Reducing the aperture increases the speckle size but at the same time reduces the overall apparent brightness.

The uniformly lit 'reference' field of the interferometer is obtained by reflection at two 45° coated surfaces within the prism block, and an adequate field of view is obtained by convergence of the originally diverging laser beam by refraction at the first spherical surface of the prism assembly. The degree of convergence is designed to make the reference field about twice as wide as the object viewing field, so as to give it as uniform a brightness as possible. The reference beam is focused through the centre of the iris diaphragm aperture so that it is not attenuated in any way.

With the object placed at a convenient distance away from the prism assembly it is found that the reference beam needs considerable attenuation to match its brightness with that of the light scattered from the object. In order to achieve this, the dielectric beam-splitters within the prism block are arranged to have a reflectivity of about 1 per cent for the parallel polarization. A polaroid disc mounted above the prism assembly can be rotated to further reduce the reference field brightness and to ensure that light of only the same polarization is transmitted from both object and reference beam. For most surfaces illuminated with polarized light, the light scattered back is depolarized, so that at most one half of it is utilized in interference. However, for objects which have a degree of specular reflection this may not be the case. A rotatable, half-wave plate which affects only the light received from the object is thus incorporated in the instrument. This rotates the plane of polarization of the polarized component reflected by the object, so that it is accepted by the analysing polaroid. The maximum brightness of object field can thus be used.

Since the interferometer is working with unequal path lengths, coherency of the interfering beams is important. By positioning the object at such a distance from the prism assembly that the difference in path lengths is equal to an even multiple of the laser resonator length, good coherence of the two beams can always be achieved. This means in practice that the object has to be placed approximately at a multiple of the resonator length away from the prism assembly.

To achieve the stability required for interferometric applications, the laser beam-expander and prism assembly are mounted on a rigid optical bench, and the components held kinematically. For alignment of the object and reference fields, the prism assembly is held in a mount capable of rotating about a vertical axis. A telescope of about $\times 3$ was found most suitable for viewing the object.

OPERATION OF LASER SPECKLE INTERFEROMETER

For observation of small movements and vibration modes of an object, it must be held rigidly with respect to the prism assembly; mounting on a steel table is convenient for small objects. When viewing, the speckle size is first adjusted by varying the iris diaphragm until the pattern is sufficiently large for easy viewing. About $\frac{1}{2}$ mm diameter for the unaided eye, or m times this value for a telescope of magnification m, is found to be optimum. This corresponds to an apparent minimum speckle size of about 5 minutes of arc. The analysing polaroid is then rotated until the reference beam is of comparable brightness to the illuminated object.

The principal use of the laser speckle interferometer is in the visual determination of the nodal regions of vibrating surfaces. These will show up as regions of high contrast speckle against a blurred background which

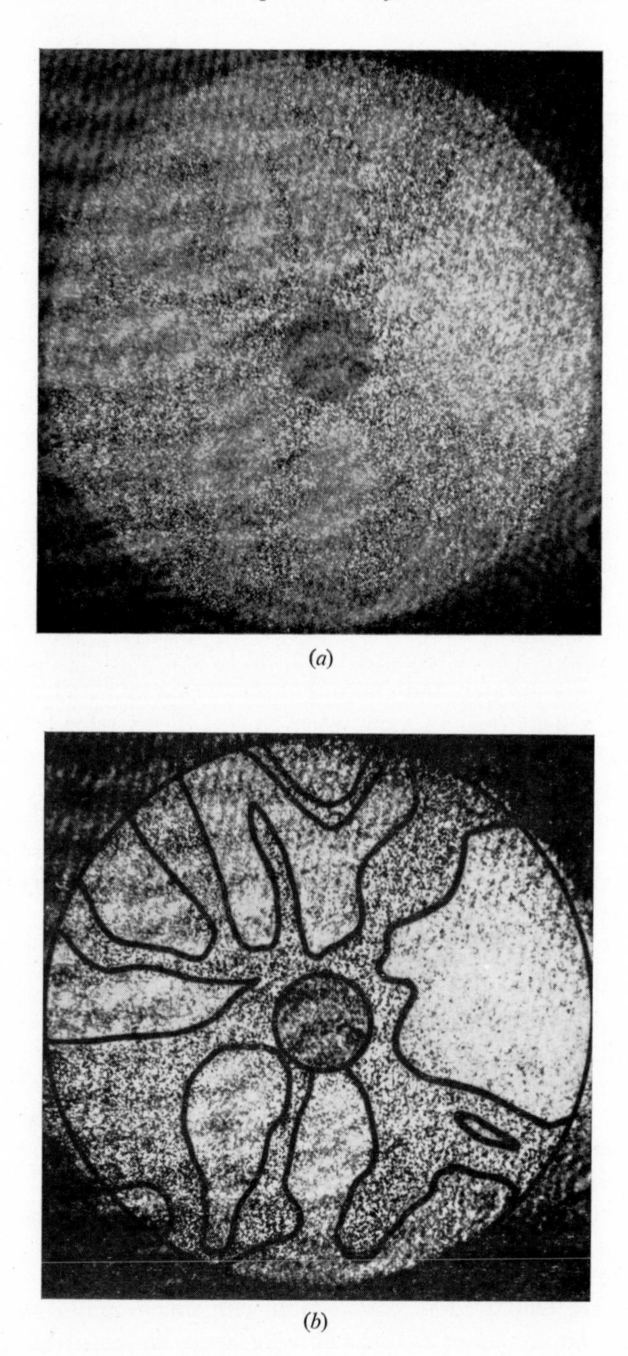

(*a*)

(*b*)

FIG. 5. (*a*) View, through interferometer, of loudspeaker cone resonating at 1·96 kHz (*b*) Similar photograph with nodal areas outlined.

exhibits no dark speckles. Figure 5*a* shows a photograph of the effect when a loudspeaker cone was energized at a frequency of 1·96 kHz. The narrow nodal regions radiating from the centre are shown more clearly when outlined as in Figure 5*b*. In Figure 6*a* the appearance of a flat brass plate, resonating at 4·28 kHz, indicates a complicated nodal pattern. The position of these nodes was confirmed by recording a time-averaged hologram of the surface vibrating under similar conditions, shown in Figure 6*b*. According to Powell & Stetson (1965), the brightest fringe in the reconstructed image will delineate the nodal regions, and these are seen to be in the same positions as the regions of high-contrast speckle of Figure 6*a*.

Instead of photographing such regions of high contrast speckle it is much easier to observe them visually, especially if they are twinkling as a result of any slow changes in path lengths of the illuminating and reference beams. If twinkle at a suitable frequency is not already present owing to atmospheric disturbance or thermal drift, it can be introduced deliberately; for example, by putting into one of the beams a gas cell the pressure of which can be varied by means of a squeeze bulb.

A simple speckle interferometer has also been constructed in which the reference beam is a speckle-pattern, derived by transmitting the reference beam through a diffuser, and omitting the convex lens which is necessary to produce a wide field of view when an undiffused reference beam is required. In this instrument the speckle brightness distribution statistics of Equation (4) apply to the vibrating regions, and there is less difference in appearance of the vibrating parts as compared with a stationary area of surface; insufficient, in fact, to be noticeable in a photograph. Only when the artificial 'twinkle' effect is added can the nodal regions be discerned visually, and then only with difficulty.

APPLICATIONS AND LIMITATIONS

The speckle-pattern interferometer can be constructed as a self-contained instrument which may be brought up to a surface to examine it for slow movement or vibration. As such, it is purely an observational instrument, requiring no photographic processing and not needing the extreme conditions of stability that are required for holography. However, in the examination of vibrating surfaces the information yielded by the speckle interferometer is limited to showing the nodal regions, whereas holographic methods (Powell & Stetson 1965; Stetson & Powell 1965) also give the amplitudes of vibration over the whole surface. If stroboscopic techniques are employed in conjunction with holography, information about the relative phases of the vibrating areas may also be obtained (Archbold & Ennos 1968; Watrasiewicz & Spicer 1968; Shajenko & Johnson 1968; Ostrovsky & Zaidel 1968).

(*a*)

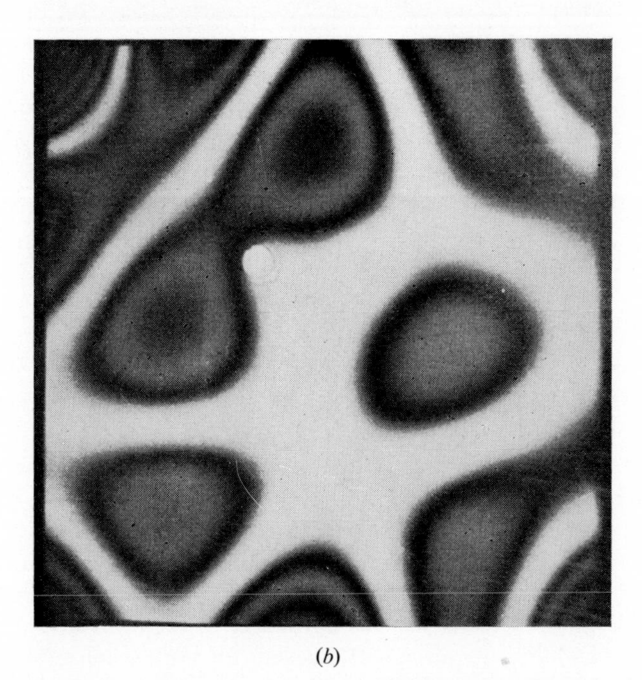

(*b*)

FIG. 6. (*a*) Square plate vibrating at resonant frequency of 4·28 kHz, viewed in speckle interferometer. (*b*) Time-averaged hologram reconstruction of same square plate vibrating under similar conditions.

When using the speckle interferometer to study vibration patterns, a steady drift in the mean position of the surface is no disadvantage; the 'twinkle' it induces in the nodal regions may indeed be useful in identifying them. On the other hand, a small drift of the surface occurring during the exposure of a time-averaged hologram will erase the fine fringe detail and result in no reconstructed image being obtained. This may rule out the use of holography for applications where the vibrating surface is made of paper, plastic, or rubber, which all tend to 'creep' with time.

The laser power required to observe the speckle effects in comfort will of course depend upon the extent of the surface illuminated and its degree of reflectivity. The apparent brightness is necessarily reduced by the need to restrict the eye pupil size so that speckles may be seen clearly. A comparatively high power laser is therefore necessary. As an example, it has been found necessary to use a 10 mW helium-neon laser to observe speckle effects on a 10 cm square plate painted white. The exposure time required to record a time-averaged hologram of a similarly illuminated object would be about 1 second, using Agfa-Gaevert 10E70 plates.

The effectiveness of the speckle interferometer depends very much on having a reference beam of uniform intensity, without additional superimposed patterns due to diffraction at specks of dust or interference between beams inter-reflected from the prism faces. The present design of prism assembly has overcome many of the sources of these unwanted patterns, but some residual effects still remain; the cause of these has been traced to mismatch in refractive index of the glass and the optical cement used to assemble the unit. A modification of the semi-reflecting film material and the optical cement should, however, overcome this.

CONCLUSIONS

A laser speckle interferometer has been described in which the speckle-pattern derived from laser light scattered at a surface is made to interfere with a uniform field of comparable brightness. The application of the instrument to detecting surface movements and vibration are outlined and compared to holographic methods. As a purely observational instrument the speckle interferometer has the advantage of high sensitivity, without the severe stability requirements needed for holography; the output information is, however, qualitative.

The work described above has been carried out at the National Physical Laboratory.

REFERENCES

ARCHBOLD, E. and ENNOS, A. E., 1968, *Nature*, **217**, 942.
OSTROVSKY, Y. I. and ZAIDEL, A. N., 1968, *Soviet, Phys. tech. Phys.*, **38**, 1824.
POWELL, R. L. and STETSON, K. A., 1965, *J. Opt. Soc. Amer.*, **55**, 1593.
RAYLEIGH Lord, 1920, *Scientific Papers*, Volume 6, Cambridge University Press, p. 565 and 610.
SHAJENKO, P. and JOHNSON, C. D., 1968, *App. Phys. Lett.*, **13**, 44.
STETSON, R. A. and POWELL, R. L., 1965, *J. Opt. Soc. Amer.* **55**, 1694.
WATRASIEWICZ, B. M. and SPICER, P. 1968, *Nature*, **217**, 1143.

Advantages of Multipass Interferometry

P. Langenbeck

Perkin-Elmer Corporation,
Norwalk, Connecticut

Abstract—Multipass interferometry is a method of multiplying the relative interferometric sensitivity. Different means are described of producing interference between a beam that has been reflected n-times (or transmitted) by the same object and a reference beam. Fringes are thus $\lambda/2n$ instead of $\lambda/2$ contour lines. Special attention is given to a modification of the well-known Fizeau interferometer in such a way that both multiple beam $\lambda/2$ fringes and $\lambda/2n$ fringes can be produced at will. The order of reflection n is selected in the focal plane of a lens, where an image of the light source is produced by each reflected beam. The luminance of the reference beam is matched to that of the nth order beam by means of controllable polarized light optics. Emphasis in this article is on various applications like measurement of thickness of thin films, dilatometry, surface mapping and homogeneity measurement. Multipass interferometry allows measurement by visual inspection of interferograms to be done with a precision that with $\lambda/2$ fringes requires densitometric evaluation.

INTRODUCTION

There is an increasing need for higher precision in the interferometric inspection of topography, typical specifications being 'The Reflected or Transmitted Wavefront to be Smooth to $\lambda/20$ to $\lambda/200$'. These tolerances occur in such common measurements as that of the homogeneity of glass plates, or angle errors of prisms, and especially, of the planeness of mirror surfaces. (All of these problems are met simultaneously, for instance, in the manufacture of triple mirror prisms.) The errors looked for are expressed in terms of optical path difference and they are met with in zones characterized by low spatial frequencies. In these cases, the microstructure of the surface and the root-mean-square departure of the real from the ideal wavefront is not so much of interest as is the exact knowledge of the maximum deformation depth in a certain area. Lateral locating accuracy must and can be sacrificed for longitudinal resolution. The precision of this type of interferometric measurement depends upon the magnification with which the depth, h, of a locally confined deformation of an otherwise plane surface can be recognized in the interference pattern (of mean fringe spacing P) as a lateral departure ΔP of the fringes from nominal straightness, $h = \lambda/2(\Delta P/P)$; ($\lambda$ =wavelength). The limit of sensitivity is reached, when $\Delta P/P$ reaches its minimum detectable value and one, therefore, has recourse to two procedures for enhancing precision: Increasing the sensitivity by magnifying the fringe spacing and improving the fringe reading accuracy.

Adjusting the spacing of the fringes to be as wide as the field of view permits, results in a decrease in the number of fringes and hence reduces the 'easily' available information in the areas between the fringes (since most fringe reading methods consider only the determination of the locations of the fringe maxima or minima). In addition, the increased fringe spacing leads to a reduction of the definition of the maximum or minimum of a fringe and the gain in sensitivity becomes meaningless as a result of decreased detectability.

On the other hand, there are sophisticated methods which allow one to determine the fringe position on an interferogram. Accuracies of $\lambda/500$ to $\lambda/1000$ in two-beam interference fringes have been reported (Gros & Roblin 1967; Lau & Krug 1967). Comparators, micro-densitometers or refined photographic processes, and a computerized analysis of the interferometric data are needed in order to obtain the desired precision; such procedures have been described in this conference by Scott (1969) and Shannon (1969).

In a majority of cases, this expensive effort would be either unnecessary or undesirable, if interferometers with higher sensitivity were available that permitted the *direct visual interpretation* of interferograms to accuracies comparable with those at present being achieved with the aid of instrumentation. In particular, those who have no easy access to a microdensitometer, would be helped by an improvement in visual interferometric testing methods.

Multipass Interferometry

'Multipass Interferometry' is a practical way of achieving higher precision in interferometric topography. In a multipass interferometer, a collimated beam is reflected repeatedly from the same test-mirror, say n times, whereby this beam adopts n times the deformation that it would have in only one reflection. Such a multipass beam is made to interfere with a neutral reference beam; this results in fringes that are interpreted as $\lambda/2n$ instead of $\lambda/2$ fringes.

A multipass beam is typically produced in a multiple beam Fizeau interferometer and the essence of multipass interferometry is to select from the entire series of beams just the one desired, say the one of the nth order of reflection. The beam is selected in the focal plane of the interferometer objective, where each beam from the multiple beam interferometer comes to a separate focus, by arranging a slit or pinhole that stops all the other beams.

A neutral reference beam can be procured in two different ways:

(*a*) The multiple-beam interferometer forms one arm of a Twyman (or Michelson) interferometer. The reference mirror in the other arm is aligned so that the beam reflected from it passes along the same path as the selected n^{th} order of reflection beam, and thus $\lambda/2n$ two-beam fringes are obtained that can be aligned for spacing and orientation by means of the reference mirror. This instrument has been described as the Multipass Twyamn Green Interferometer (TGMI, Langenbeck 1967). Omitting the focal plane stop, this is the same as the set up described by Stulla-Götz (1932).

(*b*) The beam reflected at the partially transmitting front mirror of the multiple-beam interferometer is used as the reference beam (Figure 1); to achieve this the focal plane stop has to have a second opening, through which the reference beam ($=0^{\text{th}}$ beam) may pass. Due to the necessarily large angle between the zeroth and the nth order beam, the resulting two-beam interference fringes will be quite narrowly spaced, and this interference pattern is best observed by a Moiré technique; a Moiré grating will transform the interference pattern into Moiré fringes. If the Moiré grating line spacing m equals the average fringe spacing P, the Moiré fringes (spacing M) can be interpreted as two-beam interference fringes of the same spacing: $\Delta M/M = \Delta P/P$. This instrument has been introduced as the Multipass Moiré Fizeau Interferometer (FMMI, Langenbeck 1967).

For illumination a He:Ne laser is most suitable.

Luminance Control in the Reference Beam

The luminance in the successive multipass beams decreases with the order of reflection and in order to obtain the best interference contrast the luminance of the reference beam must be controlled; a suitable method is by using polarizing optics such that the illumination is linearly polarized. In the FMMI the reference light is reflected from an uncoated glass near the polarization angle; thus its luminance is cut down to the order of that of the multiple reflected beams. The other beam is reflected from a metal coating, without any appreciable loss of its luminance. Rotation of the plane of polarization allows one to adjust the desired attenuation of the reference beam with respect to the specific n^{th} order of reflection beam.

In the TGMI there are several methods of luminance control in the reference beam, among which the use of a low reflection reference mirror is the easiest. A wide range of controllable luminance is obtained by using a prism in total internal reflection in the reference beam.

Limitations of Multipass Interferometry

The two chief causes of errors in Ml are the beam 'walkoff' and a misalignment of the focal plane stop.

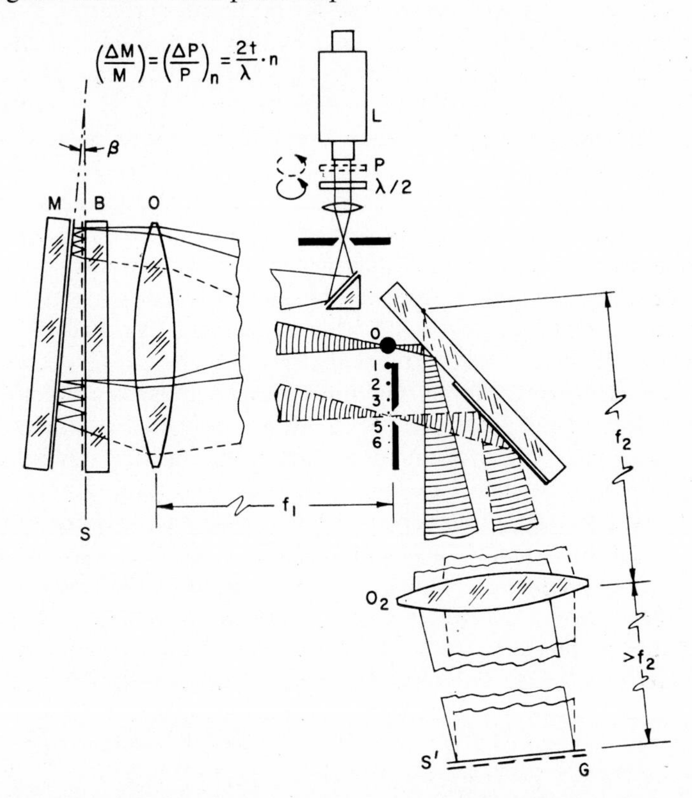

FIG. 1. Multipass Moiré Fizeau interferometer with polarized light luminance control. Interference is produced between the zero'th beam and any of the higher order reflection beams. Moiré grid *G* is used to inspect the narrowly spaced fringes. (Fringe spacing is narrower because of the large distance between light source-images.) Fringe adjustment and order selection is achieved by tilting the back mirror *M*.

The multiple reflected beam integrates the departure from planeness in two dimensions, in depth (desirable) and in width, perpendiculaı to the apex of the wedge angle (unavoidable). Misinterpretation may arise when too high an order of reflection is applied. Fringes, or fringe shifts, may be seen where there is no associated deformation of the test surface.

The walkoff W_n as a function of the order of reflection n is given by

$$W_n = [n^2 + n(n-1)]\beta t,$$

which for higher n's becomes the known expression for the walkoff in multiple-beam interferometry, $W_n = 2n^2 t\beta$, where t is the spacing of the two mirrors and β the angle between them; the dependance on n^2 is particularly disturbing. For a given interferometer the order of reflection must be limited so that the walkoff remains sufficiently smaller than the lateral extension of the defects looked for in the interferometer. For large mirror spacing, as is required for hogomeneity testing (one of the chief applications of multipass interferometry) the walkoff can amount to several mm. The domain of MI. is thus the investigation of fields with small OPD changes occurring over zones characterized by low spatial frequencies.

A technical remedy might be mentioned here, that allows the walkoff problem to be overcome to some degree; the interferometer is tilted with respect to the incoming beam so that the beam is first reflected towards the apex of the wedge and then, upon reaching either mirror normally, returns in its own path. Thus, if the 6th beam is selected, the walkoff will be effectively only that of the third beam.

Serious misinterpretation of multipass interferograms can result from a misalignment of the focal plane stop. In any case this stop acts like a spatial 'low pass' filter, eliminating from correct observation all object details of width g smaller than about $g \approx \lambda f/2\xi_0$; f=focal length, ξ_0=width of focal plane stop. It is found, however, that with practical dimensions for f and ξ_0, that the quantity g as determined by walkoff is larger by about two orders of magnitude than the g as determined by the restriction (ξ_0) in the focal plane. If, however, the stop should accidentally cut into the low spatial fiequencies, strong diffraction effects can be expected, as in Foucault or Schlieren techniques. Technically, such potential misalignment can be found out by slightly transposing the focal plane stop: only in the case of misalignment will the inteiference fringes change their appearance rapidly, otherwise the fringes will show no effect.

Typical Applications of Multipass Interferometers

Measuring Flatness. Figures 2*a* to 2*d* show some test interferograms obtained by TGMI. Obviously, it is of some advantage to use MI to check a mirror during the final stage of producing an optically flat surface. To see the advantage fully, these multipass interferograms must be compared with a normally produced interferogiam of such a mirror; this is shown in Figure 2*e*. In the Fizeau multiple beam interferogram (Figure 2*e*) local defects of up to $\lambda/4$-depth may not be discovered when they lie between the fringes; to recover this lost information one would normally switch from the axial interferogram (rings) to a normal interferogram (straight fringes) and adjust for a large number of fringes. With a few straight fringes the information could still be lost; with many fringes the sensitivity is reduced. The multipass interferogram fills the information gap, both in axial and normal alignment. In axial alignment (alignment by reference mirror), the advantage of the multipass over the Fizeau is especially important. A uniform tint multipass interferogram of, say, the fifth order will reveal almost any detail above $\lambda/100$ if only 10 per cent change in brightness is taken into account, while in the multiple beam interferogram, any detail below about $\lambda/8$ between fringes could barely be measured visually.

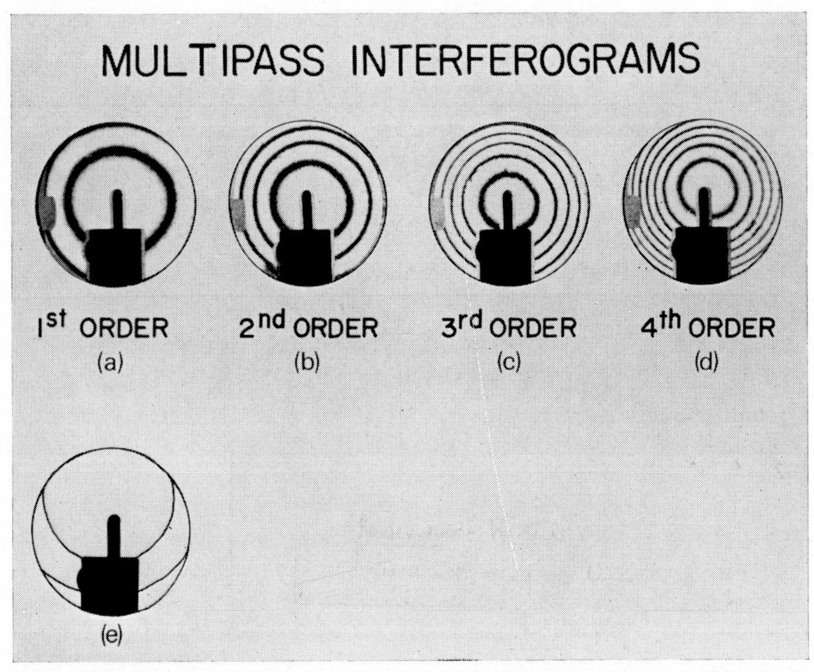

FIG. 2. (*a–d*) Multipass interferograms of a nominally plane mirror, obtained in the TGMI. Fourth order (2*d*) shows $\lambda/8$ fringes. (*e*) Multiple-beam Fizeau interferogram of the same surface ($\lambda/2$ fringes) obtained in the same interferometer, but without focal plane stop.

Grazing Incidence Interferometry. Oblique or grazing incidence is applied to the interferometrical testing of extended flat surfaces, which for lack of sufficiently large components cannot otherwise be tested. The advantage of being able to test very large surfaces interferometrically with small diameter components at grazing incidence is bought at the expense of sensitivity. The loss in sensitivity can be compensated for by employing MI. The walkoff on the large surface can be reduced using the aforementioned off-axis illumination.

Homogeneity Measurement. Measurement of homogeneity of glass plates is especially delicate if variations in index of refraction produce optical path differences on a transmitted beam of the same order of magnitude as do the departures from flatness of the two surfaces. These two contributions can be separated from each other by an interferometric method, given by Twyman and Perry (1922). The sensitivity of this method is limited because it is based upon the use of an internal Fizeau two-beam interferogram, that cannot be adjusted except by polishing the piece again.

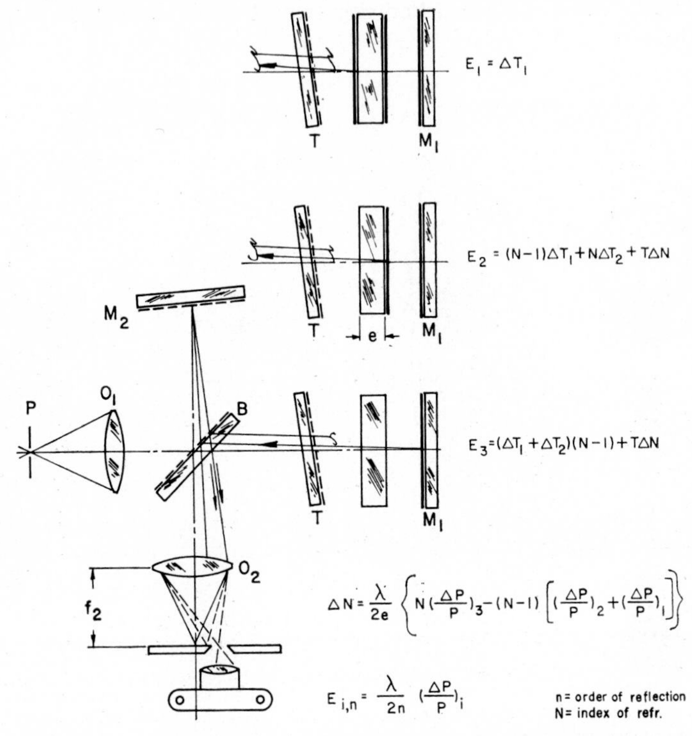

FIG. 3. Homogeneity measurement on glass plates using a method derived from Twyman and Perry (1922). TGMI is here shown, but FMMI can also be used (for larger plates). Using the known Twyman-Perry method two different interferograms are generated of a test glass which involve the surface and the interior once in a reflection relation and once in a transmission relation. This method has been modified by Langenbeck in the way outlined in this figure in order to obtain better sensitivity in homogeneity testing by the use of multipass interferometry.

The author has introduced an alternative method using multipass interferometry (Figure 3). The test plate is first coated on both surfaces. The measurement involves three steps: determining the front surface of the test glass (interferogram #1); cleaning off the front surface and taking interferogram #2 of the back surface; finally, cleaning off the back surface and taking a transmission interferogram, #3. The local fringe displacements are measured and the index variation is computed from the equation given in Figure 2. The minimum detectable ΔN is

$$\Delta N_{min} = 1{\cdot}5 \times 10^{-5}/en \times (\Delta P/P)_{min};$$

(with $\lambda = 6 \cdot 10^{-5}$ cm), e = thickness of glassplate (cm) and n = order of reflection.

Besides offering higher sensitivity, this method has other advantages over the original Twyman-Perry version; the fringes can be aligned at will, for instance, to be of the same spacing and at the same orientation in all three interferograms. This results in a convenient manual evaluation.

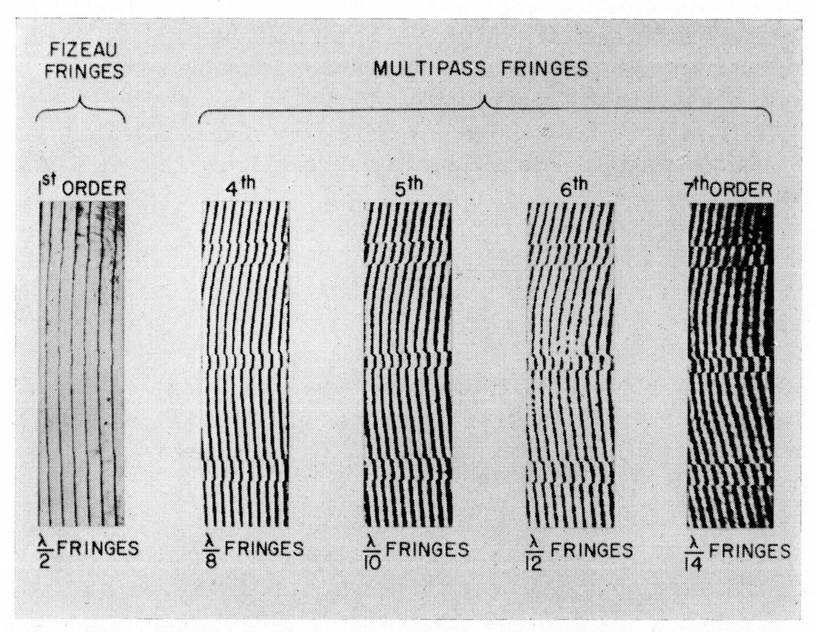

FIG. 4. Multipass interferometry applied to the measurement of thickness of thin films.

Finally, one can use the same interferometer both in a multiple-beam alignment and in a multipass alignment. Mr. E. Roberts of Perkin-Elmer has prepared a program for computerized data evaluation of homogeneity measurements using this method (1969).

Measuring Angular Stability and Differential Angles. Interference produced between an nth order of reflection beam and an independent reference beam results in fringes of spacing P, that are more sensitive to adjustment of the back mirror in Figure 1, the higher the value of n. With the reference beam in a constant position one can detect changes in angular alignment as small as $\mathrm{d}\alpha = (\lambda/nP) \times (\mathrm{d}P/P)$. With $\lambda = 6 \times 10^{-5}$ cm and $(\mathrm{d}P/P) = 0{\cdot}05$ one can detect about $\mathrm{d}\alpha = 0{\cdot}6/nP$ arc sec. between the mirrors. This implies applications such as precision interferometric pointing and especially the precise determination of 90° or 45° angles on roof prisms, Koestei prisms, pentaprisms and triple mirror prisms.

Other Applications

Multiplying sensitivity by MI is useful in some other applications. In thin-film, thickness measurement, small surface steps will appear magnified and will become easier to determine (see Figure 4). In dilatometry and distance measurement MI will provide a finer calibration; precise distance measurement also suggests better precision in determining the refractive index of glass plates. Combination of Moiré techniques with MI (as in the FMMI) leads to other useful techniques: the use of an interferogram of a standard surface (such as a Hg-mirror) as a Moiré grid in the FMMI allows Moiré interferograms to be obtained that are free from the errors of the Fizeau beam-splitter. The Moiré between an nth and an $(n+1)^{\text{th}}$ order of reflection interferogram finally allows one to separate errors such as lens aberrations or poor collimation from those of the interferometer mirrors.

The Twyman-Green of Fizeau techniques are not only means of inspecting a multipass beam; the focal plane stop, for example, can be used as a knife-edge itself, resulting in an n-times more sensitive Foucault test. Also any shearing device, such as a Wollaston, can be used to look at the multipass beam.

CONCLUSION

In multipass interferometry $\lambda/2n$ fringes replace $\lambda/2$ fringes (n =order of reflection). Multipass interferometry, however, is not the only means of obtaining higher depth resolution in interferometry: the use of Brewster's superposition fringes in its modern form (Schwider 1967) also divides the $\lambda/2$-order distance into n steps. In order to fill the gap between the $\lambda/2$ fringes with more fringes, Herriott uses multi-wavelength multiple-beam interferometry (1961).

Multipass interferometry reduces the problem of producing $\lambda/2n$ fringes to a practical operation that does not require special or new equipment; it can be used with most existing Twyman or Fizeau interferometers. The selection of the order of reflection is simple and fast. MI also provides for control of interference contrast.

Some of the MI applications have been described. In addition, it might be mentioned that multipass interferometry was an essential part in the interferometric testing of the Perkin-Elmer triple mirror prisms that are contained in the laser reflector now resting on the moon's surface.

REFERENCES

Gros, N. and Roblin, G., 1967, *Rev. Opt.*, **46**, 249.
Herriott, D. R., 1961, *J. Opt. Soc. Amer.*, **51**, 1142.
Langenbeck, P., 1967, *Appl. Optics*, **6**, 1425.
Langenbeck, P., 1969, *Appl. Optics.*, **8**, 543.
Lau, E. and Krug, W., 1967, *Equidensitometry*, Focal Press.
Roberts, E., 1969, *Appl. Optics*, **8**, 231.
Schwider, J., 1968, *Opt. Acta*, **15**, 351.
Scott, R. M., 1969, This volume, p. 63.
Shannon, R. R., 1969, this volume, p. 331.
Stulla-Götz, P., 1932, *ZS. Instrkde*, **52**, 521.
Twyman, F. and Perry, J. W., 1922, *Proc. Phys. Soc.*, **34**, 151.

T

Photographic Emulsions as Phase Objects in Coherent Optical Systems

John Mack Hood*

University of Reading,
Reading, Berks.

Abstract—The use of photographic materials for constructing phase objects such as phase filters in an optical processing system is made possible by the surface contouring effects evident in exposed portions of the emulsion. A series of gratings was produced on type 649F plates and the contouring effects as a function of exposure and spatial frequency studied. Interferometric measurements show the relationship between exposure, spatial frequency, and phase depth where phase depth refers to the phase variations induced in a plane monochromatic wave reflected from the aluminized contoured emulsion surface. At very low spatial frequency a linear relationship between exposure and phase depth is suggested by the measurements. At higher frequency a non-linear relationship is evident.

Matrix methods have been devised for the calculation of the images formed in coherent light of the phase object. The formulae derived are based on the spectrum amplitude function derived from a one-dimensional sinusoidal phase object extended to a Fourier series to allow for periodic but non-sinusoidal objects.

The matrix technique provides a means for ordering the terms in the spectrum. A compact notation of considerable generality was devised for manipulating and discussing the process of diffraction by periodic phase objects.

Images were calculated and found to be comparable to measurements made from the experimental materials.

INTRODUCTION

The contouring of a photographic emulsion surface which has been exposed and developed as a function of its exposure suggests the application of the effect to practical requirements for phase objects. A phase object can be considered to be any object which alters the phase of the incident light but not its amplitude after reflection or transmission. The questions which must be considered in making use of this phenomenon are those involving the magnitude of the effect, whether it is easily controlled, and its dependency on parameters such as spatial frequency. Hannes (1968) has made measurements which deal particularly with the phase effects on transmitted light in bleached emulsions. More easily controlled perhaps is the surface

* Present address: Naval Electronics Laboratory Center, San Diego, Calif., U.S.A.

contour effect on reflected light and that is the particular emphasis of this paper.

A theory adequate to deal with the coherent imaging of these phase objects is also needed in order to apply the effect in microscopy, holography, or coherent optical processing. Hopkins (1951) solved a portion of this problem for certain special classes of objects. Burckhardt (1966) has given the exact solution for thick, dielectric phase gratings. Here the theory is extended to a means of practical computation for objects which can be classified as one-dimensional and periodic. In principle two-dimensional non-periodic phase objects could be considered but the computational methods developed become cumbersome.

A Theory of Phase Screen Diffraction and Imaging

An object which gives to an incident plane wave of light either on reflection or transmission a form

$$f(u, v) = \exp[i\Phi(u, v)] \tag{1}$$

can be considered a phase object. If $\Phi(u, v)$ is small compared to unity then the expansion of the exponential can be truncated to the linear form

$$f(u, v) = 1 + i\Phi(u, v). \tag{2}$$

With attenuation of the constant term and quarter-wave phase shifting of the other term with respect to the constant we have the basis for the standard technique of phase microscopy (Zernike 1942).

A more general treatment of phase diffraction and imaging presents many difficulties (Berry 1966). These complexities are well known to communications scientists where frequency and phase modulation are frequently used techniques (Kharkevich 1960). If Equation (1) can be written as

$$f(u, v) = \exp[i \sum_m \beta_m \cos(2\pi m s_0 u + \phi_m)] \tag{3}$$

then each factor can be written in the Jacobi series form

$$\exp[i\beta_m \cos(2\pi m s_0 u + \phi_m)] = \sum_{n=-\infty}^{+\infty} i^n J_n(\beta_m) \exp[in(2\pi m s_0 u + \phi_m)]. \tag{4}$$

This sum is just one term of the Fourier series in the exponent in Equation (3). Taking all the terms the final expression becomes:

$$f(u, v) = \prod_{m=1}^{m} \sum_{n=-\infty}^{+\infty} i^n J_n(\beta_m) \exp[in(2\pi m s_0 u + \phi_m)] \tag{5}$$

A typical term of the expansion of this expression is

$$[i^{(n_1+n_2+\ldots+n_m)}J_{n_1}(\beta_1)J_{n_2}(\beta_2)\ldots J_{n_m}(\beta_m)]\times$$
$$\exp[i2\pi s_0u(n_1+2n_2+\ldots+mn_m)]\times$$
$$\exp[i(n_1\phi_1+n_2\phi_2+\ldots+n_m\phi_m)]. \quad (6)$$

Note that each term in Equation (5) includes a phase factor which will transform to give a delta function at a particular location in the spectrum determined by the integer sum of products,

$$\text{Spectral Order Number} = n_1+2n_2+\ldots+mn_m. \quad (7)$$

These terms might be said to be degenerate since a given integral sum called here the spectral order number may be satisfied by an infinite variety of choices of the n values. It is recollected that n is the diffraction order for a simple cosine component of the phase grating. At appreciable values of phase depth, which is the argument of the Bessel series in the Jacobi expansion, the amplitudes in the diffracted spectrum can be quite high at large values of n. Since n can be either positive or negative in Equation (6) the spectral order number can be small for various combinations of large positive and negative n and thus included in even a small aperture. This accounts for the difficulties in predicting the appearance of coherently illuminated phase objects in optical systems. For example the following might be possible.

$$\text{Spectral Order Number} = (1)(4)+(2)(-1000)+(3)(0)+(4)(500) = 4. \quad (8)$$

In summary, the properties of the diffracted spectrum from a phase grating are that each element is composed of a sum of many terms and that these are a product of many factors which individually are like the diffracted orders from a simple amplitude grating or a pure cosine phase grating.

A Qualitative View Of Phase Screen Diffraction

A simple visualization of the process is suggested. In Figure 1 the grating is represented with its individual components separated and arranged in an optical train where the highest frequency component is illuminated first and its spectrum formed by the lens P_2 in plane P_3. These spectra are collimated by P_4 and in turn illuminate from several directions the second component in plane P_5. This forms a series of superimposed spectra at P_7. The elements of the different spectra lie on each other since each successive grating is a sub-harmonic of the first. As all the various elements arise from the same illuminating wave the components at each position are added coherently to determine the phase and amplitude of the source for the next successive diffraction. The process repeats to the final output plane shown as P_{11}. Since each of the discrete spectral elements here arise from many different diffracted orders the use of a term like *spectral order number* which avoids the suggestion of *diffraction order* is justified.

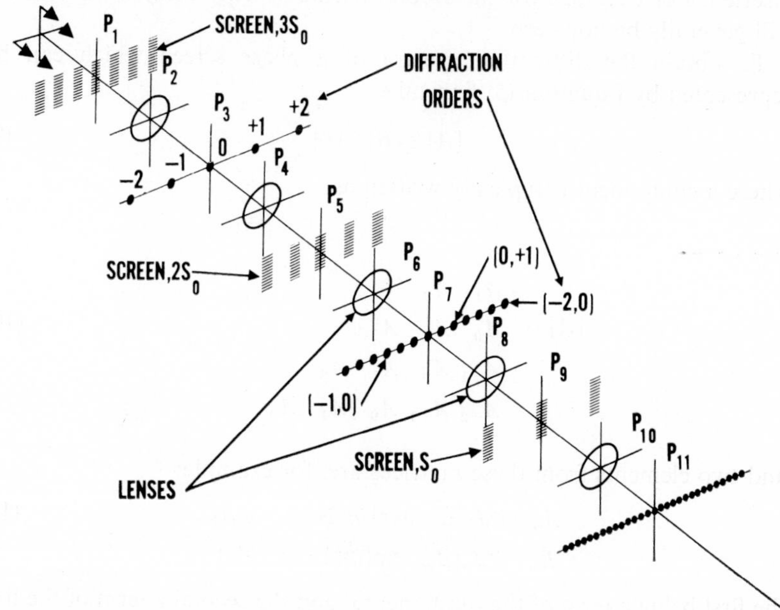

FIG. 1. Representation of phase screen diffraction as a series of successive diffractions by the individual Fourier cosine components of the original screen.

A Matrix Method for Calculating Phase Screen Diffraction Spectra and Images

The difficulty with the foregoing is that no practical computational method is evident. The form of the resulting terms of Equation (5) suggests the operation of matrix multiplication. Successive multiplications of series would build up the required sums of products for each of the final spectral elements if the terms for summing could be kept together with the correct spectral order number. From various trials it was finally discovered that a triangular matrix form for arranging the terms of the Jacobi series combined with a simple algorithm which includes successive matrix multiplications produced the required result.

The matrices are of odd order with the zero spectral order number term as the middle element of the first column. The size of the matrix is determined by the extent to which the Bessel function series gives appreciable values to the terms of the Jacobi series. The successive elements in this first column above and below the zeroth order term are the successively numbered positive and negative integers representing the spectral order numbers. Note that only the fundamental cosine grating has each element non-zero in general. For all m not equal to one, zeros are appropriately inserted. For example, for the second harmonic only every third element will generally be non-zero.

To obtain the diffraction spectra of a phase screen which can be represented by Equation (5) first take

$$[A] \times [B] = [\alpha] \tag{9}$$

where a component is typically written as

$$[A] = \begin{bmatrix} A_2 & & & & \\ A_1 & A_2 & & & \\ A_0 & A_1 & A_2 & & \\ A_{-1} & A_0 & A_1 & A_2 & \\ A_{-2} & A_{-1} & A_0 & A_1 & A_2 \end{bmatrix} \tag{10}$$

and two elements from these matrices are, for example:

$$A_n = i^n J_n(\beta_1) \exp[in(2\pi s_0 u + \phi_1)] \tag{11}$$
$$B_n = i^n J_n(\beta_2) \exp[in(4\pi s_0 u + \phi_2]$$

the first being a term of the fundamental and the second a term of the first harmonic. The index n can be either positive or negative. The α matrix in Equation (9) is a sort of partial product and if the screen could have been represented by just the two components it would be the positive half of the diffraction. An important and essential property of the matrices is that only positive values of n in the α-matrix result from the multiplication. Because the entire diffracted spectra is needed to continue the computation it is necessary to expand the α-matrix for the next step by copying the first column symetrically below α_0 but with negative n and appropriately filling out the triangular form. The operation is repeated until all components have been accounted for.

The phase term, $\exp[in(2\pi m s_0 u + \phi_m)]$, in each element transforms in the spectrum to a delta function which determines the position of the element. Finally, if it is desired to compute an image it is necessary to truncate the final spectrum by the appropriate aperture. The foregoing operations were complex so the spectrum limited to the $\pm l^{\text{th}}$ terms can be written as

$$a(x) = \sum_{-l}^{+l} (Sr + iS_i)\delta(x - ls_0) \tag{12}$$

where Sr and Si stand for the real and imaginary components of final matrix elements. The truncation cannot occur before this point because of the presence of all appreciable diffraction orders in every element of the spectrum. The synthesis of the complex amplitude in the image plane can then be written as:

$$\begin{aligned} A(u') = & S_{r,0} + 2S_{r,1} \cos 2\pi s_0 u' \\ & + 2S_{r,2} \cos 4\pi s_0 u' \\ & + \ldots \\ & + 2S_{r,l} \cos 2\pi l s_0 u' \\ + i[& S_{i,0} + 2S_{i,1} \cos 2\pi s_0 u' \\ & + 2S_{i,2} \cos 4\pi s_0 u' \\ & + \ldots \\ & + 2S_{i,l} \cos 2\pi l s_0 u']. \end{aligned} \tag{13}$$

In actually computing images by this method it is not necessary to carry out the complete formalism of matrix multiplication. One uses a complex linear array scheme which deals only with the first column of the triangular matrix with a simple indexing and iterative instruction. The details of these programs written in Algol-60 are described elsewhere (Hood 1969*a*).

Phase Gratings in Photographic Emulsions

In order to test the theoretical predictions, grating-like phase objects were made in photographic emulsions by the method described by Hood (1969*b*). Measurements were made of the grating form and depth using an interference attachment to a standard photographic microscope.* These gratings made on Kodak 649F plates had fundamental spatial frequencies of 10, 20, and 40 lines/mm and covered an exposure range of 0·25 to 2 seconds. Because of the high gamma of this emulsion as processed, $\gamma = 6$, the density range of the film was large. On each plate a standard step tablet was exposed to determine the contour effect for zero spatial frequency. Figure 2*a* is an interference photograph of one of the steps of the tablet exposure and 2*b* is a similar photograph of one of the grating exposures. The curve of Figure 3 indicates that for large areas uniformly exposed the contouring is a linear function of exposure. Figure 4 shows quite definitely a change in the effect due to the spatial frequency parameter. The only thing conclusive from these data is that the linearity is gone and the curves of higher frequency have a lesser slope. This might be logically expected on grounds that surface tension and bulk elastic effects would pull down the amplitude of the higher frequency contours.

In order to produce phase gratings for experimental use the plates were thoroughly dried in a vacuum and metallized. Plates metallized without extended vacuum drying initially had very good surface properties but later became covered with tiny blisters. The plates were coated to about 80 per cent reflection.

* Interference microscope objective, Watson and Sons, Ltd., Barnet, Herts., U.K.

If, as has been suggested by Altman (1966), contour height is proportional to density and thus to log(exposure) it is reasonable to hypothesize a log(cosine) form to the gratings. For purposes of testing the theory of

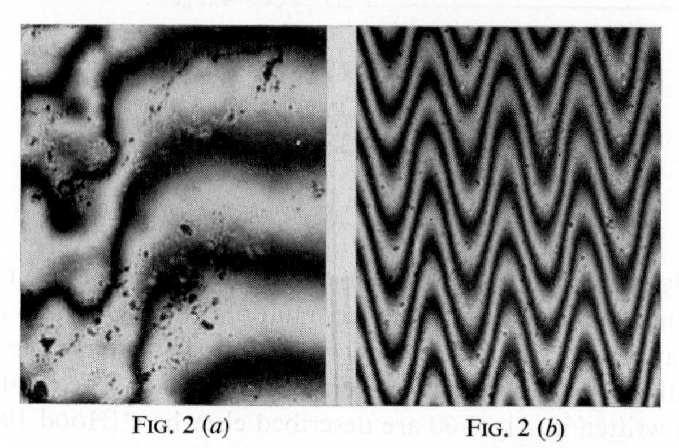

FIG. 2 (*a*) FIG. 2 (*b*)

FIG. 2. (*a*) Contour at the edge of a uniformly exposed region in Kodak type 649F emulsion. The mottled appearance is due to the contact imaging of grains in the step tablet in the high resolution emulsion. (*b*) Grating contours in Kodak type 649F emulsion as seen with an interference microscope. $\lambda = 593$ nm in both.

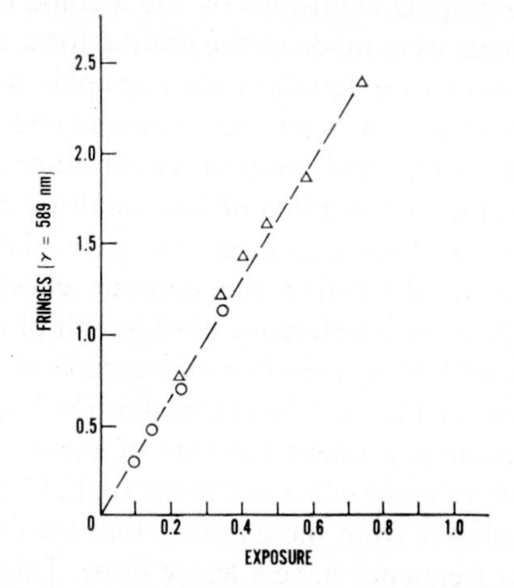

FIG. 3. Contour height of step tablet exposure on Kodak type 649F emulsion. ○, △ indicate measurements of the same plate made on two separate occasions. $\lambda = 593$ nm illuminant.

computing phase images the log(cosine) form was assumed and the phase object function accurately approximated by a five term Fourier series (Abramowitz & Stegun 1965). The phase screen images were computed, given the measured phase depth of the grating contours and compared with images formed in coherent illumination. These were measured with a sensitive, wide-dynamic-range photomultiplier photometer* behind a scanning slit. An example of these computations and measurements is illustrated in Figure 5.

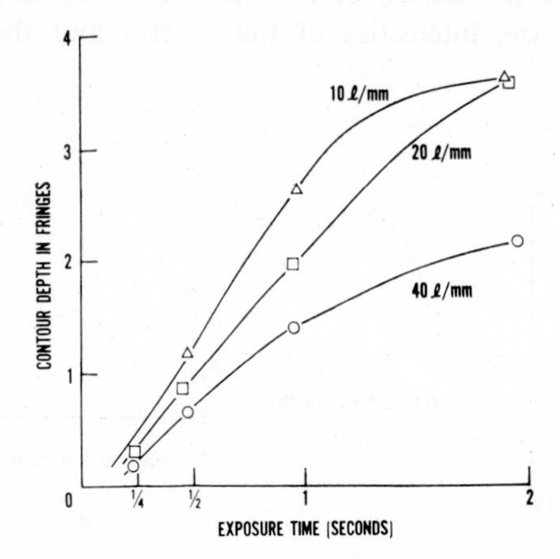

FIG. 4. Contour depth in Kodak type 649F emulsions of gratings versus relative exposure for three spatial frequencies; 10 lines/mm, 20 lines/mm, and 40 lines/mm.

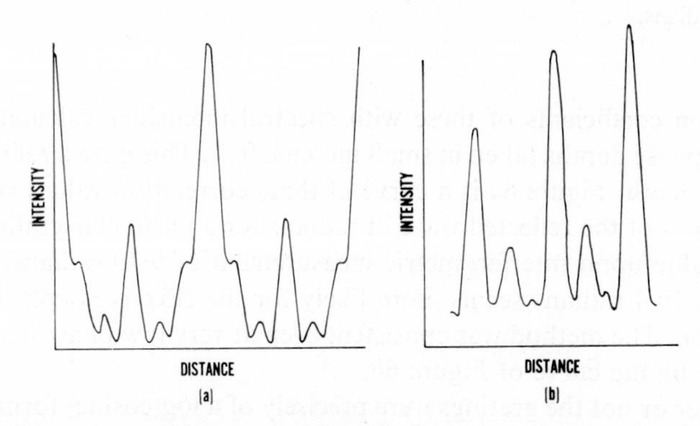

FIG. 5. (*a*) Phase grating image, calculated. (*b*) Phase grating image, measured. Phase depth = 14·70 radians. Spectral order numbers included in aperture = ±3.

Although correspondence between computed and measured values seems nominal at best it was found from an extensive series of calculations that details in the computed image were a very sensitive function of:

(*a*) the average phase depth of the grating assumed for the calculation;
(*b*) the selection of the focal plane for the measurement; and
(*c*) the precise form of the grating.

The measured phase depth was determined from photographs like Figure

* Model 700 Photometer, Gamma Scientific Inc., T.E.M. Sales Ltd., Crawley, Sussex, U.K.

2*b* which show only a small portion of each grating. Actually an amplitude averaging takes place over the whole of the illuminated surface to give an effective phase depth. Subsequently a technique was employed involving measurement of the intensities of the spectra and the calculation of

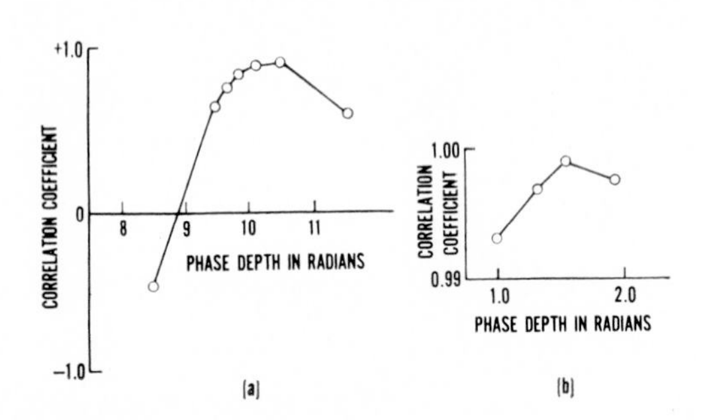

FIG. 6. Correlation coefficient of calculated versus observed spectral intensities for grating where phase depth was interferometrically measured as: (*a*) 9·70 radians; and (*b*) 1·93 radians.

correlation coefficients of these with spectral intensities calculated for a range of phase depths taken in small increments*. This gave a refined value of phase depth. Figure 6*a* is a curve of these correlation values versus the phase depth of the reflected wave in radians for a particular grating which gave a value upon interferometric measurement of 9·70 radians. A value nearer to 10·4 radians seems more likely for the effective depth from the correlation. The method was consistent even at very low phase depths as is indicated by the curve of Figure 6*b*.

Whether or not the gratings were precisely of a log(cosine) form was not proved conclusively but one observation was possible which supports the contention that they were certainly not cosinal in form. A pure cosine grating gives a diffraction spectrum which is a Jacobi series. The amplitude coefficients are just an integral order Bessel series. Zero and near-zero values occur frequently as this series oscillates about the axis with changing period. Upon squaring to get intensity these 'holes' are accentuated. The presence of these near-zero values is a sensitive indicator of the absence of components in the grating form not of the fundamental frequency. Attempts to obtain high correlation coefficients for the measured spectra using calculated spectra of appropriate pure cosine form were notably unsuccessful.

* See for example Parratt (1961) p. 141.

REFERENCES

ABRAMOWITZ, M. and STEGUN, I., 1965, *Handbook of Mathematical Functions*, Dover, New York, p. 68.
ALTMAN, J. H., 1966, *Phot. Sci. and Eng.*, **10,** 156.
BERRY, M. V., 1966, *The Diffraction of Light by Ultrasound*, Academic Press, London and New York.
BURCKHARDT, C. B., 1966, *J. Opt. Soc. Amer.*, **56,** 1502.
HANNES, H., 1968, *J. Opt. Soc. Amer.*, **58,** 140.
HOOD, J. M., 1969, Ph.D. Thesis, University of Reading.
HOOD, J. M., 1969, *Optica Acta*, **16,** 17.
HOPKINS, H. H., 1951, *Contraste de Phase*, Paris, p. 142.
PARRATT, L. G., 1961, *Probability and Experimental Errors in Science*, John Wiley, New York and London, p. 141.
ZERNIKE, F., 1942, *Physica*, **9,** 686 and 974.

Real-Time Optical Matched-Filtering

C. Atzeni and L. Pantani

Istituto di Ricerca sulle Onde Elettromagnetiche del C.N.R.,
Via Panciatichi 56, *I* 50127 *FIRENZE, Italy*

Abstract—Optical data-processing techniques devised first by Cutrona (1960) were able to process signals recorded on photographic films. The use of ultrasonic light-modulators (ULM) has allowed the real-time optical performance of signal-processing operations as Fourier transforms, spatial filtering, convolution, correlation (Reich & Slobodin 1961; Slobodin 1963). The authors have developed a new, simple technique to obtain signal correlation, and have applied this technique to matched-filtering of radar-signals.

In a first approach a reference signal was recorded as a hard-clipped optical grating, which was scanned by the ultrasonic signal propagating through the ULM. The moving interference pattern of the light diffracted by the ultrasonic beam beats with the fixed replica, causing an instantaneous variation of the emerging light. The correlation is obtained by detecting this intensity-modulated light with a photo-multiplier.

In a second approach a double-channel ULM was used; the received signal is inserted into the optical system through the first channel whereas through the second one a reference signal propagates in the opposite direction. The correlation is obtained detecting the beating of the two signals by a photomultiplier. The correlation signal so obtained exhibits an interesting feature: its duration is reduced to a half, and its carrier frequency is doubled.

Several experiments have been conducted on various waveforms: rectangular pulses, 'chirp' pulses, and Barker code sequences.

One of the chief problems in signal processing is the implementation of the matched-filter receiver (Turin 1960) which gives the maximum instantaneous signal to noise ratio.

The filter matched to a signal $s(t)$ consists of any device having an impulse response $s(-t)$. If $w(t) = s(t) + n(t)$ is the received signal, where $n(t)$ is a noise component, the output of the matched-filter will be the convolution between $w(t)$ and the impulse response:

$$u(t) = \int_{\infty}^{\infty} w(\nu)s(\nu + t)\mathrm{d}\nu \qquad (1)$$

that is the cross-correlation between the input signal $w(t)$ and the expected signal $s(t)$. A matched filter can therefore be built using a correlator fed with the received signal and some replica of the expected signal.

The processing of signals recorded on photographic films through optical correlators was introduced by Cutrona *et al.* in 1960, and real-time operation was obtained by Reich and Slobodin in 1961 by inserting an ultrasonic light modulator (ULM) in Cutrona's correlator. In 1963 Slobodin presented a simplified optical correlator whose theory was developed in 1967 by Felstead. A further simplification was independently introduced by Maloney and Meltz (1968) and by the Authors (1969;

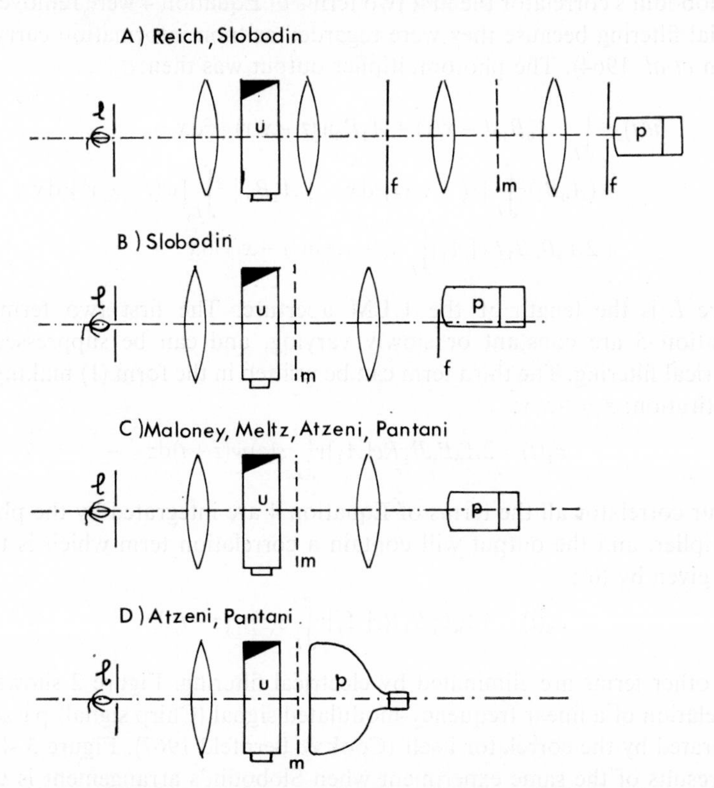

FIG. 1. Real-time optical correlators: *l*, lamp; *u*, ULM; *m*, optical mask; *p*, photomultiplier; *f*, spatial filter.

Meltz *et al.* 1969). Figure 1 shows the various models of optical correlators.

Let us consider the three simplified correlators shown in Figures 1*b*, *c* and *d*. The ULM acts like a moving optical grating and the amplitude of the light emerging from the ULM can be written (Meltz & Maloney 1968; Meltz *et al.* 1969):

$$a_1(x; t) = A_0 + A_1 w(t - x/v) \tag{2}$$

where A_0 is a real constant and A_1 a complex one whose phase depends from the depth of the modulation. Let the mask have a transparency:

$$b(x) = B_0 + B_1\, s(-x/v) \tag{3}$$

where B_0 and B_1 are real constants. As long as the distance between the ultrasonic beam and the mask is negligible the amplitude of light emerging from the mask will be:

$$a_2(x;t) = a_1(x;t)b(x) = A_0B_0 + A_1B_1w(t-x/v)s(-x/v) + A_0B_1s(-x/v) + A_1B_0w(t-x/v) \qquad (4)$$

In Slobodin's correlator the first two terms of Equation 4 were removed by spatial filtering because they were regarded as 'non information carrying' (Arm *et al.* 1964). The photomultiplier output was then:

$$\begin{aligned} u(t) &= \int_L |A_0B_1s(-x/v) + A_1B_0w(t-x/v)|^2 dx = \\ &= (A_0B_1)^2\int_L [s(-x/v)]^2 dx + |A_1B_0|^2\int_L [w(t-x/v)]^2 dx + \qquad (5) \\ &+ 2A_0B_0B_1Re[A_1]\int_L s(-x/v)w(t-x/v)dx \end{aligned}$$

where L is the length of the ULM aperture. The first two terms of Equation 5 are constant or slowly varying, and can be suppressed by electrical filtering. The third term can be written in the form (1) making the substitution: $z = -x/v$:

$$c_1(t) = 2A_0B_0B_1Re[A_1]v\int_L s(z)w(z+t)dz \qquad (6)$$

In our correlator all the terms of Equation 4 are integrated by the photomultiplier, and the output will contain a correlation term which is twice that given by (6):

$$c_2(t) = 4A_0B_0B_1Re[A_1]v\int_L s(z)w(z+t)dz \qquad (7)$$

The other terms are eliminated by electrical filtering. Figure 2 shows the correlation of a linear frequency-modulated signal (Chirp signal) passively generated by the correlator itself (Cook & Bernfeld 1967). Figure 3 shows the results of the same experiment when Slobodin's arrangement is used.

If the distance between the ultrasonic beam and the reference mask is not small, the study of ULM diffraction pattern in the near field will show that such a distance is very critical in order to obtain correlation. In the case of the Chirp signal it was recognized that the ULM acts like a moving zone grating (Gerig & Montague 1964) and therefore the diffracted waves are inherently converging ones as shown in Figure 4. As long as $z \ll z_1$ the analysis of the diffraction pattern shows that minimum changes of the $z=0$ pattern occur in planes $z = n\Lambda^2/\lambda$ where Λ is the central ultrasonic wavelength, and λ the light wavelength. If the reference mask is placed in these planes a quasi-optimum correlation will be obtained. For increasing z a progressive deterioration of the output will be observed also in these planes.

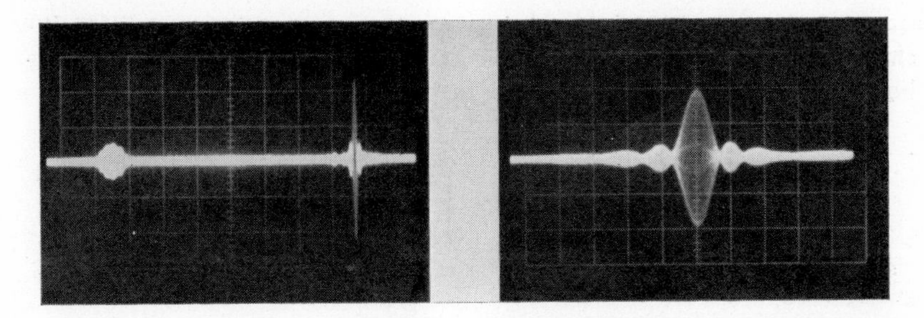

FIG. 2. Auto-correlation of a chirp pulse through the correlator of Figure 1*d*. *Left figure*: Input pulse and correlator output (20μs/div.). *Right figure*: Correlator output (2μs/div.). The input pulse has a carrier frequency of 15 MHz and a frequency sweep of 1 MHz.

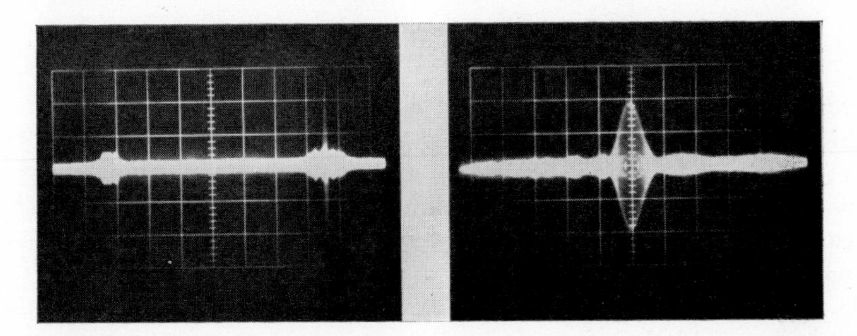

FIG. 3. Auto-correlation of a chirp pulse through the correlator of Figure 1*b*. *Left figure*: input pulse and correlator output (20μs/div.). *Right figure*: correlator output (2μs/div.). Same data as in Figure 2.

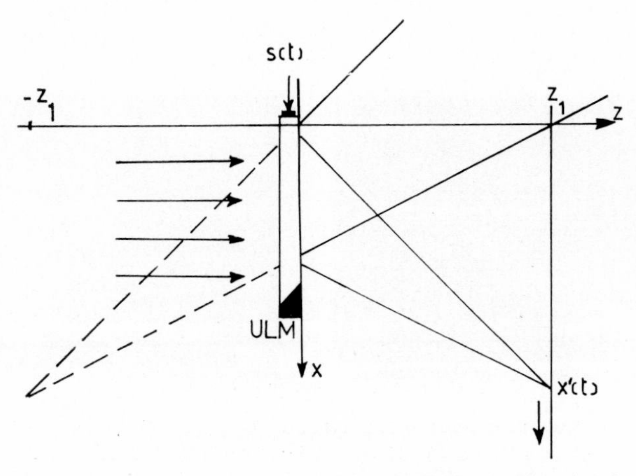

FIG. 4. Diffracted waves generated in a ULM by a chirp signal.

The use of a second ULM instead of a fixed reference mask was suggested, to our knowledge, by Di Tano in 1965 and by Browne in 1966, but no experimental results were reported in their papers. We have

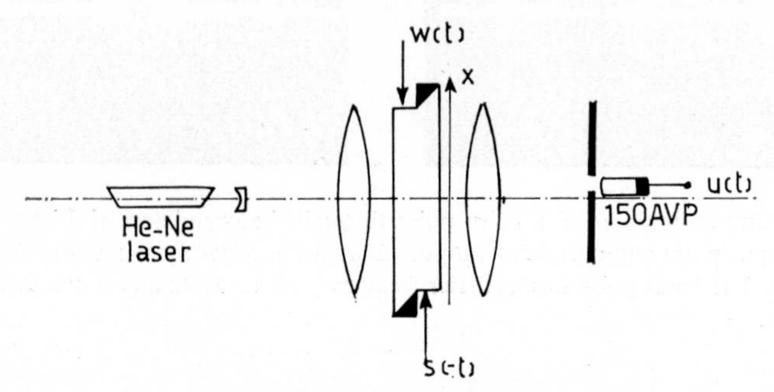

FIG. 5. Dual-channel optical correlator.

conducted some experiments using a dual-channel ULM inserted in a Slobodin's correlator (Figure 5). It is to be noted that the arrangement of Figures 1*c* and *d* cannot be used in this case because the impinging light energy is unaffected by the two ultrasonic beams; thus the correlation cannot be obtained by a time variation of the total light energy as in the previous case.

Suppose that the first channel is fed with the received signal $w(t)$ and the second one with the reference signal $s(-t)$. If the distance between the two ultrasonic beams is negligible, the amplitude distribution of light emerging from the ULM can be expressed by:

$$a(x;\, t) = [A_0' + A_1' w)t + x/v)][A_0'' + A_1'' s(-t + x/v)] \tag{8}$$

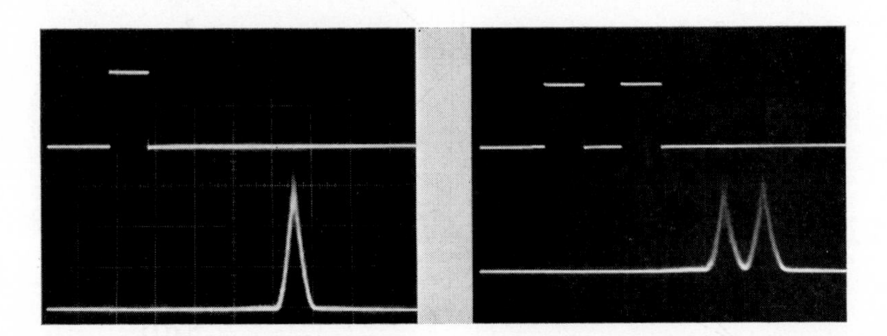

FIG. 6. *Left figure*: Rectangular pulse and its auto-correlation through a dual-channel correlator (20μs/div.). *Right figure*: correlation of two rectangular pulses having a mutual delay of 20μs (10μs/div.).

The photomultiplier output contains the correlation term:

$$c_1(2t) = 2A_o'A_o''Re[A_1'A_1'']v\int_L w(z+2t)s(z)\mathrm{d}z \tag{9}$$

where $z = -t + x/v$. From Equations (6), (7) and (9) one can deduce that the photomultiplier output presents a time scale factor 2. This means that

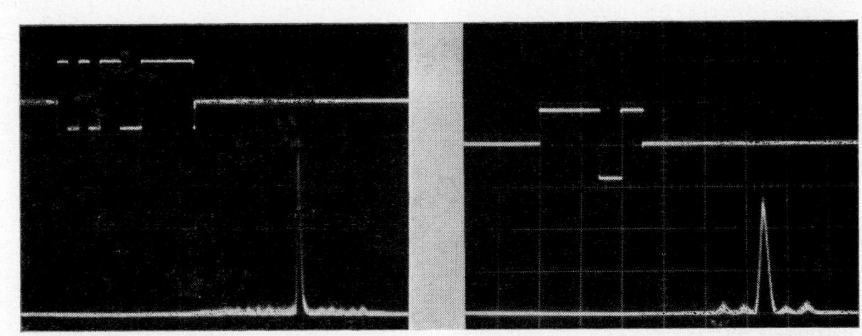

FIG. 7. Auto-correlation of Barker codes. *Right figure*: 5-bit code (10μs/div.). *Left figure*; 13-bit code (10μs/div.).

the output signal is half the output of a correlator using a fixed reference mask, and has a doubled carrier frequency. This is experimentally shown in Figure 6. Figure 7 shows some experiments conducted using Barker codes (1953).

REFERENCES

ARM, M. *et al.*, 1964, *Proc. IEEE* (*Corresp.*), **52,** 842.
ATZENI, C. and PANTANI, L., 1969, *Proc. IEEE* (*Corresp.*), **57,** 344.
BARKER, R. H., 1953, 'Group synchronizing of binary digital systems', in *Communication Theory*, W. Jackson (Ed.), London, Academic Press, Chapter 3.
BROWNE, A., 1966, 'An optical correlation technique'. *IEE Conference Publication* n. 20, p. 48.
COOK, C. E. and BERNFELD, M., 1967, *Radar Signals*, Academic Press, New York.
CUTRONA, L. J. *et al.*, 1960, IRE Trans. on Information Theory, IT-6, 386.
FELSTEAD, E. B., 1967, 'A simple real-time incoherent optical correlator'. IEEE Transactions on Aerospace and Electronics Systems, AES-3, 907.
GERIG, J. S. and MONTAGUE, H., 1964, *Proc. IEEE* (*Corresp.*), **52,** 1753.
MELTZ, G. and MALONEY, W. T., 1968, 'Optical correlation of Fresnel images'. *Appl. Optics*, **7,** 2091.
MELTZ, G., MALONEY, W. T., ATZENI, C. and PANTANI, L., 1969, *Proc. IEEE* (*Corresp.*), **57,** 1316.
REICH, A. and SLOBODIN, L., 1961, 'Optical pulse expansion/compression'. Presented at the National Aerospace Electronics Conference, Dayton, Ohio, May 8.
SLOBODIN, L., 1963, *Proc. IEEE* (*Corresp.*), **52,** 1782.
Di TANO, B., 1965, 'A more flexible optical correlator using electro-optical Doppler replicas.' Ninth Technical Meeting of the AGARD Avionics Panel on Optoelectronic Components and Devices, Paris, September.
TURIN, G. L., 1960, 'An introduction to matched-filters.' IRE Transactions on Information Theory, IT-6, 311.

La Rectification de Lentilles Asphériques par Commande Numerique

Tomochika Nakano

Laboratoire de recherche de la compagnie
Nippon Kogaku K.K., (*Tokio*)

Abstract—For grinding Schmidt correcting plates of large diameter, a numerically controlled grinding machine has been constructed. Details of this machine and the results of grinding are described.

INTRODUCTION

Généralités

Trois problèmes se présentent pour la production de lentilles asphériques.

(1) Rectification de la surface asphérique.
(2) Polissage.
(3) Contrôle.

Dans le cas de la fabrication de lentilles sphériques ces problèmes ont été résolus sans difficulté, en utilisant la particularité de la sphère. Pour fabriquer des lentilles spériques, on utilise des outils à polir également sphériques et des calibres de contrôle. Par contre, dans le cas de la fabrication de lentilles asphériques, on ne peut pas utiliser le procédé précédent. Il y a quelques ressemblances entre des machines pour fabriquer des lentilles asphériques et des machines-outils usinants les métaux. La fabrication de lentilles sphériques est un usinage par abrasif libre tandis que la fabrication de lentilles asphériques est un usinage par meule.

Tout d'abord, nous avons étudié le premier problème (la rectification de lentilles asphériques).

Si on peut rectifier des surfaces asphériques avec une haute précision, le temps de polissage diminue notablement. Jusqu'ici, la plupart des lentilles asphériques sont fabriquées sur des machines à reproduire. Mais il y a quelques défauts fondamentaux dans cette méthode. Pour éviter ces défauts, nous avons dessiné une rectifieuse à commande numérique. Evidemment l'appareil de commande numérique ne garantit pas d'obtenir des pièces de haute précision. La précision des pièces est finalement déterminée par la machine-outil elle-même. L'essentiel des mérites de l'appareil à commande numérique réside dans ces possibilités de programmation. Néanmoins, la machine-outil à commande numérique a ses propres problèmes et on ne peut pas unir, sans précaution, un appareil de commande numérique avec une machine-outil traditionnelle.

Ces problèmes sont les suivants:

(1) rigidité de la machine;
(2) résistance de frottement.

Ces points sont décrits dans les chapitres suivants.

But de la machine

Nous avons réalisé une machine destinée à la fabrication de lames correctrices de Schmidt de grand diamètre. La machine-outil à commande numérique est la mieux adaptée à la fabrication en petite série et en grande variété.

CONSTRUCTION DE LA MACHINE

La machine se compose de sept parties: une rectifieuse, un appareil de commande numérique, un centre hydraulique, une armoire d'alimentation électrique, un pupitre de commande, un générateur électrique, un système d'alimentation en huile de refroidissement. Les détails sont donnés ci-après.

Rectifieuse

Comme il est mentionné plus haut, pour dessiner la machine-outil à commande numérique, il faut insister sur l'augmentation de la rigidité de la machine et la diminution de la résistance de frottement. La Figure 1 montre une photo de la rectifieuse. La Figure 2 en donne un schèma.

La rectifieuse a 3 mètres de largeur, 1·5 mètre de longueur et 2·3 mètres de hauteur. La table circulaire sur laquelle les pièces sont fixées, a 0·83 mètre de diamètre et 0·84 mètre de hauteur par rapport au sol. La vitesse de rotation de la table est variable de façon continue de 0·5 à 8 tr/min. Pour assurer une vitesse uniforme, un guidage hydrostatique lubrifié est employé. A côté de la table circulaire, il y a deux colonnes. Le chariot porte-meule est monté sur le bras transversal qui joint les deux colonnes. Pour avoir des glissières dépourvues de frottement, le chariot porte-meule se déplace sur des guidages à rouleaux croisés. Pour éviter les poussières, les systèmes des guidages à rouleaux croisés sont couverts par des soufflets. Les parcours maximum du chariot porte-meule sont de 480 mm sur l'axe X et de 160 mm sur l'axe Z. La vitesse du chariot porte-meule est variable de 0·5 à 120 mm/min et la vitesse de déplacement rapide est 480 mm/min. Deux moteurs pas à pas (l'axe X et l'axe Z) tournent de 1·5 degrés par impulsion celle-ci provenant de l'appareil de commande numérique. Par

FIG. 1. La rectifieuse.

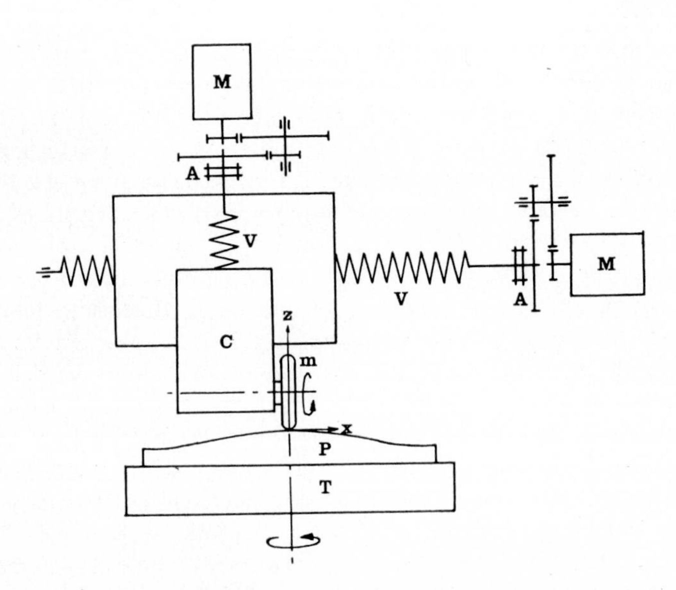

FIG. 2. Un schèma de la rectifieuse: *A*, le palier à billes; *C*, le chariot porte-meule; *M*, le moteur pas à pas; *m*, la meule; *P*, la pièce; *T*, la table circulaire; *V*, la vis-mère.

intermédiaire de roues dentées, la rotation des moteurs pas à pas est réduite en 1/25 et transmise à deux vis mères. Pour compenser les jeux, les roues dentées sont divisées en deux parties qui sont fixées avec un léger décalage tangentiel et des écrous à rattrapage de jeu sont employés. Le pas des vis mères est de 6 mm. Finalement, le chariot porte-meule se déplace de 1 μ par impulsion de l'appareil de commande numérique. Les meules ont 160 mm de diamètre et 10 mm d'épaisseur. Leur bords sont arrondis avec un rayon de 10 à 60 mm. La Figure 3 montre une meule.

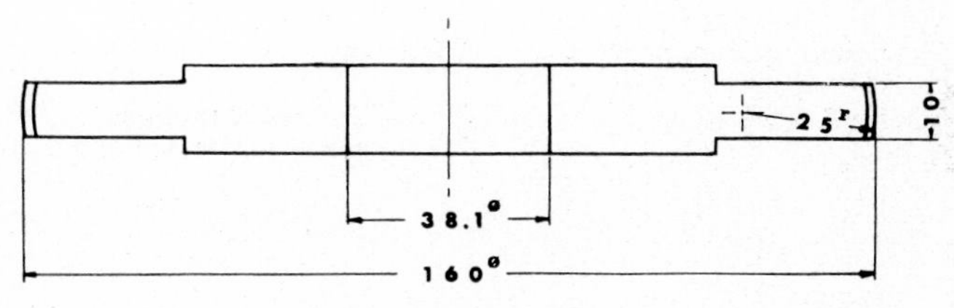

FIG. 3. Une meule.

Les grains de diamant sont de 30 μ ou de 120 μ. L'agglomérant est du cuivre jaune.

Les causes principales d'erreur d'usinage sont l'irrégularité des formes et l'usure de la meule. Pour obtenir une forme régulière, la meule est fixée sur la broche porte-meule, est dressée et arrondie par une autre meule en carborundum vert. Le centre de l'appareil d'affûtage est placé sur l'axe de la table circulaire. Après cet affûtage la meule se trouve automatiquement placée sur l'axe de la table circulaire et la détermination de l'origine est faite facilement avec précision.

Appareil de Commande Numeriques

L'appareil de commande numérique est un FANUC 280 (Fujitsu Co.). C'est un appareil qui fonctionne avec des moteurs pas à pas électro-hydraulique, en boucle de fonctionnement ouvert. Des rubans en papier noir sont perforés directement par la calculatrice-électronique ou manuellement par le perforateur électrique. Le codage est le code EIA (8 pistes). L'origine de programme peut être prise en n'importe quel point. Le Tableau 1 montre les caractéristiques de l'appareil. La Figure 4 en donne une photo.

Comme il est mentionné plus haut, le bord de la meule est arrondi. On doit donc calculer le lieu géométrique du centre du cercle qui roule sur la surface de la lentille asphérique. Le ruban est perforé d'après les résultats du calcul. Le lecteur photo-électrique lit des informations, qui sont

TABLEAU 1. Caracteristiques techniques de l'appareil de commande numeriques.

Fabricant	Fujitsu Co.
Type	FANUC 280
Domaine	1700 × 700 × 650
Entree d'information	Ruban perforé ou manuelle
Mode commande	Trajectoire
Nombre des deplacements commande numerique	2
Pas unitaire programme	0·001 mm
Vitesse Avance	Travail 0·5 ∼120 mm/min Dép lacement rapide 480 mm/min

FIG. 4. L'appareil de commande numérique.

stockées dans le registre. En même temps, l'interpolateur entre en action et les informations numériques sont transformées en impulsions dans le circuit de sortie.

Le moteur pas à pas électro-hydraulique se compose de trois parties: un moteur pas à pas électrique, un commutateur hydraulique et un moteur hydraulique. Le moteur pas à pas électrique tourne de 2·5 degrés par impulsion provenant de l'appareil de commande numérique. Suivant la rotation du moteur pas à pas électrique, un commutateur hydraulique s'ouvre et le moteur hydraulique tourne de 1·5 degré par impulsion.

Cet appareil de commande numérique a des fonctions auxiliaires. On peut commander plusieurs opérations; par exemple, la marche et l'arrêt de l'huile de refroidissement, la compensation de jeux du parcours, etc.

Autres Appareils

Centre hydraulique. Le centre hydraulique fournit de l'huile sous une pression de 70 Kg/cm^2 aux deux moteurs pas à pas et de l'huile sous une pression de 0·3 Kg/cm^2 au guidage hydrostatique lubrifié de la table circulaire.

Appareil d'alimentation en huile de refroidissement. L'huile de refroidissement est du type 'Esso-Menter 28'. La capacité du réservoir est de 120 litres. Pour aspirer l'huile vaporisée par la meule, une pompe aspirante est employée. Un papier-filtre sépare la poudre du verre de l'huile utilisée.

Générateur électrique. Pour stabiliser la tension d'alimentation de l'appareil de commande numérique, un générateur électrique est employé. La fluctuation de tension est inférieure à plus ou moins 5 pour cent.

Pupitre de commande. Le pupitre de commande sur lequel sont disposés tous les boutons, est séparé de l'autre partie et on peut le déplacer à volonté.

RESULTATS

Une lame correctrice de Schmidt pour un télescope dont la distance focale est de 1 mètre et l'ouverture est de 50 cm, a été rectifiée par cette machine. Le diamètre de la pièce est de 570 mm. Son épaisseur est de 50 mm. La matière utilisée est le crown. La Figure 5 montre la lame correctrice de Schmidt. La surépaisseur de rectification est de 0·127 mm et elle a été enlevée en une seule traversée.

La vitesse de rotation de la table circulaire est de 3 tr/min celle de la meule est de 3600 tr/min. La vitesse du chariot porte-meule est de 0·9 mm/min. La Figure 6 montre l'écart de la surface rectifiée par rapport à la surface calculée. Ces valeurs ont été mesurées par un indicateur à cadran, juste après l'usinage, la pièce étant fixée sur la table circulaire.

Le résultat obtenu est satisfaisant. La rugosité de la surface rectifiée est fine. Après le meulage la pièce a été doucie et polie par la méthode traditionnelle. La pièce polie a été contrôlée par un interféromètre à

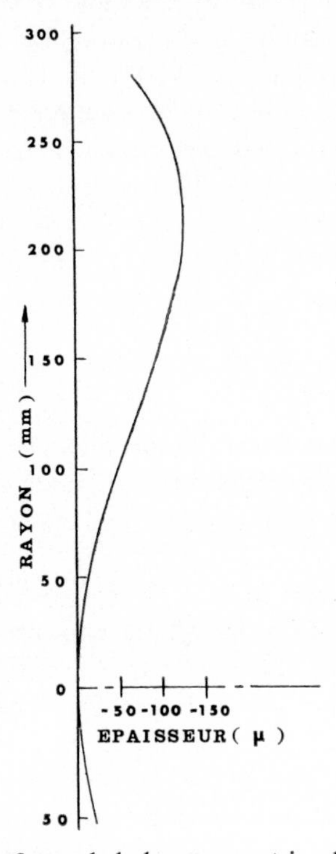

FIG. 5. La figure de la lame correctrice de Schmidt.

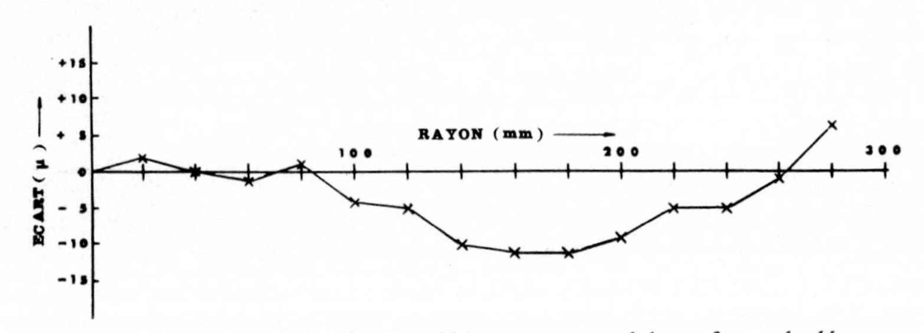

FIG. 6. L'écart de la surface rectifiée par rapport à la surface calculée.

dédoublement (T. Tsuruta 1963). Finalement elle introduit une déformation de la surface d'onde inférieure à plus ou moins $1/4\ \lambda$.

Evidemment, cette machine peut servir à la fabrication de lentilles asphériques de petit diamètre. Nous en avons déjà fabriqué quelques unes avec la même précision.

CONCLUSION

Pour rectifier des lentilles asphériques, une rectifieuse à commande numérique a été construite. Les avantages de la machine sont:

(1) Peu de temps est nécessaire pour réaliser une lentille asphérique quelconque.

(2) Les erreurs constantes (usure de la meule diamantée, flexion statique de la machine, etc.) sont compensées facilement.

(3) On peut obtenir des pièces de haute précision et de grain fin.

Une lame correctrice de Schmidt de 570 mm de diamètre a été rectifiée. La précision de l'usinage est meilleure que plus ou moins 10 μ. La machine est en cours d'amélioration. Des problèmes de polissage restent à résoudre.

BIBLIOGRAPHIE

TSURUTA, T., 1963, *J. Opt. Soc. Amer.*, **53**, 1156.

INTERVENTION DE MONSIEUR M. DETAILLE

Nous avons imaginé, avec Messieurs A. Baranne et Lemaitre de l'Observatoire de Marseille, et avec Monsieur le Professeur Courtes, une méthode originale de fabrication de lames de Schmidt.

Ces lames sont destinées à équiper des télescopes astronomiques embarquables, travaillant dans l'ultra-violet jusqu'à 200 nm.

Schmidt fabriqua ses premières lames de la façon suivante:

— il partait d'une lame à faces parallèles

— il posait sa lame sur un support dans lequel il faisait le vide

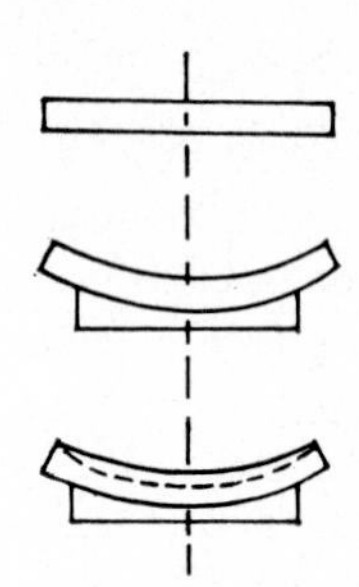

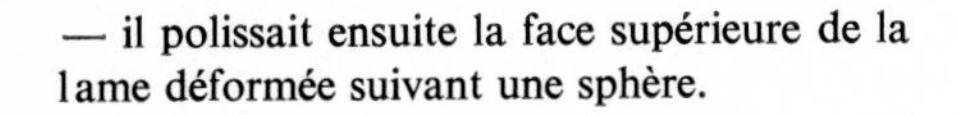

— il polissait ensuite la face supérieure de la lame déformée suivant une sphère.

Nous avons repris la même idée en utilisant deux cavités au lieu d'une. La lame ne repose que sur la couronne centrale.

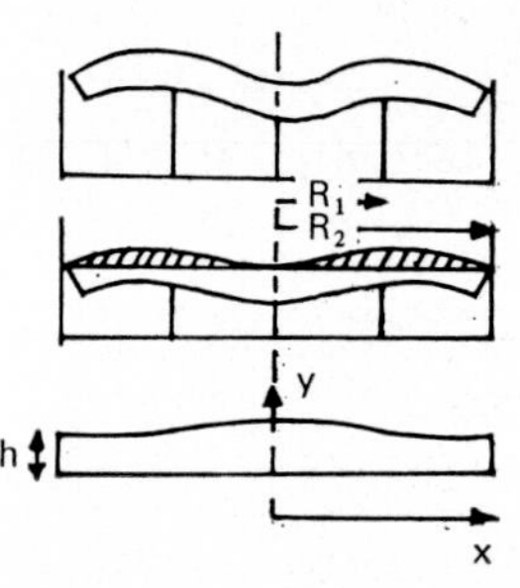

On polit ensuite la surface supérieure de la lame suivant un plan.

On obtient une lame de Kerber classique.

L'équation de la forme obtenue a été résolue par calculateur:

$$\triangle h = \frac{1-\nu^2}{E h^3} K_{R_1 R_2}(x)$$

ou ν est le coefficient de Poisson de la lame.
E est le coefficient d'élasticité de la lame.
h est l'épaisseur moyenne de la lame.

$K_{R_1 R_2}(x)$ une fonction de x qui, pour R_1 et R_2 bien choisis, a la forme de Kerber.

Pour avoir une déformation très grande, il suffit de diminuer h. Mais pour h trop petit, la lame casse dans le montage; cela se produit pour des lames corrigeant l'aberration sphérique de miroirs sphériques ouverts à F/1.

Plusieurs lames de ∅ 50 mm. d'excellente qualité ont déjà été réalisées. Nous réalisons en ce moment une lame de ∅ 220 mm.

REFERENCE

LEMAITRE, G., 1968, *Rapport de D.E.A.*, Optique Astrophysique, Faculté des Sciences de Marseille.

Fabrication de Lame de Schmidt

André Baranne, Michel Detaille et Gérard Lemaitre

Observatoire de Marseille,
Place de Verrier, Marseille

Laboratoire d'Astronomie Spatiale,
Traverse du Siphon, Les Trois Lucs, Marseille 12eme

Abstract—In experiments involving the use of optical systems attached to rockets and balloons, small cameras are required to monitor the experiment and also to photograph the field of view in ultra-violet light.

These telescopes must have very good definition and good light-gathering power, and it is found that small Cassegrain systems corrected by Schmidt plates are most suitable. The relative advantages are their achromatism, transparency to ultra-violet because of the small number of surfaces, the possibility of coating the surfaces for narrow bands of wavelength and the accessibility of the focal plane.

The manufacture of the Schmidt plates had to be undertaken in the laboratory, and the method used is described.

INTRODUCTION

Le Laboratoire d'Astronomie Spatiale place dans ses expériences fusées ou ballons, des petites caméras annexes permettant de vérifier le bon fonctionnement du pointage de l'expérience et, par la même occasion, d'avoir une photographie dans l'ultra-violet du champ à observer.

Ces télescopes doivent avoir une très bonne définition et une luminosité très importantes. On a donc été amené à construire des petits télescopes Cassegrain corrigés par une lame de Schmidt-Kerber. Les avantages de ce système sur les combinaisons telles que les Caméras Maksutov, ou objectifs dioptriques sont:

Son quasi achromatisme.

Sa bonne transparence a l'ultra-violet qui lui est donnée par son petit nombre de surface.

La possibilité de traiter les miroirs dans une bande étroite.

L'accessibilité de la surface focale.

METHODE DE FABRICATION DE LA LAME DE SCHMIDT

La méthode de réalisation est celle préconisée par Mr. Lemaitre dans son rapport de D.E.A., qu'il a fait sous la direction de Mr. Baranne à l'Observatoire de Marseille.

La fabrication de la lame a été réalisée au Laboratoire d'Astronomie Spatiale.

Nous rappelons ci-dessous les différentes méthodes de fabrication de surfaces asphériques.

A—*Méthode par Déformations Statistiques*

On part d'une surface sphérique la plus proche possible de la surface à réaliser et un choix judicieux:

(1) De la trajectoire;
(2) Du rapport diametre de l'outil/diamètre de la pièce;
permet d'approcher la forme recherchée. Cela est du au fait que l'usure locale est sensiblement proportionelle au nombre de passages de l'outil en ce point.

Lorsque la pièce à déformer est de grand diamètre, on utilise un système 'bielle-manivelle' encore appelé 'système à un bras oscillant' pour guider l'outil.

Lorsque la pièce est de taille moyenne ou petite, on peut utiliser un second système bielle-manivelle ou 'système à deux bras oscillants'. Une machine de ce type a été construite par M. D. Loomis à l'Optical Sciences Center, Tucson (Université d'Arizona). Les courbes des trajectoires obtenues sont très variées du point de vue chemin statistique ce qui laisse entrevoir quelques progrès de cette méthode.

B—*Méthode par Echauffement Thermique*

Cette méthode utilisée par A. Couder a permis d'obtenir des miroirs de télescope à méridienne parabolique en partant d'un miroir sphérique. Il faut disposer une couronne chauffante à l'extérieur du miroir, qui a pour effet de produire:

(1) Un certain gradient de température du bord vers le centre du miroir.
(2) De dilater le verre de facon plus importante au bord qu'au centre.

Pendant l'opération de chauffage, on retouche le miroir ainsi déformé à l'aide de l'outil précédent de forme sphérique. Lorsque la retouche est terminée, on laisse refroidir, et le miroir prend alors une forme voisine du paraboloïde. Cette méthode présente toutefois des difficultés, car l'outil étant en contact avec le miroir lors des retouches, empêche ce dernier d'évacuer les calories de la façon souhaitée.

C—*Méthode par Déformation Plastique*

Fabrication de la première lame de Schmidt. Cette méthode utilisée par B. Schmidt pour la première fois, permet de déformer une lame pour corriger l'aberration hors de l'axe du télescope qui porte son nom. Il procédait de la manière suivante:

En faisant le vide sous une lame appuyée sur son contour (Figure 1*a*), il remarquait que la déformée présentait une courbure plus forte sur le bord qu'en son centre. En utilisant une sphère dont le rayon est égal au

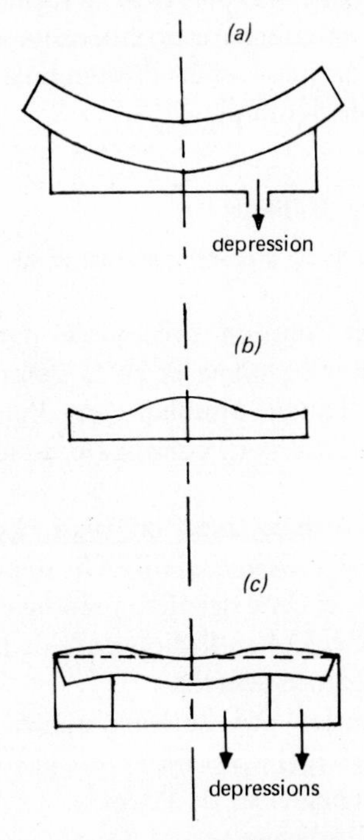

FIG. 1. Methode par déformation plastique.

rayon de courbure du bord de la déformée, il retouche la déformée jusqu'à ce que la sphère vienne coïncider tout le long de la méridienne de la déformée (voir Chretien, p. 350).

Le calcul d'élasticité montre que la déformée a pour équation:

$$z_1 = Ar^4 + Br^2$$

en retouchant avec une sphère dont le rayon est R

$$z_2 = (1/2R)r^2 \qquad (\text{si } R \gg 3).$$

Lorsque la méridienne de la sphère coïncide avec la déformée et que l'on remet le disque à la pression atmosphérique, le disque prend la forme:

$$z_1 - z_2 = Ar^4 + [B - (1/2R)]r^2.$$

Le problème est donc de choisir B et R pour que $B-(1/2R)$ soit négatif, puisque l'équation de la méridienne d'une lame de Schmidt est de la forme:

$$z = A'r^4 - B'r^2.$$

La difficulté de cette méthode provient du fait que l'outil de retouche qui est sphérique à l'origine (rayon R) se déforme lui aussi et il est très difficile de lui redonner sa courbure primitive lorsque l'on arrive à la fin des opérations ce qui a pour effet d'obtenir une surface légèrement différente de la surface de Schmidt.

D—*Etude d'une Nouvelle Méthode*

Nous avons envisagé avec Messrs. Batanne et Lemaitre la technique suivante:

Nous avons repris la solution préconisée par Schmidt, mais cette fois-ci, en partant d'une lame plane et en la déformant à l'aide de deux cavités, l'une centrale, l'autre annulaire et d'un appui situé dans le voisinage de la zone de Kerber, comme le montre la figure ci-après (Figures 1*b* and *c*).

On retouche cette déformée avec un plan. Lorsque l'opération est terminée, la déformée est devenue plane. On démonte alors la lame en supprimant la dépression, et cette dernière prend alors la forme recherchée.

Nous avons effectué le calcul en flexion pure de la déformée d'une lame appuyée sur une circonférence centrée.

Il faut résoudre l'équation aux dérivées quatrième $\Delta(\Delta W) = q/D$ avec q =dépression, et $D = ER^3/12(1-\nu^2)$ où W est la fonction des paramètres réduits r/R_1, R_2/R_1 et ν coefficient de Poisson.

Les équations de la méridienne sont:

$$\frac{r}{R_1} \epsilon [0-1], \quad Y = \frac{1}{32}\left(\frac{r}{R_1}\right)^4 + \frac{X_1}{4}\left(\frac{r}{R_1}\right)^2 + X_2$$

$$\frac{r}{R_1} \epsilon \left[1, \frac{R_2}{R_1}\right], \quad Y = \frac{1}{32}\left(\frac{r}{R_1}\right)^4 + \frac{X_3}{4}\left(\frac{r}{R_1}\right)^2 \log\frac{r}{R_1} + \frac{X_4 - X_5}{4}\left(\frac{r}{R_1}\right)^2 + X_5 \log\frac{r}{R_1} + X_6$$

où Y est la flèche réduite

$$Y = \frac{R^3 E/q}{6(1-\nu^2)R_1^4} \cdot E$$

et $X_1, X_2, \ldots, X_6$ sont les racines d'un système de 6 équations à 6 inconnues obtenues par les conditions aux limites et les conditions de continuités.

La CAB 500 a résolu ce système et nous a permis de tracer un réseau de méridiennes. (Figure 2)

On choisit ensuite sur ce réseau une courbe donnant la méridienne la plus rapprochée de la méridienne de Schmidt.

REALISATION D'UNE LAME DE SCHMIDT DIAMETRE UTILE 50 mm

Cette lame de Schmidt devait corriger la surface d'onde d'un télescope Cassegrain, formé d'un grand miroir ouvert à f/2·5.

Ce télescope devait fonctionner entre 1800 Å et 3000 Å la matière choisie pour la lame de Schmidt est donc la silice fondue spécialement traitée pour l'ultra-violet (qualité suprasil).

A—*Calcul de la déformation de la lame*

Cette équation se réduit pour un télescope Cassegrain, dont le second miroir corrige légèrement l'aberration sphérique du premier miroir.

$$(N-1)X = 8{\cdot}16y^2(1-2/3y^2)$$

N = indice de la lame

X = dépression dans le sens de la lumière, exprimée en microns

Y = distance du point considéré au centre de la lame

Y_0 = rayon utile de la lame

$y = Y/Y_0$ distance réduite du point considéré au centre de la lame (varient de 0 à 1)

Pour la zone de Kerber $y = \sqrt{3}/2$ la déformation maximum à réaliser est donnée par la formule:

$$(N-1)X = 3{\cdot}06\ \mu$$

B—*Calcul des paramètres de construction de la lame*

En reprenant les calculs de Mr. Lemaitre et en tenant compte du coefficient de Poisson que nous avons mesuré, nous avons trouvé que pour réaliser le profil voulu, on devait donner aux rayons des deux cavités:

$$2R_1 = 47{\cdot}07 \text{ mm et } 2R_2 = 69{\cdot}43 \text{ mm}$$

La valeur de l'épaisseur de la lame était:

$$h = 2{\cdot}15 \text{ mm}$$

C—*Problèmes posés par la réalisation de la lame*

La réalisation du support n'a pratiquement posé aucun problème. Le vide a été réalisé a l'aide d'une pompe primaire.

Afin de pouvoir quand même faire tourner le support autour de son axe, au cours du polissage, il était nécessaire de réaliser un joint tournant pour faire la liaison entre la pompe et le support.

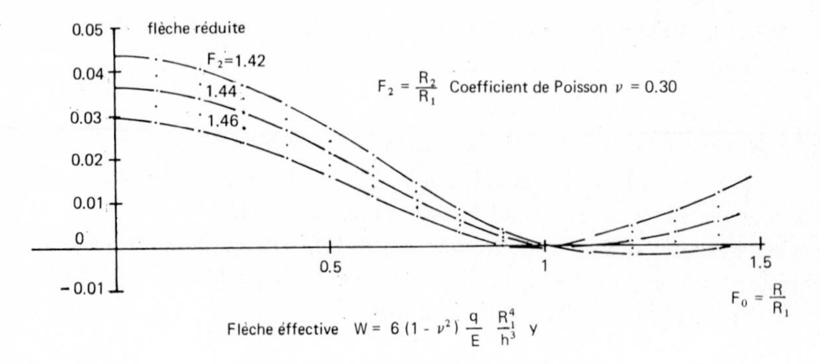

FIG. 2. Calcul theorique de la flèche.

FIG. 3. Franges d'egale epaisseur d'une lame de Schmidt à profil de Kerber équipant les caméras de champ du type CM 16. Elle corrige pour un champ nul, l'aberration sphérique longitudinale du télescope Cassegrain (ouverture du miroir primaire f/2,5).

Caracteristiques: Silice fondue (Suprasil); module d'Young $E = 7{\cdot}95 \times 10^5$ kgf.cm^{-2}; coefficient de Poisson $= 0{\cdot}16$; diamètre utile $\varnothing = 50$ mm; épaisseur $h = 2{\cdot}15$ mm.

Photographie: Source Na, filtre Schott OG3, papier Kodak super ortho; diamètre photographié $= 59$ mm; $\lambda = 589$ mμ; Pose 30 minutes.

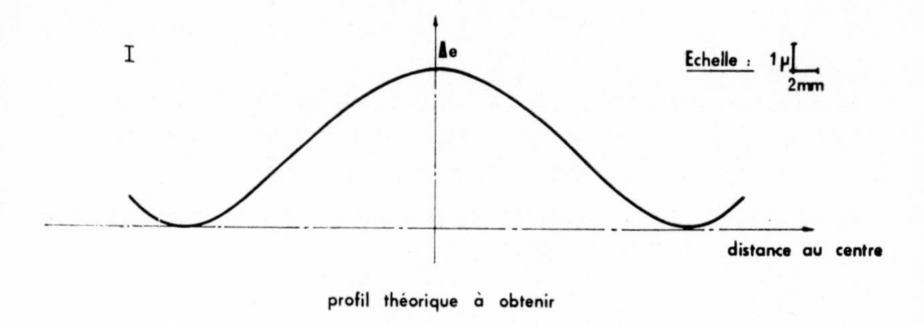

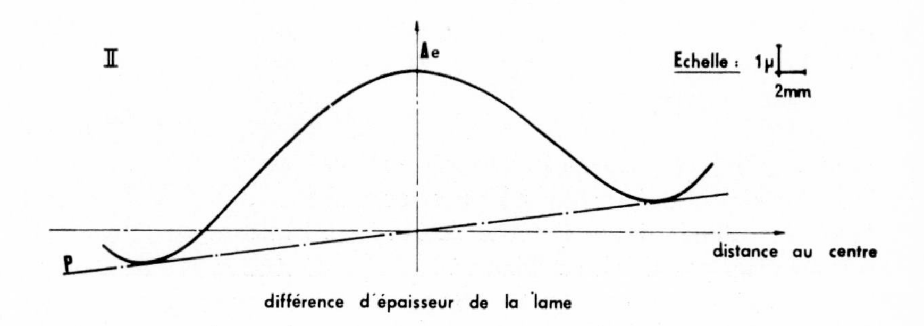

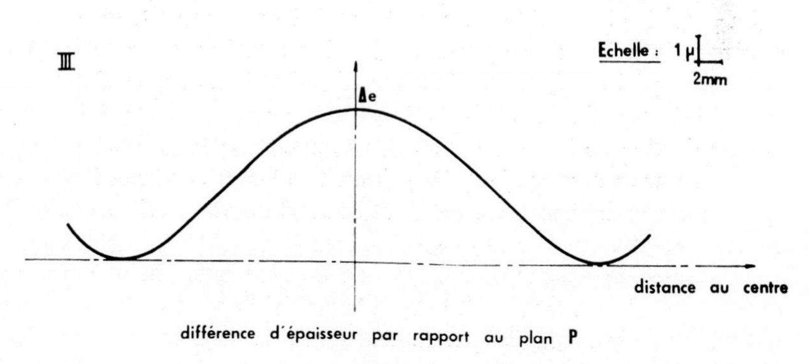

FIG. 4. Sections de lame de Schmidt. Ecarts entre la courbe I et la courbe III inférieurs a 0·1 μ.

D—*Vérification de la lame par franges d'égales épaisseurs*

La photographie (Figure 3) ci-incluse. montre les franges d'égales épaisseurs de la lame. Les franges correspondent donc à une différence de chemin optique $\Delta = 2Ne = K\lambda$; on a donc 6 fois plus de franges que si l'on mesurait directement la différence de chemin optique introduit par la lame déformée; on a donc une grande précision de lecture sur cette photographie.

L'examen de cette photographie conduit aux conclusions suivantes:

Les variations de l'épaisseur correspondent exactement aux variations de la lame de Schmidt à deux exceptions près:

(*a*) Un phénomène de coin d'angle au sommet de 10 secondes environ, se superpose à la déformation.

(*b*) Des irrégularités locales d'une frange sont visibles en deux ou trois points de la surface.

Le phénomène de coin n'a en fait aucune importance, puisque l'objet se trouve alors à l'infini par rapport à la lame. Tout revient en définitive à déplacer de 5″ d'arc le centre du champ du télescope Schmidt-Cassegrain, pour lequel la correction est parfaite, ce déplacement est négligeable pour un champ de plusieurs degrés.

Nous avons repris sur la Figure 4 le profil théorique, le profil réel de la lame et le profil ramené à une surface inclinée de 5″.

Les quelques déformations locales reviennent à déformer légèrement la surface d'onde d'une quantité de l'ordre 0·1 micron, ce qui ne sera certainement pas visible sur l'image.

CONCLUSION

La première lame obtenue par ce procédé et décrite ci-dessus a donné entière satisfaction. Elle a permis d'obtenir des taches image de dimension cinq fois supérieure à celle de la tache de diffraction, dans le plan focal du système Cassegrain dans lequel elle est montée.

Une dizaine de lames de ⌀ 50 mm et une lame de ⌀ 220 mm ont été réalisées par ce procédé jusqu'à ce jour. Ces lames corrigent l'aberration sphérique de miroirs sphériques dont l'ouverture est inférieure ou égale à F/1.

Ces lames de Kerber dont la qualité de surfaçage atteint aisément le dizième de la longueur d'onde et dont la facilité de réalisation est évidente, semblent ouvrir de grandes possibilités dans tous les domaines de l'optique et de l'astronomie.

Interference and Diffraction Phenomena Produced with Scattered Light

Solange Debrus, M. Françon and Marie May

Institut d'optique et Faculté des Sciences de Paris,
Tour 33–1er étage, 9 quai Saint Bernard, Paris Veme, France

Abstract—Some simple experiments on interference have been carried out with light scatterers; a few applications in interferometry are given.

A new interferometer is described. It consists essentially of two elements: a scatterer with a high factor of transmission, and a Gabor hologram of this scatterer. When a phase object is seen through this system, fringes of equal optical thickness of the object are seen.

These fringes can also be observed in the following way: two successive exposures are made on a photographic plate, one with the phase object and the scatterer and the other with the scatterer alone.

Both reflecting and diffusing objects can be observed by these methods. In both the cases, the reference beam is not inclined.

*Young's Fringes Obtained with a Ground Glass**

Figure 1 shows the experimental arrangement used by Burch & Tokarski (1968) to produce multiple beam fringes from photographic scatterers.

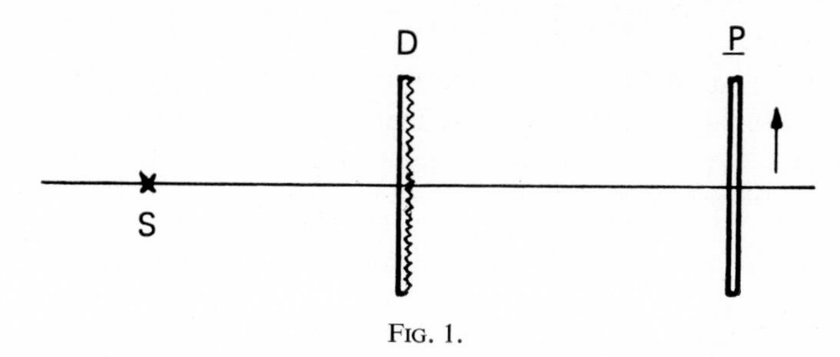

FIG. 1.

A point source S (laser source) illuminates a ground glass D and a photographic plate P records the light leaving the ground glass. Two successive exposures are made, the photographic plate having been translated in its own plane between the two exposures. After development a distant source

* (Debrus, Françon, Mallick, May & Roblin 1968*a*, 1969.)

is observed through the plate and beautiful Young's fringes are observed on an extended and practically uniform field, which is due to diffusion. The fringes are observable in white light. The source can be a slit parallel to the fringes.

Let η and ζ be the coordinates of a point on the photographic plate. During the first exposure the plate records an intensity $E(\eta, \zeta)$ at the point (η, ζ). During the second exposure the same point receives an intensity $E(\eta, \zeta-\zeta_0)$ where ζ_0 is the translation given to the plate between the two exposures. For simplicity the translation is supposed to be parallel to the ζ-axis, but obviously the direction of the translation can be arbitrary. It is assumed that the times of the two exposures are equal. The total intensity received by the plate is $E(\eta, \zeta)+E(\eta, \zeta-\zeta_0)$. The amplitude t_N transmitted by the developed plate is given by

$$t_N = t_0 - \alpha(E - E_0) \tag{1}$$

where α is the slope of the linear portion of the curve $t_N = f(E)$; t_0 is the amplitude transmitted by the plate for a mean value E_0 of the intensity. Putting $t_0 + \alpha E_0 = a$, we have

$$t_N = a - \alpha[E(\eta, \zeta) + E(\eta, \zeta-\zeta_0)] \tag{2}$$

a being a constant. Let us illuminate the plate with a parallel beam of light and study the phenomena in the focal plane of a lens L. We have to calculate the Fourier transform of Equation (2). Put $K = 2\pi/\lambda$, where λ is the wavelength of the parallel beam. The amplitude at a point (u, v) of the focal plane is given by:

$$U = a\iint \exp\,[jK(u\eta + v\zeta)]\;\mathrm{d}\eta\mathrm{d}\zeta$$
$$-\alpha\iint [E(\eta, \zeta) + E(\eta, \zeta-\zeta_0)]\exp[jK(u\eta+v\zeta)]\mathrm{d}\eta\mathrm{d}\zeta. \tag{3}$$

The first term on the right-hand side represents the Fraunhofer diffraction pattern of the lens, i.e. the image of the source. Since it is very small, we can neglect the field it occupies, i.e., we neglect the first term. Putting $\zeta - \zeta_0 = \mu$ and neglecting a constant factor, we have:

$$U = [1 + \exp(jKv\zeta_0)]\iint E(\eta, \zeta)\exp[jK(u\eta + v\zeta)]\mathrm{d}\eta\mathrm{d}\zeta. \tag{4}$$

The ground glass produces a very fine structure $E(\eta, \zeta)$ on the photographic plate, consequently the Fourier transform of $E(\eta, \zeta)$ is very much spread out which corresponds to a practically uniform field. The amplitude U and the intensity I are given by the following expressions

$$U = 1 + \exp(jKv\zeta_0) \qquad I = UU^* = \cos^2(Kv\zeta_0/2). \tag{5}$$

Young's fringes perpendicular to the direction of translation and visible in white light are obtained.

A single exposure will suffice if a birefringent plate is placed between the ground glass and the photographic plate. The birefringent plate produces the required lateral shear. The two light beams (ordinary and extraordinary) exposing simultaneously the photographic plate are polarized at right angles to each other and consequently the two exposures are independent.

Diffraction Pattern Obtained by a Continuous Displacement of the Photographic Plate During the Exposure.

Let us consider the same experiment; a series of exposures is made, the photographic plate having been translated through a small amount $\Delta\zeta$ after each exposure. The expression (5) giving the transmitted amplitude now becomes

$$U = 1 + \exp(jKv\Delta\zeta) + \exp(iKv2\Delta\zeta) + \ldots \tag{6}$$

If the plate is given a uniform translation during the exposure, we get:

$$U = \int_{o}^{\zeta_o} \exp(jKv\zeta)\mathrm{d}\zeta = \zeta_0 \exp[j(Kv\zeta_o/2)] \frac{\sin(Kv\zeta_o/2)}{(Kv\zeta_o/2)} \tag{7}$$

where ζ_0 is the total displacement of the plate. Neglecting a constant factor, the intensity is given by

$$I = UU^* = \left[\frac{\sin(Kv\zeta_o/2)}{Kv\zeta_o/2}\right]^2. \tag{8}$$

This is the diffraction pattern produced by a slit of width ζ_0, the total displacement given to the plate. Replacing the ground glass by a diaphragm placed against the photographic plate, one can obtain interesting phenomena of apodization (Debrus *et al.* 1968*b*).

Circular Fringes at Infinity with a Scatter Plate†

The scatter plates employed in all the following experiments are obtained by making a single exposure in the set-up of Figure 1. The photographic plate after development presents random variations of amplitude with very fine structure and acts as a scatter plate. This scatter plate transmits a large part of the incident light and scatters the rest. If a point source is observed through such a scatter plate, one observes an extended and practically uniform halo encircling the source. This halo represents the Fourier transform of the scatter plate. Figure 2 shows the scheme of the experiment. The scatter plate D is illuminated by a parallel beam of

† (Debrus, Françon & May 1969*b*).

monochromatic light and a photographic plate P is placed at an arbitrary distance from D. Let $F(\eta, \zeta)$ be the diffracted amplitude at a point (η, ζ) of the plate P. The light beam traversing the scatter plate without being diffracted plays the part of the coherent background. The amplitude of this coherent background is taken as unity. The plate P receives the intensity

$$E = (1+F)(1+F)^* = 1 + |F|^2 + F + F^*. \qquad (9)$$

After developing the plate under the usual conditions of linearity, the amplitude transmitted by the negative is, according to Equation (1):

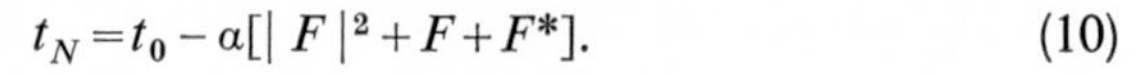

$$t_N = t_0 - \alpha[|F|^2 + F + F^*]. \qquad (10)$$

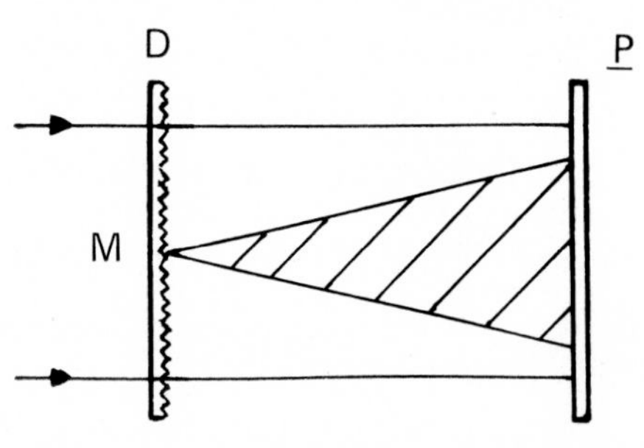

FIG. 2.

This negative is a Gabor hologram. When it is illuminated by a parallel beam two images (one real and the other virtual) of the scatter plate are reconstructed which are symmetrically situated on either side of the negative. The reconstructed scatter plates give rise to a system of circular fringes at infinity which are visible in white light.

These fringes may be observed with two identical scatter plates placed parallel to each other. It should be noted that the method of obtaining two identical scatter plates by the method indicated above is extremely simple.

Let us replace the scatter plate having a random structure by one having a fine regular structure. Taking the experiment of Figure 2 a series of exposures are made, the photographic plate having been translated after each exposure. After developing, the negative is illuminated in a parallel beam and one observes the Fraunhofer diffraction pattern of a three-dimensional structure (Laue's diagram).

Description of an Interferometer Using Scatter Plates‡

The preceeding experiments and particularly that of the previous section suggest a new type of interferometer. The proposed interferometer

‡ (Brooks, Heflinger & Wuerker 1966; Lowentral & Belvaux 1967; Gates 1968).

consists essentially of two elements: a scatter plate and a Gabor hologram of this scatter plate. We have seen in the previous section how to make a Gabor hologram of a scatter plate. Figure 2 represents the scheme of the interferometer, P being the Gabor hologram of the scatter plate D. The hologram P illuminated at O (Figure 3) by a parallel beam of monochromatic light transmitted by D, reconstitutes the virtual image D'; two rays such as IJ and $I'J'$ emerging from the hologram appear to be

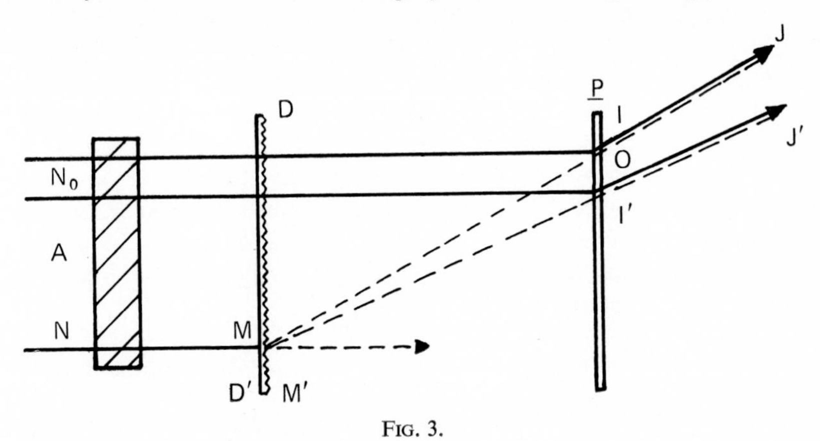

FIG. 3.

coming from a point M' of the virtual image D'. We shall call this pencil of rays [1]. The whole of the surface of the diffuser D is illuminated and the point M whose virtual image is M' scatters the light. The interferometer is adjusted when the virtual image D' (given by P) coincides with D. The pencil of rays scattered by M (pencil [2]) coincides with the pencil [1] since M and M' are superposed. On emerging from the hologram, the pencils [1] and [2] interfere. This amounts to saying that the points M and M', or in other words the scatter plates D and D', interfere with each other. The path difference being zero at all the points of the scatter plate, the eye placed at O behind the hologram sees the surface of the scatter plate uniformly illuminated. We do not take into consideration, for the time being, the real image of D.

Let us place a transparent object A before the scatter plate D. The object A is, for example, a glass plate with thickness irregularities. Now the path difference between M and M' is no longer zero since the pencil [1] has traversed A at N_0 where the thickness of A is e_0 and the pencil [2] has traversed A at N where the thickness is e.

The path difference is thus $\delta = (n-1)(e-e_0)$. Assuming that the pencils [1] and [2] have the same amplitude, the intensity I at the point M is

$$I = \sin^2\left(\frac{\pi\delta}{\lambda}\right) = \sin^2\left[\frac{\pi(n-1)(e-e_0)}{\lambda}\right] \quad (11)$$

The eye placed at O will observe fringes which represent the lines $(n-1)(e-e_0)$ =constant, as in an ordinary interferometer. But in this case it is the difference between the thickness at a point N and the thickness at N_0 which determines the state of interference. The region N_0, determined by the position of O (the pupil of the eye or that of the system of observation), fixes the origin of phases. When the position of O is altered the fringe pattern is modified. Consequently a diaphragm with a narrow aperture must be placed against the hologram. If the aperture is too large, the fringe pattern gets blurred. If the diaphragm O is situated at the edge of the hologram it is easy to visualize that the virtual and the real images do not superpose. They are just separated and if the focal length of the system of observation is small compared with the distance between D and P, the two images are simultaneously in focus. But the real image is isolated and does not give rise to any interference phenomenon. It appears to be uniform and its intensity is four times less than that of the maximum of the virtual image. The aspect of the images is represented in Figure 4.

The following remarks may be made:

(*a*) The quality of the glass of the scatter plate and that of the hologram have no effect on the interference phenomena.

(*b*) The object, placed outside the interferometer, can be of any thickness.

(*c*) The scatter plate D and the hologram P can be bleached.

(*d*) Interferometers of large dimensions are possible by juxtaposing a number of scatter plates which need not be identical. The hologram P made with this assembly of scatter plates can be of small dimensions.

(*e*) The principle of the phenomenon does not change if the object is introduced between the scatter plate and the hologram, provided the thickness of the object is small in comparison with the distance separating the diffuser from the hologram.

(*f*) When the scatter plate and its virtual image do not coincide, circular fringes analogous to those obtained with a scatter plate (see p. 321) are observed at infinity.

(*g*) The interferometer can be associated with a microscope. The whole of the surface of the hologram will be used if the details under observation cover only a small field. In this case, the real and the virtual images are not separated. However, in microscopy, a small defect of focus suffices to make the real image disappear. The real image is not troublesome since it is uniform and less intense than the virtual image.

(*h*) It is possible to introduce an optical system between the scatter plate and the hologram (this hologram was made with the above optical system). The real image of the scatter plate given by the optical system and that given by the hologram superpose and interfere. This effect may be utilized in microscopy.

(*i*) The interferometer can be used to examine reflecting and diffusing objects by modifying the experimental set up. For example, a semi-reflecting plate may be put at 45° between the diffuser and the reflecting object. For diffusing objects, a plane parallel plate may be added normal to the light beam between the semi-reflecting plate and the object; this produces the coherent background.

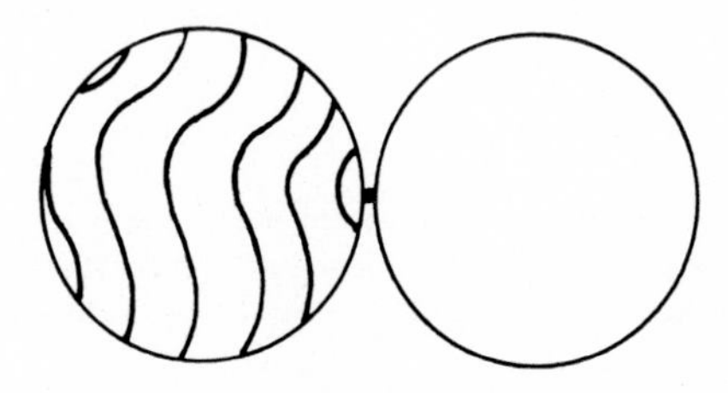

FIG. 4.

Interference Phenomena Obtained by Two Successive Exposures†

The preceeding phenomena can be studied by making two successive exposures. In the first exposure the elements are those of Figure 3. The second exposure is made after removing the transparent object *A*. After developing, a small region at the edge of the hologram is used to observe the interferogram. The images still have the disposition of Figure 4 but now they have the same intensity and the fringes are seen in both of them. It may be noted that the phenomena are observable in white light.

REFERENCES

BROOKS, R. E., HEFLINGER, L. O. and WUERKER, R. F., 1966, *IEEE J. Quantum Electron.*, **QE–2,** 275.
BURCH, J. M. and TOKARSKI, J. M. J., 1968, *Opt. Acta,* **15,** 101.
DEBRUS, S., FRANÇON, M., MALLICK, S., MAY, M. and ROBLIN, M. L., 1969, *Appl. Opt.*, **8,** 1157.
——, 1968*a*, *C.R.Ac.Sc.*, (B), **267,** 1332.
——, 1968*b*, *C.R.Ac.Sc.*, (B), **267,** 1416.
DEBRUS, S., FRANÇON, M. and MAY, M., 1969*a*, *C.R.Ac.Sc.*, (B), **268,** 317.
DEBRUS, S., FRANÇON, M. and MAY, M., 1969*b*, *Optics communications,* **1,** 89.
GATES, J. W. C., 1968, *J. Sci. Instrum.*, (*J. Phys. E.*), **1** (ser. 2), 989.
LOWENTHAL, S. and BELVAUX, Y., 1967, *Revue d'optique,* **46,** 1.

† (Debrus, Françon & May, 1969*b*.)

Visualisation des Objets de Phase en Eclairage Partiellement Coherent

Marie-Hélène Bourgeon and Gérard Fortunato

Institut d'Optique,
3 boulevard Pasteur, Paris, XV.

Abstract—During our experiments on Moiré effects, and more particularly when projecting the image of one grating on another of the same spacing and at the same orientation—we were prompted to study the interference phenomena observed.

We considered the case of replacing the first grating by two gratings of double the spacing, placed on either side of the position of the original first grating; the phase object was then placed in the position of the latter so that its image is formed in the plane of the third grating.

In this way we arrived at an interferometer which was differential, achromatic and with variable relative shear. The use of the gratings allowed us not only to study objects of large size but also to work with wavelengths ranging over the whole visible spectrum from the near-infrared to the near-ultraviolet.

Les phénomènes interférentiels observés lors de la projection de l'image d'un réseau sur un autre réseau de même orientation avaient fait l'objet de travaux de M. Vasco Ronchi (1926) et de M. Lenouvel (1924) et plus tard de calculs effectués entre autre par Messieurs Toraldo Di Francia (1940), Adachi (1960) et Theocaris (1965).

Nous avons repris les calculs dans le cas d'une source étendue éclairant le premier réseau, ce qui nous a conduit à faire intervenir la formation des images d'objets de phase en éclairage partiellement cohérent.

A la suite d'un premier montage (Figure 1) utilisant la conjugaison de deux réseaux similaires, au moyen d'un système optique bien corrigé, nous avons imaginé un second montage (Figure 2) dans lequel le premier réseau est remplacé par deux réseaux de pas double, placés symétriquement par rapport à l'objet. L'image de l'objet se trouve dans le plan du troisième réseau. Nous obtenons de cette façon un système interférométrique, du type différentiel, achromatique, à dédoublement variable.

En ce qui concerne le premier montage (Figure 1) le premier réseau est. sinusoidal en intensité, le second peut être un réseau d'amplitude ou de phase.

La méthode de calcul est la suivante: le premier réseau, de répartition d'intensité $\epsilon(u) = 1 + \alpha \cos 2\pi\omega u$, crée dans le plan de l'objet $\Omega(M)$ un certain degré de cohérence défini par le théorème de Zernike

$$\gamma(M_1 - M_2) = \int \epsilon(u) \exp [2\pi i u(M_1 - M_2)] \mathrm{d}u.$$

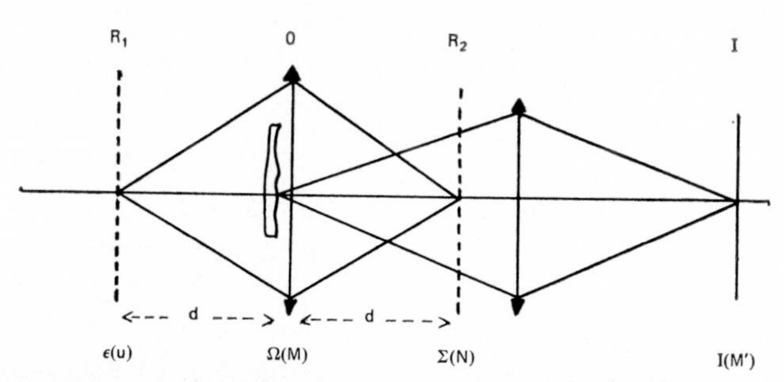

FIG. 1. Projection sur un réseau R_2 de l'image d'un réseau R_1 deformée par le présence d'un objet de phase O.

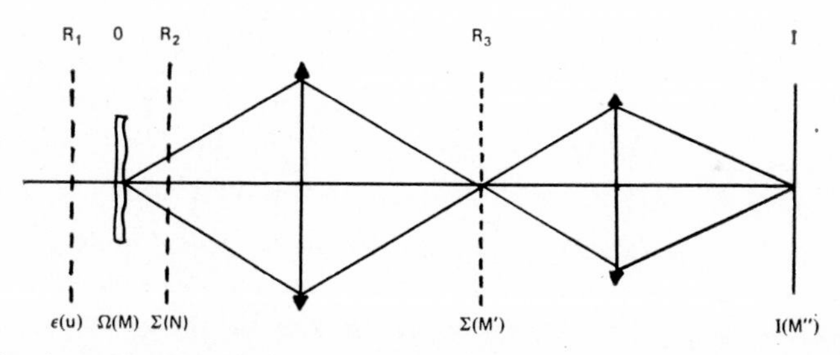

FIG. 2. Le réseau R_1 du schéma précédent est remplacé par deux réseaux, R_1 et R_2, de pas double. L'objet O se trouve à égale distance de R_1 et R_2. Son image se projette sur R_3 troisième réseau.

En appelant $\Sigma(N)$ la distribution des amplitudes du second réseau:

$$\Sigma(N) = 1 + \beta \cos 2\pi\omega N$$

le passage de l'objet à l'image fait intervenir la réponse percussionnelle du système de projection en éclairage cohérent, $E(M' - M)$. La répartition des intensités dans le plan image s'écrit:

$$I(M') = \iint \Omega(M_1)\Omega^*(M_2)\gamma(M_1 - M_2)E(M' - M_1)E^*(M' - M_2)\mathrm{d}M_1\mathrm{d}M_2$$

Dans ce cas, particulier de pupilles infinies et de résaux sinusoidaux, le degré de cohérence et la réponse percussionelle s'expriment par des sommes de pics de Dirac.

Les calculs effectués, l'intensité image est exprimée par:

$$I(M') = |\,\Omega(M')\,|^2 + \beta^2/4\,|\,\Omega(M' + \lambda\omega d)\,|^2 + \beta^2/4\,|\,\Omega(M' - \lambda\omega d)\,|^2$$
$$+ \alpha\beta/2\mathscr{R}\,\{\Omega(M')\Omega^*(M' + \lambda\omega d)\}$$
$$+ \alpha\beta/2\mathscr{R}\,\{\Omega(M')\Omega^*(M' - \lambda\omega d)\}$$

d étant le distance objet—réseau; ω étant la periode des réseaux

Dans le second montage (Figure 2) outre le dédoublement, on observe une modulation en amplitude dans le plan de l'image. Cette modulation est due à l'alignement des deux réseaux et au fait que les réseaux sont situés à des distances finies de l'objet.

La nature de cette modulation, son pas, la variation de son contraste en fonction de l'écartement des réseaux ont fait l'objet de travaux divers dont ceux de Lohman (1967) et récemment de Madame Roblin, à l'Institut d'Optique.

En introduisant dans le plan image (grandissement—1) un troisième réseau dont le pas est égal au pas de cette modulation l'objet de phase est mis en évidence.

Pour ce grandissement le pas est égal à la moitié du pas des deux premiers réseaux.

L'intensité avant le plan du troisième réseau s'exprime par:

$$I(M') = |\,\Omega(M')\,|^2 + \alpha^2/4\,|\,\Omega(M' - \lambda\omega d)\,|^2 + \alpha^2/4\,|\,\Omega(M' + \lambda\omega d)\,|^2$$
$$+ \alpha\beta/4\mathscr{R}\,\{\Omega(M')\Omega^*(M' + \lambda\omega d)\exp[-2\pi i/\lambda d(\lambda^2\omega^2 d^2 + 2\lambda d\omega M')]\}$$
$$+ \alpha\beta/4\mathscr{R}\,\{\Omega(M')\Omega(M' - \lambda\omega d)\exp[-2\pi i/\lambda d(\lambda^2\omega^2 d^2 - 2\lambda d\omega M')]\}.$$

Pour une direction de rayons incidents, une longeur d'onde déterminée, le schéma de la Figure 3 met en évidence la marche des rayons.

L'objet de phase étudié peut se mettre sous la forme exp $[i\psi(M)]$; le dédoublement $\lambda\omega d$ étant du même ordre de grandeur que la limite de résolution, il est permis d'utiliser la première, dérivée partielle sur l'objet:

$$\psi(M) - \psi(M \pm \lambda\omega d) = \pm\lambda d\omega\,(\partial\psi(M')/\partial M')$$

L'expression des intensités dans le premier interférogramme est:

$$I(M') = \lambda + \beta^2/2 + \alpha\beta\cos[(\partial\psi(M')/\partial M')\,\lambda\omega d)].$$

L'expression des intensités dans le second interférogramme est:

(*a*) avant l'introduction du troisième réseau:

$$I(M') = \lambda + \alpha^2/2 + \alpha\beta/2\cos[(\partial\psi(M')/\partial M')\,\lambda\omega d + 4\pi\omega M']\cos 2\pi\lambda d\omega^2.$$

(*b*) aprés l'introduction du troisième réseau:

$$I(M'') = (1 + \alpha^2/2) - (1 + \alpha^2/2)\cos 4\pi\omega M''$$
$$+ \alpha\beta/2\cos 2\pi\lambda d\omega^2\cos[\lambda d\omega\,(\partial\psi(M'')/\partial M'') + 4\pi\omega M'']$$
$$- \alpha\beta/4\cos 2\pi\lambda d\omega^2\cos[8\pi\omega M'' + \lambda d\omega\,(\partial\psi(M'')/\partial M'']$$
$$- \alpha\beta/4\cos[(\partial\psi(M'')/\partial M'')\,\lambda d\omega]\cos 2\pi\lambda d\omega^2.$$

Pourtant si l'aspect de l'objet est identiques pour les deux interférogrammes nous obtenons les lignes de même gradient de chemin optique, il apparait un terme de contraste chromatique: cos $2\pi\lambda d\omega^2$ dans le deuxième interférogramme.

En ce qui concerne la résolution, nous avons vu que le dédoublement doit être petit par rapport à la période de la composante spectrale dont la fréquence est la plus élevée parmi les fréquences interessantes de l'objet.

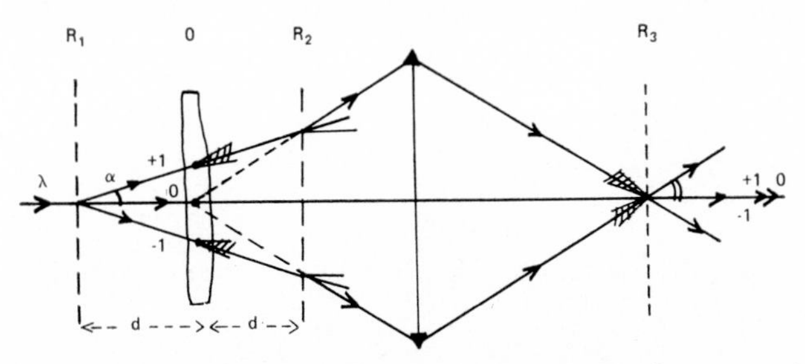

FIG. 3. Marche des rayons dans le montage interferométrique à réseaux, dans le cas d'une onde monochromatique eclairant le premier réseau.

L'avantage principal de ce montage réside dans la possibilité de modifier le dédoublement en faisant varier la distance des deux réseaux par rapport à l'objet et de choisir ainsi le dédoublement de plus approprié à l'objet étudié.

La sensibilité dépend du dédoublement; dans le cas étudié, pour des réseaux de 10 traits par millimètre, une distance réseau-objet de 40 millimètres et un minimum de contraste de 2 pour cent pour $\lambda = 0.5\ \mu$ correspond un dédoublement sur l'objet de 0·2 mm et le gradient de chemin optique décelable est théoriquement 6×10^{-5} ce qui correspond à des variations d'épaisseur de l'ordre de λ.

Le terme de contraste chromatique permet en outre d'envisager une méthode de mesure par analyse spectrale de l'image.

Utilisant des grilles de pas relativement élevés nous avons obtenu un interféromètre du type différentiel, offrant une sensibilité comparable à celle des autres interféromètres différentiels, le contraste pouvant être amélioré par la nature des grilles d'éclairage.

Parmi les applications qui peuvent être nombreuses, nous avons mis en évidence les défauts d'homogénéité dans des lames de verres, des variations d'indice en fonction de la température.

De plus l'utilisation de réseaux permet, non seulement d'étudier des objets de grande dimension, mais aussi de travailler avec des longueurs d'onde s'étendant au delà du spectre visible. (U.V., I.R., I.R. lointain).

BIBLIOGRAPHIE

ADACHI, I., 1960, *Atti della Fond. G. Ronchi*, The Diffraction Theory of the Ronchi Test.
LENOUVEL, M. L., 1924, *Revue d'Optique*, **3**.
LOHMANN, A. and BRYNGDAHL, O., 1967, *Applied Optics*, **6**, no. 11, A.
RONCHI, V., 1926, *Revue d'Optique*, no. 11.
THEOCARIS, P. S. and KOUTSABESSIS, 1965, *J. Sci. Instrument*, **42**.

Part IV

Advances in assessment and specification of performance of optical instruments

Some Recent Advances in the Specification and Assessment of Optical Images

R. R. Shannon

Optical Sciences Center,
University of Arizona,
Tucson, Arizona 85721

Abstract—The most significant development in the past three years in the field of image evaluation has been the widespread use of precise computations of the physical optical transfer function (OTF) for lens system analysis. For years it has been possible to carry out an OTF analysis in great depth, but only recently have new computation techniques made the general use of the precise OTF possible.

Advances have been made in the understanding of the behavior of optical systems in design and under various environments. Central to this advance has been the general use of the two-dimensional OTF as one part of an overall image analysis technique. From this form of analysis, an image can be evaluated to varying degrees of detail.

The major advances to be considered are in the areas of computation, measurement, and systems analysis. Some examples of each of these will be described in turn.

THEORY

The basis for computation of the transfer function for a lens system is the Kirchoff diffraction integral. The amplitude distribution at a point in the image plane is given by an integration of the wavefront over a surface enclosing the point. The most convenient surface to select is a sphere centred about the chosen image reference point, and passing through the axial intercept point of the actual exit pupil of the lens system.

The exit pupil point chosen will usually be the axial intercept point of the chief ray exiting from the lens. The chief ray should not be defined as the ray through the centre of the axial aperture stop, however, but as the central ray through the vignetted aperture.

The description of the wavefront error should be carried through in an appropriate manner in order that the transfer function theory apply to the image plane of interest. Figure 1 shows a cross-section of the imaging situation in a lens. The plane represented by (u, v) is a plane perpendicular to the optic axis at the location of the intercept of the chief ray with the optic axis (presuming a rotationally symmetric optical system). The reference image plane of interest is located a distance L from the defined pupil plane. A reference image point is taken at a height H from the axis. The amplitude distribution at points located a distance (ϵ_x, ϵ_y) about the reference image point and within the image plane is to be found.

To compute the amplitude, a spherical reference surface is chosen, centred about the reference point at H and passing through the axial pupil point. The integration will be made over this spherical reference surface, taking into account the phase of the wavefront propagated toward H as it passes through each point upon the reference sphere.

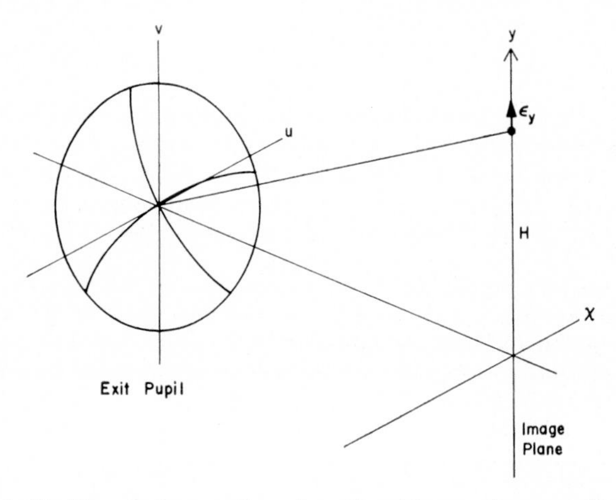

FIG. 1. Meridional plane section of pupil and image plane coordinates.

Conversion of the amplitude to intensity by squaring will allow computation of the image to be observed on the focal plane. Care must be taken that the intensity distribution computed will be that which would be observed in the chosen plane. Evaluation of the geometry shows that a particular choice of coordinates for expressing the wavefront is convenient, and adds to some understanding of the imaging process.

The diffraction integral to be evaluated is given by

$$i(\epsilon_x, \epsilon_y) = \iint_S A(u, v) \exp\left[-2\pi i\phi(u, v; \epsilon_x, \epsilon_y)\right] \mathrm{d}u\mathrm{d}v,$$

where (ϵ_x, ϵ_y) are coordinates in the image plane and (u, v) are aperture coordinates to be selected. It is required that the area element du, dv be a metric of the entering wavefront in order that a uniform mapping of the wavefront energy be obtained.

The phase error in the argument can be rewritten as

$$\phi(u, v; \epsilon_x, \epsilon_y) = W(u, v) - W'(u, v; \epsilon_x, \epsilon_y),$$

where $W(u, v)$ is a set of wavefront aberrations computed at a set of points on the reference surface through the pupil point. The function W' is the added phase error encountered when the lateral position of the (ϵ_x, ϵ_y) is varied. From Figure 2 then:

$$W' = A - B = A^2 - B^2/2R = (A^2 - B^2) \cos \theta/2L.$$

This leads to

$$i(\epsilon_x, \epsilon_y) = \int\int \exp A(u, v)[-2\pi i W(u, v; H) e 2\pi i (u\epsilon_x + v\epsilon_y)/R] du dv$$

since the u, v coordinates form a natural set of coordinates for the computation. In particular, the geometry should be noted, for it shows that the relation between the pupil amplitude and the image plane amplitude is that

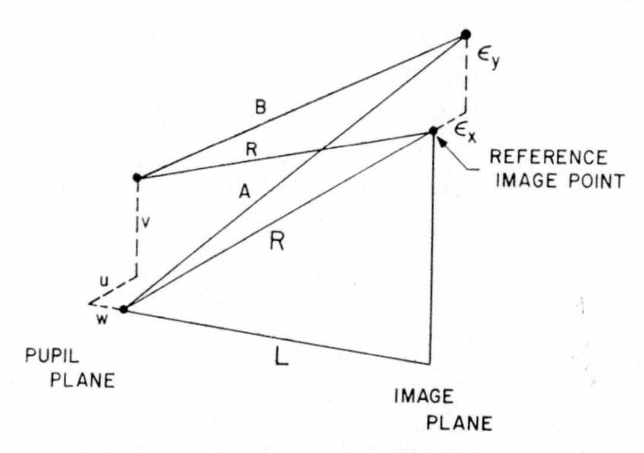

FIG. 2. Relationship between pupil and image plane.

of a Fourier transform between parallel pupil and image planes. If the amplitude image is required in a plane at some angle to the original plane, the proper transform is between planes angled through the optical axis by the proper amount, and passing through the appropriate pupil and image points.

Some additional insights are gained by rewriting the (u, v) coordinates into a form that is useful for computation. The axial paraxial exit pupil semi-diameter is defined by the quantity v_m. The region occupied axially in u, v-space can be described as a circle in the normalized region:

$$\left.\begin{aligned} u &= v' v_m \\ v &= v' v_m \end{aligned}\right\} \text{wherein } u'^2 + v'^2 \leqslant 1.$$

Inserting this into the argument for u, v and dividing by λ to obtain a phase value,

$$2\pi/\lambda R[u\epsilon_x + v\epsilon_y] = 2\pi/\lambda R[u' v_m \epsilon_x + v' v_m \epsilon_y \cos \theta] = 2\pi/\lambda[u'\epsilon_x(\lambda NA) + v'\epsilon_y(\lambda NA \cos \theta)].$$

Relative to the axial numerical aperture ($NA_0 = v_m/L$), the above function becomes

$$2\pi/\lambda[(\lambda NA_0 \cos\theta)u'\epsilon_x + (\lambda NA_0 \cos'\theta)v'\epsilon_y].$$

For convenience we will define reduced image plane coordinates as:

$$x' = \epsilon_x(\lambda NA_0 \cos\theta) \qquad y' = \epsilon_y(\lambda NA_0 \cos^2\theta).$$

The formula so obtained is then:

$$i(x', y') =$$

$$\iint A(u', v') \exp[(-2\pi i/\lambda) W(u', v')] \exp[(2\pi i/\lambda)(u'x' + v'y')]du'dv'.$$

Since this formula is in the form of a Fourier integral transform, an alternate technique for evaluation of the function at a set of sample points, as a Fourier series, is available. To do this the pupil function must be replaced by a set of sampled values. These values will be immersed in a two-dimensional matrix of samples.

For reasons to become clear later, the choice of sample values must be made in the following manner. Assume a matrix of $N \times N$ complex values. The pupil area will be sampled within a central N/2 by $N/2$ set of values. (Usually the non-zero pupil region is circular.) The size of the increment in the pupil region will then be

$$\Delta v = v_m/N/4 = 4v_m/N.$$

Since the integral form for the diffraction integral is being replaced by a Fourier transform, this increment can be considered as being located in an effective frequency parameter space. The upper limit upon the distance in this effective frequency space is given by $2v_m$.

In order to compute the image distribution, the integral is replaced by a sum. Consider increments in $\Delta\epsilon$ and Δv such that

$$x' = j\Delta\epsilon \qquad y' = k\Delta\epsilon$$

$$u' = l\Delta v \qquad v' = m\Delta v$$

Then

$$i(j, k) = \sum_{\frac{-N}{2}}^{\frac{N}{2}} \sum_{\frac{-N}{2}}^{\frac{N}{2}} A(d, m) \exp[-2\pi i W(l, m)] \exp[(2\pi/\lambda)i(jl + km)(\Delta\epsilon\Delta v/R)].$$

In order that the system of equations satisfy a Fourier series rule, the increments in frequency must be chosen to make the lowest frequency periodic in the total spatial range to be covered. Thus

$$2\pi(\Delta v/\lambda R).\ 2\epsilon_m = 2\pi.$$

Here ϵ_m is the image plane radius covered by the diffraction transform. Now:

$$\epsilon_m = 2\lambda R/\Delta\nu = N\lambda F\#.$$

Once the amplitude data in the spatial coordinate or image plane are obtained, the intensity distribution in the point image is obtained by taking the complex square of the amplitudes. This yields a data set of intensities:

$$f(j, k) = i(j, k)i^*(j, k) = R^2 + J^2.$$

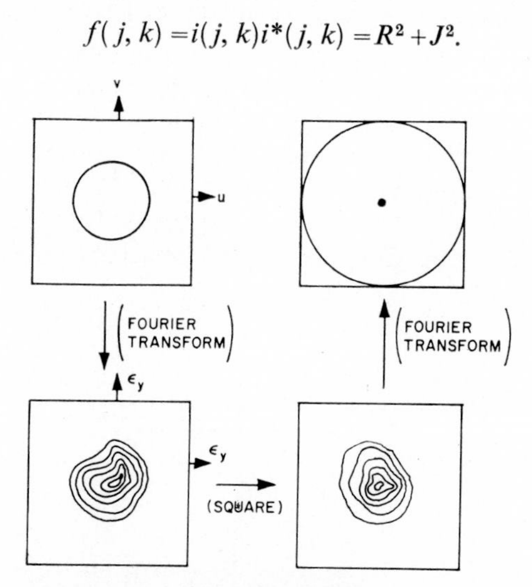

FIG. 3. Diagrammatic relationship between the matrices developed at each step of the transform process. Each square represents a $N \times N$ set of data points.

The latter form is used if the data are stored in terms of a pair of complex matrices. The transfer function can then be obtained from this matrix of data, by a Fourier series calculation.

$$F(l, m) = \sum \sum f(j, k) \exp(-2\pi i(lj + mk)\Delta\epsilon\Delta\nu).$$

The values of $\Delta\epsilon$ and $\Delta\nu$ are related again by the requirement for a Fourier series transform that $\Delta\epsilon\Delta\nu = 1$. In order to find the maximum spatial frequency included in the transform, set $2\pi(\lambda F\#/2)(2\nu_m) = 2\pi$, which results in $\nu_m = (1/\lambda F\#)$. The reason for using $2\nu_m$ in the computation is the interpretation that the spatial frequency region in the computation will extend from $-\nu_m$ to $+\nu_m$.

The physical significance of the various mathematical operations described is shown in Figure 3. The wavefront aberration is represented as a complex function within the circular region contained within half of the $N \times N$ matrix of points. The selection of the number N for the matrix must be sufficient to ensure that the wavefront is adequately represented, both in extent and shape of the aperture and in the variation of phase over the aperture.

Since the choice of the optimum size of the matrix used is related to the dimension of the aperture being analyzed, the physical area represented by the aperture matrix is always twice the linear size of the largest dimension of the aperture. In the event that the aperture is of a grossly non-symmetrical form, as may be obtained in the presence of vignetting, then an anamorphic stretch of the coordinates to fit the aperture into a circular region for the purpose of computation may be carried out. In that case a compensating 'unstretching' of the coordinates must be made after the computation is complete. This effect is related to the 'natural anamorphism' of the pupil, which takes place in any real optical system off axis.

Once the first transformation is carried out, the amplitude image is obtained in the second matrix shown in Figure 3. This amplitude matrix covers an area on the image plane determined by closeness of spacing of

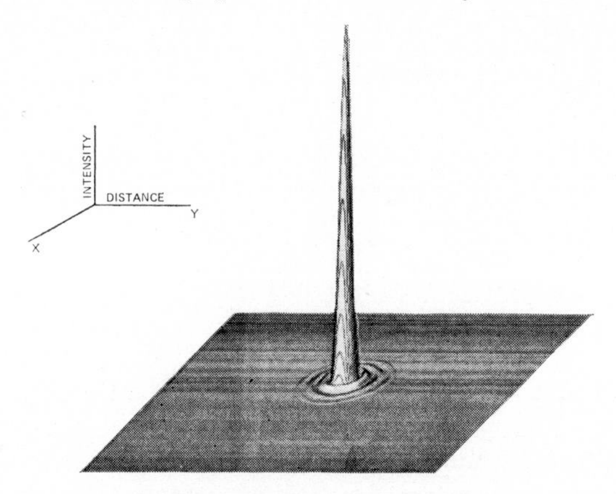

FIG. 4. Point spread function (Airy disc) for a circular aperture.

the sample points within the aperture. In other words, the spatial extent allowable upon the image plane is determined by the total number of points used in the computation of the image. In order to relate this to a convenient dimension, the period of the highest spatial frequency contained within the image is known theoretically to be equal to $1/\lambda F\#$. The spacing of the sample points on the image plane when the rule given above for determining the aperture matrix extent is used, turns out to be equal to half of the period of the highest spatial frequency. The total spatial extent of the computed image upon the image plane is then equal to N times this sample distance. (In terms of so-called 'Airy disc diameters', which may be more familiar, the sample distance is 0·21 of an Airy diameter. Thus the spatial extent is approximately $N/5$ times the Airy disc diameter.)

Values of N equal to about 100 appear to be most useful in carrying through the computation for aberrations within the range of two to three wavelengths phase error.

The intensity distribution within the image can be obtained by taking the complex square of the amplitude distribution. These data are then used for the next step of computation of the transfer function by application of a Fourier transformation operation on the matrix.

In order to understand the representation of the pupil and image plane functions as a result of a Fourier series rather than an integral computation, the diagrammatic representation of Figure 4 may be examined. To obtain a series representation, the function being evaluated must have a periodic form. Thus, although the pupil exists as, of course, only a single pupil with a set of sampled data representing the pupil and wavefront, the mathematical model which is used for the computation of the image-form considers the pupil as being periodic, with a period distance of twice the pupil diameter. Since the pupil plane can be considered as an equivalent to a spatial frequency plane, with an upper frequency cutoff of twice the pupil diameter, it is clear that this periodicity in the spatial frequency plane offers no difficulty in terms of overlap of one periodic frequency region into the next adjacent region. Indeed, the result of squaring the amplitude data to obtain the intensity is the generator of the full frequency passband that fills up the frequency matrix.

The result of carrying through the total operation upon the amplitude only, without carrying out the squaring operation for intensity, will regenerate the wavefront data, filling only half of the available spatial frequency region. This operation is equivalent to obtaining the 'coherent transfer function' and may be used in reconstructing the form of image in fully coherent optical systems.

For the case of normal incoherent imagery, however, a problem can arise from the non-linear operation of converting the amplitude wavefront, or image information, to an intensity form. The magnitude of the error that is introduced will be dependent upon the magnitude of the aberration in the pupil. For a given density of sampling points in the pupil, determined by the choice of the size of matrix, N, the spatial extent of the area on the image plane covered by the calculation is fixed. Since the representation of the pupil, in frequency space, has been chosen as an effective periodic function in order to use Fourier series techniques, the result of the Fourier operation will be another periodic function. The period distance of this image plane function will be given by the spatial extent, which is determined by the fineness of spacing (number N of samples) in the pupil plane. As the magnitude of the pupil aberration is increased, there will be a corresponding increase in the area of the image plane over which the intensity distribution in the image of a point source will be spread. Under this condition, a point will eventually be reached in which a significant amount of the light energy is spread into an adjacent region of the effective periodic areas on the image plane. This will result in a spatial frequency 'foldover' causing an error in the resulting transfer function that is proportional to the total amount of energy in the image that leaks over into an adjacent area on the image plane.

Since a single periodic region on the image plane is usually 20 or more Airy discs in diameter, this error will be small for wavefront aberrations on the order of a few wavelengths. The critical feature of the computation is that the rate of change of the aberration in the pupil shall not exceed one-half wavelength between adjacent sampling points in the pupil. Thus, the more complex in form and magnitude, the larger the number of sample

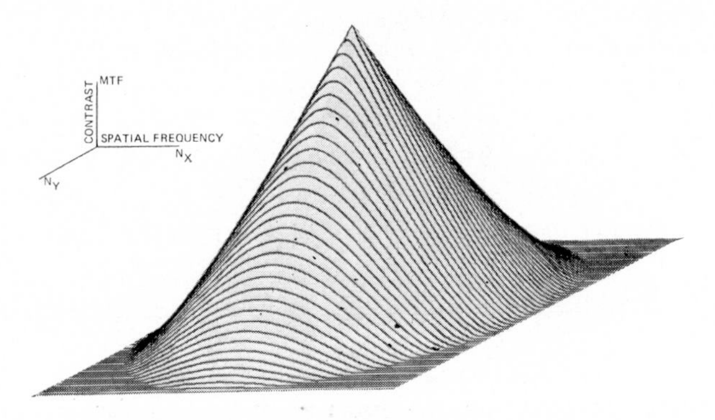

FIG. 5. Modulation transfer function for a circular aperture.

points that will be required. In fact, the system of computation is self-correcting for the foldover error. The larger the wavefront error, the more points are required to represent the error accurately, and to differentiate the error from all other possible wavefront errors of similar magnitude. The more sampling points taken, the larger N becomes, the larger the periodic area on the image plane becomes. As a result the transfer function becomes more accurate.

The greater the wavefront error, the greater will be the reduction of the transfer function at high spatial frequencies. Thus it is desirable to increase the number of sample points in the spatial frequency region in order to obtain a better representation of the transfer function at low spatial frequencies. The rules for determining the proper sampling rate are those previously described.

In summary, the Fourier series representation of the sampled data for computing the image form or transfer function does provide an accurate representation of the continuous process of image formation, if the rules regarding representation of the data samples and the rate of sampling are observed. The reason for examining this representation of the pupil and image plane of a lens system will now be shown in terms of recent developments in the high-speed computation of Fourier series transformations by Cooley-Tukey and other methods. The existence of these new mathematical techniques has made the widespread application of the previously described theory practical on digital computers.

APPLICATION

The technique can be applied to conversion of any type of pupil or wavefront to spread function, or transfer function. For example, an aberration-free circular pupil will deliver a point spread function as shown in Figure 4. The coordinates on the horizontal plane are dimensions in the image plane. The vertical coordinate is intensity.

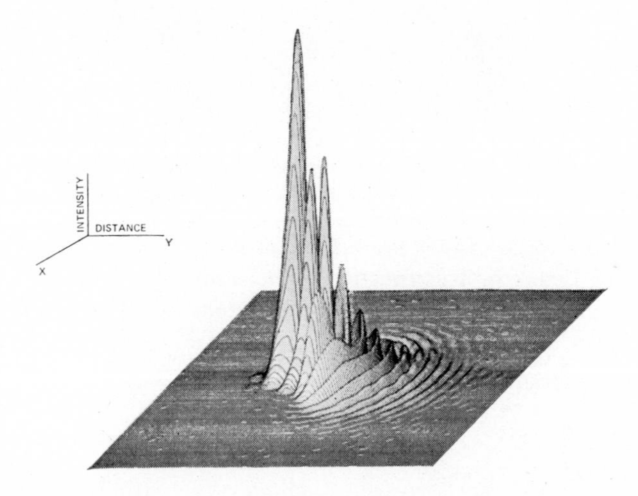

FIG. 6. Point spread function for 2·4 wavelengths (peak to valley) of cubic coma wavefront error.

The figure shown is an isometric computer plot of the matrix of data stored in computer memory after transforming from the pupil plane and squaring the amplitude image. This type of presentation of data is more useful as a device for visual presentation than in the presentation of tables of data, when two-dimensional functions are encountered. The digital data within the matrix are available for further computations when required.

A Fourier transform of the spread function yields the two-dimensional modulation transfer function plotted in Figure 5. Here the vertical coordinate is contrast, being unity at zero spatial frequency. The horizontal plane is spatial frequency, with zero at the centre and the maximum, or cut-off frequency at the edge. The familiar form of the circular aperture, 'diffraction-limited' transfer function can be seen by sighting along the edge of the transfer function surface in any direction. This surface is the maximum contrast that can be obtained for a circular aperture.

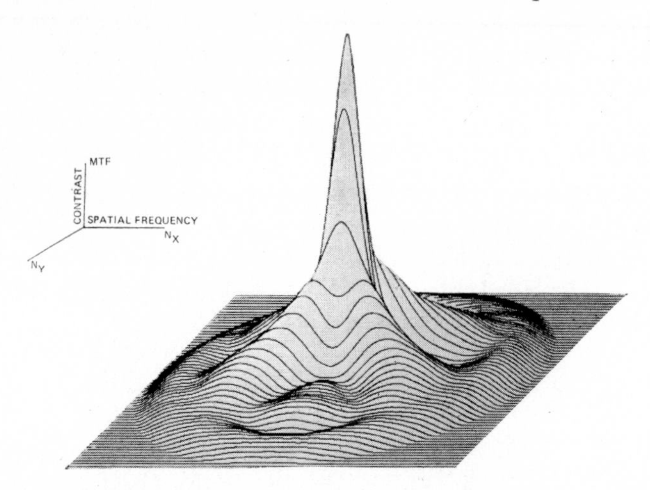

FIG. 7. Modulation transfer function corresponding to Figure 6. Note the 90° coordinate rotation in the plot.

The effect of aberrations upon the image can be seen from the spread function of Figure 6. This function is the result of converting a cubic or coma wavefront error having a peak magnitude of 2·4 wavelengths into a

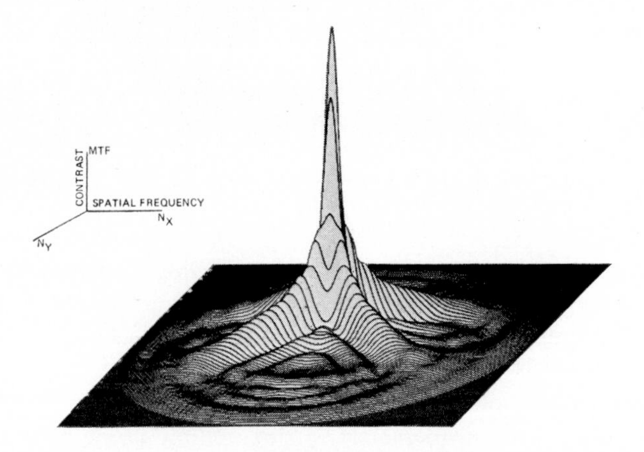

FIG. 8. Modulation transfer function for 5 wavelengths of cubic coma.

spread function. Successive Fourier transformation into an O.T.F. yields Figure 7. Some characteristics of the O.T.F. resulting from coma appear. In particular, some directions of high response appear, with lower response in the x and y directions. Calculation of the transfer function for 5 wavelengths, as in Figure 8, shows an even more predominant effect. This occurs for spatial frequencies in a direction perpendicular to the 60° caustic lines of the normal coma image.

This latter observation of a particular symmetry to the transfer function for a specific aberration can be seen for other combinations of wavefront errors. From visual observation of the two-dimensional O.T.F. surface, an intuitive feeling for the effect of aberrations on the transfer function can be developed.

Once the computation scheme is developed, the input to the wavefront error can arise in several ways. The wavefront error can be obtained from ray tracing during a lens design evaluation. After design, perturbations in the system can be carried through to evaluate the effect of assembly errors on the O.T.F. From this, tolerance limits can be established.

A very practical application is in optical system testing. An interferometer is used to gather data in terms of fringe patterns recording the two-dimensional wavefront error. The fringe pattern is then scanned, and the coordinates of the fringe locations are taken at a mesh of points over the aperture. These data are then converted into wavefront errors.

By computer, the linear and quadratic wavefront errors are removed from the data set. This amounts to a correction for lateral and longitudinal position of the measuring interferometer. Higher order errors, such as coma or astigmatic surface error, can be computed as well in order to quantify the errors in the optical surface being measured. Each time, the corrections are made in such a manner as to reduce the residual average root-mean-square wavefront error.

The residual wavefront errors remaining after a specific step of the process described above describe the wavefront produced by the lens under test. These sampled wavefront data can be corrected, by interpolation, to a new set of data, located at a rectangular mesh of points in the pupil. At this point, the Fourier transform technique can be applied to obtain the transfer function of the system.

The result of measurement of optical surfaces, especially large surfaces, is to yield a wavefront that contains a number of somewhat randomly distributed surface errors. These give rise to transfer function errors that are not simple in form, but can be observed to have a given average reduction from the best possible transfer function for a given aperture shape.

Techniques such as described above for measuring or tolerancing optical systems can be carried out to any extent desirable. The final example of this paper will show how these techniques can be used to develop an intuitive and a quantitative set of data regarding tolerance errors and their effect on the transfer function.

A random wavefront error, typical of one possible sample that may be expected from turbulence or fabrication errors, is theoretically represented in Figure 9. This wavefront has about 4:1 ratio of peak-to-valley error to root-mean-square error. The magnitude of this wavefront error was scaled to several values and the transfer functions and spread functions computed.

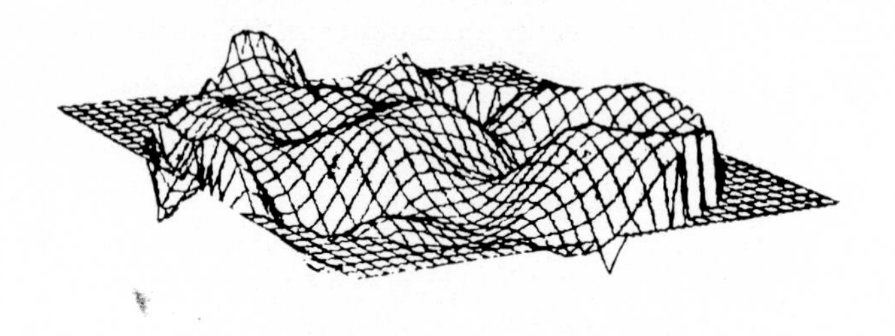

FIG. 9. Quasi-random wavefront error, theoretically developed. Vertical coordinate is optical path error; horizontal plane is pupil plane.

Figure 10 shows the effect of three different amounts of error on the spread function. Figure 11 shows the transfer functions for the same amounts of wavefront error. Visually, the effect of the random wavefront aberration can be seen to have a consistent effect upon the transfer function.

Although the details of the transfer function surface may vary, the function can be reasonably well represented by an average surface that is rotationally symmetric. If this is done for each computed transfer function, a set of average modulation transfer function curves can be computed as shown in Figure 12. These curves, which are a result of computation on one random wavefront sample, correlate well with similar curves which are a result of an average over several different smooth or 'regular' aberrations, each having the same root-mean-square wavefront error.

SUMMARY

As can be seen from the few preceding examples, the optical image can now be described numerically as well as is needed. The application of general numerical Fourier analysis methods to image analysis has led to a development in which the lens system can be viewed qualitatively in the frequency plane, allowing an intuitive development of the mind of the designer to think in a spatial frequency region.

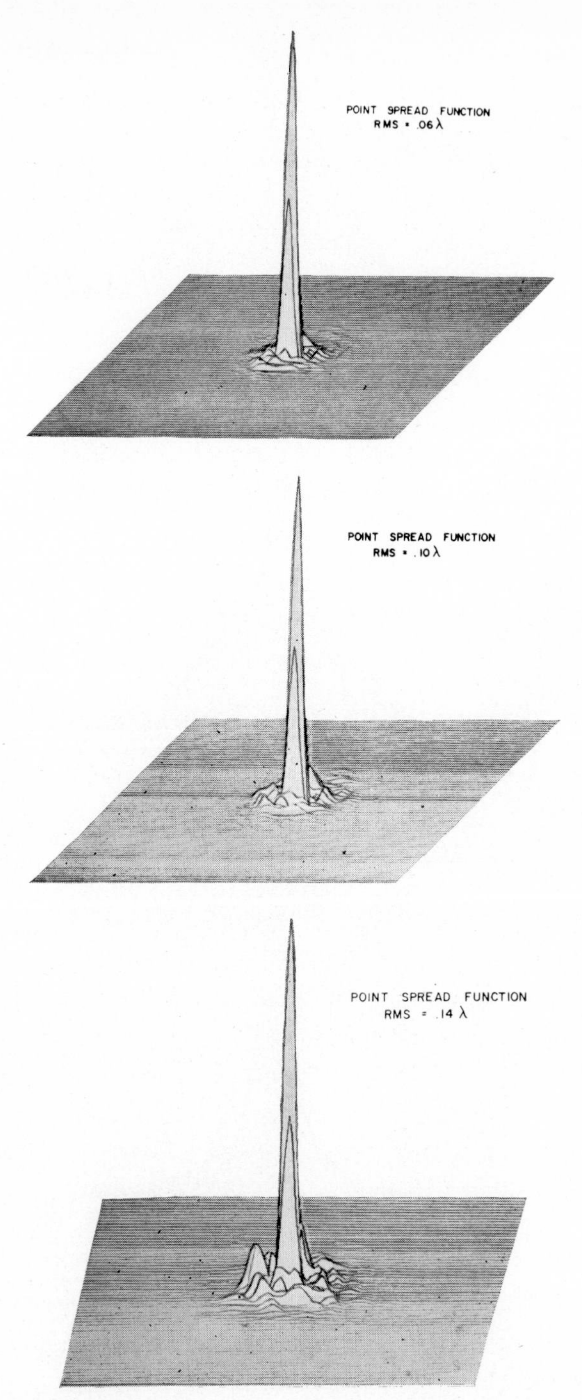

FIG. 10. Spread function for rms random wavefront error: (*a*) 0·06 wavelength; (*b*) 0·10 wavelength; (*c*) 0·14 wavelength.

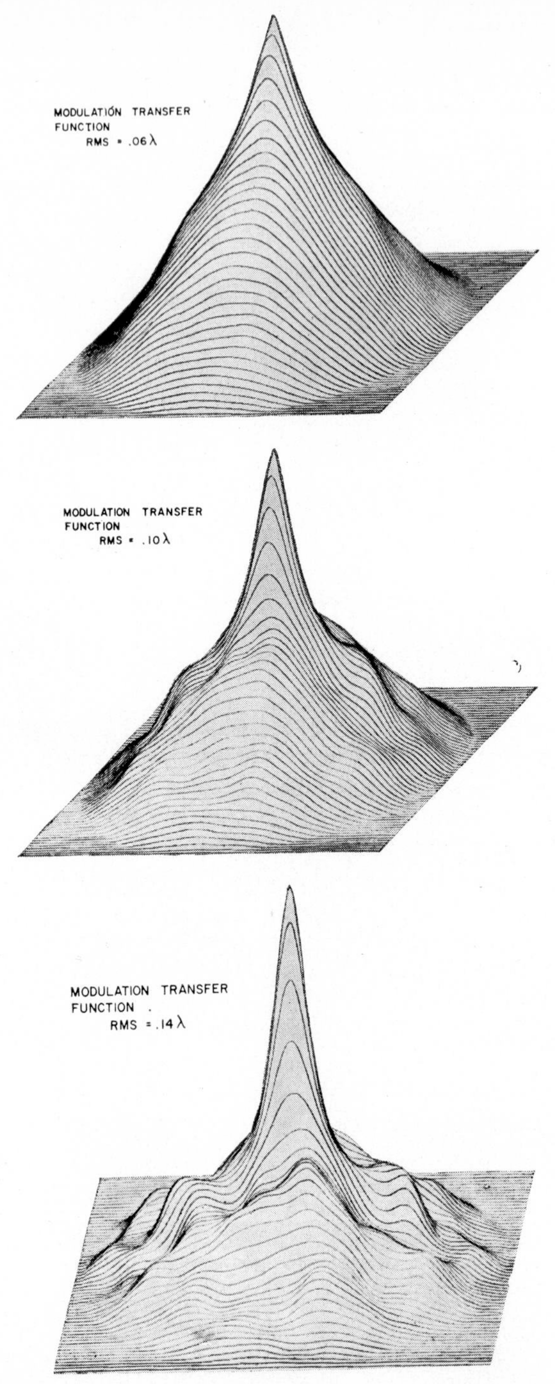

FIG. 11. Modulation transfer function for rms random wavefront error: (*a*) 0·06 wavelength; (*b*) 0·10 wavelength; (*c*) 0·14 wavelength.

Most of the developments presented here are a portion of several years work while the author was associated with Itek Corporation of Lexington,

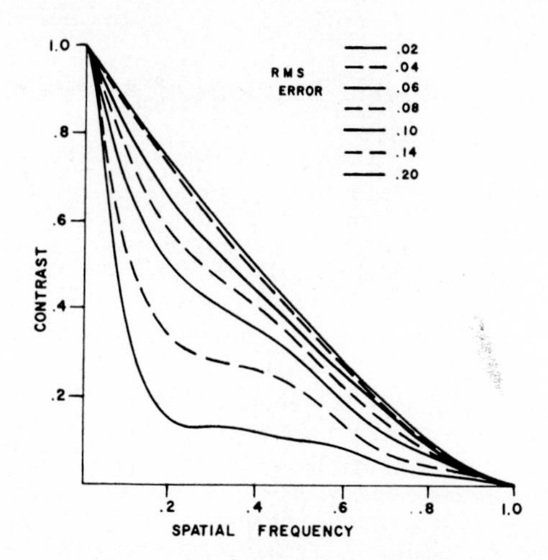

FIG. 12. Family of averaged modulation transfer functions for varying amounts of rms random wavefront error.

Massachusetts. In particular, thanks go to J. Rancourt, S. Lerman, W. Minnick and J. Clancy for their aid in computer programming.

BIBLIOGRAPHY

COOLEY, J. W. and TUKEY, J. W., 1965, *Math Comput.*, **19**, 297.
HOPKINS, H. H., 1965, *Proc. Conf. Phot. Spectroscopic Opt.*, 1964; *Japan J. Appl. Phys.*, **4**, Supp. 1, 31.
SHANNON, R. R., 1969, *IEEE Trans. Aerosp. and Elec. Sys.* V., AES-5, No. 2, p. 273.

Comparison of Different Methods of Assessing the Performance of Lenses

K. J. Rosenbruch

Physikalisch-Technische Bundesanstalt,
Braunschweig, Germany

Abstract—New methods of measuring the optical transfer function (O.T.F.) have been developed during the past few years, and some procedures have been published for correlating the results of measurement and calculation with subjective tests of image quality.

It is often more convenient, for production and design purposes, to measure other quantities depending on various parameters, such as (*a*) constructional details, radius, refractive index, dispersion, etc; (*b*) geometrico-optical aberrations; and (*c*) wave aberrations, interferometer tests, etc.

There is no difficulty in relating the results of such measurements to production and manufacturing tolerances. Image quality, however, is the final criterion and for this purpose the O.T.F. is most suitable.

Tests have therefore been made with existing lenses to show the influence of different kinds of error of measurement on the O.T.F., and a table has been compiled to show the tolerances corresponding to a just-recognizable difference of subjective image quality.

The experimental examples show that for the more complicated and large aperture systems, centring and defocusing errors are often more harmful than the random errors in the testing processes.

In recent years, several new methods of measuring the optical transfer function (O.T.F.) of lenses have been developed (Lindberg 1954; Baker 1955; Birch 1958). In addition, a number of procedures have been published for calculating this function from the measured results of other tests (Hopkins 1957; Barakat 1962; Murata 1959). Finally, there have been some successful attempts to deduce from the O.T.F. an objective measure of image quality corresponding with the subjective quality (Biedermann 1967; Heynacher 1963).

For all these reasons, the O.T.F. is a central point from which all the different methods of assessing the value of a lens, as regards image quality, the influence of aberrations, etc. Sometimes, however, it is more convenient to measure other properties of a lens which are more suitable for the processes of production or for the method of correction than the O.T.F. These properties may be:

(1) Constructional parameters such as the radii of surfaces, refractive indices and dispersions, thicknesses of lenses, distances and free apertures, centring and alignment, etc.

(2) Geometrico-optical aberrations, spherical aberration, coma, astigmatism, field curvature, defocusing, distortion, etc.

(3) Wave aberrations directly measured with the Twyman-Green-interferometer or by means of a shearing interferometer of the type of the Saunders prism, the plane plate interferometer of Murty, the grating interferometer of Ronchi, and many other similar devices.

The common purpose of testing methods for optical systems is to guarantee the image quality or to disclose noticeable deviations from the specified performance. One of the best measures for describing image quality is the O.T.F.

It is therefore shown here how large the tolerances in the above testing methods can be to prevent them from exceeding the limit which produces a perceptible change of quality. The theoretical and mathematical connection between the different methods of image evaluation is shown in Figure 1. On the left hand side of this figure, the stepwise calculation of the O.T.F. from the constructional parameters is given; on the right hand side, some measuring methods are mentioned, which can measure these values directly or indirectly.

An optical system is determined by its data of construction. Almost all the values of radii, refractive indices, thicknesses, distances and free apertures enable an optical designer to judge the image quality from his calculation. In spite of this, there are many difficulties in deducing deviations in image quality from the tolerances of the constructional data. The data for the tolerances arrived at by tedious calculation are only valid for each particular optical system, as it is known that there are more or less critical constructional parameters as, for example, the different radii of curvature in an optical system. It may be mentioned, however, that for a fully-corrected optical system, there are some critical parameters for single image aberrations, but not normally for the image quality as a whole. It should therefore be appropriate to demonstrate the influence of the usual tolerances of the radius of curvature on the O.T.F. as an example of this step.

In the second step (see Figure 1), the results of ray-tracing procedure are given and the values of the ray aberrations can be examined by a large number of experimental tests. Of these, the best known methods are those described by Wetthauer (1921) and Hartmann (1908), the spot diagram analyses given by Baker and Whyte (1965) and Williams (1968), etc. In this step, the aberrations can be measured along the paraxial ray (longitudinal aberrations, 'diapoint aberrations', Herzberger 1958) or in the image plane as lateral aberrations. A representation of the aberrations in the form of a polynominal function of the pupil coordinates is possible, for instance, from a number of rays by the least square approximation. The tolerances of the coefficients so found result also in an uncertainty in the O.T.F.

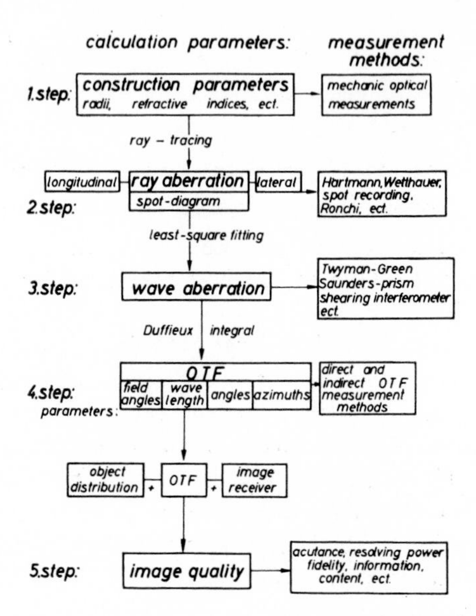

FIG. 1. Steps in the assessment of the performance of optical systems.

The next step in the row of lens-performance-tests is the wave aberration for which there are many measuring procedures. The best known is the Twyman-Green interferometer. Finally, there is the last step depending only on the optical system, the O.T.F. of which can be measured by a large number of methods. Naturally, the O.T.F. can only be used as a tool to deduce image quality, if this function is measured for a number of parameters such as wavelengths of light, angles of the field, azimuths image planes, focal steps, etc.

Finally, it is possible to combine a number of O.T.F.s in a figure of merit taking into account the structure of the object and the characteristic behaviour of the image receiver. But the effects of the object, of the receiver, and of the optical system are mixed together so that it is almost impossible to recognize the performance of the lens. Such summed-up image qualities are only useful for the comparison of lenses. This last step has therefore not been considered in the following discussion.

First let us consider the resulting errors in tne O.T.F.—according to a given scheme from the tolerances of some measuring methods. There are two ways of reaching this aim:

(1) To calculate the influence of a given variation of the performance parameter on the O.T.F.,

(2) To take the experimentally attained exact value with its errors and deduce from these errors the deviation of the O.T.F.-values.

The second method is preferred because it is better suited to the tolerances of the measuring methods.

First, the influence of the radii of curvature on the O.T.F. is ascertained. For this purpose axial rays are traced with radii changed successively by 1 per cent. It can be seen (Table 1) that this mainly gives variations in the focal length of the photographic lens, but for the best axial focus, the

TABLE 1. Changes in the O.T.F. with variation of the different radii by 1% for a lens 1:5, f=300 mm. (The transfer values are found for the best focus. The deviation from the original focal position is given in the last column.)

1‰ change of radius	O.T.F. for lines/mm						defocusing mm
	5	10	15	20	25	30	
original	0·90	0·66	0·45	0·36	0·36	0·35	0
R_1	0·89	0·63	0·41	0·31	0·32	0·34	0·7
R_2	0·90	0·67	0·46	0·35	0·34	0·33	0·1
R_3	0·86	0·58	0·36	0·32	0·36	0·34	−0·4
R_4	0·87	0·60	0·40	0·36	0·39	0·37	−0·6
R_5	0·88	0·60	0·39	0·32	0·35	0·34	−0·2
R_6	0·90	0·67	0·45	0·36	0·35	0·34	0·1
R_7	0·90	0·66	0·44	0·34	0·34	0·35	0·5

best focus for 25 lines/mm

O.T.F. shows little deviations from the original design. From the O.T.F. curves (see Figure 2) it can be seen that a change of the radii of curvature results in a variation of about 0·1.

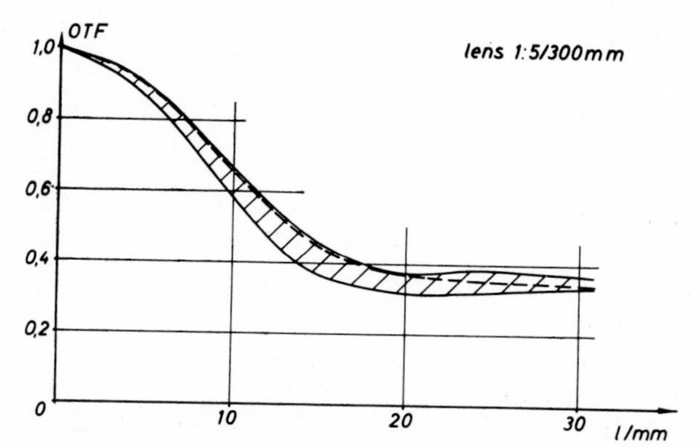

FIG. 2. Tolerance area of O.T.F. for radii variation of the lens of Table 1. The broken line shows the O.T.F. of the original lens.

To avoid any important influence on the O.T.F., the radii of the lens should be exact within 5×10^{-4}, a demand not easy to meet.

The normally attainable accuracy of measuring the ray aberrations such as longitudinal or lateral aberrations, is equivalent to their influence on the accuracy of the deduced O.T.F. This group of aberrations may therefore be represented by the measured longitudinal spherical aberration of a photographic lens, $f/2, f=50$ mm. Here the uncertainty of the measurement is about $\pm 0{\cdot}01$ mm for rays which go through the zone at half of the full aperture, and it grows to $\pm 0{\cdot}02$ mm for paraxial and full aperture rays. The spherical aberration has been varied separately in the three regions and then the relevant O.T.F.s have been calculated. Not only are

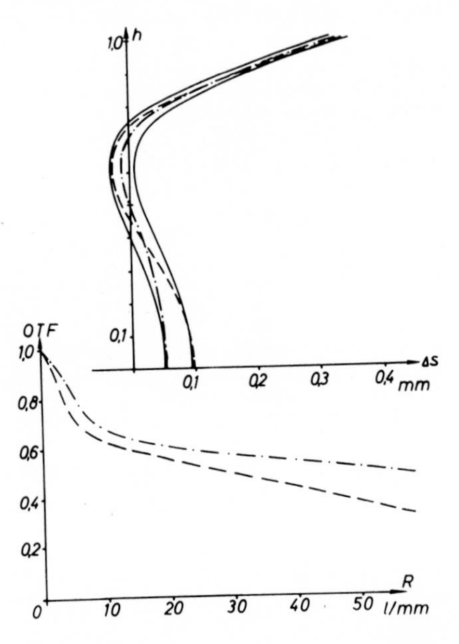

FIG. 3. Tolerance of the measurement of longitudinal spherical aberration and its influence on O.T.F.

separate parts of the spherical aberration curve uncertain, but the whole curve may deviate by about $\pm 0{\cdot}01$ mm. The influence of such a variation on the O.T.F.'s is given in Figure 3. The maximum differences of O.T.F. are 0·15 at 40 lines per mm.

To reach the same accuracy in the O.T.F., the lateral aberration y' (instead of the longitudinal aberrations $\Delta s'$) of this lens should be given within an uncertainty of 1 to 2 microns; this is normally the limit of accuracy of measurement.

This is valid because of the formula

$$y' = \Delta s' h / f$$

where f = focal length, and h = height of incidence of the rays.

The wave aberration is a further step in the assessment of lens performance which can be experimentally measured or found by ray-tracing. It is usual to display this magnitude as a polynomial expression of the coordinates of the pupil of the lens. The accuracy of the wave aberration then depends on the accuracy of the coefficients of this polynomial.

This polynomial can be deduced in different ways. Either the result of ray-tracing with 100 to 200 rays; the lateral aberration is fitted by least-square calculations to get the coefficients with their errors, or an interferogram of a Twyman-Green-interferometer is measured for 300 to 400 points over the pupil to give the best fit with these data in a polynomial

TABLE 2. Typical deviations of the polynomial coefficients.

name of coefficient	from calculated aberrations 12 surfaces lens $f/4$ for 10° field angles		measured interferogram 14 surfaces lens $f/2$ for 9° field angles	
	coeff. in wavelength	error	coeff. in wavelength	error
defocusing	−2·10	±0·020	−9·95	±0·28
spherical 1	−0·92	±0·032	−1·27	±0·71
spherical 2	1·59	±0·016	5·55	±0·48
distortion 1	3·19	±0·007	9·81	±0·09
distortion 2	−0·85	±0·010	—	—
astigmatism	2·93	±0·020	−1·25	±0·18
coma 1	0·29	±0·019	1·14	±0·34
coma 2	0·28	±0·011	−9·46	±0·32
gullstrand 1	−2·38	±0·015	12·89	±0·30
standard deviation of the wave aberration		±0·0004		±0·075

expression. The terms of this polynomial are limited by the type of lens, the exactness of the least-square fitting and the properties of symmetry which may or may not exist. According to Sunder-Plassmann (1967), either an exact adjustment of the lens is assumed or not. The normally adopted polynomial coefficients are given below

$$\begin{aligned} W(x, r) = {} & Dr^2 + S_1 r^4 + S_2 r^6 \\ & + K_1 x r^2 + K_2 x r^4 + K_3 x^3 r^2 \\ & + V_1 x + V_2 x^3 + V_3 x^5 \\ & + G_1 x^2 r^2 + G_2 x^2 r^4 + G_3 x^4 r^2 \\ & + A_1 x^2 + A_2 x^4 + A_3 x^6 \end{aligned}$$

where x, y = coordinates of pupil; $r^2 = x^2 + y^2$; D = defocusing; S_1, S_2 = first and second spherical aberration coefficients; and the first, second and third coefficients are as follows: K_1, K_2, K_3, comatic aberration; V_1, V_2, V_3, distortion; G_1, G_2, G_3, Gullstrand error; A_1, A_2, A_3, astigmatism. These are exact enough for telescopes and photographic lenses, if large centring errors are not present.

The polynomials deduced by ray-tracing from direct constructional data are of high accuracy as regards their coefficients, but the coefficients calculated from experimentally found values in the interferogram have considerably larger errors, as can be seen from Table 2. This can be easily understood, because in the ray-tracing procedure all data are exactly given and no centring errors are assumed. The only possible faults come from the limited power of the coefficients. For the evaluation of the interferogram, the errors are normally larger and, in addition, the real optical system is affected with inevitable centring errors and deviations from exact construction which results in a relatively large uncertainty in the calculated coefficients (Rosenbruch 1968).

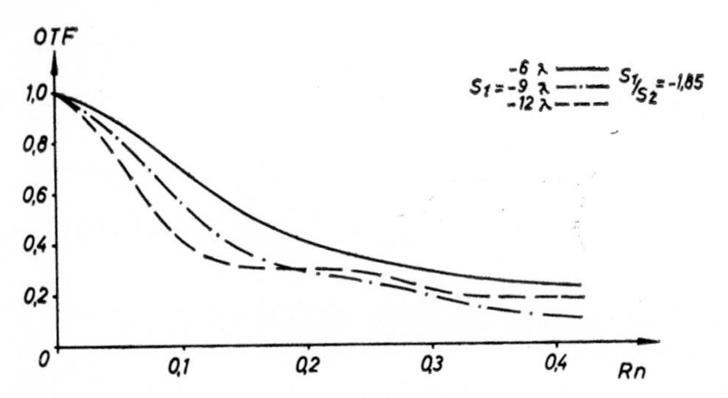

FIG. 4. Influence of variations of the axial coefficients on the O.T.F. The relation S_1/S_2 is constant.

How large can the tolerances in the coefficients be to keep the variance of the O.T.F. within a given limit of for example ±0·10? This question cannot be answered generally because there are too many variables. On the optical axis there are normally only three coefficients and one can, of course, establish exact tolerances for their influences on the O.T.F.

In Figure 4, the amount of the wave aberration is changed without varying the shape, i.e. the quotient of the first and second coefficients of spherical aberration. The defocusing coefficient is used to optimize the O.T.F. for the other two coefficients. Figure 5 gives the results if the amount is kept constant and the shape varied.

As an example of the off-axis influence of variation of the coefficients on the O.T.F., the following figures are given. For a real optical system, it is shown by how much the O.T.F. is changed, if only one of the nine coefficients of wave-aberration is varied by $\pm\lambda$ and the defocusing term is optimized. The wave aberration for the original system is given (full lines) and for the system with one coefficient varied $+\lambda$ (broken line) or $-\lambda$ (dotted line) for the best focus. For these three curves the corresponding O.T.F. curves are also shown. Figures 6(*a*) and (*b*) give the influence of the coefficient s_2 (2nd spherical aberration) for tangential and radial directions. Figure 6(*c*) describes the behaviour for K_1 (first-order comatic aberration), and Figure 6(*d*) the same for K_2 (second-order comatic aberration), both for the more important tangential direction. These examples show that the coefficients can be varied considerably before the corresponding O.T.F. is markably affected. On the other hand, there are some critical combinations where a slight change is more harmful. An exact number can only be obtained by calculating the special O.T.F. from the aberration. In spite of this, there is usually a limit to the variations of coefficients to avoid a larger change than for instance 0·05 in the O.T.F. The limit depends on:

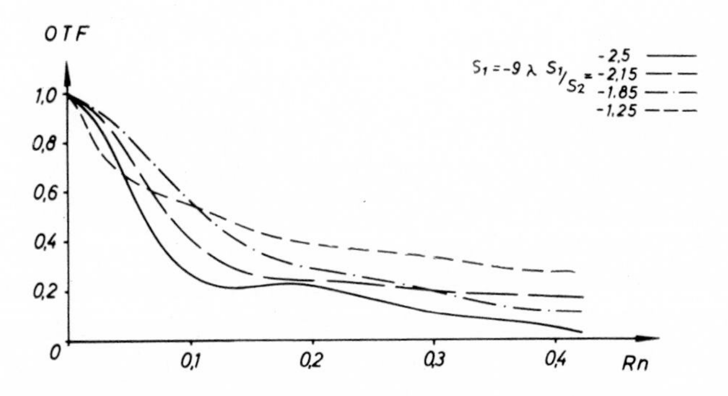

FIG. 5. Influence of variations of the axial coefficients on the O.T.F. S_1 is constant and the relation S_1/S_2 is changed.

(1) The spatial frequency region for which the O.T.F. is required;

(2) the size of wave aberration; and

(3) the shape of the aberration curve.

To reach this tolerance in O.T.F., the wave aberration in the pupil can vary from $\lambda/4$ to 2λ. Though it appears easy to fulfil this tolerance by measuring interferograms, one should not overlook the numerous difficulties at the edge of the interferogram, the measuring errors caused by inexact adjustment, and errors by incorrect positioning of the maxima and minima of fringes.

Figures 7(*a*), (*b*), and (*c*) show typical interferograms of photographic lenses, and Figure 8 gives the O.T.F. curves deduced from these interferograms (full lines) and directly measured O.T.F. curves (broken lines).

Finally, given in Table 3 in the second column, or the tolerances of the different kinds of performances which should be reached to evaluate the O.T.F. within an accuracy of $\pm 0{\cdot}10$ and, in the third column, the easily attainable measuring accuracy. This table is not valid for all optical

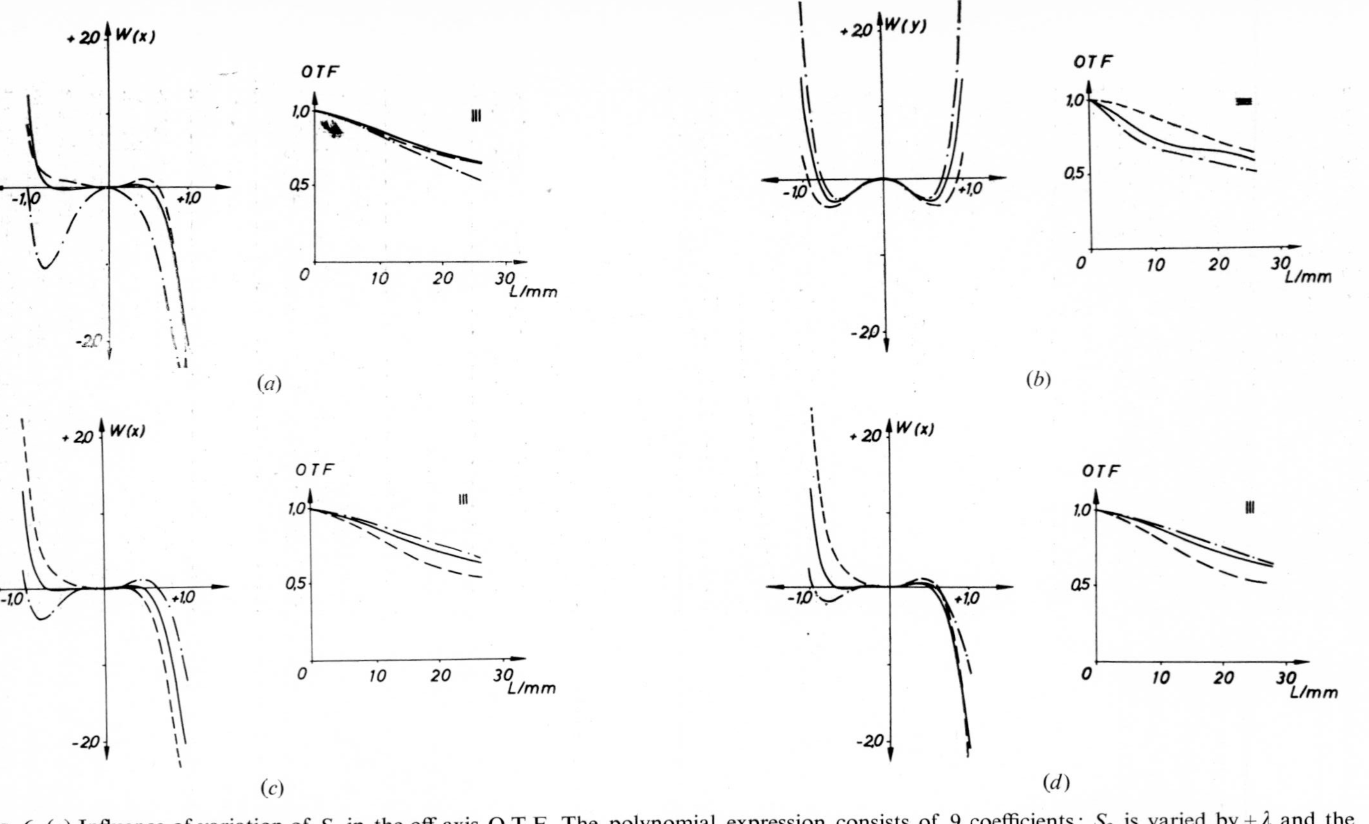

FIG. 6. (*a*) Influence of variation of S_2 in the off-axis O.T.F. The polynomial expression consists of 9 coefficients; S_2 is varied by $\pm\lambda$ and the defocusing term optimized. The wave aberration of the original system is given (full lines) and the system with one coefficient varied $+\lambda$ (broken line) or $-\lambda$ (dotted line). Also given for these 3 curves are the corresponding O.T.F. curves for tangential direction. (*b*) as for (*a*) but for the radial direction both in wave aberration and O.T.F. curves. (*c*) as for (*a*) but for variation of K_1 (comatic aberration coefficient 1). (*d*) as for (*a*) but for variation of K_2 (comatic aberration coefficient 2).

Fig. 7. Twyman-Green-interferogram:

(*a*) for an axial image point, 0·05 mm de-focused;

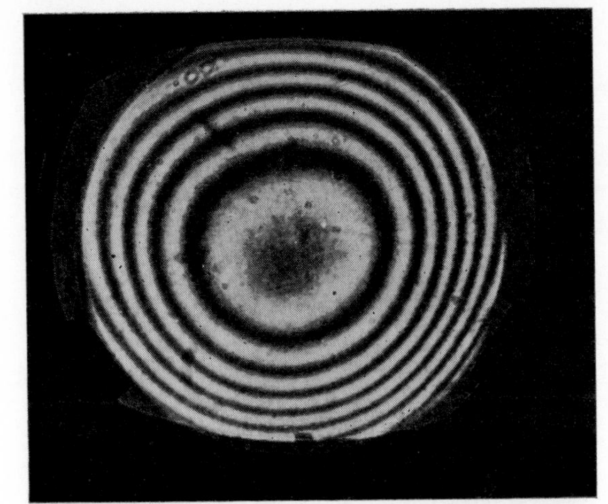

(*b*) for 4·5° field angle ;

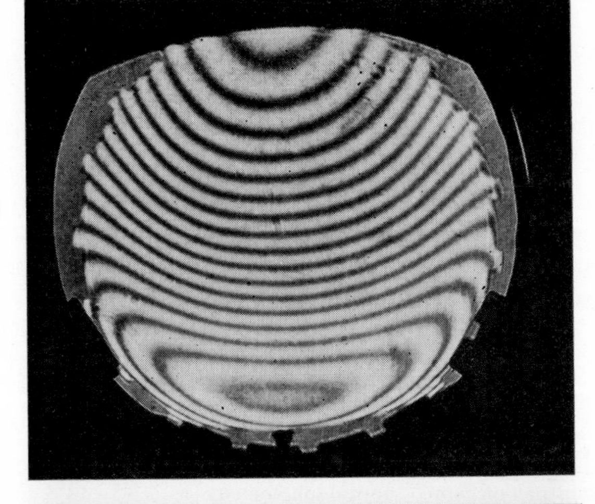

(*c*) for 6·5° field angle.

systems but gives the results of experience with photographic lenses; it can be adopted for other systems if the region of the spatial frequency of O.T.F. is suitably limited. It can be seen that no normal measuring method gives more accuracy than is necessary to find the O.T.F. to better than ±0·10.

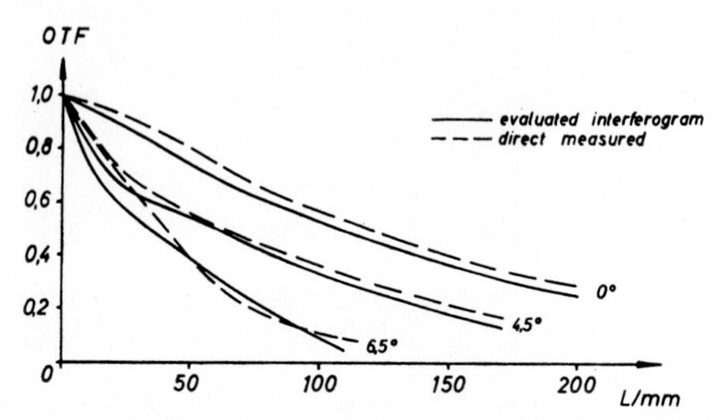

FIG. 8. O.T.F. curves of a lens 1:2 $f=100$ mm directly measured, and from the evaluated interferograms of Figure 7.

If it is necessary to know the image quality or the O.T.F. more exactly many improvements on the instruments and in the evaluation must be made. At present, however, most of the optical systems do not fulfil the

TABLE 3. Comparison of tolerances.

performance criteria	measuring method	demanded tolerance	measuring error (relative)
construction parameters p.e. radius of curvature	optical spherometer	$\pm 5 \times 10^{-4}$. R	$\pm 1\text{–}3 \times 10^{-4}$
ray-aberration p.e. spherical aberration	Hartmann or Wetthauer	for 1:2 $f=50$ mm lens ±0·02 mm in longitudinal or ±2 mikron in lateral aberration	±0·01 mm → 0·05 mm ±1 mikron
wave aberration	Twyman-Green-interferometer	$\pm\lambda/4 \rightarrow \lambda/10$	$\pm\lambda/10 \rightarrow \lambda/2$
O.T.F.	direct or indirect	±0·10	±0·01 → 0·02

necessary qualifications in stability and alignment tolerances to justify such a high standard in measuring techniques as to achieve an O.T.F. with an uncertainty less than 0·02. But this accuracy may be reached with most of the known O.T.F. measuring devices.

REFERENCES

BAKER, L. R., 1955, *Proc. Phys. Soc.*, B, **68,** 871.
BAKER, L. R. and WHYTE, J. N., 1965, *Jap. J. Appl. Phys., Suppl.*, I, **4,** 121.
BARAKAT, R., 1962, *J. Opt. Soc. Amer.*, **52,** 985.
BIEDERMANN, K., 1967, *Phot. Korr.*, **103,** 5.
BIRCH, K. G., 1958, *Optica Acta*, **5,** 271.
HARTMANN, J., 1904, *Z. Instrum. Kde.*, **24,** 1, 33, 97.
HERZBERGER, M., 1958, *Modern Geometrical Optics*, New York, Interscience.
HEYNACHER, E., 1963, *Zeiss Mitt.*, **3,** 32.
HOPKINS, H. H., 1957, *Proc. Phys. Soc.*, B, **68,** 871.
MURATA, K., 1959, *J. Appl. Phys. Japan*, **28,** 276.
ROSENBRUCH, K. J., 1968, *PTB Mitt.*, **78,** 384.
SUNDER-PLASSMANN, F. A., 1967, *Optik*, **26,** 284.
WETTHAUER, A., 1921, *Z. Instrum. Kde.*, **41,** 148.
WILLIAMS, T. L., 1968, *Optica Acta*, **15,** 553.

Use of the Optical Transfer Function for Lens Evaluation*

T. Ose

Institute of Industrial Science,
University of Tokyo

Although the method of calculation and the measurement of the optical transfer function (O.T.F.) has been well developed, its practical use does not appear to have been exploited to any great extent in lens design and testing. This is probably partly because of lack of actual experience in its practical application and also partly because of the lack of any direct correspondence between the O.T.F. and the conventional criteria for lens evaluation.

The research committee of the Optical Engineering Research Association of Japan (Chairman T. Ose) carried out an investigation to ascertain the feasibility of using the O.T.F. more directly in lens design and testing. The committee published its final report in 1968. The principle used in the investigation was the direct comparison between criteria based on the O.T.F. and the more conventional ones based on practical testing on a series of test lenses of known design parameters. The performances of five test lenses were thoroughly examined by calculation and measurement. Spot diagrams, the optical transfer functions, some image quality criteria directly related to the O.T.F. and energy contours were obtained by calculation; the O.T.F. was also obtained experimentally and related criteria calculated from these results. At the same time conventional resolution tests using bar targets and various projection tests were made. The results of these investigations were reported to the convention.

* A résumé by the author.

Use of the Modulation Transfer Function (MTF) as an Aberration-Balancing Merit Function in Automatic Lens Design

William B. King

The Perkin-Elmer Corporation,
Norwalk, Connecticut 06852, *U.S.A.*

Abstract—A new aberration-balancing merit function for automatic lens design is described in which the M.T.F. squared, with suitable weightings, serves directly as image quality criterion. Details concerning the optimization procedure are given and a new set of normal equations is described. The basic assumption is that the design parameter changes should be kept small, which is a valid approximation for the optimization technique employed. The evaluation of the coefficients in the normal equations requires extensive numerical integration; a method based on the modified Hopkins' algorithm is suggested.

INTRODUCTION

In automatic lens design, the term 'Merit Function' may be used in two senses:

(1) One sense is that of a merit function serving as a guidance function, leading to a feasible region of design parameter space.

(2) The other type of merit function is set up for the purpose of aberration-balancing during the final optimization stages of a design, and the merit function requires some criterion that is meaningful in terms of the optical performance.

The subject of this paper is concerned with a merit function of the second type.

Diffraction-based criteria of image quality in automatic lens design have been investigated by several authors (Hopkins 1966; Lenskii 1968; King 1968 a, b; King & Kitchen 1968; King 1969). Because the optimization of an optical-design merit function has direct influence on the final balancing of the residual aberrations, it is essential that the merit function be a correctly based criterion if the optimum performance is to be obtained from a given type of system. The use of the variance of the wave-aberration difference function as a useful diffraction-based merit function has already been described by Hopkins (1966). Using Hopkins' notations, we define the relative modulation $M(s, \psi)$ as

$$M(s, \psi) = \left| \frac{D(s, \psi)}{D_0(s, \psi)} \right|, \tag{1}$$

where $D(s, \psi)$ is the O.T.F. of the system, $D_0(s, \psi)$ is the O.T.F. in the absence of aberration and defocusing, and s and ψ are the reduced spatial frequency and the azimuth of the grating test object, respectively. The definition for the O.T.F. is

$$D(s,\psi) = \frac{1}{A} \iint_S \exp\left[iks\, V(x,y;\, s,\psi)\right] \mathrm{d}x\mathrm{d}y, \tag{2}$$

where

$$V(x,y;\, s,\psi) = V$$

$$= \frac{1}{s}\left[W\left(x + \frac{s}{2}, y\right) - W\left(x - \frac{s}{2}, y\right)\right] \tag{3}$$

is a wave-aberration difference function; the relative canonical pupil coordinates (x,y) are in the rotated system, with the y-axis in the direction of ψ; s is defined as

$$s = \frac{\lambda\ (\text{mm}) \times \text{number of lines per mm}}{\text{sagittal numerical aperture}}, \qquad s < 2; \tag{4}$$

$k = 2\pi/\lambda$, λ is the wavelength, A is the area of the pupil, S is the area common to two relatively displaced pupils, centred on the points $(\pm s/2, 0)$, respectively.

Hopkins (1957a) has shown that, when $M \geqslant 0{\cdot}8$, a simplified expression may be employed, namely

$$M(s,\psi) = 1 - \frac{2\pi^2 s^2}{\lambda^2} \cdot K(s,\psi), \tag{5}$$

where $K(s,\psi)$ is the variance of the wave-aberration difference function V (Equation (3)). In order for Equation (5) to be valid (i.e., $M \geqslant 0{\cdot}8$), either s or $K(s,\psi)$, or both, should be small; the implication of a small value for $K(s,\psi)$ is that the wave-aberrations of the system should also be small. These conditions are necessary in order that $K(s,\psi)$ serves usefully as a diffraction-based optical design merit function. Recent investigation by King (1969) has shown that $K(s,\psi)$ can become unreliable as a merit function when $M < 0{\cdot}6$.

Lenskii (1968) has described a merit function for automatic lens design based on an approximate expression of the O.T.F. for low spatial frequencies. However, the accuracy of the truncated exponential series expansion of the O.T.F. limits the validity of the merit function to relatively high M.T.F. values.

In optical design, there are many lens systems for which the M.T.F. falls below 0·5 at some specified spatial frequencies of interest, and the designer may still wish to optimize the system using a diffraction-based merit function. In this case, the problem area falls outside both the $M > 0{\cdot}6$ criterion and Lenskii's method. In this paper, a proposed method of applying the M.T.F. directly to the lens-optimization process is described; this new technique does not impose restrictions on the range of the M.T.F. values, and the spatial frequencies of interest need not be limited to small values. The basic assumption made is that the design-parameter changes shou ld be kept small, which is a valid approximation for the optimization technique employed.

The Differentiation of the O.T.F. with respect to the Design Parameters

Let $W(x,y)$ be the wave-aberration, which may be considered as a function of N design parameters. By using a truncated Taylor expansion, we set up the linear equation:

$$W(x,y) = W_0(x,y) + \sum_{n=1}^{N} \frac{\partial W_0(x,y)}{\partial p_n} \cdot p_n, \tag{6}$$

where $W_0(x,y)$ is the value of $W(x,y)$ for the initial point p_{n0}, the origin of our coordinate system, which we shall refer to as the current system. P_n is a first-order infinitesimal change in the nth design parameter, $n = 1, 2, \ldots, N$.

Substituting Equation (3) into (2), we may write the O.T.F. $D(s,\psi)$ as

$$D(s,\psi) = \frac{1}{A} \iint_S \exp\, ik\left[W\left(x + \frac{s}{2}, y\right) - W\left(x - \frac{s}{2}, y\right)\right] \mathrm{d}x\mathrm{d}y. \tag{7}$$

We now express $D(s,\psi)$ as a function of the design parameters. Substituting Equation (6) into (7), we have:

$$D(s,\psi) = \frac{1}{A} \iint_S \exp\, ik\left\{\left[W_0\left(x + \frac{s}{2}, y\right) + \sum_{n=1}^{N} \frac{\partial W_0(x + s/2, y)}{\partial p_n} \cdot p_n \right]\right.$$
$$\left. - \left[W_0\left(x - \frac{s}{2}, y\right) + \sum_{n=1}^{N} \frac{\partial W_0(x - s/2, y)}{\partial p_n} \cdot p_n \right]\right\} \mathrm{d}x\mathrm{d}y. \tag{8}$$

Regrouping gives:

$$D(s,\psi) = \frac{1}{A}\iint_S \exp ik\left\{\left[W_0\left(x+\frac{s}{2},y\right) - W_0\left(x-\frac{s}{2},y\right)\right]\right\}.$$

$$\exp ik\left\{\sum_{n=1}^{N}\left[\frac{\partial W_0(x+s/2,y)}{\partial p_n} - \frac{\partial W_0(x-s/2,y)}{\partial p_n}\right].p_n\right\}\mathrm{d}x\mathrm{d}y. \quad (9)$$

Let

$$\mu = \mu(x,y;\ s,\psi)$$
$$= \exp ik[W_0(x+s/2,y) - W_0(x-s/2,y)]$$
$$\sigma_n = \sigma_n(x,y;\ s,\psi)$$
$$= \left[\frac{\partial W_0(x+s/2,y)}{\partial p_n} - \frac{\partial W_0(x-s/2,y)}{\partial p_n}\right]. \quad (10)$$

In terms of μ and σ_n, Equation (9) may be written as:

$$D(s,\psi) = \frac{1}{A}\iint_S \mu \exp\left\{ik\sum_{n=1}^{N}\sigma_n p_n\right\}\mathrm{d}x\mathrm{d}y. \quad (11)$$

Exponential expansion of the second part of the integrand up to the squared term in p_n gives

$$D(s,\psi) = \frac{1}{A}\iint_S \mu\left\{1 + ik\sum_{n=1}^{N}\sigma_n p_n - \frac{k^2}{2}\left(\sum_{n=1}^{N}\sigma_n p_n\right)^2\right\}\mathrm{d}x\mathrm{d}y, \quad (12)$$

assuming that the parameter changes p_n are sufficiently small that terms higher than the square may be neglected. We thus represent the O.T.F. as a quadratic function in p. Differentiating $D(s,\psi)$ with respect to p_m, the mth parameter, $1 \leq m \leq N$, we have from Equation (12)

$$\frac{\partial D(s,\psi)}{\partial p_m} = \frac{1}{A}\iint_S \mu\left\{ik\sigma_m - k^2\left(\sum_{n=1}^{N}\sigma_n p_n\right)\sigma_m\right\}\mathrm{d}x\mathrm{d}y. \quad (13)$$

We now introduce the real and the imaginary notations, $\tilde{R}$ and $\tilde{I}$, where

$$\{\tilde{R}(m,n) + i\tilde{I}(m,n)\}_{n\neq 0} = \frac{k^2}{A}\iint_S \{\mu\sigma_m\sigma_n\}\mathrm{d}x\mathrm{d}y \quad (14)$$

$$\tilde{R}(m,0) + i\tilde{I}(m,0) = \frac{ik}{A}\iint_S \{\mu\sigma_m\}\mathrm{d}x\mathrm{d}y \quad (15)$$

with $1 \leq m \leq N$, $1 \leq n \leq N$.

In terms of $\tilde{R}$ and $\tilde{I}$, Equation (13) may be written as

$$\frac{\partial D(s,\psi)}{\partial p_m} = \tilde{R}(m,0) + i\tilde{I}(m,0) - \sum_{n=1}^{N}\{\tilde{R}(m,n) + i\tilde{I}(m,n)\}p_n \quad (16)$$

Equation (12) may be expressed as

$$D(s,\psi)=R(s,\psi)+iI(s,\psi). \tag{17}$$

Differentiating Equation (17) with respect to p_m, we have

$$\frac{\partial D(s,\psi)}{\partial p_m}=\frac{\partial R(s,\psi)}{\partial p_m}+\frac{i\,\partial I(s,\psi).}{\partial p_m} \tag{18}$$

Equating the real and the imaginary parts of Equations (16) and (18), we have:

$$\frac{\partial R(s,\psi)}{\partial p_m}=\tilde{R}(m,0)-\sum_{n=1}^{N}\tilde{R}(m,n)p_n \tag{19}$$

$$\frac{\partial I(s,\psi)}{\partial p_m}=\tilde{I}(m,0)-\sum_{n=1}^{N}\tilde{I}(m,n)p_n. \tag{20}$$

M.T.F. Related Merit Function for Optical Design

The overall merit function ψ may be defined as

$$\psi=\Phi_K+\Phi_D+\chi+Q\sum_{n=1}^{N}q^2{}_n p_n{}^2, \tag{21}$$

where Q is a positive damping factor (Wynne 1959); q_n's are the damping coefficients similar to that defined by Meiron (1965); χ is the system boundary control function; Φ_K is an image quality merit function to be defined later; Φ_D is defined as:

$$\Phi_D=\sum_{\tau}[\eta\{1-[R^2(s,\psi)+I^2(s,\psi)]\}]_\tau, \tag{22}$$

where the summation is for a range of fractional image fields τ, and η is a field-dependent, positive weighting factor. The quantity $[R^2+I^2]$ in Equation (22) represents the squared value of the M.T.F. at a given field. In practice, the summation in Equation (22) may also be extended to account for different spatial frequencies s and azimuths ψ, and for different wavelengths.

The optimization procedure (to minimize ψ) requires that the conditions

$$\frac{\partial\psi}{\partial p_m}=0, \qquad 1\leq m\leq N \tag{23}$$

be satisfied. The parameter derivatives of Φ_n from Equation (22) are given by

$$\frac{\partial\Phi_D}{\partial p_m}=-2\sum_{\tau}\left\{\eta\left[R(s,\psi)\frac{\partial R(s,\psi)}{\partial p_m}+I(s,\psi)\frac{\partial I(s,\psi)}{\partial p_m}\right]\right\}_\tau. \tag{24}$$

The values of $R(s,\psi)$ and $I(s,\psi)$ are derived from Equations (17) and (12), and by substituting Equations (19) and (20) in (24), we may show that

$$\frac{\partial \Phi_D}{\partial p_m} = \sum_{n=1}^{N} F(m,n)p_n - F(m,0), \qquad m=1,2,\ldots, N. \tag{25}$$

where

$$\{F(m,n)\}_{n\neq 0} = 2\sum_{\tau} [\eta\{\tilde{R}(0,0)\tilde{R}(m,n) - \tilde{R}(m,0)\tilde{R}(n,0) + \tilde{I}(0,0)\tilde{I}(m,n) - \tilde{I}(m,0)\tilde{I}(n,0)\}]_{\tau}, \tag{26}$$

$$F(m,0) = 2\sum_{\tau} \{\eta[\tilde{R}(m,0)\tilde{R}(0,0) + \epsilon\tilde{I}(m,0)\tilde{I}(0,0)\}]_{\tau}, \tag{27}$$

and $\tilde{R}(0,0)$ and $\tilde{I}(0,0)$ are obtained from the equation

$$\tilde{R}(0,0) + i\tilde{I}(0,0) = \frac{1}{A}\iint_{S} \mu \mathrm{d}x\mathrm{d}y. \tag{28}$$

In the derivation of Equation (25), it is assumed that the terms of $0(p^2)$ are negligible.

Referring to the merit function ψ defined in Equation (21), for convenience and for the present, we will leave out Φ_K and χ.* The optimization conditions of Equation (23) then reduce to a set of N normal equations, namely,

$$\sum_{n=1}^{N} \{F(m,n) + \omega(m,n)\}p_n - F(m,0) = 0, \tag{29}$$

$$m = 1, 2, \ldots, N,$$

where $\omega(m,n) = 0, \qquad m \neq n$

$= 2Qq_n{}^2, \qquad m = n.$

We note that the coefficients $F(m,n)$ as defined in Equation (26) have the symmetrical properties.

$$F(m,n) = F(n,m), \qquad m,n \neq 0.$$

Consequently, only those coefficients $\{F(m,n)\}_{n\neq 0}$ with $n \geqslant m$ need be computed for Equation (29).

DISCUSSION

As mentioned in the Introduction, the variance K of the wave-aberration difference function—Equation (5)—may serve usefully as a diffraction-based image quality criterion if $M > 0{\cdot}6$ (King 1969). For the range of low

* Boundary controls incorporating. Lagrangian undetermined multipliers have been described by many authors. See, for example, Meiron (1965).

spatial frequencies over which the starting design has a Relative Modulation greater than about 0·6, the image quality merit function Φ_K in Equation (21) may be defined as

$$\Phi_K = \sum_{\lambda} \sum_{s} \{\alpha_s K_{axial}(s)\}_{\lambda} + \sum_{\lambda} \sum_{\tau} \{\sum_{s} \sum_{\psi} [\beta_{s,\psi} \mathrm{K}_E(s,\psi) + \gamma_{s,\psi} K_0(s,\psi)]\}_{\gamma,\tau},$$

where α, β, and γ are suitable weighting factors, τ being the fractional field; K_{axial}, K_E, and K_O are the axial, extra-axial even, and the extra-axial odd components of the variance K. Details concerning the technique of computing Φ_K directly from the wave aberrations have already been published (Hopkins 1966; King & Kitchen 1968); therefore, it is sufficient to say that, by using a set of universal coefficients in conjunction with the elliptical pupil approximation (Hopkins 1966; King 1968c), both the merit function Φ_K and its parameter derivatives may be computed with great efficiency. Consequently, it is of practical advantage to incorporate Φ_K in the definition of ψ, Equation (21), for those spatial frequencies $s \leq 0{\cdot}2$ and $M(s,\psi) > 0{\cdot}6$. For those spatial frequencies over which $M(s,\psi) < 0{\cdot}6$, Φ_K tends to become unreliable (King 1969); in these cases, we should employ the image quality merit function Φ_D, Equation (22). Thus, by combining Φ_D and Φ_K as defined in Equation (21), the overall merit function ψ remains physically meaningful over a wide range of spatial frequencies during the aberration-balancing stage of performance optimization.

A lens-design programme based on the merit function ψ, Equation (21), is now under development for the IBM 360-67 computer.

Acknowledgement

The author wishes to express his appreciation to Professor H. H. Hopkins for calling attention to the permissibility of differentiating the O.T.F. under the integral sign, and to Abe Offner for discussions.

Appendix I

The computation of the $F(m,n)$ coefficients requires the evaluation of $\tilde{R}(m,n)$ and $\tilde{I}(m,n)$ (Equation (14)). For this purpose, we may use the numerical integration technique based on a modified version of Hopkins' algorithm (Hopkins 1957b) which requires the pupil to be divided into elementary rectangles. The modification to Hopkins' formula consists of multiplying the contribution from each one of the rectangles by the product $(\sigma_m \sigma_n)$ evaluated at the centre of that rectangle. The definition of σ_n is given in Equation (10).

Appendix II

Although the method outlined for optimizing the merit function Φ_D (Equations (22) to (29)) can be applied to systems with arbitrarily shaped pupils, the use of the elliptical pupil approximation (Hopkins 1966; King 1968c) greatly improves the computational efficiency. Thus, for a given field, we first define a suitable wave aberration polynomial $W(x,y)$ in conjunction with the elliptical pupil; differentiating $W(x,y)$ with respect to the design parameters, p_m, results in a new polynomial, $\partial W/\partial p_m$. The parameter derivatives C_p of the wave aberration coefficients are next calculated from the values of $\partial W/\partial p_m$ for sampled rays. The values of $\partial W/\partial p_m$ at any other points within the region of integration (see Appendix I) may then be computed from the $\partial W/\partial p_m$ polynomial, using the predetermined C_p coefficients.

REFERENCES

Hopkins, H. H., 1957*a*, *Proc. Phys. Soc.*, B, **70**, 449.
Hopkins, H. H., 1957*b*, *Proc. Phys. Soc.*, B, **70**, 1002.
Hopkins, H. H., 1966, *Optica Acta*, **13**, 343.
King, W. B., 1968*a*, *J. Opt. Soc. Amer.*, **58**, 655.
King, W. B., 1968*b*, *Appl. Optics*, **7**, 489.
King, W. B., 1968*c*, *Appl. Optics*, **7**, 197.
King, W. B., 1969, *J. Opt. Soc. Amer.*, **59**, 692.
King, W. B. and Kitchen, Jane, 1968, *Appl. Optics*, **7**, 1193.
Lenskii, A. V., 1968, *Optics & Spectrosc.*, **24**, 229.
Meiron, J., 1965, *J. Opt. Soc. Amer.*, **55**, 1105.
Wynne, C. G., 1959, *Proc. Phys. Soc.*, **73**, 777.

Extended Range Diffraction-Based Merit Function for Least Squares Type Optimization

Abe Offner

The Perkin-Elmer Corporation,
77 Danbury Road,
South Wilton, Connecticut 06897, *U.S.A.*

Abstract—Diffraction-based criteria of image quality for use in balancing aberrations in automatically optimized optical systems have been described by H. H. Hopkins. These result in merit functions which are useful when the Strehl ratio, or the relative MTF of the optimized system is greater than about 0·7. A diffraction-based merit function that is applicable to systems with larger aberrations has been developed. It is based on the method of computing the OTF in which H. H. Hopkins' algorithm is employed to evaluate the auto-correlation integral. After minor modifications, the damped least squares optimisation procedure can be applied to minimize this extended range merit function.

The pupil function is described by a set of Taylor expansions around sampled values that are obtained by ray tracing. Use of a differential ray trace results in the accurate description of complicated wave forms on the basis of data obtained from relatively few ray traces.

In programmes used for the 'automatic' design of optical systems, a criterion of progress must be established. Most commonly this takes the form of a defect vector whose components are weighted aberrations of the system. Although the term 'aberrations' has been broadened to include quantities such as cost of manufacture, sensitivity to misalignment, and other quantities that are not defects of imagery, in this paper we will consider only those components of the defect vector that result in image deterioration. The square of the modulus of the defect vector is called the merit function. An automatic design programme minimizes the merit function subject to the constraints necessary for realizability and any additional constraints that are associated with the particular design problem.

In the early stage of a design, the merit function should guide the design to a region of parameter space in which a solution is feasible. It is neither necessary nor generally desirable that the merit function be an accurate measure of performance of the system at this stage (Glatzel & Wilson 1968). In the final stage of a design, however, the system should reach a point in the feasible region of parameter space in which the residual aberrations are balanced to optimize system performance. A merit function for this purpose is the subject of the present paper.

Hopkins (1966) has pointed out the importance of basing an aberration-balancing merit function on diffraction criteria. He has shown how, by using Maréchal's (1947) approximation to the Strehl definition or Hopkins' (1957*a*) approximation to the MTF, a merit function suitable for incorporation in an automatic optical design programme can be set up and used for optimizing the optical performance in systems in which these approximations are valid. The optimization procedure was first worked out for systems whose pupil boundaries are either circular, or can be distorted by anamorphotic scaling of the pupil coordinates to become circular (Hopkins 1966). Use of a numerical integration method enabled Meiron (1968) to employ Hopkins' type merit functions in systems in which the shape of the pupil boundary is arbitrary and in which the pupil may be partially obscured. In the case of MTF-based merit functions, however, Meiron's implementation requires the data from a troublesomely large number of ray traces.

In both Maréchal's approximation to the Strehl definition and Hopkins' approximation to the MTF, a truncated power series expansion is substituted for a complex exponential function. Because of the oscillatory nature of this function, the approximation is valid for only small values of its argument. In practice, this limits the usefulness of the merit functions that are based on these approximations to systems in which the relative M.T.F. at the spatial frequencies of interest, or the Strehl definition is greater than about 0·7 (Hopkins 1966).

An approximation to the MTF with an extended range of validity can be derived by making use of a well-known algorithm of Hopkins (1957*b*) which is widely used in the computation of the O.T.F. In order to implement the method in systems with pupils of arbitrary shape, the pupil function is described in terms of its value at sampled points. The number of ray traces required is kept manageable by making use of a differential ray trace method (Feder 1968) to obtain, for each ray traced, the second derivatives of the wave aberration with respect to the pupil coordinates. Truncated Taylor expansions about each pupil point through which a ray is traced are used to obtain the additional data required for the difference function and its derivatives. A modification of a previously described method (Offner 1969) permits the use of a coarse pupil sampling grid for merit functions related to the M.T.F. at any desired spatial frequency and azimuth without adding the difficulty of defining the arbitrarily shaped pupil in terms of a rotated coordinate system. The normal equations required for a least squares solution corresponding to this merit function will be derived.

Both the Strehl ratio and the O.T.F. are related to an integral which can be evaluated numerically in the form:

$$f(\theta) = (1/R)\sum_{j=1}^{N} \exp\{i(\theta_j - \bar{\theta})\}\Delta R_j, \tag{1}$$

where

$$\bar{\theta} = (1/R)\sum_{j=1}^{N} \theta_j \Delta R_j \tag{2}$$

and ΔR_j is the area of an element of the region R whose area is also denoted by R. The complex exponential can be approximated by the truncated Taylor expansion

$$\exp\{i(\theta - \bar{\theta})\} = 1 + i(\theta - \bar{\theta}) - 1/2(\theta - \bar{\theta})^2. \tag{3}$$

If we substitute Equation (3) into Equation (1) and carry out the summation, an approximation to Equation (1) is obtained in the form

$$f(\theta) = 1 - (1/2)R\sum_{j=1}^{N} (\theta_j - \bar{\theta})^2 \Delta R_j. \tag{4}$$

The last term of Equation (4) is the variance of the function θ in the region R. This relation is the basis of both Maréchal's approximation to the Strehl ratio and Hopkins' approximation to the M.T.F.

In accordance with Equation (4), the contribution to $f(\theta)$ from the element ΔR_j is

$$f_j(\theta) = (1/R)[1 - (1/2)(\theta_j - \bar{\theta})^2]\Delta R_j. \tag{5}$$

The quantity in the square brackets results from the approximation (3). It is actually an expression for the value of the real part of exp $i(\theta_j - \bar{\theta})$ whose value cannot be less than -1. When $(\theta_j - \bar{\theta})$ is large, this limit is exceeded, and the value of $f(\theta)$ obtained by the use of Equation (4) is a poor approximation to its true value.

In order to increase the range values of $(\theta_j - \bar{\theta})$ for which a good approximation to $f(\theta)$ is obtained, we have added a weighting function to the term $(\theta_j - \bar{\theta})^2$ and a lower limit to the contribution from any element. The weighting function we have chosen is the one that Hopkins (1957*b*) used in his computation of the O.T.F.

Using a modification of Hopkins' notation, for properly chosen (Hopkins, 1964) rectangular coordinates (u,v) in the exit (or entrance) pupil of the system, we may compute the O.T.F. as the integral

$$D(\Delta u/\lambda R, \Delta v/\lambda R) = (1/A)\iint_S \exp\{ik(W_+ - W_-)\}\mathrm{d}u\mathrm{d}v \tag{6}$$

where

$$W_{+} = W[u+(\Delta u/2),\ v+(\Delta v/2)] \tag{7}$$

$$W_{-} = W[u-(\Delta u/2),\ v-(\Delta v/2)] \tag{8}$$

$W(u,v)$ is the distance from the point (u,v) of the reference sphere of radius R to the actual wave front along a radius of the sphere, and k is equal to $2\pi/\lambda$. The region S is the overlapping area of two pupils symmetrically displaced so that their centres are at $(\Delta u/2, \Delta v/2)$ and $(-\Delta u/2, -\Delta v/2)$, respectively; A is the area of the pupil, and $(\Delta u/\lambda R, \Delta v/\lambda R)$ are the coordinates in the two-dimensional frequency space of the spatial frequency whose magnitude is $[(\Delta u/\lambda R)^2+(\Delta v/\lambda R)^2]^{\frac{1}{2}}$.

The pupil is divided into a grid of rectangular elements whose dimensions are $2\epsilon_u \times 2\epsilon_v$. The integral pertaining to Equation (1) can then be expressed as the sum of contributions from the grid elements, or portions of grid elements, that are in the region S.

$$D(\Delta u/\lambda R, \Delta v/\lambda R) = (1/A)\underbrace{\sum_m \sum_n}_{S} \iint_{\boxed{m,\,n}} \exp\{ikV(u_m+h, v_n+j)\}\,\mathrm{d}h\,\mathrm{d}j, \tag{9}$$

where $\boxed{m,n}$ is the portion of the grid element that is in S; (u_m, v_n) are the coordinates of the midpoint of this portion; and V is the difference function $(W_{+}-W_{-})$. For a grid sufficiently fine so that the departure of V from linearity within a grid element is not great, the sum of integrals (9) may be evaluated in the form of a weighted sum (Hopkins 1957*b*).

$$D(\Delta u/\lambda R, \Delta v/\lambda R) = (1/A)\underbrace{\sum \sum}_{\substack{mn \\ S}} 4\epsilon_u\epsilon_v\alpha_{mn} \exp\,(ik\,V_{mn}), \tag{10}$$

where

$$V_{mn} = V(u_m, v_n) \tag{11}$$

and

$$\alpha_{mn} = \operatorname{sinc}\,\{k\epsilon_u(V_u)_{mn}\}\ \operatorname{sinc}\,\{k\epsilon_v(V_v)_{mn}\}. \tag{12}$$

The function $\operatorname{sinc} x = (\sin x)/x$, and $(V_u)_{mn}$ and $(V_v)_{mn}$ are the partial derivatives of V with respect to the variable in the subscript evaluated at the point (u_m, v_n).

For an element that is only partly within the region S, the semi-dimensions ϵ_u' and ϵ_v' of the portion of the element within S replace ϵ_u and ϵ_v, respectively, in the weighting factor α. The area $4\epsilon_u\epsilon_v$ is replaced by $4\beta_{mn}\epsilon_u\epsilon_v$, where β_{mn} is the fraction of the grid element area that is in S.

The expression (5) for the O.T.F. takes the form

$$D(\Delta u/\lambda R, \Delta v/\lambda R) = (1/A) \underbrace{\sum_m \sum_n}_{S} 4\epsilon_u \epsilon_v \alpha_{mn} \beta_{mn} \exp(ikV_{mn}). \qquad (13)$$

When Equation (13) is used for aberration balancing, a rather coarse grid may be set up. The values of W and its first and second derivatives with respect to the pupil coordinates are determined accurately by tracing rays through the centres of each grid element. The additional values required in determining the difference function for a given spatial frequency are then obtained from the truncated Taylor expansion:

$$W(u_m + h, v_n + j) = W_{mn} + h(W_u)_{mn} + j(W_v)_{mn} + (1/2)h^2(W_{uu})_{mn} + (1/2)j^2(W_{vv})_{mn} + hj(W_{uv})_{mn} \qquad (14)$$

where again the subscripts indicate partial differentiation with respect to the variables in the subscript, so that, for example

$$W_{uv} = \partial^2 W / \partial u \, \partial v. \qquad (15)$$

The values of the first derivatives of W at any required points are obtained in a similar manner and used in conjunction with evaluations (14) to compute the values of the difference function and its derivatives that occur in Equation (13).

The right-hand member of Equation (13) may be multiplied by the unit factor $\exp(ik\bar{V}) \exp(-ik\bar{V})$, where $\bar{V}$ is the weighted average

$$\bar{V} = \left(\underbrace{\sum_m \sum_n}_{S} \alpha_{mn} \beta_{mn} V_{mn} \right) \Big/ \underbrace{\sum_m \sum_n}_{S} \alpha_{mn} \beta_{mn}. \qquad (16)$$

Equation (13) then takes the form

$$D(\Delta u/\lambda R, \Delta v/\lambda R) = (1/A) \exp(ik\bar{V}) \underbrace{\sum_m \sum_n}_{S} 4\epsilon_u \epsilon_v \alpha_{mn} \beta_{mn} \exp\{ik(V_{mn} - \bar{V})\} \qquad (17)$$

Upon replacing the complex exponentials of Equation (17) by the truncated Taylor expansion (3), we may put the expression for the O.T.F. into a form that is similar to Hopkins' (1957*a*) approximation.

$$D(\Delta u/\lambda R), \Delta v/\lambda R) = (1/A) \exp(ik\bar{V}) \underbrace{\sum_m \sum_n}_{S} [4\epsilon_u \epsilon_v \beta_{mn} - 4\epsilon_u \epsilon_v \beta_{mn}(1 - \alpha_{mn}) - 4\epsilon_u \epsilon_v \alpha_{mn} \beta_{mn} (k^2/2)(V_{mn} - \bar{V})^2] \qquad (18)$$

Hopkins' variance of the difference function is replaced by the third term of Equation (18), which is a weighted variance of the difference function, and the second term, which preserves the deteriorating effect of a large value of the difference function when the associated weight, α_{mn}, is small. The truncation of the Taylor series expansion of the complex exponential may still result in large false negative contributions to the modulus of D when $(V_{mn} - \bar{V})$ is too large. This is prevented by limiting kV_{mn} and $k(V_{mn} - \bar{V})$ to a maximum value of $\sqrt{3}$ so that the minimum contribution to the modulus of D from any grid element is $-0{\cdot}5$. Equation (18) can be put in the more manageable form

$$D(\Delta u/\lambda R, \Delta v/\lambda R)$$
$$= \exp(ik\bar{V})\{S/A - (1/N)\sum_{j=1}^{P}[\delta_j\beta_j + \alpha_j\beta_j(k^2/2)(V_j - \bar{V})^2]\} \quad (19)$$

where the single subscript, j, has replaced the double subscripts, m,n; S is used for the area of the region S; P is the number of grid elements in the region S; N is the number of grid elements in the entire pupil; and $\delta_j = 1 - \alpha_j$. The following relations hold:

$$A = 4\epsilon_u\epsilon_v N \quad (20)$$

and

$$S = 4\epsilon_u\epsilon_v\sum_{j=1}^{P}\beta_j. \quad (21)$$

The M.T.F. can then be computed from the expiession

$$M(\Delta u/\lambda R, \Delta v/\lambda R) = |D(\Delta u/\lambda R, \Delta v/\lambda R)| = M_0 - (1/\mathrm{N})\phi \quad (22)$$

where M_0 is the aperture-limited value of the M.T.F.;

$$M_0 = S/A \quad (23)$$

The deteriorating effect of the aberrations is given by the merit function ϕ.

$$\phi = \sum_{j=1}^{P}[\delta_j\beta_j + \alpha_j\beta_j(k^2/2)(V_j - \bar{V})^2]. \quad (24)$$

A practical merit function will invariably include other components of the form (24) corresponding to the different spatial frequencies, azimuths, field angles, and/or colours. These may be weighted and combined to form a single merit function of the form (24) with appropriate values of $\bar{V}$ used for each component. The resulting merit function differs from the usual quadratic one by virtue of the presence of the quantities α and δ. When a damped least squares optimization procedure is used, the damping factors must be chosen so as to restrain the solution to a region in which both α and V are adequately described by linear functions of the parameters.

Since the merit function contains (positive) linear as well as quadratic terms, the procedure for obtaining the normal equations must be modified. For this purpose, it is useful to set up the following arrays:

$$g = \begin{bmatrix} k(\alpha_1\beta_1)^{\frac{1}{2}}(V_1 - \bar{V}) \\ \cdot \\ \cdot \\ \cdot \\ k(\alpha_p\beta_p)^{\frac{1}{2}}(V_p - \bar{V}) \end{bmatrix} \qquad d = \begin{bmatrix} \delta_1\beta_1 \\ \cdot \\ \cdot \\ \cdot \\ \delta_p\beta_p \end{bmatrix} \tag{25}$$

$$x = \begin{bmatrix} x_1 \\ \cdot \\ \cdot \\ \cdot \\ x_R \end{bmatrix} \qquad b = \begin{bmatrix} 1 \\ \cdot \\ \cdot \\ \cdot \\ 1 \end{bmatrix} \updownarrow R \tag{26}$$

$$A = \begin{bmatrix} \partial g_1/\partial x_1 \ldots \partial g_1/\partial x_R \\ \cdot \quad \ldots \quad \cdot \\ \cdot \quad \ldots \quad \cdot \\ \cdot \quad \ldots \quad \cdot \\ \partial g_P/\partial x_1 \ldots \partial g_P/\partial x_R \end{bmatrix} \qquad D = \begin{bmatrix} \partial d_1/\partial x_1 \ldots \partial d_1/\partial x_R \\ \cdot \quad \ldots \quad \cdot \\ \cdot \quad \ldots \quad \cdot \\ \cdot \quad \ldots \quad \cdot \\ \partial d_P/\partial x_1 \ldots \partial d_P/\partial x_R \end{bmatrix} \tag{27}$$

The components of the vector x are the parameters of the system that are being varied to optimize it. We take the origin in parameter space at the point corresponding to the system at the start of a cycle. The components of x are then the changes in the parameters.

The merit function (24) can now be put in the form

$$\phi = \sum_{i=l}^{P} d_i + (1/2) \sum_{i=l}^{P} g_i^2. \tag{28}$$

Within the domain of linearity,

$$g = g_0 + Ax \tag{29}$$

and

$$d = d_0 + Dx \tag{30}$$

where g_0 and d_0 are the values of these vectors at the start of a cycle.

Upon substituting Equations (29) and (30) into Equation (28) and setting the components of the gradient of ϕ equal to zero, we obtain the normal equations for the minimization of ϕ in the form

$$A^T Ax = -(A^T g_0 + D^T b). \tag{31}$$

Equation (31) is easily modified to include a damping factor that restricts the parameter change vector, x, to the region in which the vectors g and d are sufficiently linear functions of x (Levenberg 1944).

A system in which the aberrations have been balanced by the use of the extended range diffraction-based merit function may suffer from too great a departure from linearity of the relation between the phase and the spatial frequency of the O.T.F. in the region of interest. Since this defect is always associated with asymmetry of the wave aberration, it can be reduced by augmenting the merit function with terms that measure this asymmetry. Data for this purpose are available from the rays already traced and the derivatives (or differences) previously computed. This corrective measure should be used carefully so that its effect on the aberration balance is minimal.

REFERENCES

Feder, D. P., 1968, *J. Opt. Soc. Amer.*, **58,** 1494.
Glatzel, E. and Wilson, R., 1968, *Appl. Optics*, **7,** 265.
Hopkins, H. H., 1965, *J. Appl. Phys.*, **4,** Suppl. 1, 31.
—— 1966, *Optica Acta*, **13,** 343.
—— 1957*a*, *Proc. Phys Soc.* B, **70,** 449.
—— 1957*b*, *Proc. Phys. Soc.* B, **70,** 1002.
Levenberg, K., 1944, *Quart. Appl. Math.*, **2,** 164.
Maréchal, A., 1947, *Rev. Opt.*, **26,** 257.
Meiron, J., 1968, *Appl. Optics*, **7,** 667.
Offner, A., 1969, *Seminar Proc. Soc. Photo-Optical Instrumentation Eng.* (U.S.A.), **13,** 79.

Wavefront Determination Resulting from Foucault Test as Applied to the Human Eye and Visual Instruments

Francoise Berny and Serge Slansky

Institut d'Optique,
3 *Bd. Pasteur, Paris* 15, *France*

Abstract—The aberrations of the human eye are studied by an experimental method based on the principle of the Foucault test. Some numerical results concerning the wavefront shape with its irregular defects are given and light distribution in the diffraction pattern is deduced for different pupil diameters. When the human eye is associated with optical instruments, having an instrumental exit pupil not exceeding the entrance pupil of the eye, the final image quality is essentially determined by instrumental aberrations as long as the eye pupil is smaller than 3 mm.

INTRODUCTION

To measure the aberrations of an optical system, Philbert (1967) suggested a method derived from the classical method of Foucault's shadows. Its interest lies in giving directly the transverse aberration from photometric measures. A similar procedure shown in this paper is used to study the aberrations of the human eye which is not generally a system of revolution. The retina not being accessible, a double-path method has been set up. The eye observes a source and the retinal image is used as a secondary source in a Foucault-test arrangement for the outgoing light. After a short description of the optical apparatus, the validity of the method and the approximations made are investigated for trivial cases. Some results relating the wavefront shape of the human eye are given. Knowledge of this wavefront makes it possible to study the eye when associated with optical instruments.

PRINCIPLE OF THE METHOD

The principle consists in using a wide slit *DRB* as an object (Figure 1) and a knife-edge at a given point *E* of the caustic of the image given by the optical system W'' under study. The knife-edge is parallel to the larger side of the object slit having a uniform luminance. A lens *O*, placed against the knife edge, images W'' at W'''. The illumination *I* is required for any point H_1 on the diameter of W''' which is perpendicular to the knife-edge. In the presence of aberrations, the wave is locally distorted and the elementary image of the object slit R_2TR_1 associated with the wavefront element $\mathrm{d}W''$ is translated in the knife-edge plane. The change in illumination at point H_1, the centre of the elementary surface $\mathrm{d}W'''$, is therefore proportional to the component ET of the translation of the image R_2TR_1.

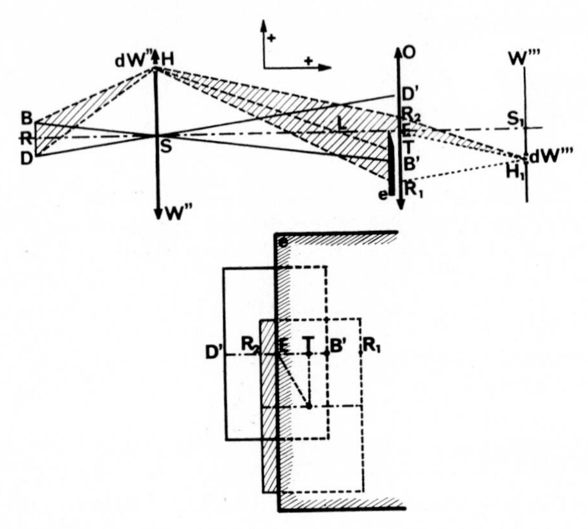

FIG. 1. Principle of the wide slit Foucault's method: $B'D' = R_1R_2 = 2a$ = slit width; $t = ET$ = transverse aberration; E = intersection of the knife-edge e and the optical axis; W'' = optical system to be studied; W''' = image given by lens O.

This component is perpendicular to the knife-edge and located in its plane. It represents the tangential deviation or transverse aberration $t = ET$ of the light ray. If I_0 is the illumination at point H_1 when the knife is removed from the image slit, and $2a$ is the geometrical slit-width, we can write

$$\frac{I}{I_0} = \frac{a \pm (t-x)}{2a} \tag{1}$$

where x = deviation which may exist between the optical axis and knife-edge. The positive sign refers to the case in which the knife is in the direction $t < O$, and the negative sign to the converse case.

The relation (1) is only valid if the slit image is never completely masked or unmasked by the knife, for the extreme deviations, i.e.

$$|t - x| < a. \tag{2}$$

In fact, the expression (1) is an approximation because it has been deduced from geometrical optics. A more rigorous study would need the introduction of diffraction theory as shown by Linfoot (1955) and Dupuy (1964) for the classical Foucault test; but the expressions obtained are too complicated to be used practically. Formula (1), however, is a sufficient approximation: it has been experimentally checked by applying the method to systems having aberrations known by some other means.

The wave aberration Δ which represents the deviation between the actual wavefront and a reference spherical wavefront centred in E is easily obtained by integrating the tangential profile t.

OPTICAL APPARATUS

The principle is realized as follows. The source (Figure 2) used is the image R of a wide slit M: this image, formed on an orthotropic diffuser (magnesia), or on the retina (r), is transmitted again by the optical system W'' and forms a real image in point E. A Foucault's knife edge e

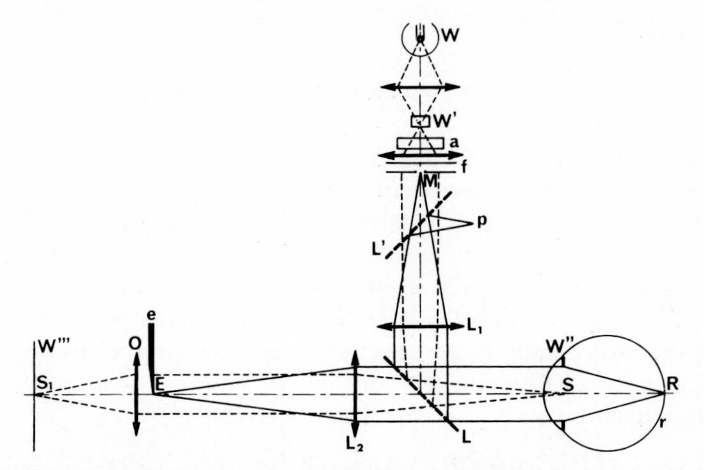

FIG. 2. Optical apparatus: W, light source; W', electronic flash; W'', subject pupil; W''', photographic plate; a, water cell; f, filter; M, wide slit; p, fixation test; r, retina; e, Foucault's knife-edge; O, photographic lens; L, beam-splitter; L_1L_2, focusing lens system for the variation of the subject's accommodation.

and a photographic lens O which images W'' at W''', are placed in E. In the case of the human eye, a fixation object p is placed in the plane of M. The focusing lenses L_1L_2 are used to allow for variations of the subject's accommodation. The source used is an electronic flash W' which is imaged at W''.

Consideration of the photographic images of W'', along a direction perpendicular to the knife-edge, after calibration of the photographic emulsion used, allow the calculation of t as well as the normalized wave-aberration $N = \Delta/\gamma$ referred to a sphere centred on the point E.

Double Passage of the Optical System W″

The double passage of the system W'' is necessary because in the case of the eye, the retinal image is not accessible. The retina can be considered as a perfect diffuser as a result of various experimental studies, such as those reported by Arnulf and Dupuy (1956). They were recently confirmed by Campbell and Gubish (1965). On the other hand it has been established that the degree of partial coherence γ (Hopkins 1953; Maréchal & Françon 1960) is cancelled out for two very close points in the object slit plane M.

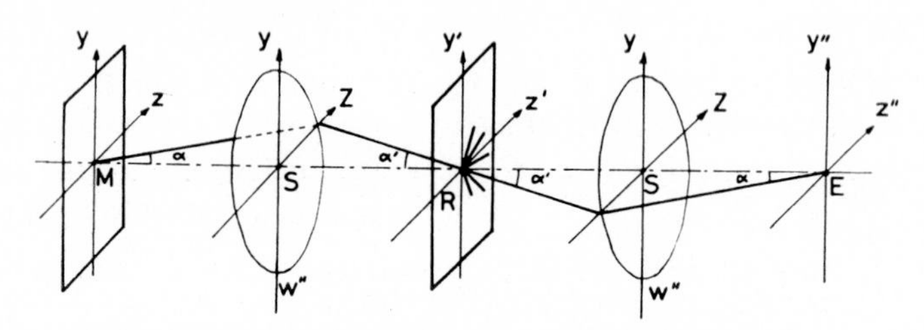

FIG. 3. Double crossing of the optical system W''.

Under our experimental conditions (Figure 3) the light used is a quasi-monochromatic light produced either by a mercury vapour lamp or thermal source fitted with a red filter (Schott). It has been found that $\gamma(z) = 0$ when $z = 2\mu$, a distance ten times smaller than the radius of the central spot of the diffraction pattern given by W''. Consequently (Françon & Slansky 1965) the image is practically the same as in incoherent light. If O is the luminance distribution of the object slit, E and D the respective amplitude and intensity distribution over the image $(y'z')$ of point $M(yz)$, the incoherent slit image is given by:

$$O'(y'z') = O \otimes D \quad \text{where} \quad D(y'z') = E(y'z')E^*(y'z'). \tag{3}$$

A diffuser put in the plane $(y'z')$ wipes out the phase correlation in the image of the point M. The phases of the diffused light are random; i.e. the image around R is an incoherent image given by (4):

$$D' = D \otimes G \tag{4}$$

$G(y'z')$ is the diffusion function due to a magnesia diffuser or the retina (Dupuy 1968).

The light source $O''(y'z')$ acting as an object, is defined by:

$$O''(y'z') = O \otimes D \otimes G. \tag{5}$$

The intensity distribution in the image $(y''z'')$ may be written:

$$O'''(y''z'') = O'' \otimes D. \tag{6}$$

As a consequence it appears that, firstly, the aberrations studied are only due to the second passage of the lens W'', and secondly, the light source O″ becomes wider owing to aberrations in the first passage of W'' and by diffusion phenomena. In the presence of aberrations it has been shown (Berny 1968) that the decrease in sensitivity of the method can be estimated to within 2 per cent with respect to the case of a source represented by $O(yz)$.

Check Experiments

The preceding results have been checked by two different methods:

(*a*) *Comparison of the experimental wave aberrations of a plano-convex lens with its theoretical value.*

The spherical aberration of a plano-convex lens has been measured with the optical apparatus shown in Figure 2. This experiment uses the double crossing of the lens under study W'' and reflexion at a

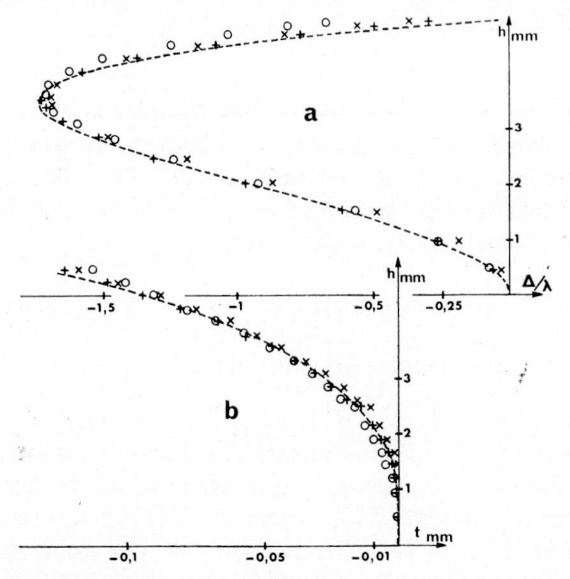

FIG. 4. Normal profile Δ/λ corrected for maximal aperture (curves *a*) and transverse aberration *t* (curves *b*) of a plano-convex lens (○ × +, experimental curves; ------, calculated curves).

magnesia screen (*r*). It is shown (Figure 4) that very good agreement exists between experimental results and theoretical computations; the aberration measured is therefore due only to one passage of the lens. On the other hand, the accuracy of the method is about $\lambda/10$ on the wave-aberration Δ. (Scatter of experimental curves *a*.)

(*b*) *Formation of a double slit on a mirror or a diffuser.*

Another experiment consists of carrying out Young's interference experiment in the following way (Figure 5). A double-slit is imaged with a lens on to a mirror or a diffuser. The phenomenon is observed

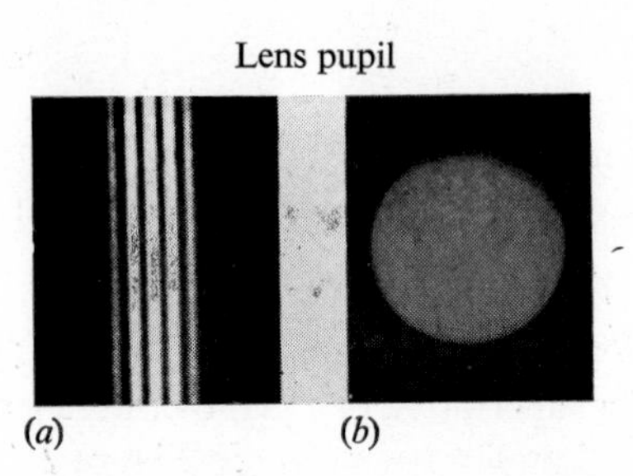

FIG. 5. Double slit imaged on a mirror or an orthotropic diffuser by a lens Q: the phenomenon is observed in the lens pupil: (1) mirror; *a*, interference fringes; (2) magnesia diffuser; *b*, uniform field.

in the lens pupil; the absence of interference fringes when a double-slit is imaged on the magnesia diffuser *b* shows that the degree of coherence of the light diffused by magnesia is weak enough to consider that the source formed on magnesia is incoherent, for the quasi-monochromatic sources used.

OPTICAL SYSTEM OF THE EYE

After the previous check, the human eye is studied by using the experimental device of Figure 3. The Foucault's shadows are photographed for two directions of the knife-edge perpendicular to each other (Figure 6): A_1A_2, vertical knife-edge; B_1B_2, horizontal knife-edge. On each photographic image, only the components of the deviation perpendicular to the knife-edge are recorded. It is possible to observe shadows which are symmetrical with reference to the knife-edge as a result of spherical aberration, and grooves due to the stress of ciliary muscle fibre and other irregular defects. Sets of photographs A_1A_2 or B_1B_2, taken at different times, shows micro-fluctuations in the accommodation of the eye (Arnulf & Dupuy 1960b). From negatives A_1B_1 (or A_2B_2) it is possible to find the components of transverse aberration deviations, $t_{y''}(YZ)$ and $t_{z''}(YZ)$ where Y, Z are normalized Cartesian coordinates in the pupil W'' of the eye (Figure 7).

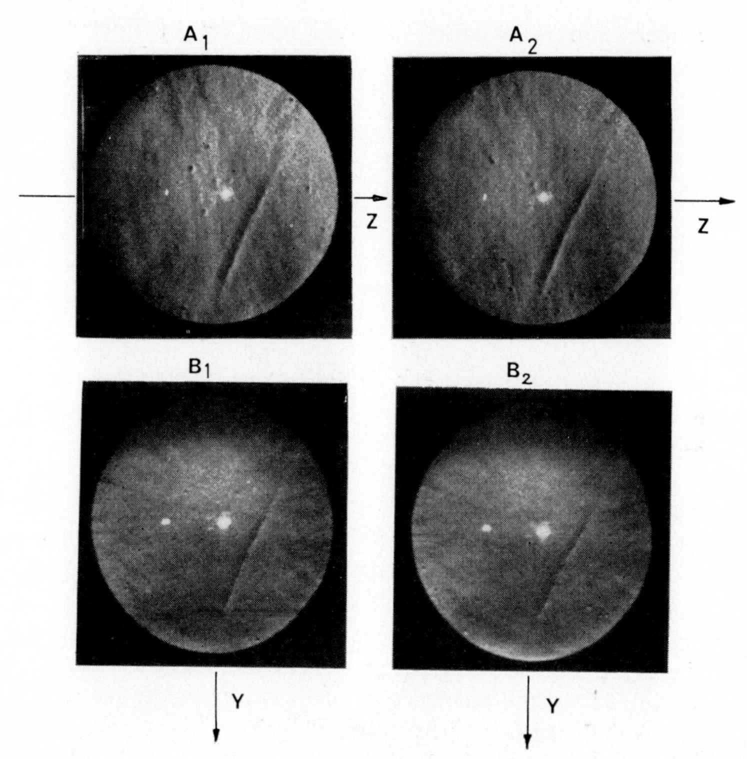

FIG. 6. Foucault's shadows in the pupil, showing deviation components which are perpendicular to the knife-edge. Accommodation, 1 dioptre (δ); A_1A_2, vertical knife-edge; B_1B_2, horizontal knife-edge.

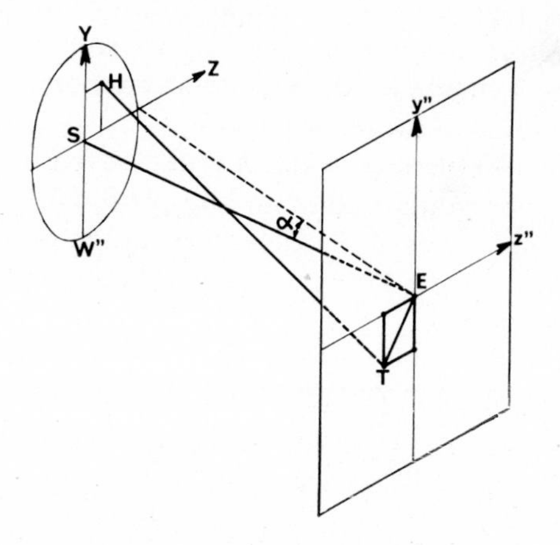

FIG. 7. Transverse aberration ET related to normal deviation at point $H(YZ)$.

Wave aberration is obtained at every point H by integrating the Equations:

$$\left| \begin{aligned} \frac{\partial N}{\partial Y} &= -\frac{h}{F\lambda} t_{y''}(YZ) \\ \frac{\partial N}{\partial Z} &= -\frac{h}{F\lambda} t_{z''}(YZ) \end{aligned} \right. \tag{7}$$

They give either of the equations

$$\left[\begin{aligned} N(YZ) &= -\frac{h}{F\lambda}\left[\int_0^Y t_{y''}(YZ)\mathrm{d}Y + \int_0^Z t_{z''}(OZ)\mathrm{d}Z\right] \\ N(YZ) &= -\frac{h}{F\lambda}\left[\int_0^Z t_{z''}(YZ)\mathrm{d}Z + \int_0^Y t_{y''}(YO)\mathrm{d}Y\right] \end{aligned} \right. \tag{8}$$

$N(YZ)$ is obtained with respect to reference point E; $2h/F$ is the maximum aperture of W'', λ being the wavelength.

To derive $N(YZ)$ it is sufficient to have the recordings of $t_{z''}(YZ)$ carried out along lines which are parallel to axis Z on the first photograph A_1, the knife-edge being perpendicular to Z; but furthermore it is necessary to have the recording of $t_{y'}(YO)$ carried out on the second negative (B_1) along axis Y, the knife-edge being perpendicular to Y.

RESULTS

Wavefront Given by Equal Phase Curves

A dimensioned diagram of wavefront distortion in the pupil can then be drawn. The wavefront is represented by equal phase curves referred to a sphere centred on a chosen focal point, at intersection E of the knife-edge and an axis very close to the optical axis.

Figure 8 shows two pictures A and B of the wavefront of the same eye obtained from photographs taken in the pupil for 1 dioptre accommodation at different times. They show irregularities of the optical system of the eye. Pictures A and B have been obtained respectively with 1257 and 6400 experimental points.

The application of suitable correcting curves (Yvon 1926) to these wavefronts makes it possible to obtain the wave aberrations referred to the best focus E_0. The position E_0 is such that the intensity distribution is a maximum at E_0, the centre of the diffraction pattern, image of a point (Maréchal 1947). This best focus will be obtained when using a longitudinal and lateral displacement of the centre of the reference sphere. Calculations have been made by numerical integration. For *A*, a CAB 500 has been used by taking 1257 points on the wavefront. For *B*, a bigger computer (UNIVAC 1108) has been used; which allowed for the integration of 6400 points.

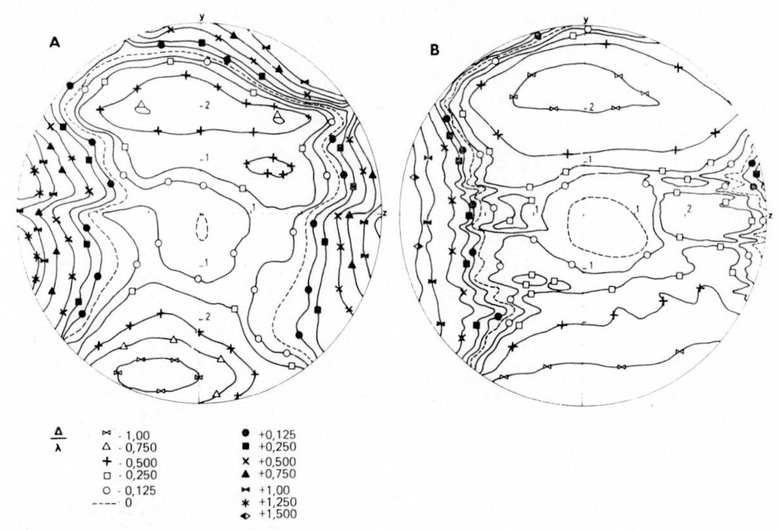

FIG. 8. Wavefront Δ of the optical system of the eye. Level curves in terms of wave number (Δ/λ): , ⋈ −1·000; Δ, −0·750; +, −0·500; □, −0·250; ○, −0·125; -----, 0·000; ●, +0·125; ■, +0·250; ×, +0·500; ▲, +0·750; ⧓ ,+1·000; ✱, +1·250; ◆, +1·500. Reference point *E*. Pupil diameter 7 mm. Accommodation 1 δ. *A*, 1257 experimental points; *B*, 6400 experimental points.

Figure 9 shows two appearances, *A* and *B*, of the wavefront, referred to the best focus E_0, for a pupil diameter $2h = 7$ mm. Although the aberrations are irregular, the Rayleigh's zone where Δ_0 remains stationary ($\Delta_0 \leqslant \lambda/4$) is observed for $2h \leqslant 3$ mm. When $2h$ increases, the pattern becomes progressively impaired and reaches values up to 2 to 3λ at the pupil edge.

Diffraction Pattern at the Best Focus

The evolution of the space diffraction pattern at the best focus, for pupils ranging from 7 to 2 mm in diameter is studied. This was carried out for *B* in which case results are more accurate (Figure 10.).

The intensity distribution given by (9):

$$D(y''z'') = D(\rho\theta) = E(\rho\theta)E^*(\rho\theta) \tag{9}$$

where

$$\left[\begin{array}{l} E(\rho\theta) = \displaystyle\int_0^1 \int_0^{2\pi} \exp\left[j\frac{2\pi\Delta(H\phi)}{\lambda}\right] \exp\left[-j\frac{2\pi h\rho}{\lambda F}\cos(\phi-\theta)\right] HdHd\phi \\ H\cos\phi = Z;\ H\sin\phi = Y;\ \rho\cos\theta = z'';\ \rho\sin\theta = y'' \end{array}\right. \tag{10}$$

is studied as a function of the distance ρ (in mm) from the centre of the diffraction pattern in the plane $y''z''$ using different azimuths θ (in degrees). The values are normalized by taking as unity the central intensity distribution of a perfect instrument of the same aperture. For large pupils the spatial diffraction pattern is not a solid of revolution about the optic axis; this shows the presence of astigmatism on the axis for the eye under study.

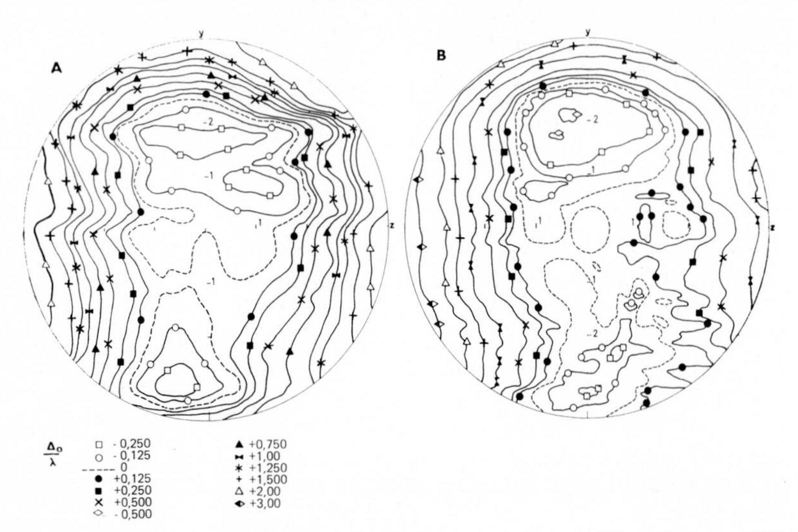

FIG. 9. Wavefront Δ_0 of the optical system of the eye. Level curves in terms of wave number (Δ_0/λ): ◇, −0·500; □, −0·250; ○, −0·125; – – – – – –, 0·000; ●, +0·125; ■, +0·250; ×, +0·500; ▲, +0·750; ⧓, +1·000; ∗, +1·250; +, +1·500; △, +2·000; ◐, +3·000. Reference point E_0. Pupil diameter 7 mm. Accommodation 1 δ. *A*, 1257 experimental points; *B*, 6400 experimental points.

The central intensity (b:$2h$ = 4 mm, D = 0·40; c:$2h$ = 2 mm, D = 0·94) is improved when $2h$ decreases. When $2h = 1$ mm, the Airy disk is again found and when $2h \leqslant 3$ mm, $D \geqslant 0{\cdot}75$ (Table 1). The quality of this optical system having irregular defects is equivalent to that of a well-corrected instrument as long as $2h \leqslant 3$ mm, i.e. in photopic vision. For larger pupils the optical quality is impaired. This confirms earlier results (Arnulf & Dupuy 1960a; Berny 1969).

The present conclusions raise the possibility of investigating an association of the eye with an almost perfect or slightly under-corrected optical instrument, e.g. binoculars.

Three cases must be considered:

(*a*) When associated with instruments having exit pupil not exceeding 2 mm in diameter, the eye brings the image to its best focus and realises practically the Rayleigh tolerance. The final wavefront (eye and instrument) is that of the instrument alone if any aberration remains.

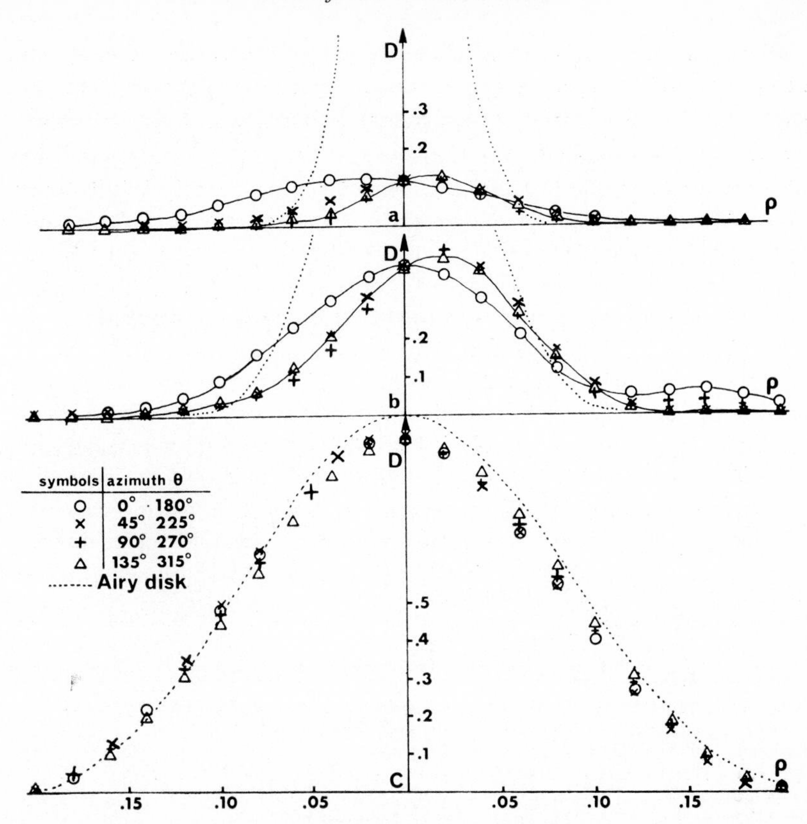

FIG. 10. Intensity distribution $D(\rho\theta)$ in image E_0. Different pupil diameter $2h$: (*a*) 6 mm; (*b*) 4 mm; (*c*) 2 mm.

TABLE 1. Intensity distribution $D(\rho\theta)$ in image E_0; $2h = 3$ mm.

	θ (degree)							
ρ (mm)	0	45	90	135	180	225	270	315
0·000	0·7523	0·7523	0·7523	0·7523	0·7523	0·7523	0·7523	0·7523
0·010	0·6982	0·7332	0·7541	0·7480	0·7186	0·6842	0·6659	0·6712
0·020	0·5704	0·6318	0·6676	0·6589	0·6099	0·5499	0·5236	0·5286
0·030	0·4049	0·4762	0·5167	0·5100	0·4589	0·3861	0·3653	0·3631
0·040	0·2449	0·3075	0·3437	0·3406	0·3048	0·2319	0·2268	0·2128
0·050	0·1248	0·1634	0·1918	0·1892	0·1783	0·1143	0·1275	0·1025
0·060	0·0577	0·0659	0·0892	0·0811	0·0938	0·0425	0·0687	0·0381
0·070	0·0356	0·0172	0·0408	0·0219	0·0488	0·0095	0·0391	0·0103
0·080	0·0373	0·0037	0·0316	0·0018	0·0307	0·0004	0·0247	0·0031
0·090	0·0418	0·0068	0·0384	0·0032	0·0252	0·0007	0·0157	0·0028
0·100	0·0376	0·0116	0·0424	0·0101	0·0223	0·0019	0·0085	0·0023
0·110	0·0249	0·0114	0·0363	0·0135	0·0180	0·0014	0·0040	0·0010
0·120	0·0112	0·0071	0·0235	0·0117	0·0126	0·0004	0·0034	0·0006
0·130	0·0036	0·0027	0·0119	0·0071	0·0770	0·0004	0·0059	0·0022
0·140	0·0041	0·0007	0·0073	0·0029	0·0043	0·0015	0·0090	0·0049
0·150	0·0092	0·0012	0·0096	0·0007	0·0025	0·0027	0·0103	0·0065
0·160	0·0132	0·0024	0·0145	0·0000	0·0015	0·0030	0·0090	0·0060
0·170	0·0128	0·0028	0·0171	0·0001	0·0008	0·0022	0·0060	0·0038
0·180	0·0085	0·0019	0·0151	0·0002	0·0003	0·0010	0·0030	0·0016
0·190	0·0035	0·0007	0·0098	0·0000	0·0000	0·0003	0·0012	0·0005
0·200	0·0010	0·0004	0·0043	0·0000	0·0000	0·0004	0·0005	0·0008

(*b*) When associated with instruments having eye circles ranging from 2 to 4 mm, the irregular but rather under-corrected wavefront of the eye shows weak aberrations which are comparable to those of the instrument.

(*c*) When associated with instruments having exit pupil exceeding 4 mm in diameter (scotopic vision), the irregular wavefront is largely over-corrected near the edge. Over-correction of the eye may partially balance under-correction of the instrument. However in scotopic vision the defects of the eye seem to dominate those of optical instruments.

We remind the reader that this investigation has been made with only one subject. An experimental investigation of a wavefront made by exploring a set of photographs of the pupil requires several months, because of the photometric examination operations. The present results are therefore given as examples. They do not take into consideration the micro-fluctuations in accommodation. We think that by improving the present apparatus it will be possible to make a statistical study of a sufficiently large number of subjects and to draw general conclusions.

REFERENCES

ARNULF, A. and DUPUY, O., 1956, *Problems in Contemporary Optics*, Inst. Naz. Ottica, Arcetri-Firenze, p. 330.
ARNULF, A. and DUPUY, O., 1960*a*, *C.r. Seanc. Acad. Sci.*, Paris, **250,** 2757.
ARNULF, A. and DUPUY, O., 1960*b*, *Revue Opt. Theor. Instrum.*, **39,** 195.
BERNY, F., 1968, Thèse, Paris.
BERNY, F., 1969, *Vision Res.*, **9,** 977.
CAMPBELL, F. W. and GUBISH, R. W., 1966, *J. Physiol.*, **186,** 558.
DUPUY, O., 1964, *Revue Opt. Théor. Instrum.*, **43,** 217 and 282.
DUPUY, O., 1968, *Vision Res.*, **8,** 1507.
FRANÇON, M. and SLANSKY, S., 1965, *Cohérence en Optique*, Ed. Centre National de la Recherche Scientifique, Paris, p. 37.
HOPKINS, H. J., 1953, *Proc. Roy. Soc.*, A **217,** 418.
LINFOOT, E. H., 1955, *Recent advances in Optics*, Oxford, Clarendon Press, p. 128.
MARECHAL, A., 1947, *Revue Opt. Théor. Instrum.*, **26,** 257.
MARECHAL, A. and FRANÇON, M., 1960, *Diffraction, Structure des Images*, Ed. *Revue Opt. Théor. Instrum.*, Paris, p. 81.
PHILBERT, M., 1967, *Optica Acta*, **14,** 169.
YVON, G., 1926, *Controle des Surfaces Optiques*, Ed. *Revue Opt. Théor. Instrum*, Paris, pp. 17, 48 and 142.

The Optical Evaluation of High Resolution Grating Spectrographs

Eugene van Rooyen and Arthur J. Boettcher

National Physical Research Laboratory,
P.O. Box 395, *Pretoria,*
Republic of South Africa

Abstract—A method, which makes use of wavefront shearing interferometry, for taking interferograms of long focal length plane grating spectrographs, is described. The deviations of the wavefront from the best-fitting reference sphere are then calculated. The necessity of using a small shear and a large number of points becomes evident from the calculated wavefronts. The spread functions and optical transfer functions are calculated by means of Fourier transform methods at a number of wavelengths and are discussed in detail.

INTRODUCTION

Spectrographs have been, and are being used to establish the energy level structures of atomic and molecular systems possessing spectra in the ultraviolet, visible and infrared regions (Shenstone 1960). These instruments have also been used to obtain information on the shape of lines, but in nearly all cases the distortion due to the transfer properties of the instrument was minimized by using special techniques such as pressure broadening (Penner 1959). The application of these techniques are not always possible and so the spread function is determined at a single wavelength using a spectral line having a small width in comparison to the width of the spread function. It is generally assumed that the spread function will not vary significantly with wavelength. This may be applicable to prism infrared spectrometers. However, in large grating spectrographs, the width of the spread pattern in the visible and ultraviolet is comparable to the width of many lines which could possibly be used. In such cases laser lines have been used (Sica 1967) but care has to be exercised in the interpretation of the results due to the larger coherence length of laser light.

The spread function and optical transfer function of an optical instrument can be calculated from the instrument's pupil function by Fourier transform methods (Smith 1963). Using this method Stroke (1960; 1963) studied the effect of ruling errors on diffraction pattern imperfections. However, to date no quantitative pupil functions of plane grating spectrographs have been determined and the purpose of this paper is to describe a method to achieve this. From this data the spread and transfer functions were computed, assuming incoherent slit illumination. Convolution with real slit widths were not considered.

EXPERIMENTAL

The spectrograph used in this study was an off-plane Littrow mount. The optical components were a well corrected f/45 lens of 6 metre focal length and a B & L plane grating (300 grooves/mm) blased at 63° 24′.

Interferograms were taken at a number of wavelengths using an interferometer with variable shear (van Rooyen & van Houten 1969). Furthermore, because this interferometer was adjusted for white light compensation this assured that the fringe visibility depended only on the shear ratio and on the width of the spread function or spectral line used· In a study of the ruling errors of concave grating spectrographs Birch (1966) used a He:Ne laser source and a plane parallel plate shearing interferometer. When wavefront aberrations are required at other than laser wavelengths this method is clearly not usable and a white light compensated interferometer is necessary. In this work it was found that the resolved isotopic structure of the Hg lines 4358 Å and 5461 Å did not affect the visibility to the extent that a single isotope lamp was required. Moreover, a Th electrodeless discharge lamp could be used, should data be required at other wavelengths.

To facilitate the initial adjustments of the interferometer the entrance slit of the spectrograph was opened to about 1 mm and the grating set to the normal position. The interferometer was next set to zero shear and white light compensation and adjusted so that 5 to 7 horizontal fringes crossed the field. The slit was then closed and the grating rotated so that the spectral line was transmitted through the interferometer. The shear was then set to a suitable value and the slit closed to strike a compromise between intensity and fringe visibility.

A lens with a focal length of 800 mm was used to image the interferograms on a suitable film. The lens and film-holder unit was adjusted, using the Boyes' points technique (Taylor & Thompson 1957), so that the film plane was parallel to the pupil plane of the spectrograph. Since the aberrations of the lens can affect the position of the fringes a fine grid, placed at the pupil distance, was photographed. Measurements indicated that the rather small deviations would have no serious effect on the accuracy of measurement.

Interferograms were obtained for 6328 Å He:Ne laser light on Adox R17 film using an exposure time of $1\frac{1}{2}$ minutes. Ilford HP4 film was used for photographing the interferograms in 5461 Å and 4358 Å light. Exposure times for these ranged up to 30 minutes.

In order to analyse the interferograms, ten times enlargements of the negatives were made. The enlarger lens and enlarger alignment were checked in the same way as described above.

Saunders (1961) described a method whereby the wavefront deviations from a reference sphere could be obtained at equally spaced intervals across the pupil. These intervals could be equal to the shear distance or could be sets or families of points spaced 1/2 or 1/3, etc., of the shear distance (Saunders 1964). To compute the wavefront deviations a reference line was drawn across the centre of each enlargement and perpendicular to this reference line a number of parallel lines, separated by the shear distance. The fractional order between the reference line and a fringe was then calculated. The deviations from a close-fitting reference sphere could be computed from the fractional orders for more than a hundred points across the pupil in a few seconds by means of a suitable computer programme. Such a programme was specially written for the available IBM 360 computer.

Since the wavefront aberrations of gratings are not smooth the deviations have to be determined at a large number of points, ~ 100 or more. This means that the shear must be small or that a number of families must be used. It was found that for a large number of points the method described by Saunders, whereby the deviations of two families were referred to the same reference sphere, did not give correct values but that the computed wavefront deviations would oscillate when the aberrations were small compared to the coefficients of the condition equation. All the results reported in this paper were therefore based on the measurements from one set of points. In spite of the lower possible accuracy when such small shear ratios were used, results agreed to within $\lambda/10$. The absolute value of the deviations were checked using scatter fringe interferometry (Scott 1967).

Contrary to expectations no serious problems were encountered in obtaining the required fringe stability. This was due to the fact that the shear was generally small and that the interferometer was rigidly clamped to the spectrograph. Relatively stable atmospheric conditions during the time when this experimental work was performed also helped to prevent fringe oscillations.

THEORETICAL

A Fourier transform relationship exists between the pupil function, $g(\eta)$, and the complex amplitude distribution of the spread function (Stroke 1963) and is given by

$$B(\beta) = \int_{-1}^{1} g(\eta) \exp{(i\beta\eta)} d\eta$$

where η is the normalized length across the pupil and β is the angular coordinate. Since modern gratings are ruled interferometrically with the subsequent elimination of fan errors, it is only necessary to perform a one-dimensional transform across the pupil to obtain the spread function. β is a function of the linear coordinate x_n and is so chosen that the first diffraction minimum is at $(\lambda/A)f$ for $g(\eta)=1$ where f is the focal length, λ is the wavelength and A the aperture.

The Fourier transform of the intensity distribution of the spread function yields the optical transfer function $H(\omega)$ which is given by (Smith 1963)

$$H(\omega) = \frac{\int_a^b |B(x_n)|^2 \exp(i\omega x_n)\mathrm{d}x_n}{\int_a^b |B(x_n)|^2 \mathrm{d}x_n},$$

where ω is the spatial frequency, the limits of the integration being the limits of the x_n. From the real and imaginary parts of $H(\omega)$ the modulation transfer function $T(\omega)$ and the phase function $\Phi(\omega)$ can be calculated.

The computer programme which was written to calculate the aberration function numerically, also calculated the pupil function from the aberration function; from the pupil function, the spread function, and finally from the spread function, the modulus and phase of the optical transfer function. Provision was made for adding a defocusing term to the aberration function so as to study the influence of this aberration on the spread and transfer functions. The numerical integrations in the Fourier transforms were performed by means of the method of overlapping parabolas (Secrest 1965) and yielded very good results.

RESULTS AND DISCUSSION

Determination of the Wavefront Aberrations

Figure 1 shows three interferograms of the 6-metre spectrograph taken in situ. The experimental conditions used are given in Table 1. Columns 4 and 5 show detail of interferograms, not reproduced here, taken at different shear values and at different wavelengths. Since the wavefront determinations were made across the pupil centre, the vignetting of the grating corners by the lens did not affect the wavefront deviation calculations.

Figures 2 and 3 show the wavefront deviations from a close-fitting reference sphere calculated from the interferograms. Figure 2*a* shows the wavefront calculated from the interferogram shown in Figure 1*a*.

Figures 2*b*, 3*a* and 3*b* show the wavefronts obtained from interferograms, details of which are listed in columns 2 and 4, column 3 and column 5 of

Table 1, respectively. In Figure 2*a* spherical aberration of the lens is clearly shown. The asymmetry is possibly due to lens centring errors. From Figures 2*b*, 3*a* and 3*b* it is clear that if the number of points across the wavefront is increased, greater detail becomes evident. This is especially

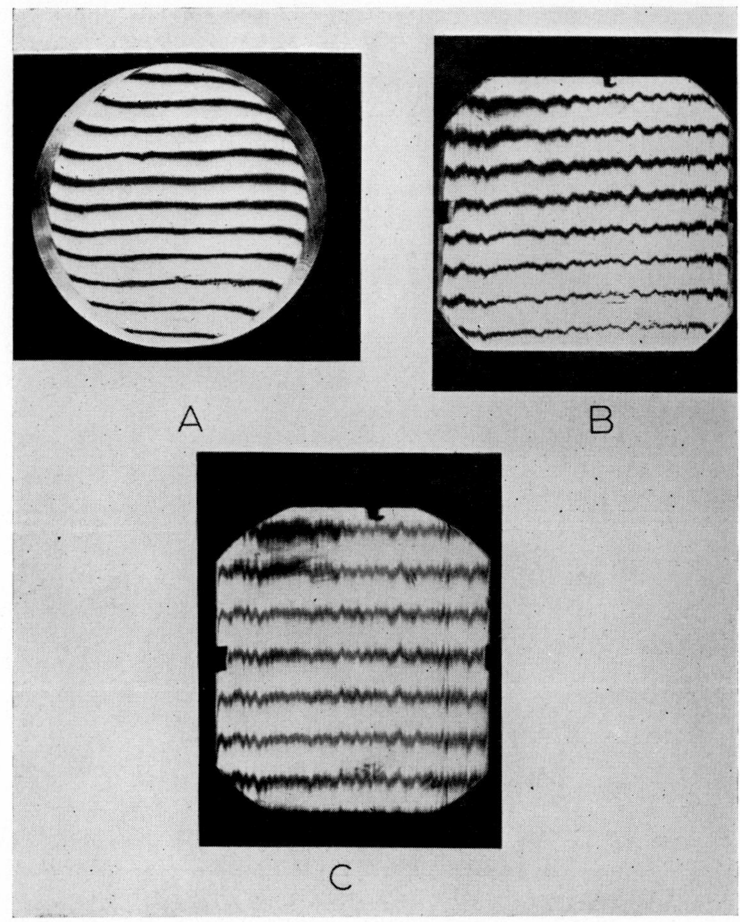

FIG. 1. Sheared interferograms of the spectrograph's optics photographed at the following wavelengths: (*a*) lens only in 6329 Å laser light (normal incidence); (*b*) lens and grating in 6328 Å laser light (9th order); (*c*) lens and grating in 5461 Å mercury light (11th order).

important when abrupt changes are present. Such a change is very clearly seen in Figure 3*a* but is less clearly indicated in Figures 2*b* and 3*b*.

Since the problems involved in the use of a larger number of families have already been discussed the use of a small shear ratio is indicated. Moreover, the results indicate that a shear ratio of smaller than 1 in 100 should be used in order to obtain maximum information.

Referring to Figure 2*b* where the wavefronts obtained at shear ratios of 1 in 33 and 1 in 57 are depicted, it is clear that in spite of the fact that some detail is lost due to the lower number of reference points employed the differences between the curves are smaller than $\lambda/10$. Also the edge aberration of the lens alone seen in Figure 2*a* does not contribute significantly to the aberration shown in Figures 2*b*, 3*a* and 3*b* since the grating width is smaller than the lens diameter.

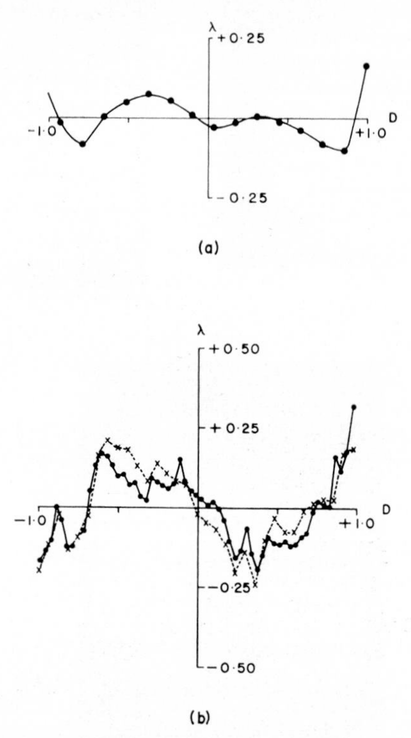

FIG. 2. Wavefront deviations calculated for the: (*a*) lens only at 6328 Å (normal incidence); (*b*) spectrograph at 6328 Å (9th order) where —●——●— is for the shear ratio 1 in 57 and —×——×— is for the shear ratio 1 in 33.

Calculated Spread Functions

Figure 4 shows the normalized spread functions calculated from the wavefront aberration functions shown in Figures 2 and 3. From Figure 4*a* it can be seen that the spread function for the lens alone is somewhat asymmetrical due to the asymmetrical wavefront. The halfwidth of the spread function of the lens is also broader (25·0 μ) than for the aberration-free case (22·0 μ). In the wings of the spread function some asymmetry is found as well.

Figures 4*b*. 4*c* and 4*d* show that the spectral line will be broadened even more when the effect of the grating aberrations are taken into account. The actual values are indicated in Table 2 which also lists the positions and intensities of the central maxima and principal ghosts. The effect of these grating imperfections is the appearance of a strong ghost on the shorter wavelength side of the main line. Table 2 shows that the distance of this ghost from the central maximum increases with decreasing wavelength and that the ghost intensity decreases with a decrease in wavelength. However, when the spread functions were computed for both the wave-

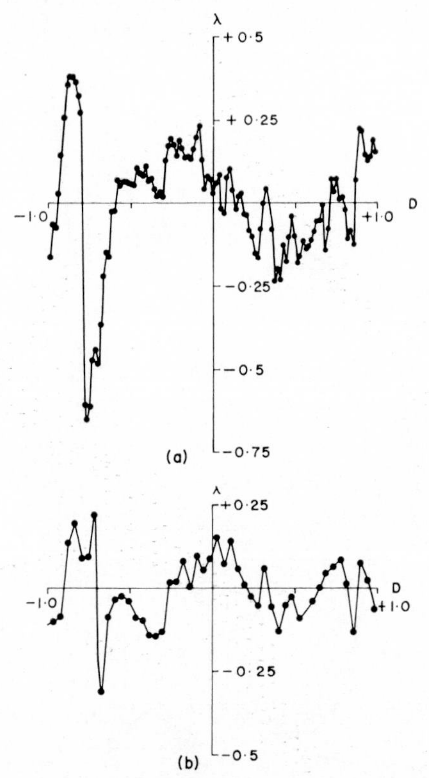

FIG. 3. Wavefront deviations calculated for the: (*a*) spectrograph at 5461 Å (11th order); (*b*) spectrograph at 4358 Å (14th order).

fronts shown in Figure 2*b* it was found that the positions of the ghost and of the central maximum did not change significantly but that the peak intensity of the ghost changed from 41·5 per cent to 61 per cent even though the computed wavefronts differed by less than $\lambda/10$. This indicates that the intensity of the ghost and satellite lines are therefore very sensitive to the accuracy with which the wavefront is determined.

The addition of a 0·25 λ defocusing term to the wavefront results in a broadening and a shift of the central maximum as is shown in Table 2.

However, the position of the ghost is changed more with defocusing than that of the central maximum. Since small differences in the wavefront's aberrations, as was shown above, have a marked effect on ghost line intensities, a defocusing aberration of $\pm 0{\cdot}25\ \lambda$ added to the wavefront affects the intensities even more. This is clearly indicated in Table 2.

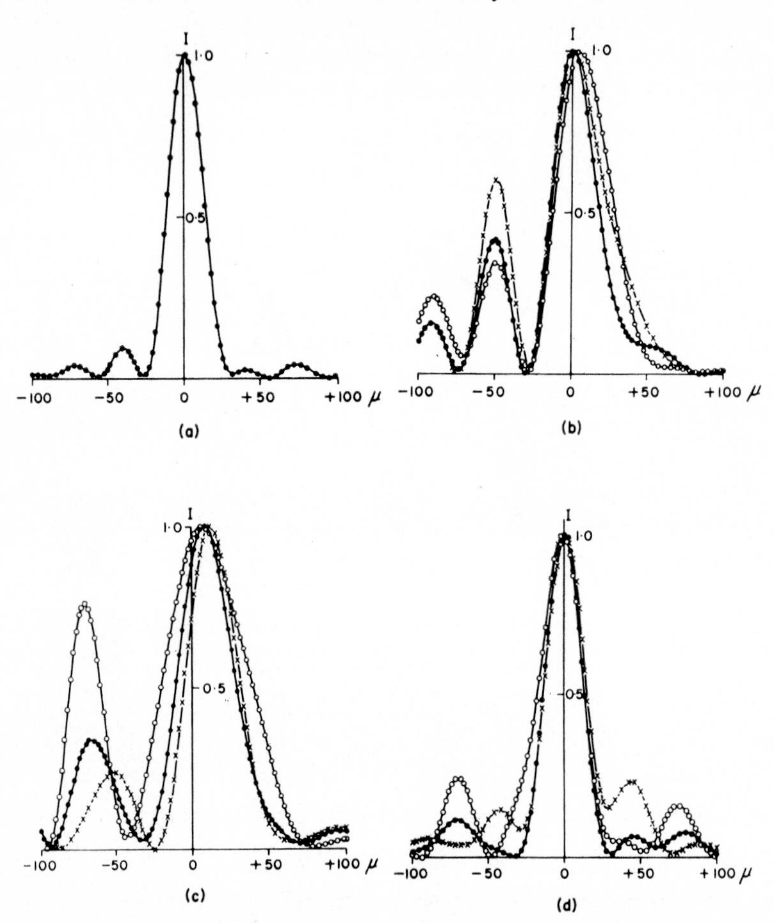

FIG. 4. Spread functions calculated for the: (*a*) lens only at 6328 Å; (*b*) spectrograph at 6328 Å; (*c*) spectrograph at 5461 Å; (*d*) spectrograph at 4358 Å; where —●—●— is for zero defocusing, —×—×— is for an added $-0{\cdot}25\ \lambda$ defocussing and —○—○— is for an added $+0{\cdot}25\ \lambda$ defocusing.

It is further clear that, should the above grating be used for wavelength measurements with coarse-grained film, errors in the wavelength determination will result since the position of the ghost is wavelength dependent. In order to verify the above calculations photographic exposures were made of the spread functions of the lens alone and of the spectrograph at

6328 Å (van Rooyen 1966). A single mode He:Ne laser was used for this purpose. The results are shown in Figure 5. Taking into account the response and background of the film and also the slit width of the

TABLE 1. Summary of experimental conditions.

Interferogram taken of	Lens		Lens plus grating		
Wavelength (Å)	6328		5461	6328	4358
Order	Normal incidence	9	11	9	14
Shear ratio	1 in 15	1 in 57	1 in 117	1 in 33	1 in 48
Pupil width (mm)	135·8	112·7	94·4	112·7	87·6
No. of points across interferogram	15	57	117	33	48

instrument (~10 μ), the results show the calculated asymmetric spread function for the lens alone and also the position of the principal ghost at approximately $-50\ \mu$ for the lens and grating. The intensity of the ghost is in qualitative agreement with the calculations.

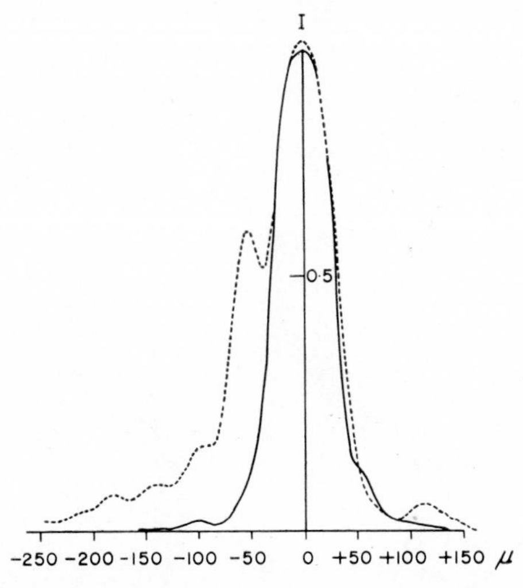

FIG. 5. Experimentally determined spread functions for the lens only (solid line) and for the spectrograph (dotted line) at 6328 Å.

Optical Transfer Functions

The modulation transfer function (M.T.F.) and the phase transfer function (P.T.F.), that is, the optical transfer function (O.T.F.) were computed from the spread functions.

In order to determine the effect of the ghost on the O.T.F., the spread function for zero defocusing shown in Figure 4*b* was used. The O.T.F. was computed for the spread function, first without the ghost ($-29\ \mu$ to $86\ \mu$);

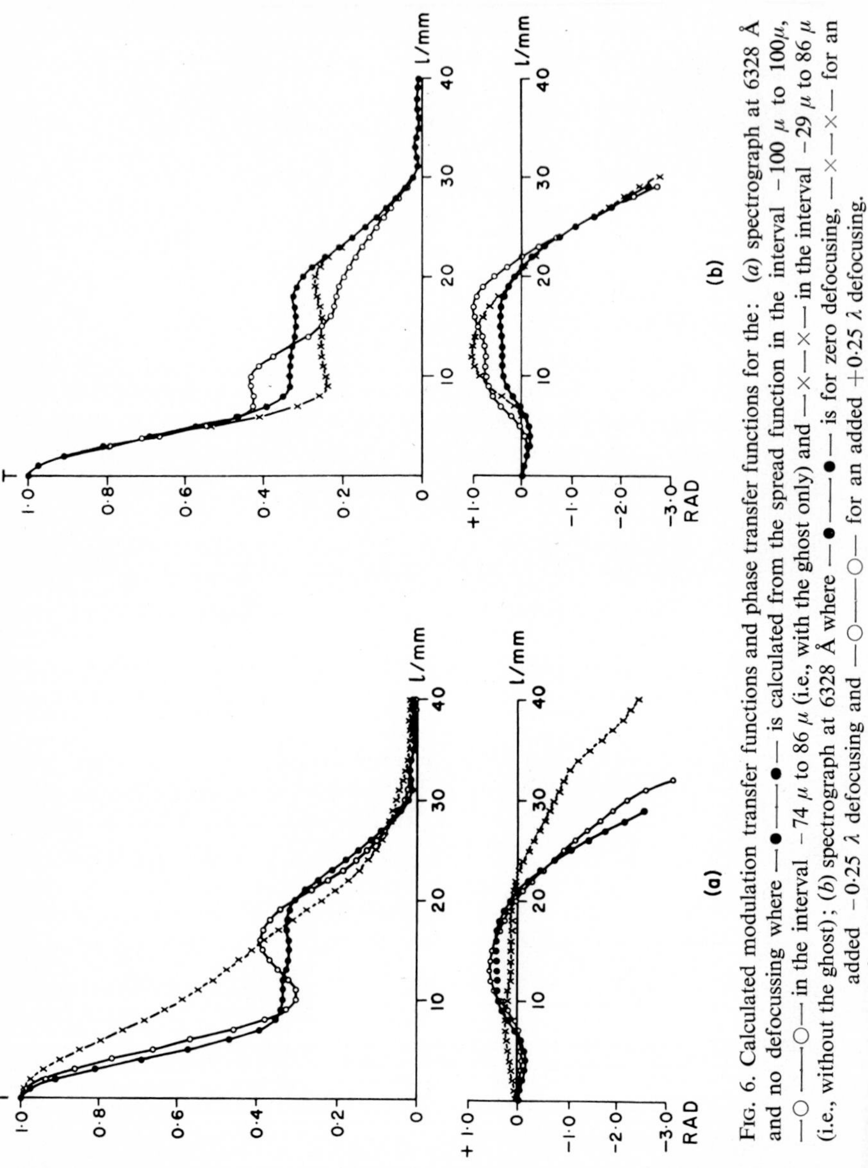

FIG. 6. Calculated modulation transfer functions and phase transfer functions for the: (*a*) spectrograph at 6328 Å and no defocussing where —●——●— is calculated from the spread function in the interval $-100\ \mu$ to 100μ, —○——○— in the interval $-74\ \mu$ to $86\ \mu$ (i.e., with the ghost only) and —×—×— in the interval $-29\ \mu$ to $86\ \mu$ (i.e., without the ghost); (*b*) spectrograph at 6328 Å where —●——●— is for zero defocusing, —×—×— for an added $-0{\cdot}25\ \lambda$ defocusing and —○——○— for an added $+0{\cdot}25\ \lambda$ defocusing.

then with the ghost ($-74\ \mu$ to $86\ \mu$) and finally for the spread function over the full $200\ \mu$ width. These results are shown in Figure 6*a*. The response of the M.T.F. for the central part of the spread function ($-29\ \mu$ to $86\ \mu$) is

TABLE 2. Dependence of spread pattern positions, intensities and halfwidths on defocusing and wavelength.

	Lens	Spectrograph			Spectrograph			Spectrograph		
Wavelength (Å)	6328	6328			5461			4358		
Defocusing (λ)	0	0	−·25	+·25	0	−·25	+·25	0	−·25	+·25
Position of central maximum (μ)	+2	0	+2	+6	+7·5	+13	+6	0	+1	−1
Halfwidth of central maximum (μ)	25·2	34·5	39	40	40	38	60	27·5	27·5	33
First minimum position (positive) (μ)	32	86	82	>100	71	56	77	31	30	32
First minimum position (negative) (μ)	−26	−29	−27	−28	−32	−24	−42	−33	−29	−48
Principal ghost maximum position (μ)	−40	−50	−50	−59	−66	−52	−71	−20	−42	−70
Principal ghost peak intensity (%)	9·3	41·5	60	34·5	34	23·8	76·5	12·8	14·5	25
Halfwidth of principal ghost (μ)	17	22·5	23	22·9	31·5	32·5	26	27	14	21·5

smooth, having a cut-off at 40 lines/mm. When the M.T.F. is calculated for the central part plus the ghost ($-74\ \mu$ to $86\ \mu$), the M.T.F. exhibits

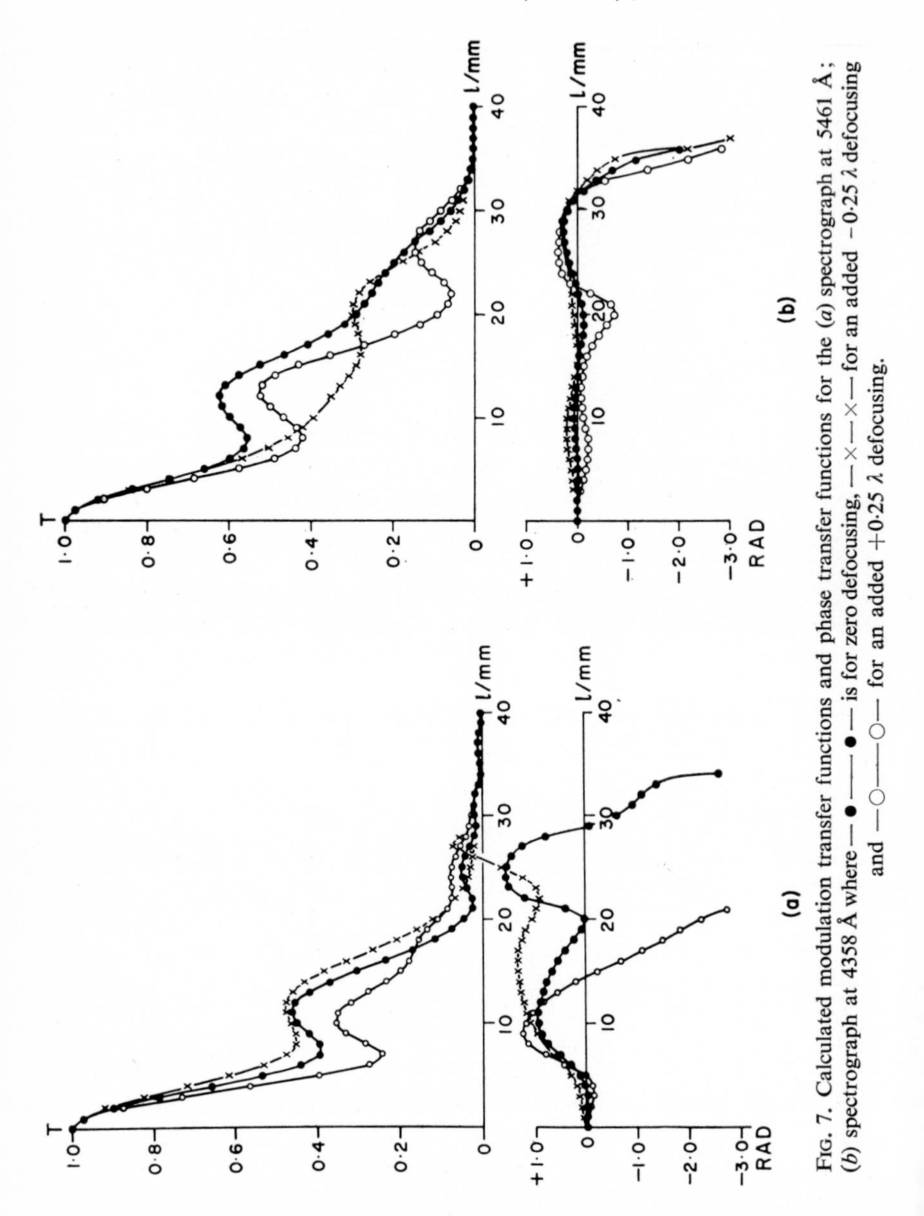

FIG. 7. Calculated modulation transfer functions and phase transfer functions for the (*a*) spectrograph at 5461 Å; (*b*) spectrograph at 4358 Å where —●——●— is for zero defocusing, —×—×— for an added $-0{\cdot}25\ \lambda$ defocusing and —○——○— for an added $+0{\cdot}25\ \lambda$ defocusing.

a minimum at about 10 lines/mm and a maximum at 16 lines/mm. The M.T.F., when calculated for the $-100\ \mu$ to $+100\ \mu$ range, tends to flatten in the central frequency range due to the irregularities on the other side of the line. Similarly the P.T.F. will show changes due to the asymmetries present and the shift of the central maximum.

Figures 6*b*, 7*a* and 7*b* show the O.T.F. calculated from the spread functions shown in Figures 4*b*, 4*c* and 4*d*. The range for the calculations was chosen to be $-100\ \mu$ to $100\ \mu$. It is clearly seen that a large satellite will tend to form a hump in the response but if other satellites are present the response will tend to flatten out in the central frequencies. Asymmetry and positional shifts are reflected in the phase functions and dips in the M.T.F. curves are also reflected in the P.T.F. curves. These results are in agreement with the calculations done by Perrin (1960).

CONCLUSIONS

Extreme accuracy is required in the determination of the wavefront aberrations of plane grating spectrographs, should these data be required for evaluation purposes. This is because the intensity of a ghost in the spread function depends on the accuracy with which the wavefront deviations have been determined and thus the optical transfer function as well.

The application of wavefront shearing interferometry to the quantitative determination of wavefront aberrations of grating spectrographs has been demonstrated and it was further shown that the shear ratio should be as small as possible.

In the present study the effect of slit functions or partial coherence slit illumination has not been considered. For the application of this technique in practical line shape analysis the effect of the above, however, will have to be evaluated. Since the spread function is wavelength dependent and is in general narrow in the case of grating spectrographs, a laser source can only be used to scan the spread function if it is very highly stabilized, if its coherence length is known and if it is used in single mode. In the case of a He:Ne laser it is not very difficult to stabilize it well and obtain a single mode, but this is more difficult in the case of an A^+ laser if spread functions are required at the other wavelengths. Taking interferograms at different wavelengths eliminates these difficulties which might be considerable when the spread functions of long focal length grating spectrographs are evaluated. It is also important to know the focal plane of the instrument accurately since it was shown that small defocusing errors introduce a shift both of the central maximum and of the ghosts as well as changes in the relative intensities of the ghosts.

The nature of the aberrations of gratings presents a problem to the evaluation of spectrographs but it is shown that these can be solved to obtain reliable spread and optical transfer functions.

REFERENCES

BIRCH, K. G., 1966, *J. Sci. Instr.*, **43**, 243.
PENNER, S. S., 1959, *Quantitative Molecular Spectroscopy and Gas Emissivities*, Addison-Wesley, Reading, U.S.A.
PERRIN, F. H., 1960, *J. Soc. Mot. Pict. and Tel. Eng.*, **69**, 239.
SAUNDERS, J. B., 1961, *J. Res. N.B.S.*, **65B**, 239.
——, 1964, *J. Res. N.B.S.*, **68C**, 155.
SCOTT, R. M., 1967, *New Developments in Interferometry*, Perkin-Elmer Corp., Optical Group, Norwalk, Connecticut.
SECREST, D., 1965, *J. SIAM. Numer. Anal., Ser. B*, **2**, 52.
SHENSTONE, A. G., 1960, Discussion of Paper 5–4, *National Physical Laboratory Symposium No. 11, Interferometry*, London, Her Majesty's Stationery Office, p. 431.
SICA, L., 1966, Ph.D. dissertation, The Johns Hopkins University, Baltimore.
SMITH, F. D., 1963, *Appl. Opt.*, **2**, 335.
STROKE, G. W., 1960, *Revue D'Optique*, **39**, 291.
——, 1963, *Progress in Optics*, **2**, 3.
TAYLOR, C. A, and THOMPSON, B. J., 1957, *J. Sci. Instr.*, **34**, 439.
VAN ROOYEN, E., 1966, D.Sc. dissertation, University of Pretoria, R.S.A.
VAN ROOYEN, E. and VAN HOUTEN, H. G., 1969, *Appl. Opt.*, **8**, 91.

Measurement of the Transfer Function for Interference-Phase-Contrast under Coherent Illumination

D. Haina and W. Waidelich

Gesellschaft für Strahlenforschung,
m.b.H München/Neuherberg
and
I. Physikalisches Institut der Technischen Hochschule,
Darmstadt

Abstract—Low spatial frequencies of the object are lost in the image, when using Zernike's phase-contrast method. The transfer function has the form of a band-pass.

The interference-phase-contrast method does not compress the low frequencies. For this method a recording interferometer, operating with spherical waves, is described. From the intensity distribution of the image, the transfer function is calculated. The linearity and the dependence on coherent illumination of the transfer function are discussed.

In Zernike's phase contrast method for objects with large dimensions and high phase differences the image no longer looks like the object. The low spatial frequencies of the object's phase distribution lie in the pupil plane of the microscope objective in the space surrounding the image of the light source and because of the finite size of the phase ring will not be changed into intensities. For small phase differences the phase contrast function $B(R)$ is linear (Hauser 1962) and takes the form of a band pass.

For high object modulations and partially coherent or coherent illumination the transfer is nonlinear. Weingärtner and Menzel (1967) defined a pseudo-transfer-function $T(R)$ for this case in analogy to the modulation transfer function M.T.F.

$$T(R) = \frac{\text{Fourier transform of the image}}{\text{Fourier transform of the object}}$$

R is the spatial frequency. If an antisymmetrical object is used for measurement and it is transferred by a symmetrical optical channel there is antisymmetrical part of the image as well as a symmetrical one, and the latter can depend only on the non-linear transfer properties. If there is only an antisymmetrical part of the image the transfer function is not necessarily linear. The pseudo-transfer-function is split into two parts,

$$T(R) = h_1(A)E(R) + h_2(A)G(R)$$

where A refers to the object modulation, $G(R)$ describes that part of the transfer function which comes from the symmetrical part of the image intensity and $E(R)$ that which comes from the antisymmetrical part. For Zernike's phase contrast and microscopic bright field illumination (Hirscher, Menzel & Weingärtner 1968) it was shown that $E(R)$ describes only the linear part of the transfer. The factor $h_1(A)$ tends to 1 and the factor $h_2(A)$ tends to zero for small object modulations. This means that for Zernike's phase contrast method $E(R)$ will become identical with the phase contrast function $B(R)$ and for bright field illumination identical with the M.T.F.

For the observation of extended objects it is better to use the interference contrast method. In this method a homogeneous reference wave is superimposed on the object wave in an interferometer. If the wave-fronts are parallel, no interference fringes are generated in the image, but the intensity distribution of the image is approximately equivalent to the phase distribution in the object. The transfer function will have the shape of a band-pass. In interference contrast with coherent illumination the transfer function must be non-linear because in this case the transfer can only be linear with respect to light amplitudes, but we are interested in intensities.

To measure the transfer function the image intensity corresponding to the known phase distribution of the object must be recorded. Therefore an interferometer of Mach-Zehnder type was constructed which in contrast to the known apparatus works with the coherent light of a He:Ne laser. Figure 1 shows the principle of the optical arrangement. The mirrors $S1$ and $S2$ together with the beam-splitters $T1$ and $T2$ form the actual Mach-Zehnder interferometer. From a plane wave incident at $T1$ the microscope objectives $O2$ and $O3$ produce spherical waves, which interfere behind the beam-splitter $T2$. If the distances of the foci of $O2$ and $O3$ with respect to the beam-splitting plane of $T2$ and the angle of incidence are made equal, the interference fringes disappear and an equally illuminated field of view arises.

A phase object OB in the object plane is transformed by this arrangement into an intensity image, which can be observed by means of the mirror $S4$ in the eyepiece $OK2$ or projected by the eyepiece $OK1$ to the slit SP. The objective lens $O1$ with the pinhole $B1$ works as a spatial filter in the same way as $O3$ and $B3$. An additional arrangement—indicated by broken lines—illuminated by a conventional source and a ground glass screen makes it possible to illuminate the object incoherently. In order to record the image intensity with minimum aberration the object table is moving with a constant velocity of 1 mm/h, keeping the small 12 μm scanning slit fixed. The light intensity is measured by a multiplier PEV through the prism PR. The alternating part of the signal is registered by a recorder.

The use of coherent laser light has the advantage that an intense light source is available, which can be directly modulated. The microscope objectives do not have to be identical nor is there any need for a similar microscope slide to be put into the reference beam to equalize the paths, as is the case in apparatus using partially coherent light.

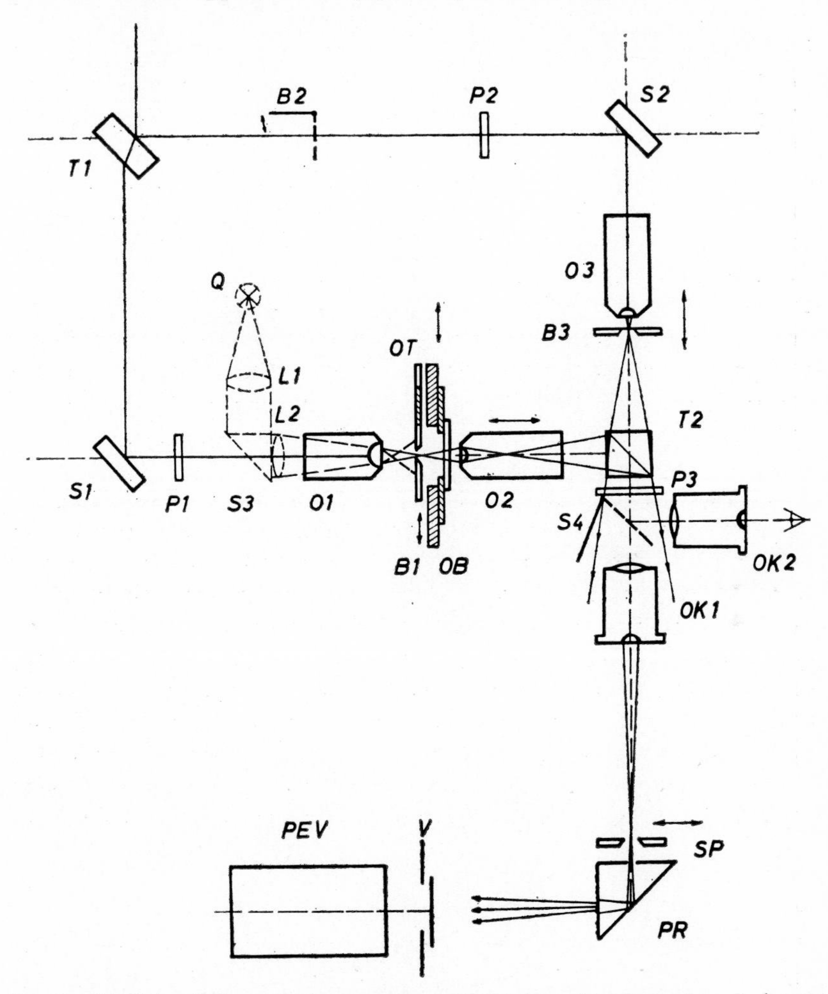

FIG. 1. Principle of the optical arrangement of the laser interferometer: *L*, lenses; *Q*, light source; *P*, polarizers to equalize the light amplitudes in the two interferometer paths; *V*, optical shutter. (The other abbreviations are explained in text).

The objects used were phase edges, the path difference ϕ of which were measured with a three-slit interference microscope. Errors in the calculation of the transfer function from measurements arise from three causes: the exact shape of the edges used is not exactly known; a slight defocusing has a large effect on the intensity distribution in the image; and there is an uncertainty in the position of the phase jump in the object corresponding to the intensity distribution of the image.

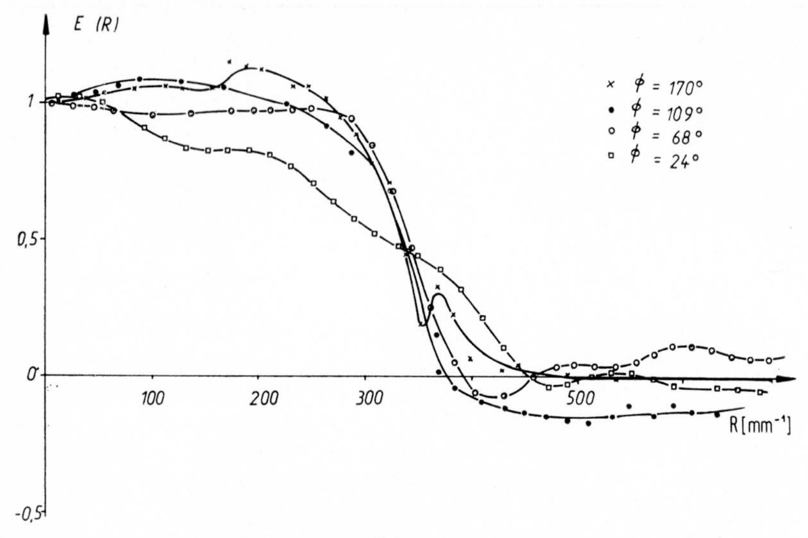

FIG. 2. Functions $E(R)$ with phase difference ϕ of the phase edges as parameter.

The calculations were carried out on an IBM 7040 computer. Figure 2 shows the functions $E(R)$ due to the antisymmetrical and normalized part

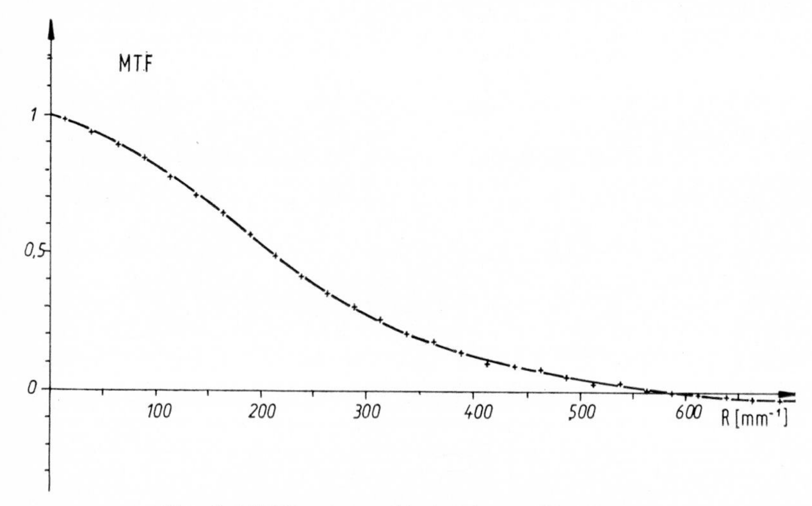

FIG. 3. M.T.F. measured in incoherent illumination.

of the image for different object phase angles ϕ. These functions are not independent of modulation A with respect to ϕ, which means that $E(R)$ in interference contrast, contrary to Zernike's method, includes not only the linear transfer but also non-linear parts. For small phase differences ϕ for which $E(R)$ is identical with the phase contrast function $B(R)$, it will correspond, except for a constant, to the M.T.F. shown in Figure 3, which is measured under incoherent illumination in the apparatus described above.

REFERENCES

HAUSER, H., 1962, *Optica Acta*, **9**, 121 and 141.
HISCHER, H., MENZEL, E. and WEINGARTNER, I., 1968, *Optik*, **27**, 42.
WEINGARTNER, I. and MENZEL, E., 1967, *Optik*, **26**, 442.

Part V.

Image forming systems of essentially novel design

New Developments in Photographic Objectives

E. Glatzel

Carl Zeiss,
7082 *Oberkochen,*
West Germany

Abstract—There are many possible incentives for developing new optical systems. The most important are:

1. human interest in research;
2. utilization of new possibilities offered by modern technology;
3. utilization of new techniques in optical computation;
4. more extreme requirements of the users.

They are illustrated here by means of examples. The examples refer primarily to one of the above incentives, although in practice there are always several which stimulate new lens development.

1. HUMAN INTEREST IN RESEARCH

The oldest and most important reason is human interest in research for its own sake. It induces men to follow up an idea without worrying about a possible application.

The development of a triplet with negative refractive powers in the outer lenses may serve as an example. It has long been known from the

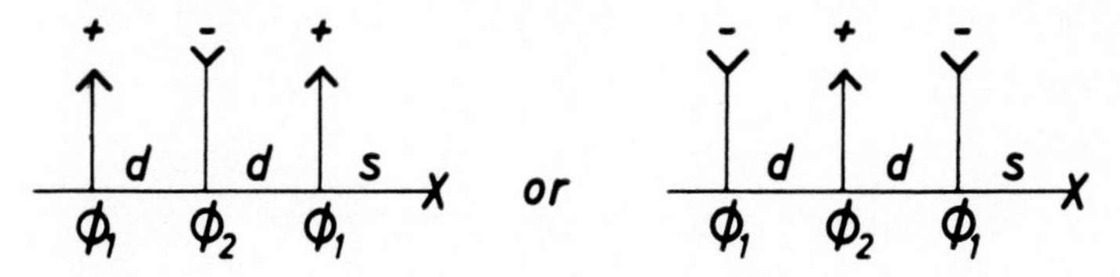

FIG. 1.

theory of thin lenses that this solution is possible. Let us now derive this solution. Since we are only interested in the basic principle, a number of simplifications are permitted. We set up a symmetrical triplet with identical refractive indices for all three lens elements (Figure 1). There are three unknowns: two refractive powers (ϕ_1 and ϕ_2) and one lens separation (*d*). The fourth unknown, the refractive index, is eliminated when the Petzval sum is reduced to zero. We now impose the conditions:

1. The equivalent focal length is normalized to 1.
2. The back focal distance will be supposed fixed. It would, however, not complicate the problem if this back focal distance were included in the computation as a parameter *s*.
3. The Petzval sum is to be eliminated.

We cam now set up the following three equations for the determination of the three unknowns:

$$\phi_1+\phi_2+\phi_1-d\phi_1(\phi_2+\phi_1)-d\phi_1(\phi_1+\phi_2)+dd\phi_1\phi_2\phi_1=1$$
$$1-d\phi_1=d(\phi_1+\phi_2)+dd\phi_1\phi_2=s$$
$$\phi_1+\phi_2+\phi_1=0$$

This mathematical problem can be solved explicitly. We obtain a quadratic equation and thus two solutions (Figure 1).

$$\phi_1=\pm\sqrt{\frac{1-s}{2}}\cdot\frac{1}{d}$$

$$\phi_2=-2\phi_1$$

$$d=\pm(1-s)\sqrt{\frac{1-s}{2}}.$$

Apart from triplets with *positive* refractive powers in the outer lens elements there are triplets with *negative* refractive powers in the outer lens elements. Let us now deal with the two solutions for a fixed value of *s*.

for example: s= 0.878

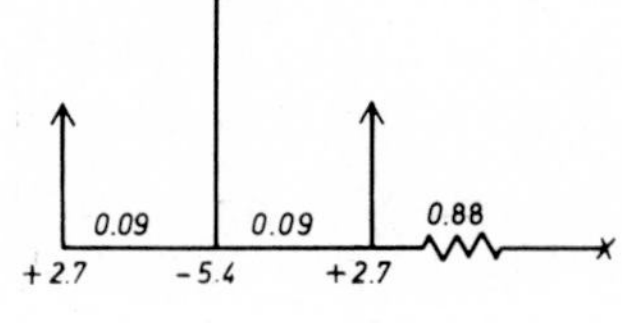

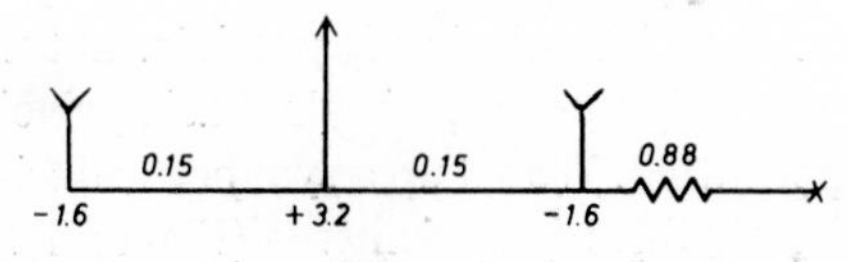

FIG. 2.

Figure 2 shows the separations and refractive powers for $s = 0{\cdot}878$ to full scale. These results apply to thin lenses, without regard to lens bendings.

For the format of 35 mm. photography it is known that it is possible to obtain apertures up to f/2·8 and angular fields up to $\pm 25°$ for triplets with positive outer lens elements; provided, however, the lens bendings are 'normal' and the lens thicknesses are as small as possible. The Petzval sum can only be further improved by using very high refractive indices. The correction is never perfect.

The situation is different with triplets with negative outer lens elements. Here, meniscus-shaped bendings and great thicknesses are permissible and necessary for the outer lens elements for the correction of the spherical and oblique spherical aberrations. This means such a gain in the Petzval sum that it can be corrected at will for triplets with negative outer lens elements. It is possible to obtain angular fields up to $\pm 60°$ at f-numbers between 8 and 5·6 with this type of lens.

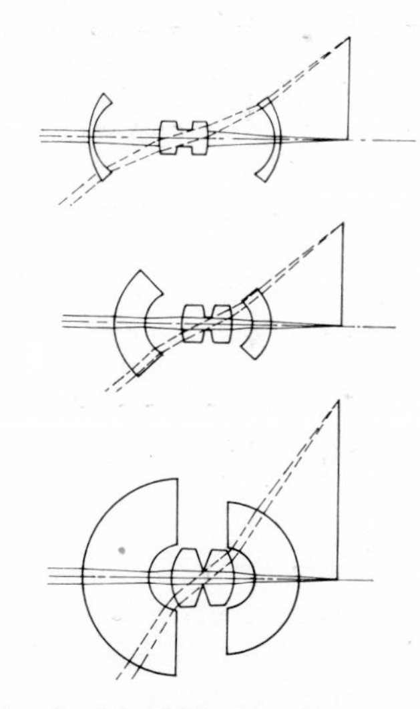

FIG. 3. Zeiss Hologon: Design steps.

The disadvantage of this new triplet type is that the longitudinal chromatic aberration cannot easily be corrected. As can be seen in Figure 2, the individual refractive powers are considerably smaller than for the other solution. This remains true even when the back foci are not the same for the two solutions. The low values of the individual refracting powers bring considerable advantages but render the correction of the longitudinal chromatic aberration more difficult.

Because of the symmetrical design, distortion, chromatic difference of magnification and coma can be completely corrected. Correction of astigmatism is possible by adjusting the thickness of the centre lens. Figure 3 shows the correction process. We started from an arbitrarily selected system and tried at first the correction for f/11 and $\pm 40°$ with considerable vignetting, since we had no idea for what aperture and field this triplet type would be suitable; we had to formulate the problem during the design. Of course, it was clear that this type of lens would be more suitable for large angular fields rather than for large relative apertures, but the optimum balance between these was not known.

From the point of view of the optical designer the correction process

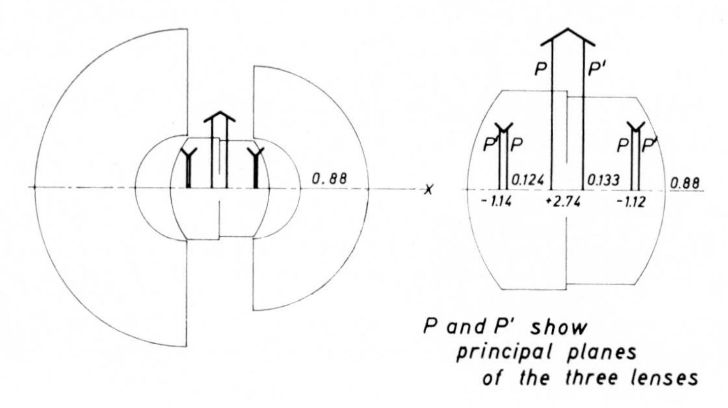

In practice: Petzval sum = -0.030

In theory: $\frac{\phi_1}{n_1} + \frac{\phi_2}{n_2} + \frac{\phi_1}{n_1} = 0.345$

FIG. 4. Zeiss-Hologon: In practice and theory.

was very long, especially since the range of linearity of the aberration derivatives relative to the parameters was very small. This applies in particular to the third lens section in Figure 3. Here the correction was made for $\pm 55°$ at f/8 without vignetting. This seems to be the most favourable balance. We wished to produce a design with acceptable illumination in the corners without a correction filter. Furthermore, one must bear in mind that a stop is not acceptable as it lies in glass. From the point of view of aberrations, this lens type can also be corrected for $\pm 60°$ at f/5·6. The correction process showed that all aberrations, including higher orders, can be corrected, except for the oblique spherical aberration. The applications of this new objective type are limited by this aberration alone.

Figure 4 shows a comparison of theory and practice. The front and back principal planes of the three lenses, the refractive powers and principal plane distances are entered in the lens section in such a way that they can be compared with the values from the theory of thin lenses (cf. Figure 2). The objective itself is no longer exactly symmetrical and as the Hologon f/8, 15 mm it is manufactured for a 35 mm. camera (Deutche Patentschrift 1 241 637).

Figure 5 shows the diapoint diagram of this objective. We have introduced spherical undercorrection to compensate for the oblique spherical over correction.

More or less complex designs derived from this basic principle are well known and are used as distortion-free wide-angle objectives. The Hologon can be considered as the progenitor of a whole group of such lenses.

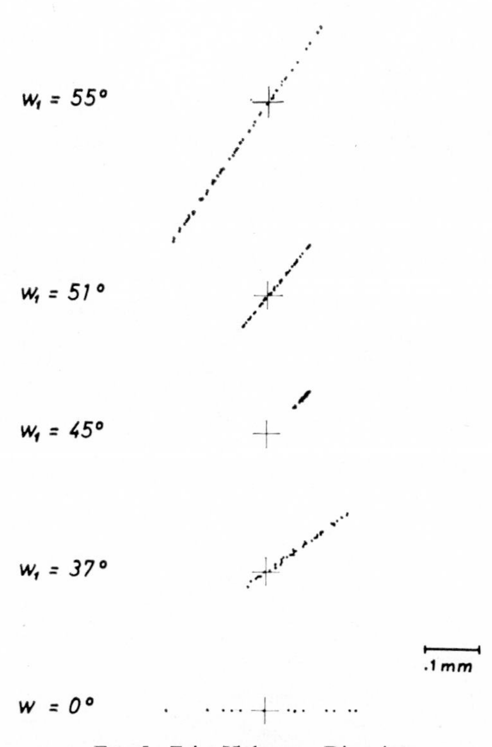

FIG. 5. Zeiss-Hologon; Diapoints.

Figure 6 shows the Hologon together with a number of these wide-angle lenses. These are, from top to bottom: Russar (Roossinov 1946), an objective designed by Rudolf Steinheil in 1901 (British patent 21 211), Pleogon (Richter 1955), and S-Pleogon (Roos, Winzer 1968). On the right side, from top to bottom: Super-Angulon (Klemt 1954), Biogon (Bertele 1951), Aviogon (Bertele 1950), and Super-Aviogon (Bertele 1956). The Hologon (Glatzel, Schulz 1966) is in the centre.

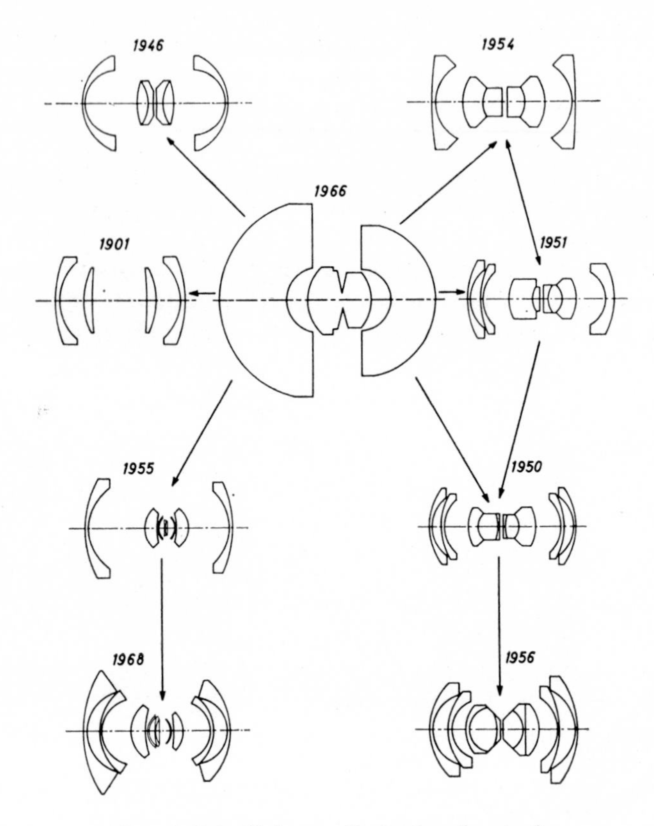

FIG. 6. Zeiss-Hologon; Derivatives (see text).

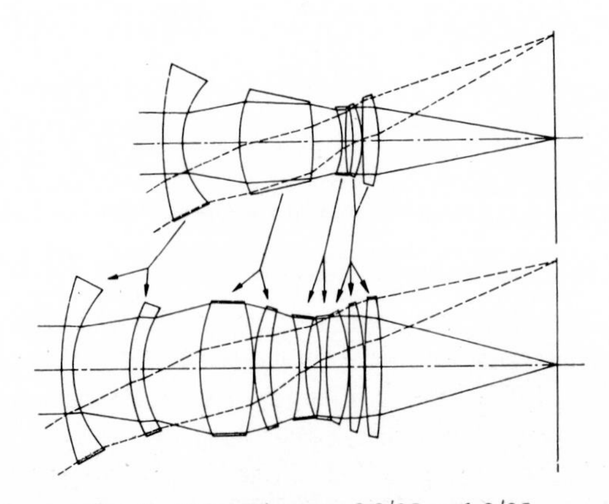

FIG. 7. Zeiss-Distagon 2.8/35 and 2/35.

2. UTILIZATION OF NEW POSSIBILITIES OFFERED BY MODERN TECHNOLOGY

2.1. *Anti-reflection Coatings*

The possibility of providing optical surfaces with anti-reflection coatings opened up completely new fields for the designer of optical systems: there is now practically no limit to the number of lens elements in an optical system. I would like to explain by means of two examples how the performance of lenses can be increased by means of this procedure. Retro-focus lenses, which have gained considerable importance in the course of the last fifteen years, illustrate this development. These were the objective types which made the use of mirror-reflex cameras possible. Two examples are given. In one the f/number is increased and in the other the angular field.

Figure 7 shows Distagon lenses with focal lengths 35 mm and angular fields $\pm 32{\cdot}5°$ for the 35-mm format. The f/number of the upper lens section is 2·8. This aperture can just be attained with 5 lens elements. If

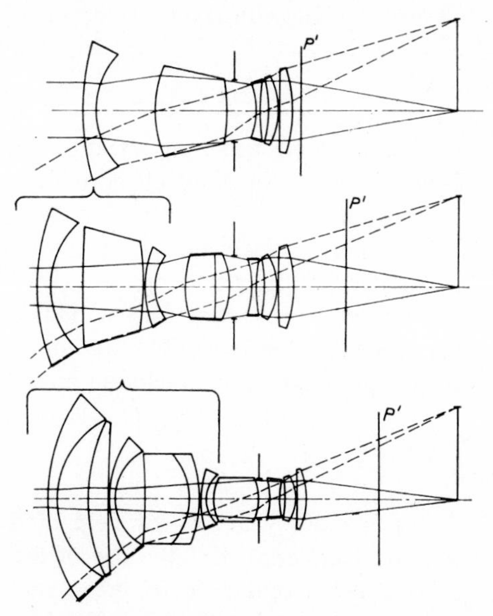

FIG. 8. Zeis-Distagon 2·8/35, 2·8/25 and 4/18.

the individual refractive powers of these lenses are spread over several lens elements the difficulties with regard to the spherical aberration can be reduced. It is then possible to reach an f/number of 2 and excellent correction with 9 lens elements.

Figure 8 demonstrates the increase in angular field attainable by the use of an increasing number of lens elements for the retro-focus objectives.

The upper lens section is identical to that of the previous illustration. It is just possible to attain an angular field of $\pm 32{\cdot}5°$ with *one* lens element in the dispersive front component. The rear principal plane lies just behind the last lens surface. With a focal length of 35 mm there is just sufficient space for moving the mirror. For an increase in the ratio of back focal distance to focal length (which means further shift of the rear principal plane) the influence of the front negative refractive power must be increased. This shift of the rear principal plane is required if the space for the mirror motion is to be maintained and a larger angular field is to be covered simultaneously. The diameters can be kept small only by an increase in the front negative refractive powers. In order to avoid large incidence angles, which would increase the aberrations, the required negative refractive power must be spread over several lens elements. This has been done (the lens section in the centre and, even more marked, in the lower one).

In the lens section in the centre the negative refractive powers were spread over *two* dispersive lens elements. For better correction of distortion and chromatic difference of magnification, a collective lens element has been introduced between the two. We obtained an angular field of $\pm 40°$ at a focal length of 25 mm.

The lower lens section shows *three* dispersive front lens elements with one single and one cemented collective component inserted between them. The collective components have the same effect as in the lens section in the centre. The cemented surface is required in order to correct at these large angular fields higher-order aberration in the chromatic difference of magnification. The angular field thus obtained is $\pm 50°$ which results in a focal length of 18 mm for the 35-mm format. However, whereas the f/number in the two upper lens section is 2·8, the f/number is only 4.

Note the amount of glass material contained in these lenses.

2.2 *Aspheric Surfaces*

In the course of the last few years a new field of research has been opened up for the optical designer, i.e., the use of aspheric surfaces in 'standard lenses'. This new freedom is of great value since aspheric surfaces can be applied to influence either a single aberration or a group of aberrations. A large matrix is thus split up into several smaller ones whose consequences can be more readily understood.

It is, for instance, possible to use an aspheric surface in the vicinity of the diaphragm so that it affects spherical aberration alone. This aspheric surface can be designed in such a way that it eliminates the aberrations of all orders.

If an aspheric surface is provided immediately in front of the image plane or a long way in front of the stop, it influences above all the distortion. There are many possible solutions between these two extremes.

The use of two aspheric surfaces arranged approximately symmetrical to the diaphragm offers particularly interesting possibilities.

With this arrangement it is possible to influence both the axial spherical aberration and the field aberrations (coma and oblique spherical aberration).

Figure 9 gives an example of latest developments. The first and the last surface of a double Gauss lens are aspheric surfaces. This lens has become well-known under the designation Leitz-Noctilux. Consisting of only 6 lens elements, this lens has an f/number of 1·2 and an angular field of ±23° (British Patent 1 088 192).

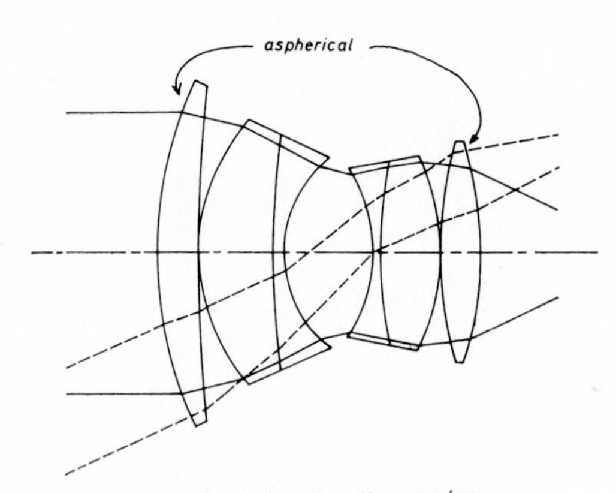

FIG. 9. Leitz-Noctilux 1,2/50.

2.3. *New Glass Types*

In the course of the last decades new fields have been opened up for optical designers by the development and manufacture of new types of optical glass. The new LaSF and SF glass types with refractive indices between 1·85 and 1·96 and Abbe numbers between 40 and 20 are types which have been marketed only recently. Let us take again the Noctilux as an example. It consists of three types of glasses in the aforementioned range.

Today, optical designers are in a position to increase, for a number of photographic objectives, the aperture by another 1/2 stop without having to increase the number of lens elements, merely by using these highly refractive glasses.

It is impossible at the moment to make a judgement on the future development and expansion in this specific field, but a judgement is even more difficult with regard to the applications of special glass types with abnormal partial dispersions and crystals, e.g., fluorite in photographic objectives. Unfortunately I cannot deal with this topic here because it would go beyond the scope of this paper.

3. UTILIZATION OF NEW TECHNIQUES IN OPTICAL COMPUTATION

The enormous progress in the computation of optical systems has only been possible by making use of electronic computers. New techniques of computation have enabled the theoretical possibilities—especially the use of large numbers of lens elements—to be fully utilized in a manner which

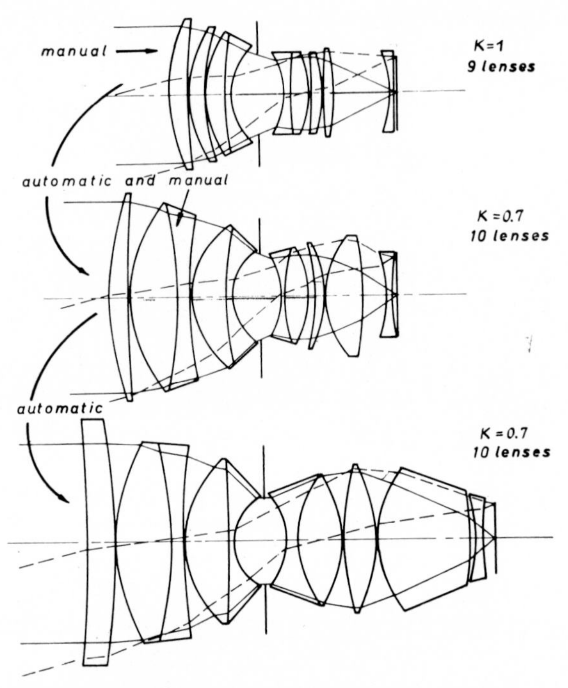

FIG. 10. Zeiss-Planar 0·7/50 ($\pm 15^{\circ}$); Development.

was not possible without computers. I would like to emphasize the following three advantages of automatic correction.

We are in a position to carry out research on a very wide front: The optical designer is able to try out many of his ideas, and changes his starting system frequently. He gets the results while the ideas are still fresh. It is possible today to undertake systematic research of objective types and/or problems in lens computation.

It is possible to achieve optimum correction for optical systems: Large matrices, even if they contain thousands of elements which change continuously, can be fully exploited with electronic computers. Even when well trained, the efficiency of the human brain is limited in this respect.

Gathering of experience and training in finding most favourable starting systems are speeded up and become more reliable: This is the consequence of the two above-mentioned advantages. In addition to this, automatic programmes can recognize and indicate unknown correlations of aberrations and parameters. In spite of this, automation covers only part of the entire problem. It is absolutely necessary to carry the development of automatic programmes further.

I would now like to illustrate by means of an example the correction

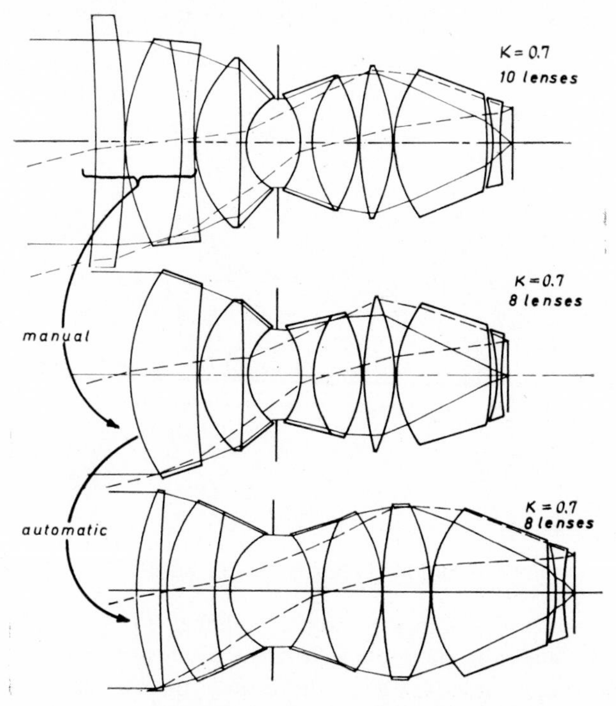

FIG. 11. Zeiss-Planar 0·7/50 (±15°): Development.

procedure for a lens by means of a method (Glatzel 1961, Glatzel & Wilson 1968), I have developed.

The task was to compute a lens with an f-number of 0·7 and an angular field of ±15° for a focal length of 50 mm. The first attempt did not, however, achieve the desired result in one computer pass. The initial system and the changed systems emerging during the optimization were not always appropriate to the problem. The resulting objective remained a well-known type, but it is just for these reasons that it is well suited for demonstration purposes.

For the starting system we had considered that for this angular field a double Gauss lens would be most suitable; and because of the aperture we had to distribute the positive refractive powers over several lens elements. We extended the system by an image flattening lens. This arrangement is shown in the upper lens section in Figure 10. The first automatic correction was at first limited to an f/number of 1·0, but we reduced the incidence angles of the central bundle in order to be able to use a higher relative aperture later. The result corresponds approximately to the lens section in the centre of Figure 10. However, a cemented surface was introduced by hand into the second lens from the front which had to be made thicker on this account. Due to this provision for improved

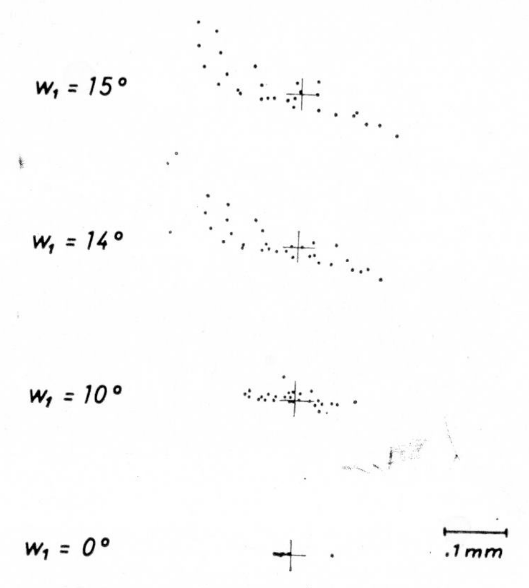

FIG. 12. Planar 0·7/50; Diapoints

correction, it is now possible to use the f/number 0·7. We hoped that this cemented surface in front would relieve the dispersive surface behind the diaphragm space with respect to the central bundle. At first the total negative contribution of the Seidel sum I lay on this surface. However, in the second automatic correction step this contribution automatically distributed itself uniformly over the two dispersive surfaces surrounding the diaphragm space but did not appear at the cemented surface we had introduced for this purpose. On the contrary, this surface was used less and less and had finally (lower lens section in Figure 10 and upper lens section in Figure 11) no influence at all. The first lens element, too, has no effect except for a small share in the Seidel sum II. We could thus remove the cemented surface and the first lens element. This was done as is shown in the lens section in the centre of Figure 11. With the last automatic correction step we could relieve the incidence angles of the objective somewhat more, partly made possible by reduction in lens thickness at the front. The result is shown in the lower lens section in

Figure 11. It is the well-known X-ray screen objective of the type from Jena which permitted in that design, however, an f/number of only 0·85. In spite of the higher relative aperture the diaphragm space is now much larger and all spaces behind the diaphragm space are filled with glass.

Figure 12 shows the diapoint diagram of the new Planar f/0·7, 50 mm. The deformation error is clearly recognizable.

Figure 13 shows the geometrical-optical modulation transfer functions for integral illumination [daylight as defined elsewhere (Glatzel 1967)].

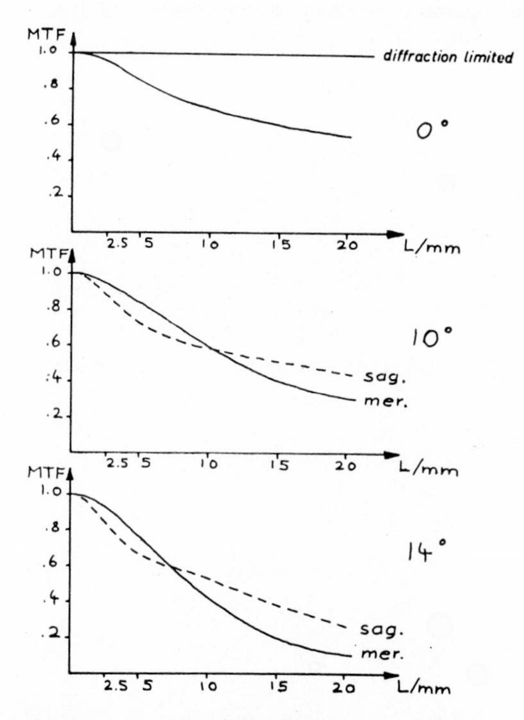

FIG. 13. Planar 0·7/50; MTF for integral light.

Finally I would like to refer to a trend which has accompanied automatic correction. Modern objectives which are the result of automatic correction contain much more glass than previous types. I have just illustrated this by means of the Planar f/0·7, 50 mm. The same applies to the retro-focus lenses and to the Hologon. This fact may perhaps prove that with automatic programmes results are obtainable which were never possible before. With the previous form of correction it was often impossible to notice that greater glass thickness would bring about improvements at a certain point in the design, because these improvements were much too small to be realized. Automatic programmes, however, can recognize and utilize every possible freedom allowed by the boundary conditions.

4. MORE EXTREME REQUIREMENTS OF THE USERS

These requirements increase so considerably and are so manifold that it is possible only to deal with a few of them.

4·1 *Modified Double Gauss Objectives*

This objective has been the subject of intensive research in the course of the last few years. Many modifications have been suggested. One suggestion

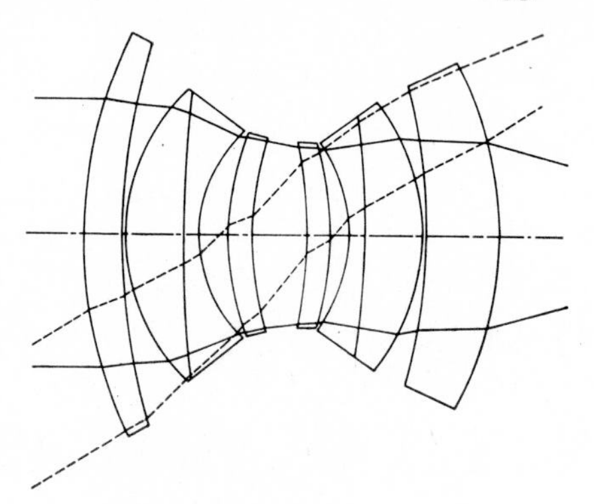

FIG. 14. Leitz-Summicron 2/35.

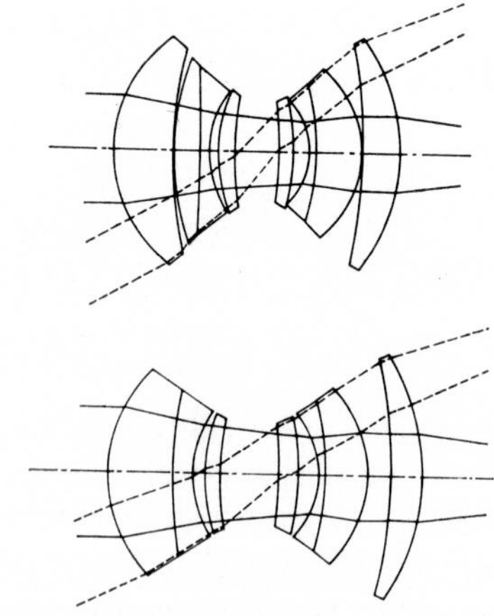

FIG. 15. Zeiss-S-Orthoplanar 4/50 and 4/60.

was made by W. Mandler (U.S. Patent 3 003 249) and is shown in Figure 14. This modified type combines the advantages of two older objective types: the standard double Gauss lens and the Orthometar. The result is a new objective type which is suitable for relatively large f/numbers at relatively large angular fields. This lens has become known under the name Leitz-Summicron 2/35 for 35-mm photography. However, it also meets the requirements for high resolution at f/numbers of 4 and angular fields up to $\pm 30°$, for instance in micro-documentation.

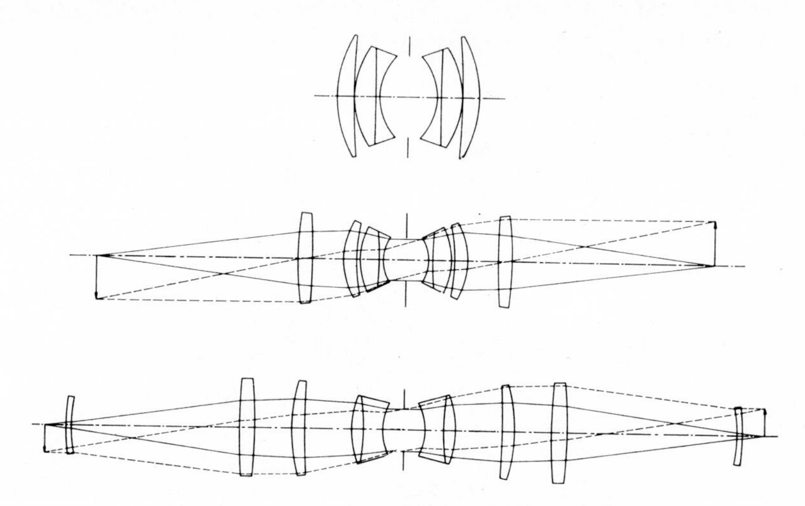

FIG. 16. Development of lenses for projection printing, 1 : 1.

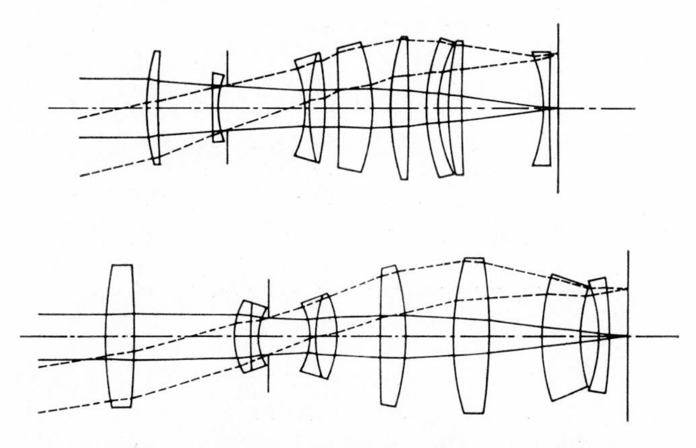

FIG. 17. Lenses for projection printing 1 : 20 and 1 ; 2.

Figure 15 shows the *S*-Orthoplanar lenses f/4, 50 mm (upper lens section) and f/4, 60 mm (lower lens section) which are specially corrected for this purpose. They yield a resolution of 120–150 line-pairs/mm over the entire field at an aperture of f/5·6, for which the theoretical resolution is 300 line-pairs/mm.

However, projection masking requires today even higher resolutions.

Modified double Gauss lenses are specially suitable for this purpose at magnifications of about 1:1.

Figure 16 illustrates one of the possible design approaches. Starting from a symmetrical system, a positive lens element is introduced in front and behind the objective. They are placed at considerable distances from the lens. The lens section in the centre shows the *S*-Planar f/2,

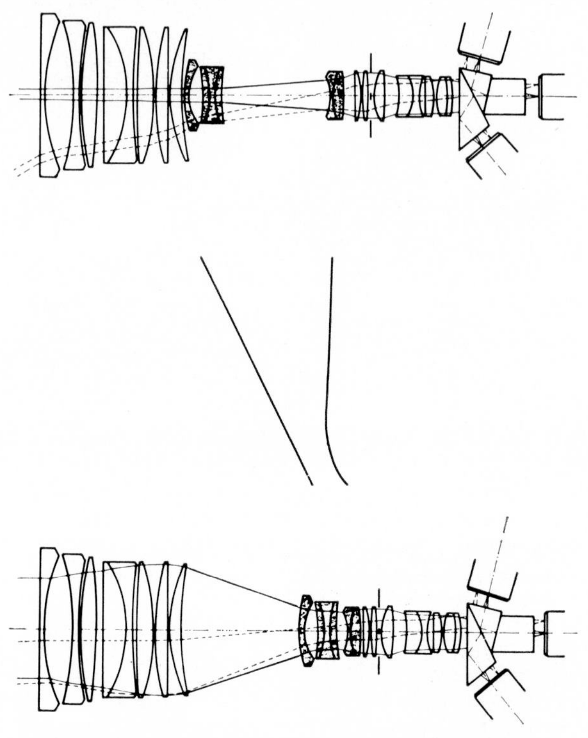

FIG. 18. Schneider-TV-Variogon 1 : 2·1/18–200 mm.

160 mm, *NA* =0,125. This lens type can be corrected so that 200 to 250 line-pairs/mm are resolved (at least monochromatically) over image fields up to 70 mm diameter. The theoretical resolution limit in this case is 570 line-pairs/mm. The lower lens section shows an objective from the next stage of the design development. Here a field flattening lens was added both in the vicinity of object and of the image. This objective type can, for instance, be corrected so that within an image field of 58 mm diameter the aberrations remain below the Rayleigh limit for *NA* =0·125, which means that in practice structures of about 3 μ can be reproduced.

However, for projection masking at reduction ratios of 1:2 completely asymmetrical arrangements are much more suitable. Figure 17 shows two lenses for such purposes: the upper lens section is the *S*-Planar f/4, 190 mm, for an image circle diameter of 90 mm at a magnification of 1:20 and the lower lens section the *S*-Planar f/2·8, 125 mm, for an image circle diameter of 50 mm at a magnification of 1:2.

4·2 *Zoom Lenses*

Let me select two items from the great number of new designs. Figure 18 shows the Schneider-TV-Variogon f/2·1, 18–200 (lens section) (Deutsche

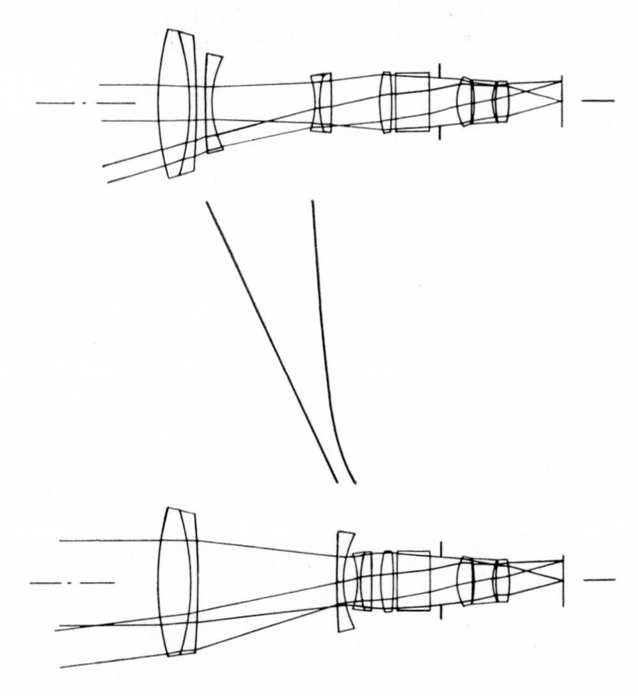

FIG. 19. Zeiss-Vario-Sonnar 1·9/12–30.

Auslegeschrift 1 279 962). This lens is corrected for colour TV and a Plumbikon-format. It can also be used for studio photography in cramped surroundings. Owing to the special design of the front component a distance setting down to 0·7 m is possible, whereby the front lens diameter is kept relatively small. The first three lens elements are shifted for focusing. For this reason these three lens elements are achromatized.

The 'dernier cri' is represented by small lenses for pocket cameras. Figure 19 illustrates the design of a Vario-Sonnar f/1·9, 12–30. With this lens a front lens diameter of 24 mm can be obtained for the super-8 format. This success is the result of special arrangements for the moving elements, so that the cemented surface required for chromatic correction could be displaced from the first moving component into the second one, thus reducing the diameter.

4·3 *Asymmetrical Wide-Angle Designs*

Figure 20 shows an unusual design of this type, namely a wide-angle lens for the 3·7 m bubble-chamber at CERN with an angular field of $\pm 53°$ and a

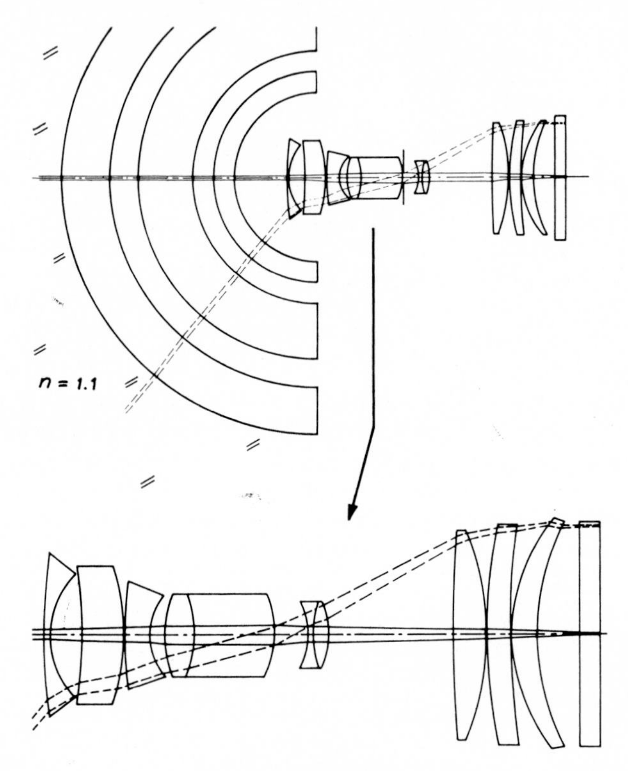

FIG. 20. Zeiss-S-Distagon.

telecentric ray path of the principal rays in the image space. The bubble-chamber is imaged through three concentric meniscus windows. This lens type is related to both the retro-focus and the fish-eye objectives. At an angular field of $\pm 50°$ the distortion is 20 per cent, but the f-number is only 11, because of the required depth of focus.

4·4 *Quasi-Symmetrical Wide-Angle Lenses*

There is a steadily increasing requirement for wide-angle lenses with high relative aperture, distortion correction and angular fields of about

±45°. Bertele (Schweizer Patentschrift 449 995)) has recently made a considerable step forward in this respect. Figure 21 shows the lens sections of two versions of this new lens type, namely lenses with f/numbers 2·8 and angular fields of ±45°. For the correction of the fifth-order spherical aberration both contain a thin almost concentric air lens between the second and third lens elements. In the upper lens section this air lens is much more strongly bent with respect to the central pencil than in the lower one, whereas in the lower lens section the incidence angles of the oblique bundles are higher.

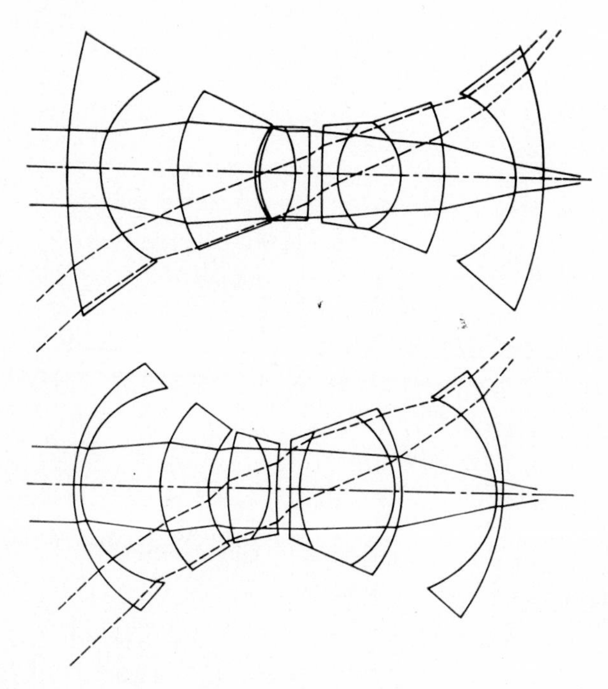

FIG. 21. Zeiss-Biogon 2·8/38.

4·5 *UV- and IR-lenses*

Let me now refer briefly to the lens designs for other spectral ranges. In the UV range the optical designer has to manage with relatively small refractive indices (below 1·5) and small differences in dispersion (e.g., quartz glass and fluorite).

In the IR range the refractive indices are very high (up to 3 or even 4). Consequently, the available objective designs so assiduously investigated and developed in the course of the last decades can rarely be used for these new spectral ranges. On the other hand, completely new fields are opened up for research and development.

As an example of UV-objectives let me consider the design of the UV-Sonnar f/4, 100 mm, made of quartz glass and fluorite, with an angular field of ±20·5°. The basic type permits the chromatic correction of several components of the objective during automatic correction. Since the Petzval sum cannot be corrected by an increase in refractive indices,

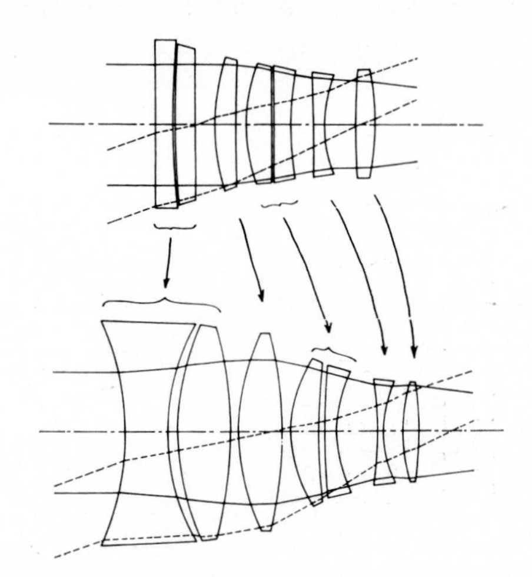

FIG. 22. Development of UV-Sonnar 4/100.

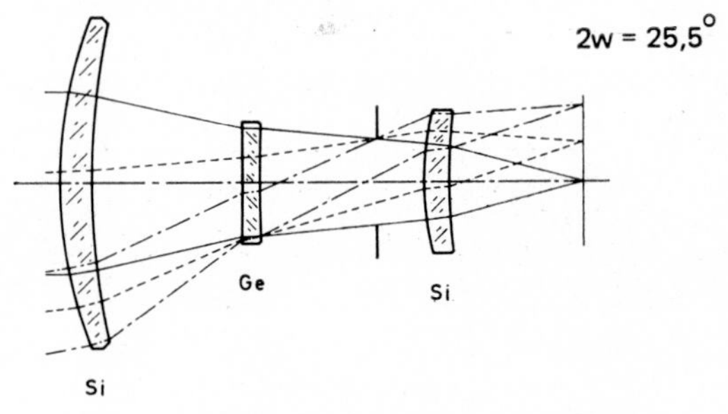

FIG. 23. Rodenstock-IR-Objective 1·9/57.

we provided in the arrangement for the possibility that lens components may bend to a meniscus-shape, if this is required for the correction of the Petzval sum.

Figure 22 shows in the upper lens section the starting system and in the lower one the final result. All correction parameters we introduced have been fully utilized. The objective can be used for the wavelength range between 210 and 650 nm (Glatzel & Scheid 1970). It is manufactured as the UV-Sonnar f/4·3, 105 mm, for the 6×6 format. We hope that it will prove its worth in space applications.

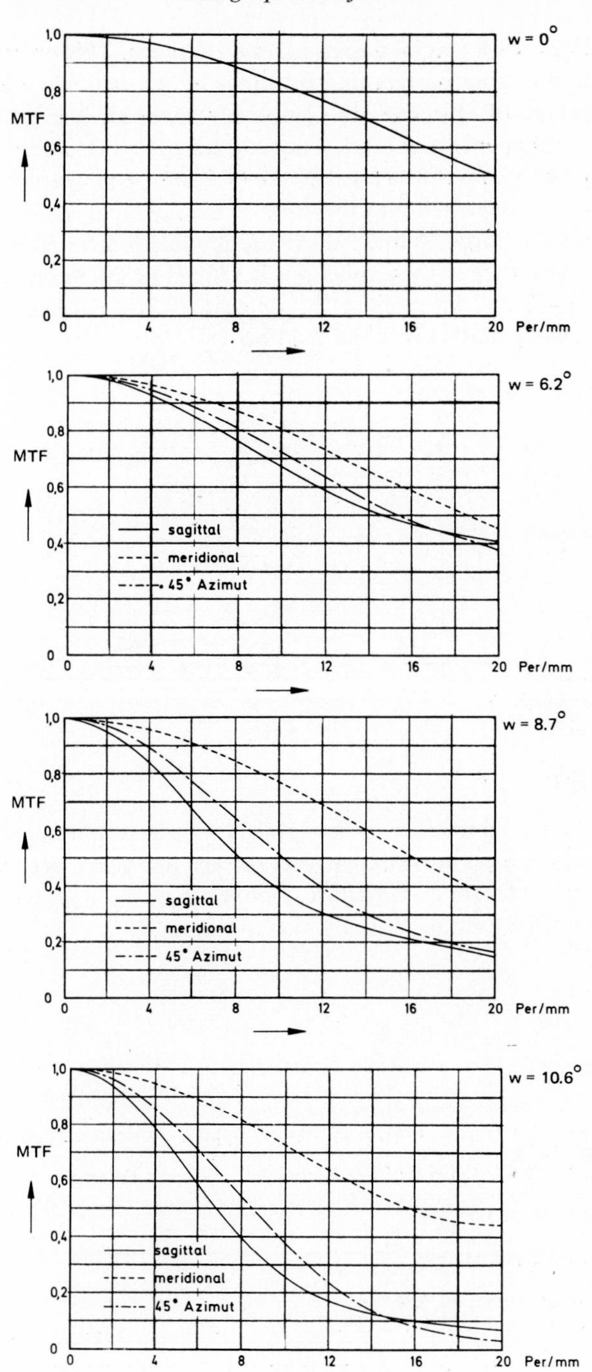

FIGS. 24, 25, 26, 27. Modulation Transfer Functions IR-Objective 1·9/57 (see Fig. 28 p. 248).

Figure 23 shows a special design for the IR-range. This example shows us how the refracting powers and bendings of the individual lenses in a well-known type of objective are completely changed. It is quite easy to reach a relative aperture of f/1·9. The field here is $\pm 13°$ and the spectral range from 2 μ to 2·7 μ .For a focal length of 57 mm, a resolution of 20 line-pairs/mm is obtained well into the zone of the field and 12 line-pairs/mm in the corners. Figures 24 to 28 show the modulation transfer functions. The 'white light' used in these functions is defined by the three wavelengths 2·1 μ, 2·7 μ and 3·3 μ, with weights 60 per cent, 20 per cent and 20 per cent (Deutsche Offenlegungsschrift 1472 189).

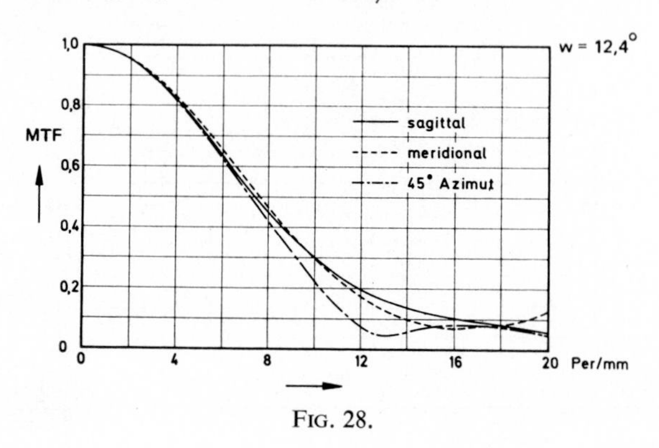

FIG. 28.

CONCLUSION

I have described a number of new objectives which represent, however, only a tiny fraction of those designed in the last few years. We may expect a further rapid increase in the future. I hope I have been able to give you a glimpse into the exciting future which awaits us.

REFERENCES

Deutsche Patentschrift 1 241 637.
British Patent 1 088 192.
GLATZEL, E., 1961, Ein neues Verfahren zur automatischen Korrektion optischer Systeme mit elektronischen Rechenmaschinen. *Optik*, **18**, 577.
GLATZEL, E. and WILSON, R., 1968, Adaptive Automatic Correction in Optical Design, *Appl. Opt.*, **7**, 265.
GLATZEL, E., 1967, UV-Objektiv 2/50, Bildwinkel = 30°, aus Quarzglas und Flussspat. Korrektionsgang und Korrektionszustand, *Optik*, **26**, S. 411, p. 420.
United States Patent 3 006 249.
Deutsche Auslegeschrift 1 279 962.
Schweizer Patentschrift 449 995.
GLATZEL, E., und SCHEID, G., 1970, UV-Sonnar 4/00, Bildwinkel=41°, aus Quarzglas und Flussspat, *Optik*, **30**, 354.
Deutsche Offenlegungsschrift 1 472 189.

A Unit-Power Telescope for Projection Copying*

C. G. Wynne

Optical Design Group,
Imperial College of Science and Technology,
London, S.W.7.

Abstract—For projection printing at unit magnification, e.g., in making micro-circuits, resolution requirements in some cases necessitate numerical apertures of 0·35 or greater, and the need to conserve scale to close limits makes it desirable that the system should be telecentric at the object and image, i.e. that the system be a unit-power telescope.

Any set of concentric refracting or reflecting surfaces whose Petzval sum is zero is paraxially telescopic, and such a system can give very high aberration correction over an extended field of view over quite a wide spectral range. One such system consists of a spherical mirror, a meniscus lens and a plano-convex lens, the curved surfaces all being concentric and the plane surface, through the common centre of curvature, lying at the object and image planes. These can be separated by an inclined beam-splitting surface in the plano-convex lens. If the curvatures be chosen to make the Petzval sum zero, then all primary aberrations are zero, and in addition there is perfect correction on axis, and for sagittal aberrations, of all orders. Chromatic difference of Seidel tangential field curvature can be corrected by an appropriate choice of the dispersions of the two glasses. An appropriate choice of their refractive indices gives a high level of correction of spherical aberration of principal rays (and hence of telecentricity), and this condition at the same time corrects the main higher order meridian plane aberrations of oblique imagery, and also makes the primary aberrations substantially invariant with conjugates.

An air space may be required between the optical system and the object and image planes. The aberrations of a plane parallel air space in this (telecentric) region may be corrected by using a thinner plano-convex lens. and adding behind its plane surface a further plane parallel glass plate of higher refractive index and dispersion.

An example is given of a system covering a field of 3·8 cm diameter, with aberrations within $\pm 0{\cdot}06$ wave lengths over this field monochromatically and within $\pm 0{\cdot}10$ wave lengths over the spectral range 405 to 546 mm.

In the manufacture of micro-circuit systems a silicon slice coated with a photo-resist is printed with an array of masks, and then processed, and a

* The National Research Development Corporation have made patent applications covering the systems described in this paper.

sequence of ten or more such exposures, followed by processing, may be needed to build up the final circuits. The silicon slice is typically 3 to 4 cm in diameter and has many hundreds of identical circuits formed upon it; lines of one or two microns width must be resolved, and registration between successive masks is needed to micron precision, all over the silicon slice. The individual circuit masks are first reduced photographically from large master drawings to an intermediate size, and then subjected to a final reduction (typically to 1 or 2 mm square) in a step and repeat camera, which gives an array covering the silicon slice. The final transfer of these arrays of masks is generally made by contact printing. Because of problems of registration, and of wear of the printing masks, it would be better if projection printing could be used in this final set of operations, instead of contact printing. This paper is concerned with unit magnification optical systems for this purpose.

The optical requirements are rather severe. To allow for the likely future reduction in size of microcircuits, numerical apertures of 0·25 to 0·35 at the object and image are needed, with a high degree of aberration correction. Correction over a considerable spectral range is desirable, so that registration may be obtained in a spectral range to which the photo-resist is insensitive, followed by exposure at an appropriate shorter wave length range. And to ensure accurate registration of successive exposures, which may have slightly different focal settings, it is desirable that the imaging system should be telecentric in both object and image spaces—i.e. the system should be a unit-power telescope.

It seems unlikely that conventional forms of lens system can meet all these requirements, so less usual systems have been investigated. Any system of concentric refracting or reflecting surfaces gives perfect correction of all orders of spherical aberration, sine condition and chromatic aberration for an object at the common centre of curvature, and of all orders of sagittal field curvature over an image plane passing through this centre; and any monocentric system with zero Petzval sum has its Seidel tangential field curvature zero, and is paraxially telescopic. The simplest such system consists of a plano-convex lens, with its axial thickness equal to its radius, together with a spherical mirror, concentric with the curved lens surface, the two radii being chosen so that the Petzval condition is satisfied. This system (Figure 1) has been discussed by Dyson (1959). The plane surface of the lens is imaged, inverted, on itself. Dyson used this system to image one half of the plane surface onto the other half. The use of a semi-reflecting surface would enable the object and image surfaces to be separated, without sacrificing field size. This involves the loss of three-quarters of the imaging light, but for the micro-circuit application this is not important; a large numerical aperture is needed for resolution, not to obtain short exposures. In practice the system would be made by cementing a beam-splitting cube to the plano-convex lens (Figure 2).

The simple Dyson system (and equally the system with beam-splitter),

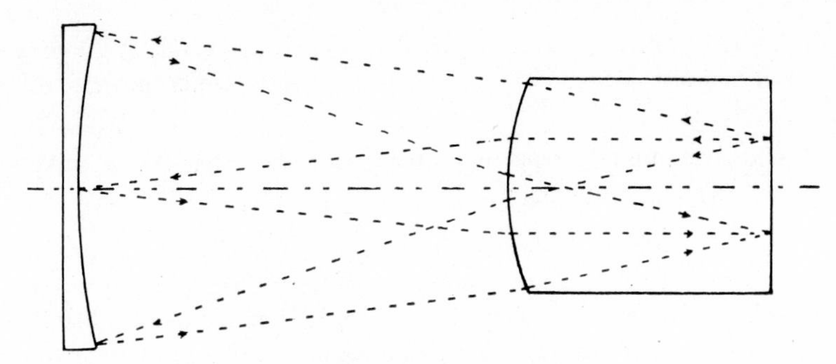

FIG. 1. The Dyson imaging system.

being symmetrical about a central stop on the mirror, is free from all orders of odd power aberrations (coma, distortion, chromatic difference of magnification): being monocentric it has perfect imagery on axis, and in a sagittal plane off axis. It has a chromatic difference of Petzval sum (P), and hence of tangential field curvature and telecentricity. These chromatic

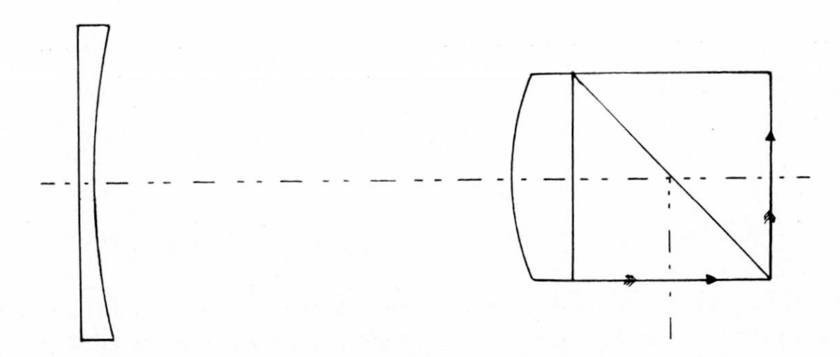

FIG. 2. Imaging system with beam-splitter to separate object and image planes.

differences can be corrected by compounding the plano-convex lens into a monocentric meniscus and a plano-convex lens, making the meniscus of glass of higher dispersion than the positive lens (Figure 3); the residual secondary spectrum of Petzval curvature is extremely small. This chromatic correction of P could be carried out using two glasses of the same mean refractive index but different dispersions.

If this is done, the remaining monochromatic aberrations in the system are:

(*a*) A higher order tangential field curvature (depending on the fourth and higher powers of the field size).

(*b*) A higher order spherical aberration of oblique pencils in the tangential section (depending on the fourth and higher powers of the aperture).

(*c*) An aberrational departure from telecentricity, i.e. a spherical aberration of principal rays.

Now it follows consideration of the ray geometry that (*a*) and (*b*) are in fact a consequence of (*c*); and the elimination of (*c*) is equivalent to correction of the spherical aberration of the glass block (the meniscus plus the plano-convex lens) for rays normal to the plane surface. If this condition could be exactly satisfied, there would be perfect monochromatic imagery for any object distance, in the medium of the rear lens. The aberration theory of monocentric systems, such as this glass block, is amenable to detailed analysis (Herzberger 1958). Applying this theory, it can be shown quite

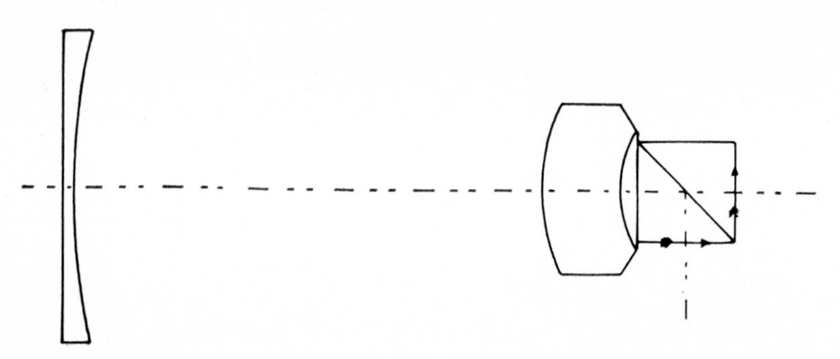

FIG. 3. Imaging system compounded to allow correction of higher order field aberrations.

generally that no system of concentric surfaces having a finite power can be perfectly corrected for spherical aberration for a parallel pencil of rays. But correction of Seidel spherical aberration is possible, for a monocentric meniscus whose concave surface is cemented to a glass of lower refractive index, for parallel light within this lower index space (Wynne & King 1964). By an appropriate choice of the refractive indices and dispersions for the rear member of the system shown in Figure 3, it is therefore possible to ensure a high degree of telecentricity, and hence a high degree of correction of the higher order aberrations of the system for oblique imagery in a meridian plane; there is, of course, still perfect correction in a secondary plane.

There is one further modification that is desirable to the system shown in Figure 3, namely to remove the object and image planes from the surfaces of the beam-splitter block. One could either introduce a plane parallel air space directly between the glass surfaces and the focal planes, or what is more convenient for the micro-circuitry application, between the beam-splitter block and plane parallel cover glasses, whose outer surfaces coincide with the object and image planes (Figure 4). If the beam-splitter block is made of the same material as the cover plates, the optical problem is the same in the two cases. Since both object and image are in telecentric spaces, a parallel air space introduces the same aberrations at any point in the field as it does on axis, viz., spherical aberration and

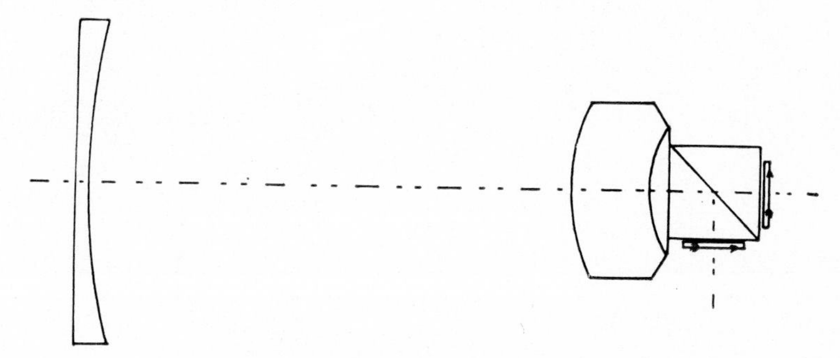

FIG. 4. Complete system with air spaces introduced.

chromatic difference of focus. For a converging beam in glass, the spherical aberration caused by introducing a plane parallel air space is of opposite sign to that caused by introducing a plane parallel plate of higher refractive index. The Seidel spherical aberration of the plane parallel air space in Figure 4 can therefore be corrected by making the beam-splitter cube of slightly higher refractive index than that of the plano-convex lens to which it is cemented; and by an appropriate choice of the glass dispersions, chromatic aberration can be corrected at the same time. The higher orders of spherical aberration cannot be corrected, but for small air spaces, the uncorrected residuals are very small.

The level of aberration correction that it is possible to achieve, in systems of this kind, is higher than seems necessary for the micro-circuitry application; and pressing correction to the limits attainable, in particular the interconnected ones listed as (*a*), (*b*) and (*c*) above, involves increasing the physical size of the system for a given object and image diameter. It would therefore seem that the most useful compromise is one in which the Seidel spherical aberration of principal rays is not fully corrected, but is balanced, across the field, against a small departure from the exact Petzval condition.

In a typical compromise design, with a numerical aperture at object and image of 0·35, a field diameter of 3·8 cm, a spectral range from 546 nm to 405 nm, and a mirror diameter of 27·5 cm, the aberrations can be reduced to within about $\pm 0{\cdot}1$ wavelength at the worst corrected part of the field.

REFERENCES

DYSON, J., 1959, *J. Opt. Soc. Amer.*, **49**, 713.
HERZBERGER, M., 1958, *Modern Geometrical Optics*, New York: Interscience Publishers, pp. 193–212.
WYNNE, C. G. and KING, W. B., 1964, *Optica Acta*, **11**, 107.

A New Lens System for use in Optical Data-Processing

Brian A. F. Blandford

Department of Applied Physical Sciences,
University of Reading, Berks.

Abstract—The conditions are given for a coherent optical data-processing system to give exactly the Fourier spectrum in the filter plane. It is also desirable to reduce the large overall length of the conventional arrangement. To satisfy these requirements it has been found convenient to use a symmetrical lens system. A four-component system is described which reduces the overall length of the processor from $4F$ to $1{\cdot}4F$, and has aberrations nowhere greater than $\lambda/8$. The primary aberrations of such a system are analysed, and details given of the final performance of a practical design.

INTRODUCTION

In previous work several factors affecting the design of lenses to be used for spatial filtering appear to have been ignored. The arrangement frequently employed for spatial filtering is shown in Figure 1. The object,

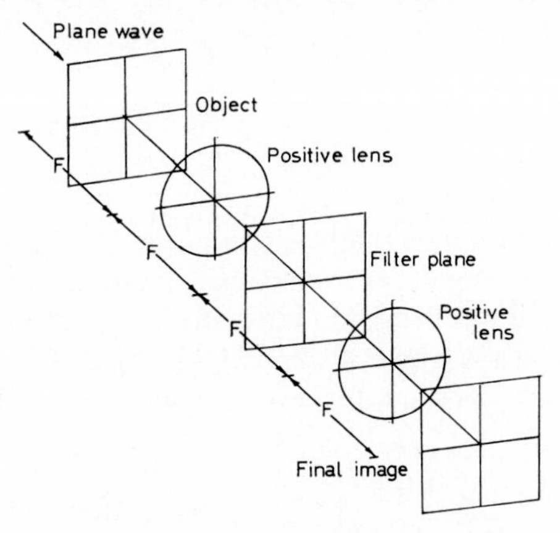

FIG. 1. A conventional arrangement for optical data-processing using single-component ens systems.

a transparency which may have variations in phase as well as amplitude transmittance, is illuminated by a beam of coherent monochromatic light. If this is a collimated beam, the entrance pupil of the lens system is at infinity, with the principal rays parallel to the axis in the object space.

For many applications, the Fraunhofer diffraction pattern is required to be given exactly by the complex Fourier transform of the transmittance of the object. To achieve this, not only must the system be free of all the usual aberrations, but the optical path lengths along the principal rays must also be considered. This factor does not seem previously to have been taken into account, and it has a decisive effect on the design of the system.

The overall length is another important factor. From the diagram it can be seen that the total distance from object to final image will be equal to $4F$ in a system using single lenses of focal length F. Lenses of long focal length are generally necessary to give an adequate spread of the lower spatial frequencies in the transform plane. It is therefore desirable to employ a 'double telephoto' type of system, so as to reduce the distance between the focal planes.

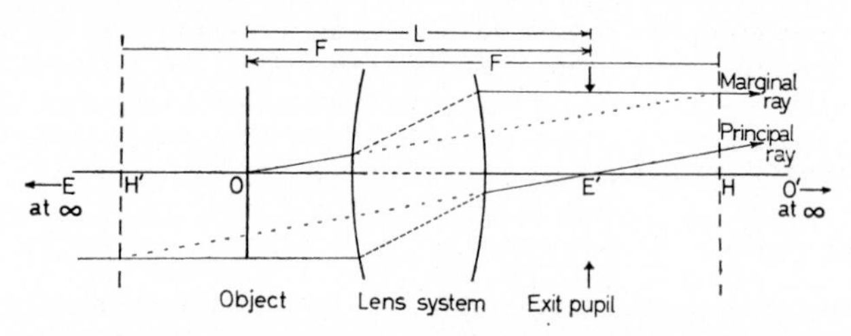

FIG. 2. The geometrical optics of a lens system for data-processing.

If the complex amplitude of the light leaving the object plane is $A(\xi, \eta)$ the complex amplitude at the point (X, Y) in the exit pupil, E', will be given by

$$a(X, Y) = \iint \frac{A(\xi, \eta)}{F} \exp\,[ik(LX+MY)]\exp\,[-ikQ(\xi, \eta)]\,\mathrm{d}\xi\mathrm{d}\eta]$$

where $Q(\xi, \eta)$ is the optical path length along the principal ray from a point having coordinates (ξ, η) in the object plane to the axial point E' of the exit pupil. If the sine condition is satisfied, the direction cosines of the principal ray will be given by $L = -\xi/F, M = -\eta/F$. If, in addition, the path lengths $Q(\xi, \eta)$ are constant for every point in the object plane, the expression becomes

$$a(X, Y) = \frac{\exp\,[-ikQ(0, 0)]}{F} \iint A(\xi, \eta)\exp\,[-i2\pi(\xi X+\eta Y)/\lambda F]\mathrm{d}\xi\mathrm{d}\eta$$

which has, apart from a constant factor, exactly the form of the complex Fourier transform.

The final requirement is that the light diffracted from the object must form a series of plane waves in the final image space. Thus the imagery of both the object and the entrance pupil by the system is required to be aberration-free. A system consisting of a single element cannot be corrected for astigmatism or field curvature for two sets of conjugates, so the use of separated elements is required. The fact that the entrance pupil is at infinity in one direction, and the image at infinity in the other suggests that a symmetrical lens system is most suited to the task. In this case, if the object is imaged without aberrations, the entrance pupil will be also, and the object and exit pupil will be at the principal focal planes, symmetrically located about the lens. This is illustrated in Figure 2.

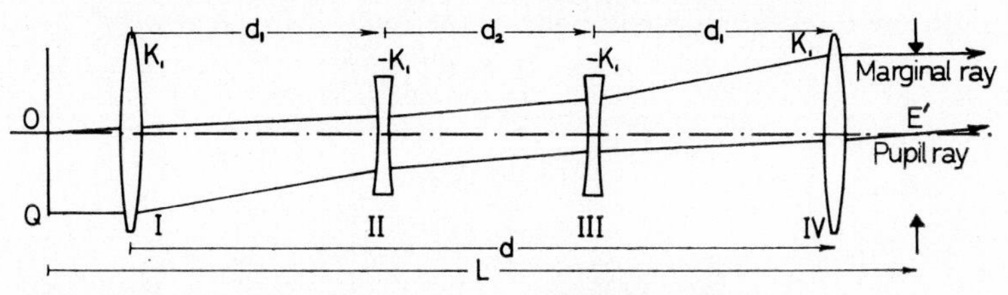

FIG. 3. A four-component symmetrical system which is corrected for field curvature, and which gives a reduction in the distance between its principal focal planes.

GAUSSIAN DESIGN

In addition to the requirements of symmetry and the telephoto effect, the Gaussian design should satisfy one other condition—namely that the Petzval sum $\Sigma(K/\mu)$, be zero, to ensure that the image is free of field curvature. A system satisfying these requirements must necessarily have three or more elements. A four-lens system of this type is shown in Figure 3. The outer components have power K_1 equal and opposite to that of the inner components. In the following analysis of the thin lens system, the ratio of the separation of the first two components to the total length of the lens system is denoted by $d_1/d = \gamma$ and the product dK_1 by β. The three-lens system is treated as the special case of the four-lens system for which $\gamma = 0{\cdot}5$.

Following the notation of Hopkins (1950) the passage of a ray through the system is given by the paraxial ray trace formulae:

$$u' = u + hK; \qquad h_{+1} = h - du'.$$

Applying these to a ray for which $u_1 = 0$, $h_1 = 1$ gives the following values for the successive paraxial angles and incidence heights:

$$\begin{aligned}
u_2 &= \beta/d \\
h_2 &= 1 - \gamma\beta \\
u_3 &= \gamma\beta^2/d \\
h_3 &= 1 - \gamma\beta - \gamma(1-2\gamma)\beta^2 \\
u_4 &= [-\beta + (1-\gamma)\beta^2 + (1-2\gamma)\beta^3]/d \\
h_4 &= 1 - \gamma\beta^2 - \gamma^2(1-2\gamma)\beta^3 \\
u_4' &= [2\gamma\beta^2 - 2\gamma^2\beta^3 - \gamma^2(1-2\gamma)\beta^4]/d.
\end{aligned}$$

In this case, however, we have $h'_4 = \epsilon$, the ratio of the back focal length to the equivalent focal length of the system. Also $u'_4 = h_1/F$, giving:

$$d/F = 2\gamma\beta^2 - 2\gamma^2\beta^3 - \gamma^2(1-2\gamma)\beta^4,$$

and

$$L/F = 2\epsilon + d/F = 2 - 4\gamma^2(1-\gamma)\beta^3 - \gamma^2(1-2\gamma)\beta^4.$$

The ratios d/F and L/F are shown in Figure 4 as a function of $\beta = dK_1$ for both the three-lens system, and for the four-lens system with $\gamma = 0{\cdot}35$.

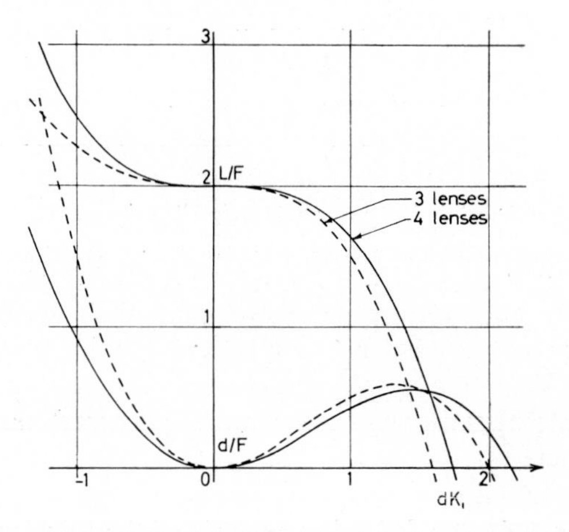

FIG. 4. The ratios of the interfocal length, L, and the length of the lens system, d, to the equivalent focal length, F, of the system shown in Figure 3. $d_1/d = 0{\cdot}35$ for the four-lens system, 0·5 for the three-lens system.

The principal focal planes are outside the system when $L \geqslant d$. In the former case $L = d$ when $L = 0{\cdot}586F$. This minimum length is reduced in the case of the four-lens system for a wide range of values of γ, the minimum being $L = 0{\cdot}554F$ when $\gamma = 0{\cdot}33$. In the usual configuration for spatial filtering this corresponds to a total distance from object to final image of $1{\cdot}11F$, which is a considerable reduction on the value of $4F$ required for systems consisting of single components. This distance must, of course, be slightly increased to allow for the finite thicknesses of the lenses, and to give reasonable clearances between the object and transform planes and the lens system.

Although the advantage of a four-lens system over a three-lens system is small as far as the reduction in length is concerned, it is preferable to employ a four-lens system in any practical design. The reason is that the introduction of the additional lens provides extra parameters which are useful for the correction of the aberrations of the system.

PRIMARY ABERRATIONS

Each of the primary aberrations must now be considered in turn. As a consequence of the symmetry of the system, if the total spherical aberration and coma of the image, ΣS_I and ΣS_{II}, are corrected, those of the pupil, $\Sigma \bar{S}_I$ and $\Sigma \bar{S}_{II}$ will also be corrected. The remaining monochromatic aberrations of the image and pupil are related by the equations:

$$\sum \bar{S}_{III} - \Sigma S_{III} = H(u'_K \bar{u}_K' - u_1 \bar{u}_1)$$
$$\sum \bar{S}_{II} - \Sigma S_V = H(\bar{u}_K'^2 - \bar{u}_1^2)$$
$$\sum \bar{S}_V - \Sigma S_{II} = H(u_K'^2 - u_1^2)$$

Here, H is the optical invariant, and u_1, u'_K and $\bar{u}_1$, $\bar{u}'_K$ refer to the angles of the paraxial marginal and pupil rays in the object and image spaces. In the present case, $u_1 = \bar{u}'_K$ and both $\bar{u}_1$ and u'_K are zero. Thus, if the astigmatism of the image, ΣS_{III} is zero, that of the pupil $\Sigma \bar{S}_{III}$ will also be zero. However, if the coma in both the image and the pupil are made to be zero, there will be distortion in both these planes, given by:

$$\sum S_V = \sum \bar{S}_V = -Hu_1^2 = -h_1^4/F^3$$

This is negligible in systems such as this which have a large focal length, and the symmetry of the overall system ensures that the distortion in the final image is zero.

As a system of this type is designed to be used only in monochromatic light, it may be thought that no account need be taken of the chromatic aberrations. If, however, it is to be used in light of a wavelength other than that for which it was initially designed, the correction of the monochromatic aberrations at the new wavelength will not be adequate if there is large longitudinal chromatic aberration. It is therefore desirable to make this as small as possible and this will be considered later.

It can be seen from this analysis that the only monochromatic primary aberrations of the system which must be considered are the spherical aberration, coma and astigmatism, ΣS_I, ΣS_{II} and ΣS_{III}.

It will now be shown that, as a result of the symmetry of the system, it is possible to express these quantities as linear combinations of the spherical aberration and central coma of the last two elements. For the purpose of this analysis all the lenses are assumed to have zero thickness.

The contributions to the total aberrations from the fourth element are given directly by the stop-shift formulae:

$$S_I = (S_I)_4$$
$$S_{II} = (S_{II})_4 + E_4(S_I)_4$$
$$S_{III} = H^2K_4 + 2E_4(S_{II})_4 + E_4{}^2(S_I)_4$$

where $(S_I)_4$, $(S_{II})_4$ and H^2K_4 are the central values of spherical aberration, coma and astigmatism for this element and where E_4, the stop-shift factor, is the ratio of the incidence heights of the paraxial pupil and paraxial marginal rays. In the present case, writing α_4 for h_1/h_4 and noting that $\bar{h}_4 = -h_1$, we have $E_4 = \bar{h}_4/h_4 = -\alpha_4$.

The aberrations of the first element can also be expressed in terms of the same quantities. First, if the fourth element is reversed, the central coma will be merely reversed in sign, the spherical aberration and astigmatism remaining unchanged.

The object and image positions are now shifted to those required for the first element. The angles at which the marginal ray enters and leaves the lens will thus be changed by the same amount, namely $\delta u = h_4 u_1/h_1 - (u_4') = -H/(\alpha_4 h_4)$ where $H = \bar{h}_1 u_1 - h_1 \bar{u}_1$. The resulting changes in the spherical aberration and central coma will be:

$$\delta S_I = -(4(S_{II})/H + u'^2 - u^2)(h\delta u) + (3 + 2/\mu)K(h\delta u)^2$$

$$\delta S_{II} = -HK(2 + 1/\mu)(h\delta u)$$

the central astigmatism being unaffected. The aperture is now reduced to that of the first element, i.e. by a factor of $(h_1/h_4) = \alpha_4$, giving the following expressions for the central aberrations of this element

$$(S_I)_1 = [(S_I)_4 + \delta S_I]\alpha_4^4$$

$$(S_{II})_1 = [-(S_{II})_4 + \delta S_{II}]\alpha_4^2$$

$$(S_{III})_1 = H^2K_4$$

Substituting for δS_I and δS_{II} in these expressions and applying the stop-shift formulae with $E_1 = -1/\alpha_4$ gives immediately the contributions of the first component to the total aberrations of the system. Completely analogous arguments can be used to find the contributions of the second and third elements, in this case with α_3 for h_2/h_3, and putting $\mu_3 = \mu_4 = \mu$ to a first approximation.

Adding together the results for the four lenses gives the following expressions for the total spherical aberration, coma and astigmatism of the system, as linear combinations of the spherical aberration and central coma of the last two components:

$$\sum S_I = (\alpha_4^4 + 1)(S_I)_4 + (\alpha_3^4 + 1)(S_I)_3 - 4\alpha_4^3(S_{II})_4 - 4\alpha_3^3(S_{II})_3 + H(-\alpha_4^3(-u_4^2) - \alpha_3^3(u_4^2 - u_3^2)) + (3 + 2/\mu)H^2(\alpha_4^2K_4 + \alpha_3^2K_3)$$

$$\sum S_{II} = (-\alpha_4^3 - \alpha_4)(S_I)_4 + (-\alpha_3^3 - \alpha_3)(S_I)_3 + (3\alpha_4^2 + 1)(S_{II})_4 + (3\alpha_3^2 + 1)(S_{II})_3 + H(\alpha_4^2(-u_4^2) + \alpha_3^2(u_4^2 - u_3^2)) + (-1 - 1/\mu)H^2(\alpha_4K_4 + \alpha_3K_3)$$

$$\sum S_{III} = 2\alpha_4^2(S_I)_4 + 2\alpha_3^2(S_I)_3 - 4\alpha_4(S_{II})_4 - 4\alpha_3(S_{II})_3 + H(-\alpha_4(-u_4^2) - \alpha_3(u_4^2 - u_3^2)) + H^2(K_4 + K_3)$$

These three total aberrations are required to be zero. This condition gives three simultaneous equations whose coefficients are constants depending only on the Gaussian properties of the system.

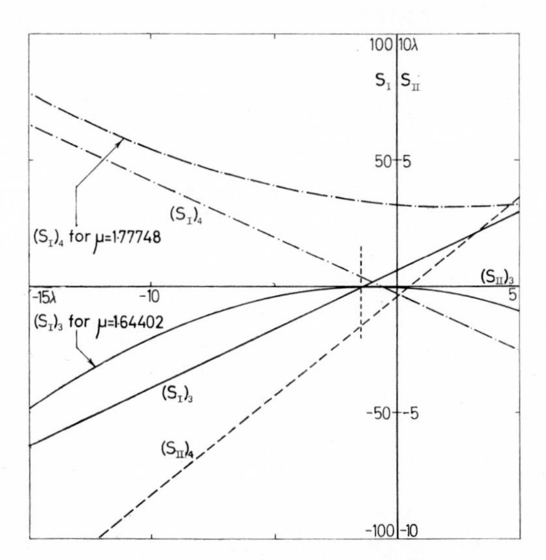

FIG. 5. Solutions of the simultaneous aberration equations for the system with $\beta=1{\cdot}5368$, $\gamma=0{\cdot}35$, and the spherical aberrations of singlet lenses as the third and fourth components. The dotted vertical line indicates the solution used for the system shown in Figure 6.

The problem is now reduced to one of finding lenses for the last two components which will have values of spherical aberration and central coma that satisfy these equations. The straight lines in Figure 5 give, for a typical case, the necessary values of $(S_I)_4$, $(S_{II})_4$, and $(S_I)_3$ for each value of $(S_{II})_3$.

Because a doublet can be designed to have, within reason, any desired values of spherical aberration and coma, it would not be difficult to find two doublets that meet these requirements. In the interests of economy, howevei, it is desirable to use a singlet for one of the two components. As the third lens works at only half the aperture of the fourth, it would be preferable to have the doublet in this position. This, however, is not possible, for reasons which will now be given.

The spherical aberration of a thin singlet is given as a function of its coma by the expression:

$$\frac{S_I}{h^4K^3}=\frac{\mu(\mu+2)}{(\mu+1)^2}\left(\frac{S_{II}}{Hh^2K^2}\right)^2+\frac{\mu Y}{(\mu+1)^2}\left(\frac{S_{II}}{Hh^2K^2}\right)-\frac{\mu^2Y^2}{4(\mu+1)^2}+\frac{\mu^2}{4(\mu-1)^2}$$

where $Y = (u' + u)/hK$ is the magnification factor of the lens. This function is shown in Figure 5 for a singlet in the position of the fourth component. In this particular case, the parabola does not intersect the solution line for $(S_I)_4$ even when a glass of very high refractive index is used. A solution of this type can be achieved only within a limited range of Gaussian arrangements and these are not satisfactory for any practical system. The reason is that the values of primary spherical aberration required by the solution are so large (over 60 wavelengths) that they would induce a large amount of aberration of higher orders. It is therefore necessary to use a doublet for the fourth component.

The spherical aberration of a singlet used as the third component is shown on Figure 5 as a continuous curve, which intersects the solution line for $(S_I)_3$ at a point very close to the axis. The aberrations required to satisfy the equations will thus be acceptably small.

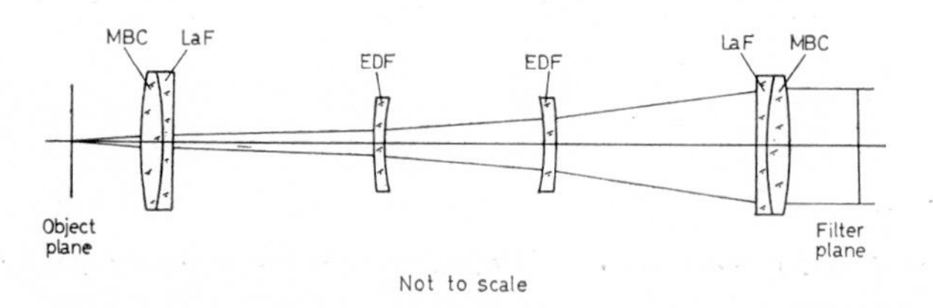

FIG. 6. A practical four-lens system for optical data-processing.

It now remains only to specify the doublet to be used as the fourth component of the system. Hopkins (private communication) has developed a technique for selecting the glass types to give a doublet with specified values of spherical aberration and central coma. This method gives a wide choice of possible solutions for the present case. If, however, for reasons already given, it is desired to correct for the longitudinal chromatic aberration of the system, ΣC_L, the choice of glass types will be considerably restricted. For, in addition to S_I and S_{II}, the value of C_L of the doublet must be specified to compensate for that of the other components. If this can be achieved, the system may be used in light of any wavelength.

To realize a practicable system it is now only necessary to introduce finite thicknesses for the lenses, and the first order design is complete.

A PRACTICAL SYSTEM

In designing a practical system for spatial filtering, the focal length will be governed by the range of frequencies required in the filter plane for a given field size. If the pupil has radius h, the spatial frequency which corresponds to the edge of the filter will be $N_{max} = h/\lambda F$.

A system designed by the author is shown in Figure 6. The equivalent focal length was one metre. For object and exit pupil of radius 30 mm the maximum spatial frequency was thus 47 lines/mm when used with light from a helium-neon laser (0·6328 μm). The lens system had a length of 576 mm with a clearance of 65 mm at each end, giving an overall length of 1·412 metres from object to final image for the complete data-processing system.

It was not necessary to introduce any finite primary aberrations to correct for higher-order aberrations, as these were found to be negligible. This was to be expected, for reasons already given. The maximum value of the total wave aberration at the edge of the pupil was only 1/8th of a wavelength on axis, reducing to 1/20th of a wavelength at full field.

Acknowledgements

I should like to express my gratitude to Professor H. H. Hopkins for his help and advice at all stages of this research, and to the Atomic Weapons Research Establishment at Aldermaston for providing financial support for this work.

REFERENCE

HOPKINS, H. H., 1950, *Wave Theory of Aberrations*, Oxford, Clarendon Press.

2-Conjugate Zoom Systems

H. H. Hopkins

Department of Applied Physical Sciences,
The University, Reading

Abstract—A 2-conjugate zoom system is defined as one which maintains stable focus in fixed image planes of two separate object planes, namely that of the object and that of the entrance pupil of the zoom system. Such a system demands the use of at least three moving elements. Formulae are given to ensure the exact fulfilment of the conditions and a typical system is described.

INTRODUCTION

In existing zoom systems, the focal lengths and movements of the separate lens elements are arranged to form the image of a fixed object plane in an image plane which remains fixed during operation of the zoom. The aperture stop is customarily placed to the rear of the moving elements, so that a fixed position is obtained for the exit pupil. By contrast the position of the entrance pupil is found to vary widely for different regions of the zoom range. For a camera objective, this wandering of the entrance pupil is usually of no disadvantage. When, however, a zoom system exists as an intermediate stage of a complete system, it is desirable, and in many cases essential, to have the zoom system to image the fixed exit pupil of the system preceding it at the fixed position of the entrance pupil of the system following it. Such a zoom system must thus maintain stable focus in fixed image planes of two separate object planes, namely that of the object and that of the entrance pupil of the zoom system. It is proposed to denote such systems as 2-conjugate zoom systems.

To design a 1-conjugate zoom system in which focus is accurately maintained requires a minimum of two moving components. It has been shown that a 2-conjugate zoom system demands the use of at least three moving components if focus is to be held accurately for two pairs of conjugate planes. Given three moving elements, formulae may be obtained which ensure exact fulfilment of the conditions for a 2-conjugate zoom system.

In a particular case, a negative lens is placed between two positive lenses. It is shown that each of these lenses must move during operation of the zoom, and further that the separation of the outer lenses may not remain constant. As a further special case of this structure, the two outer positive lenses may be made of equal focal length, and arranged so that in the mean position each of these lenses, and also the middle negative lens, work at -1 magnification. The object and entrance pupil must then, in this case, be placed at the positions giving overall magnification of $+1$ and -1. The theory to be described gives explicit values of $d = d_{12} + d_{23}$ and $\beta = d_{12}d_{23}$, where d_{12} and d_{23} are the inter-lens separations.

The Gaussian optics of a large number of such systems, and also of those comprising a positive lens between two negative lenses, has to be explored. Zoom ratios as large as 20:1 are easily obtainable. A typical system will be described, together with the manner in which it is used to obtain a large-range, zoom phase-contrast microscope.

GENERAL CONSIDERATIONS

In the past zoom systems seem to have been applied primarily to camera objectives, both for photography and television. The zoom principle has also found use in microscopes, in particular for stereoscopic microscopes and as variable magnification eyepieces in ordinary microscopes. When used to provide a camera objective, the aperture stop is placed in the rear part of the system and is followed only by fixed lens elements. In this way, the *F*/No. is automatically held constant during operating the zoom; and, additionally, the exit pupil of the system remains fixed in position. Calculation shows that, in contrast, the position of the entrance pupil varies widely as the objective is made to vary from the short to the long focal length parts of the zoom range. This wandering position of the entrance pupil is of no disadvantage in a camera objective; indeed, it can be turned to advantage by arranging that the entrance pupil lie more towards the front of the objective for its short focal length range, thereby reducing the incidence heights of the pupil rays at the front of the system.

In the case of a zoom stereoscopic microscope, and also in that of a zoom eyepiece, there is a small, but tolerable, wandering of the eyepoint. Again, this is of no practical consequence for small zoom ratios. For larger zoom ratios, and for phase-contrast systems, the wandering of the pupil is a serious disadvantage, and the purpose of the present study was to investigate the design possibilities of 2-conjugate zoom systems. This description is suggested for any zoom system in which two object planes, and their conjugate image planes, both remain exactly constant in position during operation of the zoom. Both the image and exit pupil then 'hold focus'. A 2-conjugate zoom system is also necessary in cases when it is used as a relay system, where the fixed image and exit pupil of a preceding system have both to be imaged at the object and entrance pupil of a following system.

In an ideal microscope a given diameter of the object is illuminated with cones of suitable aperture for any given magnification. If the magnification is changed, the object diameter and aperture angle should change in, respectively, inverse and direct proportion to the magnification. The illuminating system should also, in principle, change correspondingly. In this way, both the level of illumination in the final image and the ratio of the limit of resolution to the diameter of object viewed remain constant. Such an arrangement can clearly be obtained if the microscope has both a zoom objective and a reciprocally operating zoom illuminator.

If a zoom microscope of this kind is to work over a large zoom ratio, or if it is to be used as a phase-contrast system, both the zoom objective and the zoom illuminator must be 2-conjugate systems. In this case, the annular stop is placed in the object space of the illuminator, and is imaged at infinity in the object space of the zoom objective: it is then finally re-imaged, with constant focus and size in the image space of the zoom objective. The result is that, as the two zoom systems are operated, a varying size of the object is illuminated with hollow-cone illumination, whose inner and outer radii remain constant fractions of the (varying) full aperture angle of the illuminator. This is exactly the condition that is demanded for a zoom phase-contrast microscope, and it clearly requires that both the zoom condenser and zoom objective be designed as 2- conjugate zoom systems.

GENERAL DESIGN FORMULAE

In Figure 1 the object O is imaged at O' with a magnification M. A second object point, E, is imaged at E', where these points are the entrance

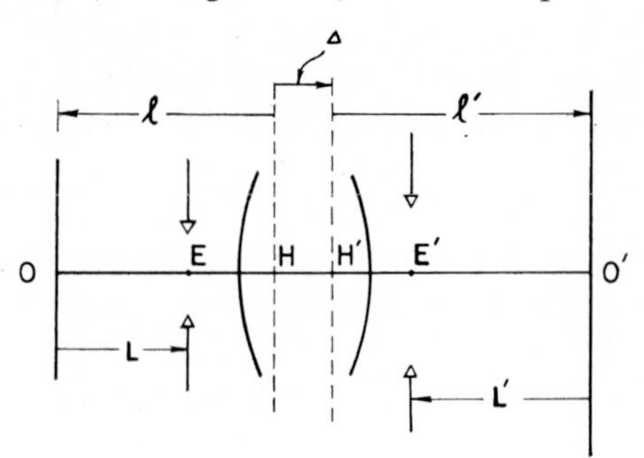

FIG. 1. Diagram of the Gaussian optics.

and exit pupils of the system. If K is the equivalent power of the image-forming system, the general form of the conjugate equation gives

$$\frac{M}{L'} - \frac{1}{ML} = K \tag{1}$$

where $L = OE$ and $L' = O'E'$. In equation (1) it is assumed that the initial object and final image spaces have unity refractive index. Alternatively, the distances L and L' can be regarded as reduced lengths. If the points E, E' as well as O, O' are to remain constant when the zoom is operated, and thus varying the magnification M, equation (1) shows that K cannot be zero, and that its value must change as M changes, since both L and L' remain constant. This excludes, of course, the cases when one or other of O or E, and also one or other of O' or E', are at infinity. The condition (1) then states that the object/entrance pupil and the image/exit pupil distances remain constant.

To ensure that all four of the points O, O', E and E' remain fixed in space requires, in addition, that the object/image distance, $P = OO'$ also remains constant. It has been seen that the system cannot be of zero equivalent power; and there will therefore always exist principal planes. Let these be H, H', and employ the notation $l = HO$, $l' = H'O'$ and $\Delta = HH'$. The object/image distance is then given by

$$P = l' - l + \Delta$$

or, using the formulae $l' = (1 - M)F$, $l = \left(\frac{1}{M} - 1\right)F$, where $F = 1/K$ is the equivalent focal length,

$$\Delta K = PK - \left[2 - M - \frac{1}{M}\right] \tag{2}$$

and this relation, together with equation (1), specifies completely the gaussian properties of the complete system. Thus, for given values of L and L', equation (1) determines the equivalent power of the system for any magnification M; and, then, equation (2) determines the Gaussian thickness, Δ, of the system. Two independent conditions have thus to be satisfied by the system for any zoom position.

For a system consisting of two elements, of powers K_1 and K_2, with a separation d_{12}, the equivalent power and Gaussian thickness are given by:

$$\begin{aligned} K &= K_1 + K_2 - d_{12}K_1K_2 \\ \Delta K &= -d_{12}^2 K_1 K_2 \end{aligned} \tag{3}$$

and both conditions cannot be satisfied simultaneously for a range of different values of the single variable d_{12}. It follows that a 2-conjugate zoom system cannot be designed using only two independently moving components. It is thus necessary to use a minimum of three moving components to design a 2-conjugate zoom system.

For a separated system of three components, with powers K_1, K_2 and K_3, and inter-lens separations d_{12} and d_{23}, the formulae corresponding to Equation (3) are:

$$K = K_1 + K_2 + K_3 - dK_1K_3 - (d_{21}^{2}K_1 + d_{23}^{2}K_3)K_2 + d_{12}d_{23}K_1K_2K_3 \quad (4)$$

and

$$\Delta K = -d^2K_1K_3 - (d_{12}^{2}K_1 + d_{23}^{2}K_3)K_2 + dd_{12}d_{23}K_1K_2K_3 \quad (5)$$

where $d = d_{12} + d_{23}$ is the overall length of the system. Given the values of the powers K_1, K_2 and K_3, the separations d_{12} and d_{23} giving desired values of K and ΔK may be found from equations (4) and (5) regarded as simultaneous quadratic equations in the variables d_{12} and d_{23}. Real systems exist when both of these quantities are greater than zero. The problem is thus essentially solved by using equations (1) and (2) in conjunction with equations (4) and (5).

A Special Case Having Symmetry

A case of particular interest and simplicity is that in which the system is symmetrical as regards the powers and movements of the three elements, and where, to obtain a zoom ration R, the numerical value of the magnification is made to vary from $|M| = \sqrt{R}$ to $|M = 1/\sqrt{R}$. The equivalent power of the system then has the same value for any given magnification and its reciprocal. Equation (1) then gives

$$K = \frac{M}{L'} - \frac{1}{ML} = \frac{1}{ML'} - \frac{M}{L}$$

and this requires that $L' = -L$. For this case, therefore, equation (1) becomes

$$K = \frac{-1}{L}\left(M + \frac{1}{M}\right) \quad (6)$$

Now, if $\bar{M}$ is the pupil magnification, that is between E and E', the distances L and L' are given by

$$L = \left(\frac{1}{\bar{M}} - \frac{1}{M}\right)F$$

$$L' = (M - \bar{M})F$$

and, because now $L' = -L$, it follows that either $\bar{M} = -1/M$ or $\bar{M} = M$. Since $\bar{M} = M$ merely gives $L = L' = 0$, the pupil and object/image magnifications must satisfy the former, and must, therefore, be numerically reciprocal and of opposite sign. It may further be seen that, at the mean position, either $M = -1$ and $\bar{M} = +1$, or conversely $M = +1$ and $\bar{M} = -1$. For any given powers, and for given separations in the mean position, the values of L, M and $\bar{M}$ are thus no longer free parameters when the above conditions of symmetry are imposed.

A symmetrical system of the type under discussion has $K_3 = K_1$, and the power of the middle component may usefully be specified by the ratio

$$\gamma = -K_1/K_2 = -F_2/F_1 \tag{7}$$

Let, now, M_0 denote the object/image magnification of the first component when in the mean position, assuming that $M = -1$, so that $\bar{M} = +1$: the magnifications of the second and third components are then respectively -1 and $1/M_0$. In this position the object and image distances for the three components are thus given by:

$$\begin{aligned} l_1 &= \left(\frac{1}{M_0} - 1\right)F_1 \\ l'_1 &= (1 - M_0)F_1 \\ l_2 &= -2F_2 = 2\gamma F_1 \\ l'_2 &= +2F_2 = -2\gamma F_1 \\ l_3 &= (M_0 - 1)F_1 \\ l'_3 &= \left(1 - \frac{1}{M_0}\right)F_1 \end{aligned} \tag{8}$$

From these formulae, it easily follows that the inter-lens separations are given by:

$$d_{12} = d_{23} = (1 - M_0 - 2\gamma)F_1 \tag{9}$$

and the object to image distance is given by

$$P = 2\left\{\left(2 - M_0 - \frac{1}{M_0}\right) - 2\gamma\right\}F_1 \tag{10}$$

To find the value of L, the distance of the entrance pupil from the first element, $\bar{l}_1$, is needed. This is found from the condition that $\bar{M} = +1$, so that the pupil ray crosses the axis at the middle component. Thus,

$$\frac{1}{\bar{l}_1} = \frac{1}{d_{12}} - \frac{1}{F_1}$$

or, substituting from equation (9),

$$\bar{l}_1 = \frac{(1 - M_0 - 2\gamma)F_1}{(M_0 + 2\gamma)} \tag{11}$$

and, by symmetry, $\bar{l}'_3 = -\bar{l}_1$. The value of $L = \bar{l}_1 - l_1$ is found using Equations (8) and (11), giving after some reduction:

$$L = -\frac{2F_1}{M_0(M_0 + 2\gamma)} \tag{12}$$

Since F_1 merely appears as a scale factor in equations (8), (9), (10), (11) and (12), it is usefully taken to be equal to unity in a practical case, so that the system has effectively two design parameters, namely γ and M_0. For any values of these two quantities, equations (8) and (9) give the object and image positions and the inter-lens separations, while equation (11) gives the entrance and exit pupil positions. The system is then completely specified in the mean position of the zoom run.

To calculate the positions of the three lens elements which give magnifications other than $M = -1$, the values of L and P from equations (12) and (10) are used in equations (1) and (2) respectively to give the necessary values of K and ΔK.

The formulae (4) and (5) also simplify when $K_3 = K_1$. Thus, noting that $d_{12} + d_{23} = d$, and using equation (7), equation (4) becomes

$$K = \left\{\left(2 - \frac{1}{\gamma}\right) - \left(1 - \frac{1}{\gamma}\right)dK_1 - \frac{1}{\gamma}d_{12}d_{23}K_1^2\right\}K_1 \tag{13}$$

and equation (5) becomes, after some reduction,

$$\Delta K = -\left(1 - \frac{1}{\gamma}\right)(dK_1)^2 - \frac{1}{\gamma}dd_{12}d_{23}K_1^2 - \frac{2}{\gamma}d_{12}d_{23}K_1^2 \tag{14}$$

Again noting that F_1 appears as a scale factor, and defining

$$\rho = dK_1 = \frac{d}{F_1} \tag{15}$$

$$\beta = d_{12}d_{23}K_1^2 = \left(\frac{d_{12}}{F_1}\right)\left(\frac{d_{23}}{F_1}\right). \tag{16}$$

equations (13) and (14) become

$$\frac{F_1}{F} = \frac{K}{K_1} = \left(2 - \frac{1}{\gamma}\right) - \left(1 - \frac{1}{\gamma}\right)\rho - \frac{\beta}{\gamma} \tag{17}$$

and

$$\Delta K = \rho^2\left(\frac{1}{\gamma} - 1\right) - \frac{\beta}{\gamma}(2 + \rho). \tag{18}$$

These last two equations are easily solved to give, eliminating first β/γ,

$$\rho = \frac{\Delta K - 2\left(\frac{K}{K_1}\right) + 2\left(2 - \frac{1}{\gamma}\right)}{\left(\frac{K}{K_1}\right) - \frac{1}{\gamma}} \tag{19}$$

and then, from equation (17),

$$\beta = \gamma\left\{\left(2-\frac{1}{\gamma}\right)-\left(1-\frac{1}{\gamma}\right)\rho-\left(\frac{K}{K_1}\right)\right\}. \tag{20}$$

Again the scaling factor F_1 may be put equal to unity; and, the values of K and ΔK being known from equations (1) and (2) in terms of M_0 and γ, the formulae (19) and (20) solve the problem of finding the necessary inter-lens separations. In explicit terms, the overall length of the system is given by

$$d = \rho F_1 \tag{21}$$

and, writing $d_{23} = d - d_{12}$, equation (16) gives the quadratic equation

$$d_{12}^2 - dd_{12} + \beta F_1^2 = 0 \tag{22}$$

whose roots, having the sum d, are the values of the two separations d_{12} and d_{23}. Clearly d_{12} and d_{23} may be interchanged and still give a solution.

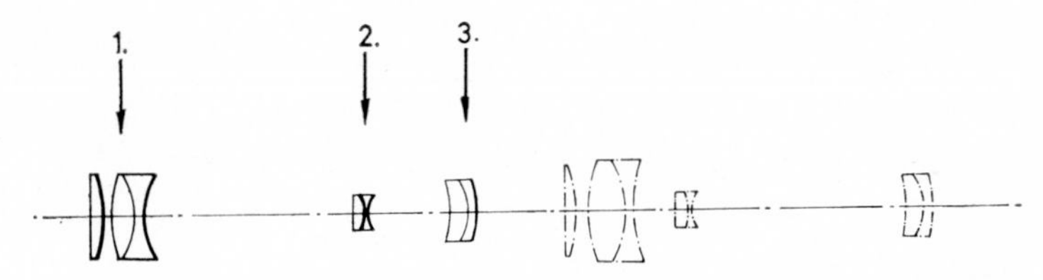

FIG. 2. Design for a 20:1, 2-conjugate zoom system.

A computer programme was written to investigate the family of systems corresponding to different values of the parameters γ and M_0. This yielded a large variety of different systems, in some of which the overall length, d, showed considerable variation. In all cases, the maximum obtainable zoom range, R, was limited by the condition that one or other of the inter-lens separations became negative. From among the many possibly useful systems, one was chosen having a zoom ratio of 20:1, and components were designed to establish that stable correction of the aberrations was also possible. This particular system, described in more detail elsewhere (Hopkins 1968), is shown in Figure 2, the full and broken lines showing respectively the two extreme positions of the three components. The movements of the three components are shown in Figure 3, with the logarithm of the magnification as ordinate.

In a zoom microscope, the system described here is used to provide a variable magnification relay system in both the illuminator and for the observing side. The former is followed by a fixed focus substage condenser, and the latter is preceded by a fixed focus objective. The resulting system allows a rapid change between normal and phase-contrast observation. Since the entrance pupil of the illuminator and the conjugate exit pupil of the observing system remain exactly constant in size and also hold focus

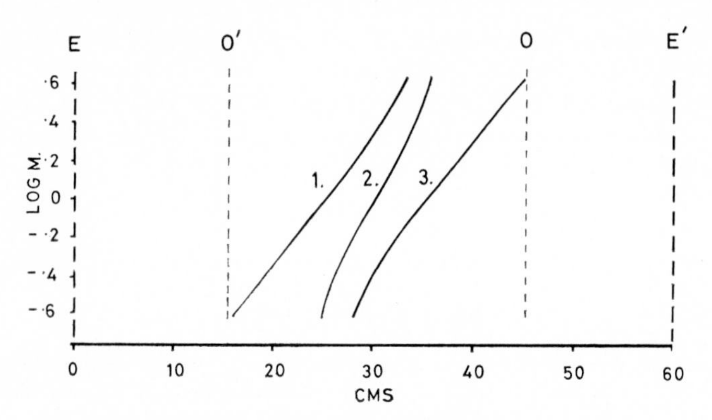

FIG. 3. Lens movements 20:1, 2-conjugate zoom system.

exactly, phase-contrast observation can be made continuously during operation of the zoom. For the same reason many other forms of illumination and observation, for example Rheinberg's differential colour method, are easily made interchangeably available.

Acknowledgements

This investigation was undertaken in conjunction with W. Watson & Sons, and it is a pleasure to acknowledge the assistance of Miss J. M. Drewitt.

REFERENCE

HOPKINS, 1968, British Patent Application No. 4026/68.

A New Ultrasonic Imaging Technique

Takuso Sato and Mitsuhiro Ueda

Research Laboratory of Precision Machinery and Electronics,
Tokyo Institute of Technology,
O-Okayama, Meguro-ku, Tokyo, Japan

Abstract—Work in the Institute of Technology in Tokyo has shown that it is possible to use the property of the diffraction of light by ultrasonic waves to obtain holograms with possible applications in the fields of diagnostic medicine or non-destructive testing of materials. The ultrasonic waves are in the frequency range of 1 to 10 MHz and this is combined with laser light to obtain the holograms. It has also been found possible to use three different ultrasonic frequencies and to obtain coloured images by projection through tri-colour filters. The theory of the method and the necessary techniques are described.

INTRODUCTION

Various kinds of acoustic imaging techniques, mainly holographic, have been reported recently (Korpel 1966; Kreuzer 1967; Marom 1968; Massey 1967; Metherell 1967; Muller 1966). Noteworthy among them is Korpel's method, in which the amplitude and phase distribution of ultrasonic waves can be reconstructed instantaneously by using the Bragg's diffraction. This method, however, has been applied only in the case of ultrasonic waves of higher frequency than a few tens of MHz because the Bragg's diffraction was necessarily used.

The ultrasonic imaging technique in the frequency range of 1 to 10 MHz is especially useful for medical diagnoses and non-destructive testings because of the relatively good transmissivity of waves and the sufficient resolution (Saneyoshi, Kikuchi & Nomoto 1966).

In this paper a new ultrasonic imaging technique which is suitable to this frequency range is presented. Diffraction of light by ultrasonic waves in the above-mentioned range can be put to use by adopting an acoustical slit and an optical filtering device, and the visible image is displayed on a screen in quasi-real time. The resolving power of this device is derived theoretically and verified experimentally.

By using three different frequencies of ultrasonic waves, coloured images can easily be obtained in this system, as verified experimentally.

PRINCIPLE

In Figure 1 the object, which is for the moment considered to be of one dimension, is illuminated by plane ultrasonic waves.

Let us assume that the amplitude and the phase of the ultrasonic waves at point P on the hologram plane (this term is used since the amplitude and phase of this plane is reconstructed) are given by:

$$u_2(x_2) \exp [jb(x_2)] \tag{1}$$

where the term depending on time is neglected for simplicity.

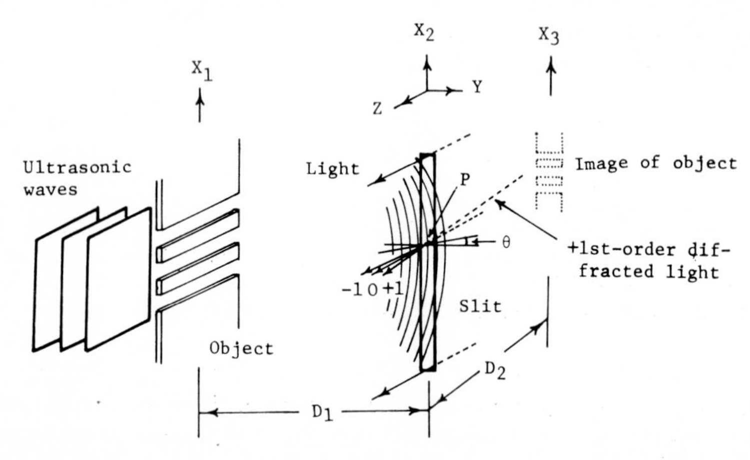

FIG. 1. Reconstruction of the amplitude and phase distribution of ultrasonic waves by means of diffracted light.

A narrow window is placed along the hologram plane and collimated laser light is passed through it in the direction z. If the width of the window is so small as to be negligible in relation to the change of $u(x_2)$ and $b(x_2)$ but wide enough to produce the diffraction of light, the light in the vicinity of the point P is diffracted by the plane waves which are given as follows:

$$U_p(x_2, y) = u(P) \exp [jb(P)] \exp [-jK(x_2 \sin \theta + y \cos \theta - P \sin \theta)] \tag{2}$$

By the phase-lattice theory (Saneyoshi *et al.* 1966) the amplitude and phase distribution of the nth order diffracted light A_n is given by

$$A_n = J_n[v(P)] \exp [jnb(P)]$$
$$\exp [-j(k_n Z + nKx_2 \sin \theta + nKy \cos \theta - nKP \sin \theta)] \tag{3}$$
$$k_n^2 = k^2 - (nK)^2, \qquad k = 2\pi/\lambda, \qquad K = 2\pi/\Lambda$$

where $v(P)$ is the Raman-Nath parameter at the point P, and is proportional to the amplitude of the ultrasonic waves at the point, J_n is the

n^{th} order Bessel function, Λ is the wavelength of ultrasonic waves and λ is that of the light.

In the case of a weak ultrasonic disturbance the following relations can be applied:

$$J_{+1}(x) \simeq \frac{x}{2}; \qquad J_{-1}(x) \simeq -\frac{x}{2}; \qquad x \leqslant \frac{1}{2} \tag{4}$$

hence the amplitude and phase of the $\pm$1st-order diffracted light is derived as follows:

$$A_{\pm 1} = \pm cu(P) \exp [\pm jb(P)] \exp [-j(k_1 Z \mp Kx_2 \sin \theta \mp Ky \cos \theta \pm KP \sin \theta)] \tag{5}$$

where c is a constant.

Since this relation holds for any point on the hologram plane, we can see that the amplitude and the phase of the ultrasonic waves on the hologram plane are converted to those of the light passing through this plane.

Then, according to the principle of holography, the reconstructed real image of the object is formed by this modulated light at a distance D_2 from the hologram plane, where D_1 and D_2 are related by

$$D_1 \Lambda = D_2 \lambda. \tag{6}$$

If a wide window is used and an object which has depth along the direction of propagation of the ultrasonic waves is considered, the reconstructed image of the cross-section parallel to the x_2y-plane of the object is visible because of the relation of reconstruction.

So far the object is restricted to a homogeneous one along the coordinate z, since the disturbance of the ultrasonic waves along this direction affects the diffraction of the light cumulatively and we cannot discriminate the change in this direction. To see this change a slit made of sound-absorbent material is placed behind the object and the system is arranged so that ultrasonic waves scattered from the part of the object corresponding to the aperture of the slit diffract the light.

Thus a sliced image of the object is seen. By moving the object in the z-direction the reconstructed image of the whole cross-section parallel to the zx_2-plane can be seen.

To select the modulated light from the unmodulated light a concave mirror is used and the +1st-order diffracted light is separated by a slit placed at the focal plane. The reconstructed image is magnified onto a screen by a lens.

It is also possible to focus any plane parallel to the x_2z-plane by adjusting the focus of the magnifying lens, and the image focused on this plane can then be seen on the screen.

CONSTRUCTION OF ACOUSTIC IMAGING DEVICE

The ultrasonic imaging device based on the foregoing principle is shown schematically in Figure 2. The object is immersed in water and illuminated by ultrasonic waves. To move the object and the camera synchronously

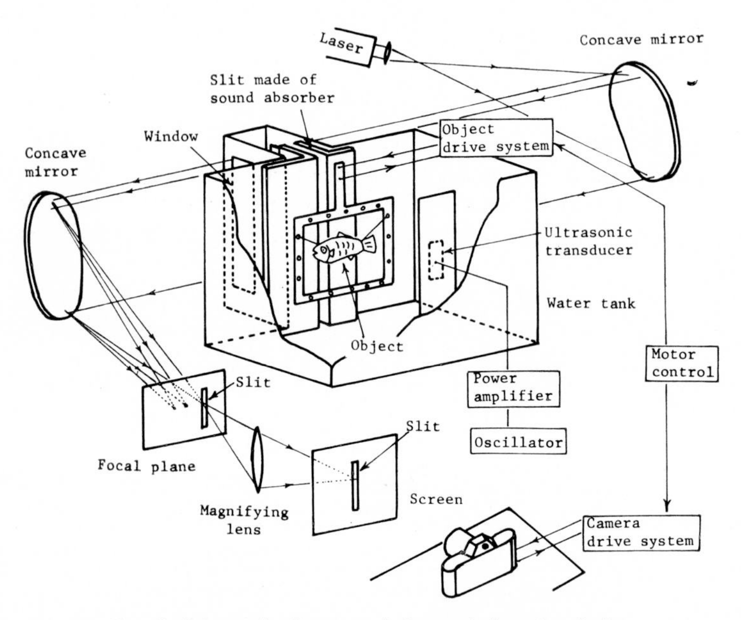

FIG. 2. Schematic diagram of ultrasonic imaging device.

two stepping motors are used, and to take the picture of the reconstructed image of the cross-section parallel to the x_2z-plane a slit is placed on the screen and the segment of the image of the cross-section parallel to the x_2y-plane is picked up.

RESOLUTIONS OF THE SYSTEM

Resolution in the Direction x

When the transparency of the object for ultrasonic waves is denoted by $O(x_1)$ then the amplitude and phase distribution of the modulated $+1$st-order light $G(x_3)$ on the plane at a distance $D_2(D_1\Lambda = D_2\lambda)$ from the hologram plane is given by the equation:

$$G(x_3) = -c^2\left(\frac{K}{D_1}\right)^2 \int_{-\infty}^{\infty} O(x_1) \exp\left\{ j\frac{K}{2D_1}(x_1{}^2 - x_3{}^2)\right\} \times \frac{2 \sin Ka(x_3 - x_1)/D_1}{K(x_3 - x_1)/D_1} \mathrm{d}x_1 \quad (7)$$

where $2a$ is the length of the window placed along the hologram plane. By using the Fourier transforms, this may be rewritten as follows:

$$\hat{G}(\omega) = \int_{-\infty}^{\infty} G(x) \exp\left[j\frac{K}{2D_1}x^2 \right] \exp\left[-j\omega x \mathrm{d}x\right]$$

$$\hat{G}(\omega) = \hat{H}(\omega)\hat{O}(\omega) \qquad \hat{O}(\omega) = \int_{-\infty}^{\infty} O(x) \exp\left[j\frac{K}{2D_1}x^2 \right] \exp\left[-j\omega x \mathrm{d}x\right] \quad (8)$$

$$\hat{H}(\omega) = -c^2\left(\frac{K}{D_2}\right)^2 \int_{-\infty}^{\infty} \frac{2 \sin Kax/D_1}{Kx/D_1} \exp\left[j\omega x \mathrm{d}x\right]$$

From this result we can say that the least resolvable spacing s of a grating is given by $s \geqslant \Lambda D_1/a$.

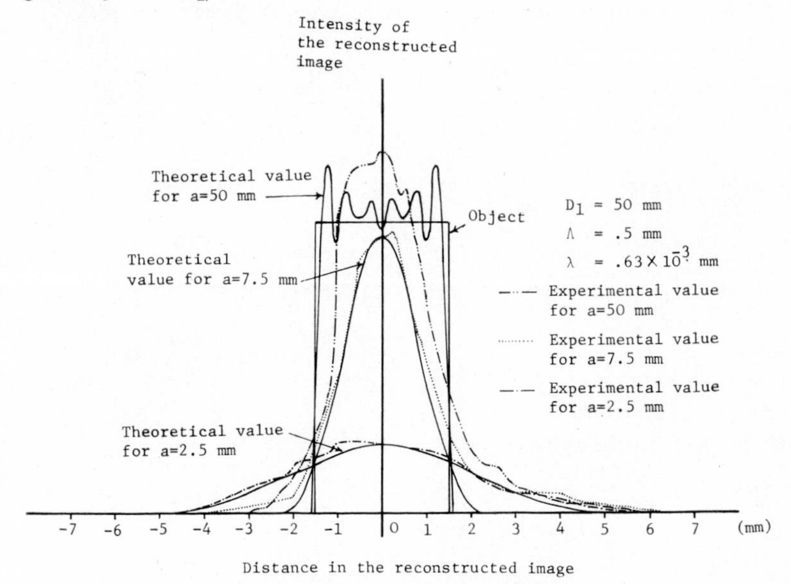

FIG. 3. Intensity distributions of the reconstructed images of a slit.

In Figure 3 the variation of the resolution is shown as a function of a. The object is a slit whose aperture is 3 mm.

Resolution in the Direction z

The slit made of sound absorber is placed at a distance d from the object as shown in Figure 4, where 2ρ is the width of the slit and l is its length. If the coordinates z_1 and z_2 are taken as shown in the Figure 4 and a point sound source is placed at the point B, then the sound pressure $P(b)$

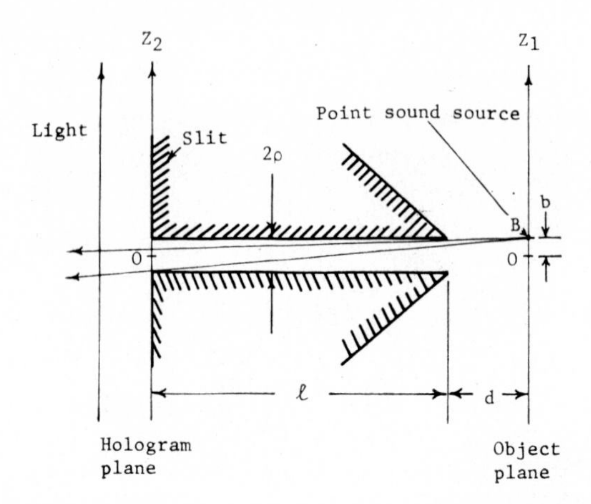

FIG. 4. Schematic diagram of a slit made of sound absorbent material.

which contributes to the diffraction of the light at the outlet of the slit is derived as follows:

$$
\begin{aligned}
b) &= \frac{c}{l+d}\int_{-\rho}^{\rho} \exp\left\{ j\frac{K}{2(l+d)}(z_2-b)^2 \right\} dz_2; && (O \leqslant b \leqslant \rho) \\
&= \frac{c}{l+d}\int_{-\rho}^{\frac{\rho(l+d)-lb}{d}} \exp\left\{ j\frac{K}{2(l+d)}(z_2-b)^2 \right\} dz_2; && \left(\rho \leqslant b \leqslant \rho+\frac{2\rho d}{l}\right) \\
&= 0 && \left(\rho+\frac{2\rho d}{l} < O\right).
\end{aligned}
\qquad (9)
$$

where c is a constant.

By using the inequality $l \gg \rho$ this is calculated as follows:

$$
\begin{aligned}
P(b) &= \frac{2c\rho}{l+d} && \text{if } O \leqslant b \leqslant \rho \\
&= \frac{c}{l+d}\left\{ 2\rho + \frac{l(\rho-b)}{d} \right\} && \text{if } \rho \leqslant b \leqslant \rho + \frac{2\rho d}{l}.
\end{aligned}
\qquad (10)
$$

Then the variation of sound pressure $G(m)$ which contributes to the diffraction of light is given by:

$$
G(m) = \int_{-\infty}^{\infty} O(z_1 - m)P(z_1)dz_1 \qquad (11)
$$

where m indicates the distance of movement of the object O.

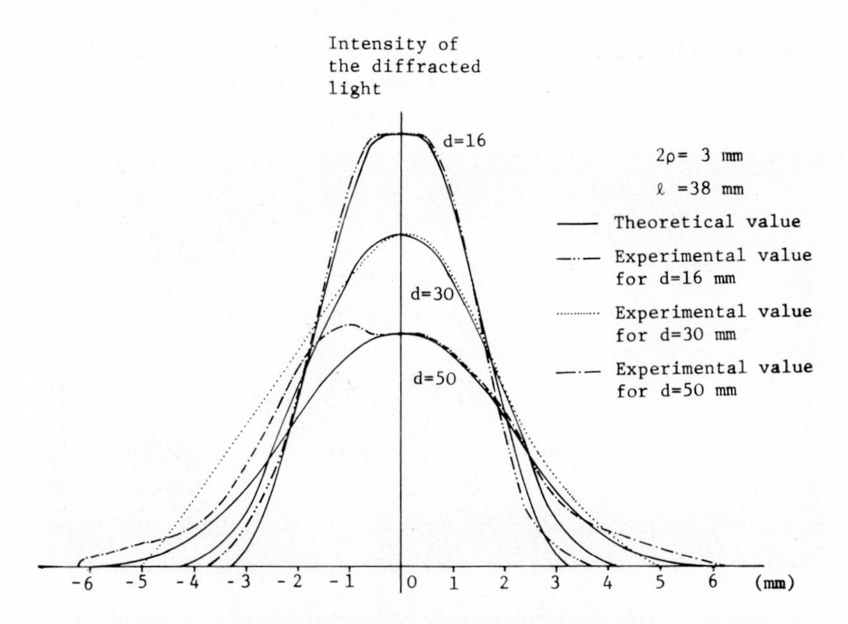

FIG. 5. Intensity variation of the light diffracted by ultrasonic waves as the point sound source is moved in the direction z.

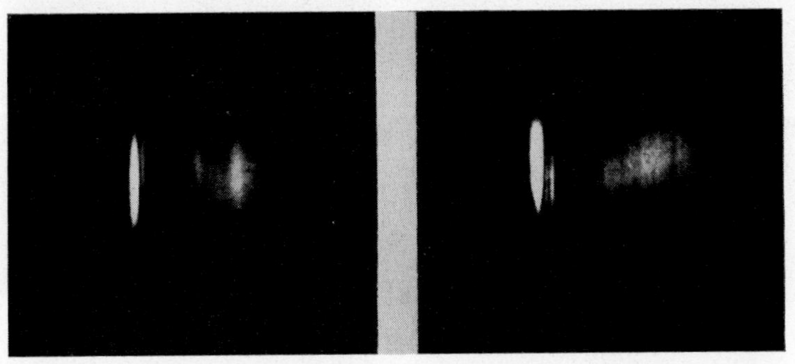

FIG. 6. The reconstructed images of slits placed at (*a*) 38 mm, and (*b*) 63 mm from the hologram plane. If one of the slits is focused the other becomes blurred.

Now the Fourier transform of $P(z_1)$ is given by

$$\hat{P}(\omega)=\frac{cd}{\omega^4(l+d)}\sin\frac{\rho d\omega}{l}\sin\rho\left(1+\frac{d}{l}\right)\omega \qquad (12)$$

hence if the pass band of $\hat{P}(\omega)$ is taken up to the first zero-point, we can see that the grating whose interval is smaller than $2\rho[1+(d/l)]\omega$ cannot be resolved on the screen.

For example, a slit of $2\rho = 3$ mm and $l = 38$ mm is placed behind the object, and the object is illuminated by 3·1 MHz ultrasonic waves. The variation of the intensity of the diffracted light is shown in Figure 5 as the

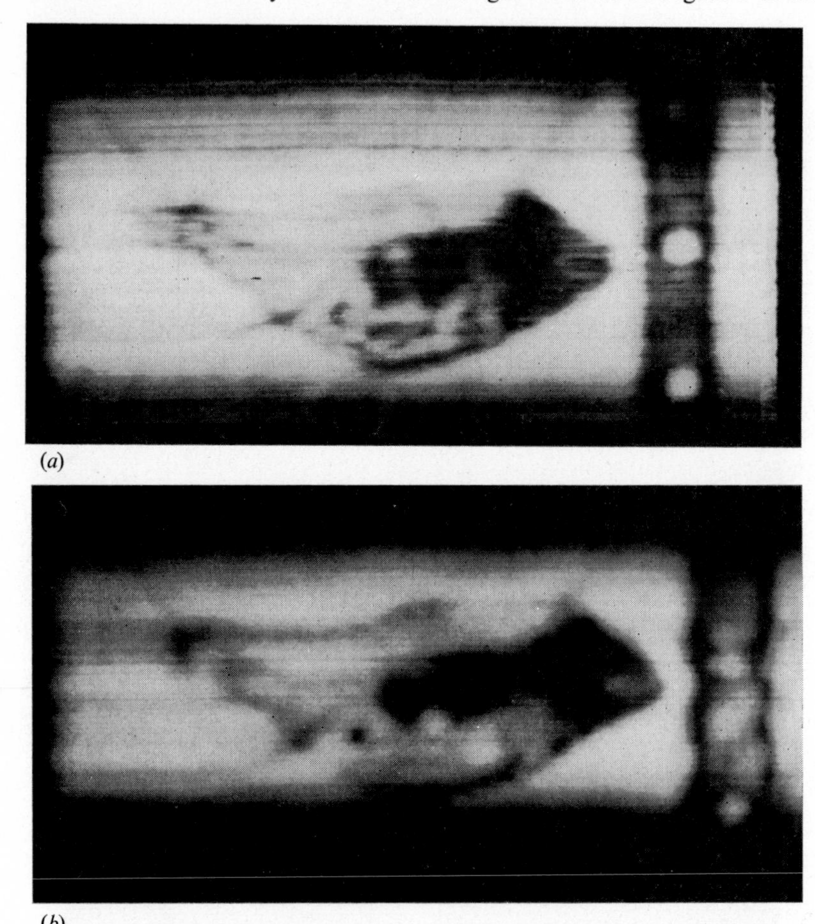

(*a*)

(*b*)

FIG. 7. The reconstructed image of a goldfish using ultrasonic waves of frequency: (*a*) 5·7 MHz; (*b*) 3·2 MHz.

object is moved in the direction of z_1. The theoretical estimation is in good agreement with experimental result.

EXPERIMENTS

The object immersed in water is illuminated with 3·1 MHz ultrasonic waves from a 30×15 mm P.Z.T. transducer. The wavelength in water is 0·5 mm. A He:Ne gas laser is used as a light-source and the light is collimated by using a 150 mm-diameter concave mirror.

To show that it is possible to discern the objects placed at different distances from the hologram plane by adjusting the focus of the magnifying lens, two slits of 2 mm in width are placed at distances of 38 mm and 63 mm from the hologram plane. The reconstructed images of the slits are shown in Figure 6. We can see that if one of the slits is focused the other becomes blurred.

The reconstructed images of a goldfish are shown in Figure 7. The black region in the belly corresponds to the bladder.

We can see the difference of the images for two different frequencies of ultrasonic waves. By using three different frequencies of ultrasonic waves, three images may be taken and projected on the same screen with red, green and blue lights, thus obtaining coloured ultrasonic images.*

CONCLUSIONS

Theoretical and experimental studies are described concerning a new ultrasonic imaging technique which displays the optically reconstructed image of an acoustically illuminated object in quasi-real time. Ultrasonic waves in the frequency range 1 to 10 MHz can be used, adopting the acoustical slit and an optical filtering device, accordingly and the reconstructed image is obtained as the sum of sliced images.

Because of the range of the frequency used this device has many applications, especially in medical diagnoses and non-destructive testing, and it is hoped that the method may be used effectively in these fields.

Acknowledgements

The authors wish to express their sincere thanks to Professors J. Ikebe, E. Mori, J. Tsujiuchi and M. Okujima and Mr. S. Ohba of the Tokyo Institute of Technology for their support and advice. Thanks are also due to the Matsunaga Science Foundation for their generous support.

REFERENCES

KORPEL, A., 1966, *Appl. Phys. Letters*, **9**, 425.
KREUZER, J. L., 1967, *Modern Optics*, Polytechnic Press, p. 91.
MAROM, E., 1968, *Appl. Phys. Letters*, **12**, 26.
MASSEY, G. A., 1967, *Proc. I.E.E.E.*, **55**, 1115.
METHERELL, A. F., 1967, *J. Acoust. Soc. Amer.*, **42**, 733.
MULLER, R. K., 1966, *Appl. Phys. Letters*, **9**, 32.
SANEYOSHI, J., KIKUCHI, Y. and NOMOTO, O., 1966, *Handbook of Ultrasonic Techniques*, Nikkan Kogyosha Press, p. 178 (Japanese).

---

*Unfortunately, it is not possible to reproduce colour prints in these Proceedings.

New Systems of Precision Optical Pyrometry

V. V. Kandyba

Committee for Standards, Measure and Measuring Instruments,
Kwartal Yugo-Zapad 38,
189 *Moscow*

Abstract—Progress in engineering science and the development of new branches of technology require a radical improvement in high temperature measurements. This in its turn requires the improvement of the accuracy of reproduction and transfer of the International Practical Temperature Scale (I.P.T.S.), i.e. to improve the accuracy of high temperature standards.

Up till now visual optical pyrometers with disappearing filaments have been used as standards for high temperature measurements. A classical optical scheme of such pyrometers is shown in Figure 1 (Riband 1934).

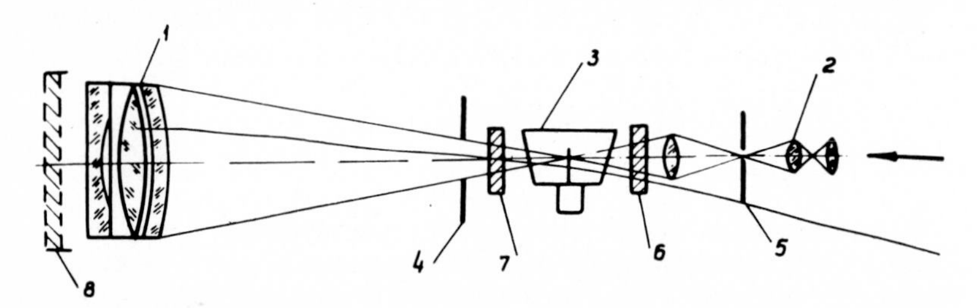

FIG. 1. Classical optical system of disappearing filament pyrometer: 1, objective lens; 2, eyepiece system; 3, pyrometer lamp; 4, input diaphragm; 5, output diaphragm; 6, light filter; 7, absorbing glass; 8, additional absorber for high temperatures 6000–10,000°C.

One of the best visual optical pyrometers, the pyrometer of the ЗОП-51-type has a standard temperature deviation equal to 0·5°C at the 'gold point' (Kandyba 1951). The standard deviation of reproduction of the temperature scale established by means of this pyrometer is given in Table 1 (Phinkelstein & Kandyba 1959).

A further increase of accuracy by means of visual optical pyrometers does not seem possible due to the limitations of the human eye. Considerable efforts have been made to replace subjective methods of measurement by objective ones using photocells and photomultipliers.

A first model of this pyrometer (Figure 2) was patented by us (Kandyba & Kovalevskiï 1956). Here the human eye is replaced by a photocell but the scheme remains the classical one (Figure 1) the only difference being that the pyrometer lamp is movable. During the measurement the lamp is either placed on the optical axis of the apparatus or moved aside (in the diagram—downwards). This causes the modulation of the light beam incident upon the photocell.

TABLE 1.

Temperature (°C)	Standard deviation (°C)
1063	0·5
1400	0·8
1800	1·0
2000	1·5
2500	2·0
2800	3·3

When the brightness of the lamp filament and that of the object measured are equal, the movement of the lamp will not change the intensity of the beam incident upon the photocell. Unequal brightness causes the appearance of an alternating photo-current component. The temperature

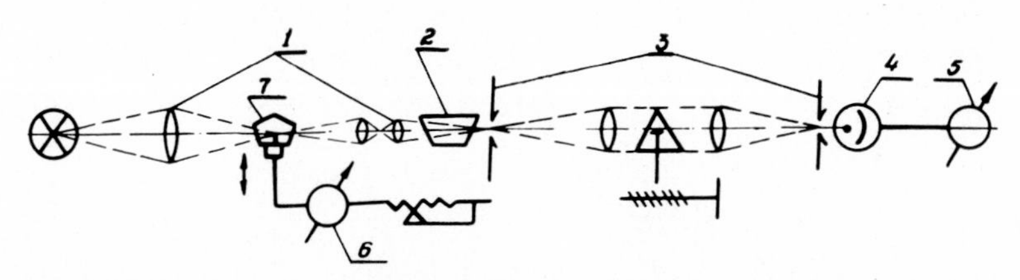

FIG. 2. Classical system but with photocell in place of eye: 1, objective lens and the eyepiece of the pyrometer; 2, Dove prism; 3, diaphragm; 4, photocell; 5, indicator; 6, ammeter; 7, pyrometer lamp.

measurement (photometry) here consists in matching both the filament and object brightness until the variable component vanishes. Therefore the method was called a zero modulation method. Subsequently our method was made use of by R. D. Lee (1962) in his NBS photoelectric pyrometer. According to Lee (1966)the application of this method provided an increase in accuracy presented in Table 2.

Usually the reproduction and transfer of the temperature scale are carried out by means of temperature lamps (band lamps). So, the systems described are designed for the use of such lamps.

In our investigations we also used the principle of matching the brightness of two band lamps—the method of direct temperature comparison (Kandyba & Kovalevskii 1957). A photoelectric spectropyrometer-comparator СПК (Kandyba & Kovalevskii 1956) based on this principle

TABLE 2.

Temperature (°C)	I.P.T.S. realization error (°C)
1063	0·06
1250	0·12
1650	0·27
2330	0·7

has the following optical arrangement (Figure 3). The pyrometer lamp is here missing and instead a direct comparison of the brightness of two band lamps is made. The main element of the scheme is a modulator. This is a switching device which directs the light beams from the two lamps alternately onto the monochromator slit. When the brightness of the lamps is equal the substitution of one beam for another results in no change of the photomultiplier current. If this is not the case an alternating photo-current component appears which is amplified and rectified for indication. The block-diagram of the СПК spectro-comparator is shown in Figure 4.

TABLE 3.

Temperature (°C)	Standard deviation (°C)
1063·0	0·05
1200	0·11
1400	0·2
1800	0·45
2000	0·55
2500	0·8
2800	1·0

The advantage of the system realized in the СПК as compared to that of the NBS pyrometer consists in the elimination of the error of calibration of the pyrometer lamp. Generally in any pyrometer containing a lamp the error appears twice: the first time when it is calibrated against a reference lamp and the second time when the lamp to be measured is calibrated against it. As the temperature of both lamps in the СПК is measured simultaneously, the error appears only once and thus the I.P.T.S. transfer error is minimized.

The СПК attains a threshold sensitivity of 0·02°C with a standard deviation of the temperature reproduction at the 'gold point' of 0·05°C (Kovalevskii, Ioselson & Kandyba 1956). Thus, the accuracy of the СПК is one order higher than that of the best visual optical pyrometers.

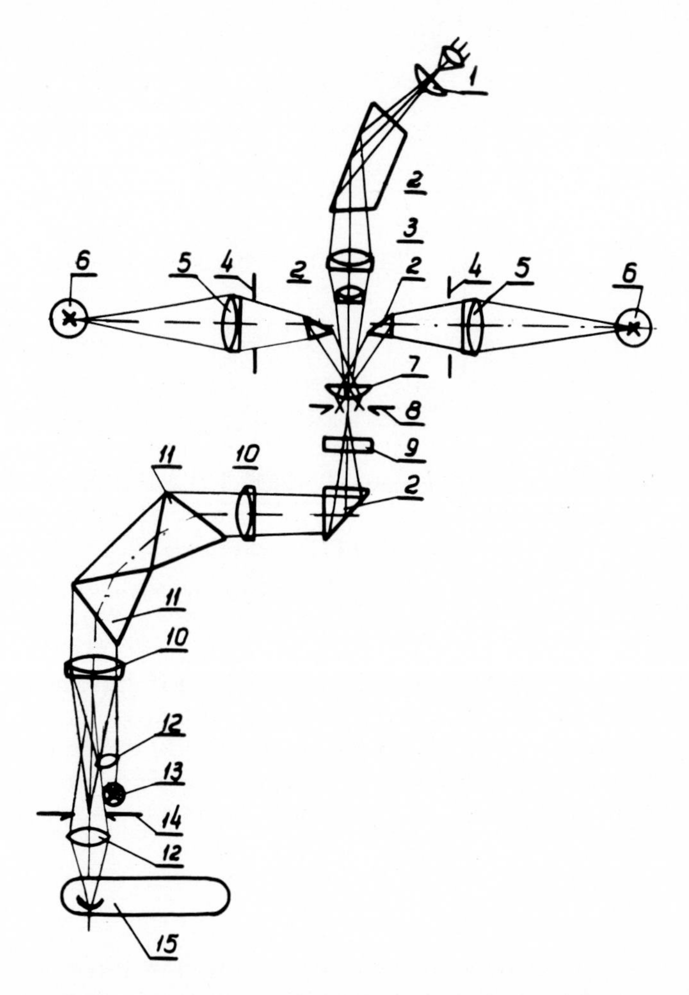

FIG. 3. The СПК Spectrocomparator-pyrometer: 1, microscope eyepiece; 2, reflecting prism; 3, microscope objective; 4, diaphragm; 5, objective of external optics; 6, temperature lamp; 7, modulator prism; 8, entrance slit; 9, light filter holder; 10, monochromator objective; 11, dispersive prism; 12, lens; 13, backward-beam lamp; 14, exit slit; 15, photomultiplier.

The accuracy of the I.P.T.S. realization attained by means of the СПК is given in Table 3 (Kirenkov 1964).

The data refer not only to the apparatus but include also the I.P.T.S. realization errors. In practice the instrumental errors of the СПК are much lower than indicated.

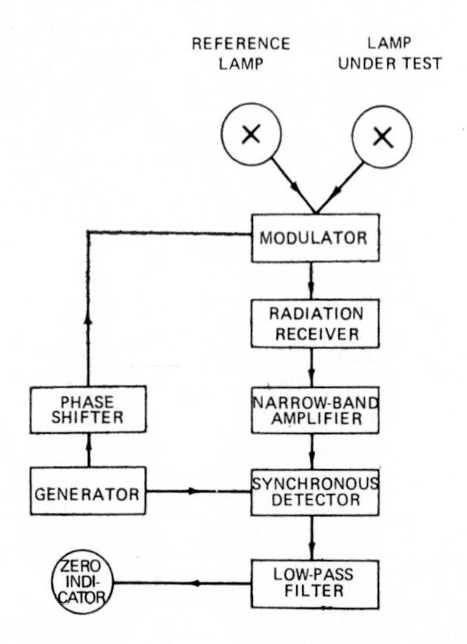

FIG. 4. Block diagram of СПК spectro-comparator.

SUMMARY

A new optical system realized in the СПК spectro-pyrometric comparator ensures a considerable improvement in accuracy for the reproduction and transfer of the I.P.T.S. Taking into account the great prospects of objective methods in optical pyrometry it would be desirable to organize international comparisons of band lamps by means of photoelectric pyrometers.

REFERENCES

KANDYBA, V. V., 1951, Standard optical pyrometers, Zavodskaya Laboratoriya, No. I. p. 36.

KOVALEVSKII, V. A., IOSELSON, G. L. and KANDYBA, V. V., 1956, Objective spectro-pyrometer apparatus СПК-I, *Izmeritelnaya Technika*, No. 2, p. 16.

KANDYBA, V. V. and KOVALEVSKII, V. A., 1956*a*, A photoelectric spectro-pyrometer of high precision, *Doklady Akad. Nauk S.S.S.R.*, **108,** 633.

KANDYBA, V. V. and KOVALEVSKII, V. A., 1956*b*, A method of temperature measurements by means of an optical pyrometer with a disappearing filament. Author Certificate U.S.S.R., No. 102953.

KANDYBA, V. V. and KOVALEVSKII, V. A., 1957, A method of calibrating temperature lamps in precision work on reproduction of International Centigrade Temperature Scale and a spectrometer for its realization. Author Certificate U.S.S.R., No. 105268.

Kirenkov, I. I., 1964, Elaboration and study of methods of precision absolute measurements and high temperature standards, *Dissertation Abstract*, Leningrad.

Lee, R. D., 1962, The NBS photoelectric pyrometer of 1961. Temperature, its measurements in science and industry, vol. 3, part I, New York, p. 507.

Lee, R. D., 1966, The NBS photoelectric pyrometer and its use in realizing the International Practical Temperature Scale above 1063°C, *Metrologia*, **2**, 150.

Phinkelstein, V. E. and Kandyba, V. V., 1959, Nouvelle methode pour l'etalonnage des pyrometres optiques. Nouveau pyrometre optique de precision. Proces-Verbaux des Seances, Comite Consultatif de Thermometrie (Comite International des Poids et Mesures), T-142, Paris.

Ribaud, G., 1934, *Optical pyrometry*, GTTI, M.

An Apparatus for Measuring Aspherical Surfaces

J. Hajda

Institute of Measurement Theory of the Slovakian Academy of Sciences, Bratislava-Patronka, Czechoslovakia

Abstract—A method of measuring aspherical surfaces is described in which two optical benches are used, one fixed and the second pivoting about a vertical axis intersecting the first. The surface to be measured is mounted on the fixed bench and adjusted by means of an auto-collimating microscope until the vertex centre of curvature of the surface is made to coincide with the rotation axis. The rotating or swinging bench is then used to measure, by auto-collimation, the deviations from the vertex sphere at various angles of incidence. Slightly different techniques are used for concave and convex surfaces and for increasing and decreasing curvatures.

In 1967 a method of measuring aspherical surfaces was described. In this method two separate instruments were used, each giving discontinuous measurements only. Since that time this method has been further developed and it was found that one measuring instrument only is required.

The apparatus mentioned, consists of two optical benches: one bench Z is fixed and immobile. The other bench, rotating about a vertical axis, is situated above the first one. The axis passes through the centre of both benches. The aspherical surface to be measured is set up on an adjustable holder on the fixed bench. Both benches are fitted with collimation microscopes (W, S) pointing at the surface to be measured; the position of the microscopes may be read on the scales by means of reading microscopes V and P with an accuracy of 1/100 mm.

Figure 1 shows the layout of the apparatus.

The method here used is an improvement on the well-known auto-collimation method. The aspherical surface is first adjusted in its holder centrally with the microscope of the fixed bench. This microscope is then pointed to the vertex and to the centre of curvature of the surface, in this way determining the radius of curvature of the vertex. Then the holder, with the surface to be measured, is moved so that the axis of rotation of the upper bench passes exactly through the centre of curvature of the vertex. The microscope of the fixed bench is then removed and the microscope on the upper rotating bench is pointed at the surface at a small angle (3°–5°). In general, the microscope in this displaced position will not be normal to the surface. By observing the exit pupil, the holder of the surface is roughly rotated and moved until the pupil is well illuminated. The position of the holder is then adjusted with great care until the auto-collimation maık in the microscope is situated in the centre of the field of view. The distance between the initial and the final position of the holder gives the point where the normal to the surface to be measured intersects the rotational axis. At the same time the reading of the microscope position on the revolving bench gives the accurate position of this part of the aspherical surface. This operation is repeated for further rotational angles of the upper bench so as to measure the whole aspherical surface from spot to spot.

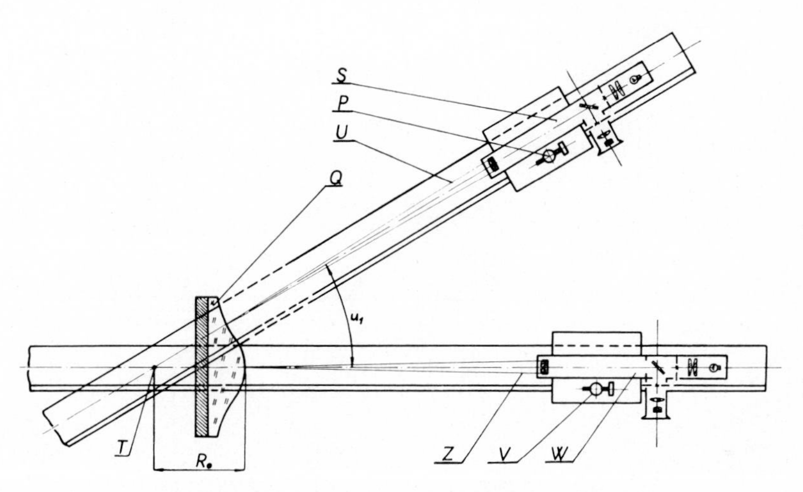

FIG. 1. Schematic view of the measuring instrument.

If a spherical surface is being measured, the centre of which is on the turning axis of the upper bench, the case would be very simple. As shown in Figure 2 it is possible to adjust the microscope on the rotating bench at any angle, so that the surface to be measured would reflect the rays coming from the reticle of the auto-collimation microscope back along their path and thus to the eyepiece of the microscope.

In aspherical surfaces there may exist four cases, depending upon whether the surface to be measured is concave or convex and whether it is more or less curved than the spherical surface corresponding to the vertex. In a convex surface flattening from the centre to the edge, on turning the upper bench the entrance pupil is only partially filled by the reflected light in the direction of an increasing angle of rotation u. When the microscope is moved so that an image of the reticle, projected by the microscope, is on the measured surface, an image mirrored by the surface is seen in the microscope eyepiece. This image is displaced from the centre of the reticle.

The shift may be removed by moving the holder of the surface to be measured G along the fixed bench away from the axis of rotation T.

In Figure 4 the opposite case of a surface more curved towards the edge is shown. The illumination of the entrance pupil of the objective now moves in the opposite direction.

Similar effects occur with concave surfaces, but the direction of shift

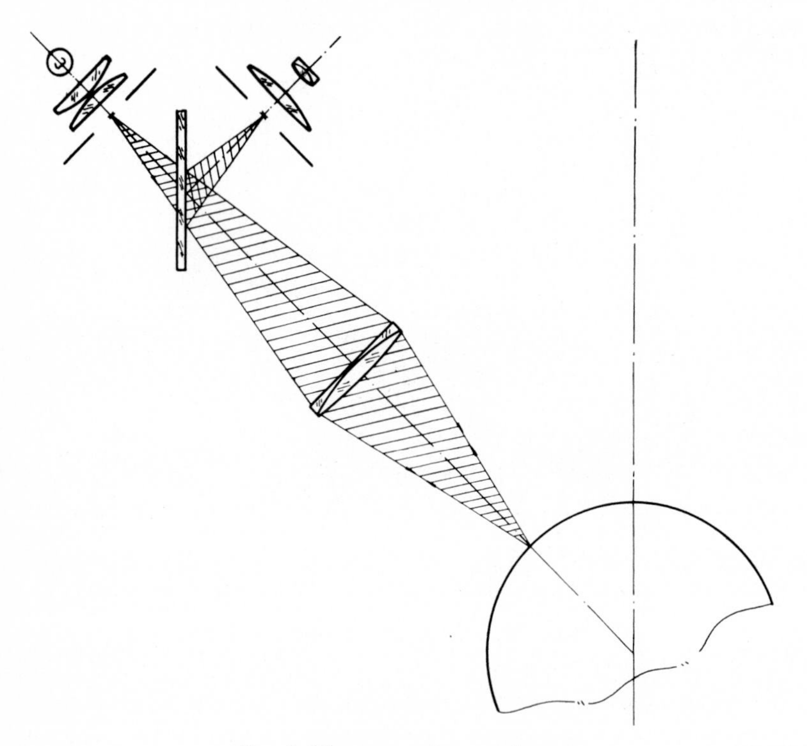

FIG. 2. Measurement on a sphere.

of the illumination in the entrance pupil of the microscope is in the opposite direction.

On the instrument three readings can be made for each turning angle of the upper bench:

1. The angle u;
2. The shift of the holder on the fixed bench with respect to the position corresponding to the vertex;
3. The position of the intersecting point of the normal with the surface to be measured.

The precision of indication of angle is connected with the reading of the movement of the holder. The smallest angle, that can be observed by shifting the holder, to obtain coincidence of the mirrored image with the reticle cross, corresponds to the angular width of the scale lines on the reticle.

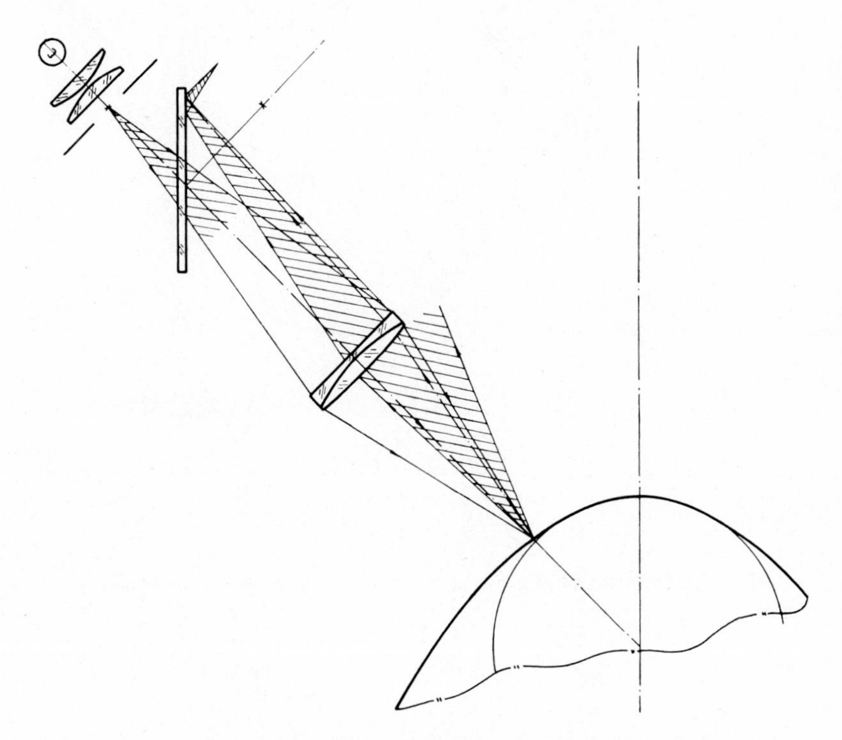

FIG. 3. Measurement of a surface flatter than the vertex sphere.

Thus we say

$$\Delta u = \tan^{-1} \frac{s}{f}$$

where s = width of the scale lines, and f = focal length of the microscope objective. As shown in Figure 5, an error Δu corresponds to an error in the position of the holder of

$$x = N \tan (\Delta u) \sin u,$$

where N = the length of the normal. In our case, when $f = 100$ and $s = 0{\cdot}01$ mm,

$$\Delta u = \tan^{-1} 10^{-4} \simeq 10^{-4}.$$

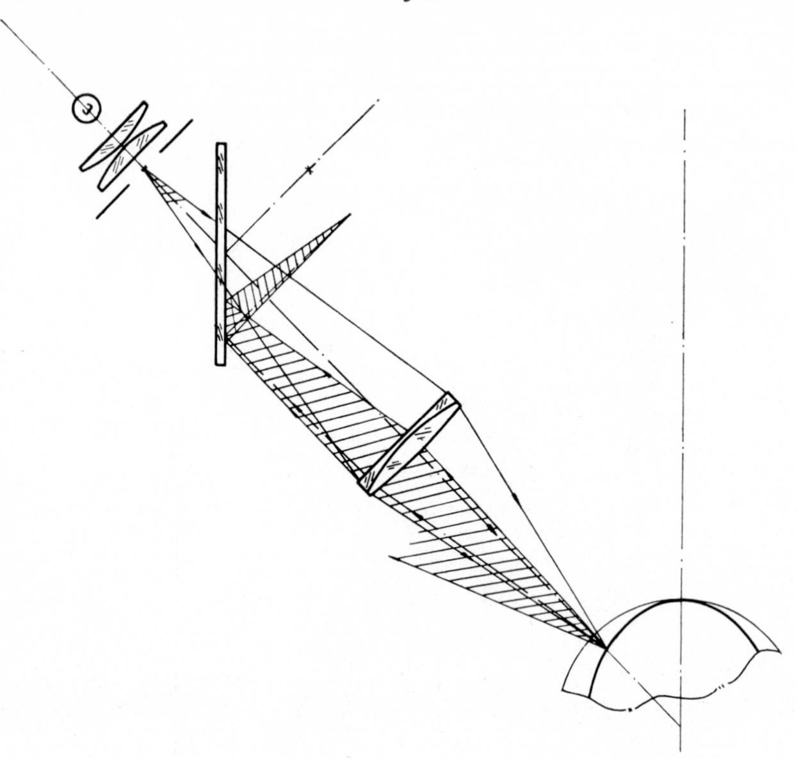

FIG. 4. Measurement of a surface more curved than the vertex sphere.

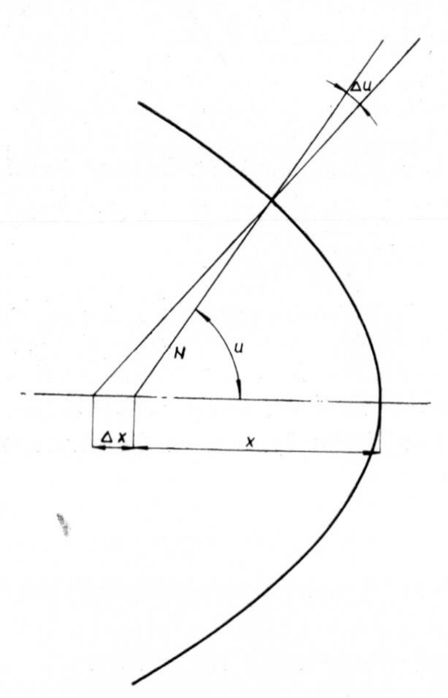

FIG. 5. Ratio of $\triangle u$ and Δx.

Then we have

$$\Delta x = N \sin u 10^{-4}.$$

When $N = 100$ mm and $u = 10°$ we have

$$\Delta x = \sin u 10^{-2} = 0{\cdot}0017 \text{ mm}.$$

Thus under these circumstances a scale graduated in 1/100 mm on the fixed bench is appropriate.

The revolving bench was equipped with the same scale. With a numerical aperture of the microscope of 0·3 it is possible to measure a distance $\Delta x = 3\ \mu$m, using the well-known equation derived from the $\lambda/4$ Raleigh law:

$$\Delta x = \lambda/8 \sin^2 (\theta/2) \qquad (\theta = \text{aperture angle}).$$

For small deviations from the spherical shape one may proceed by pointing the microscope of the revolving bench at the surface of the sphere

FIG. 6. General view of the instrument.

for increasing angles, and reading on the scale of this bench the deviations from the spherical shape. This method may be useful also for measuring convex spherical surfaces. In this case microscope objectives with long focal length must be used, resulting in diminished accuracy of measurement. Using a microscope objective of short focus and wider aperture the spherical shape of the surface can be measured with greater accuracy by revolving the upper bench.

A further development of the apparatus using electro-optical sensors and the possibility of partial automation is planned.

REFERENCE

HAJDA, J., 1967, *J. Sci., Instr.*, **44,** 1005.

Part VI.
Systems design of astronomical instruments

Design of Astronomical Telescope Systems

P. B. Fellgett

Department of Applied Physical Sciences,
University of Reading, U.K.

'A good idea is worth more than a large telescope'—R. O. Redman.

Abstract—The total astronomical observing system must be optimized, not just the telescope, and it is obsolete to express a project in terms of constructing 'an x-hundred inch telescope'. The system includes the atmosphere, collecting optics, analysing equipment, receptors, measuring equipment and (in modern work) digital processors. Classification into flux-collecting and field-imaging methods demonstrates that field imaging may be 10^2–10^4 times faster, and that Schmidts are much more effective than other current configurations. Isoplanatism to better than 0″ ·1 arc is desirable.

1. INTRODUCTION

A distinguished theoretical optician once suggested to me that the principal optical problems of telescopes concern the design of field-correcting lenses. This is rather like asserting that the most important requirement in running a railway is to have good re-railing equipment for rolling stock that has gone off the line. One begins with a telescope, such as a paraboloid or Cassegrain system, which is stigmatic only on the optic axis, and then attempts to correct this defect by additional equipment. An excellent elementary guide to the problems which are then involved has been given by Gascoigne (1968). As a *mise en scéne*, consider the problem of making a parabolic telescope into an anastigmat; that is to say, one free from Seidel spherical error, coma and astigmatism. A direct method of bringing about the necessary Seidel corrections is to place aspheric plates in the beam converging on the focal surface. A single plate evidently gives insufficient freedom to cancel all three Seidel aberrations, but two aspheric plates appear at first sight to be ample for the purpose, since the positions and strengths of the plates provide four disposable parameters.

Following the method of plate-diagram analysis (Burch 1942; Linfoot 1955) the real plates are regarded as imaged into star-space by Gaussian optics at the parabolic mirror (see Figure 1). The condition for the spherical error, coma and astigmatism to vanish is given in Seidel approximation by the relations:

Spherical: $$S_1+S_2=0 \tag{1}$$

Coma: $$-2f+S_1d_1+S_2d_2=0 \tag{2}$$

Astigmatism: $$-4f^2+S_1d_1^2+S_2d_2^2=0. \tag{3}$$

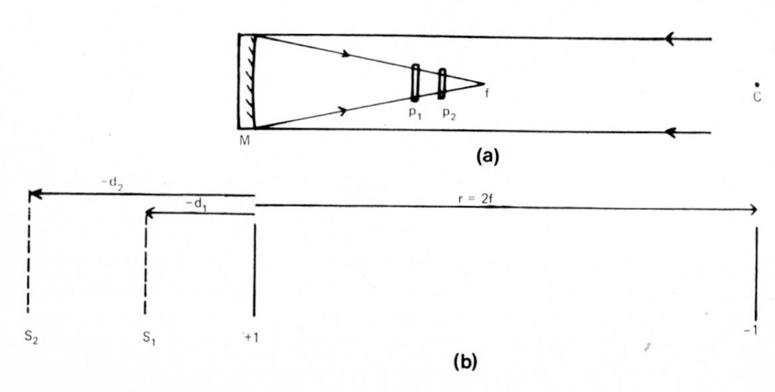

FIG. 1. Plate diagram analysis of field-corrector for parabolic telescope. (*a*) Physical layout of field-corrected telescope: *M*, parabolic primary mirror; p_1 and p_2 aspheric corrector plates; *f*, focus; *C*, centre of paraxial curvature. (*b*) Plate diagram: S_1 and S_2 Gauss images of plates p_1 and p_2.

Substituting Equation (1) into Equations (2) and (3) yields, respectively:

$$S_1(d_1-d_2)=2f \tag{4}$$

$$S_1(d_1{}^2-d_2{}^2)=4f^2 \tag{5}$$

and dividing Equation (5) by (4):

$$d_1+d_2=2f. \tag{6}$$

Equation (6) shows that it is not possible for d_1 and d_2 to be both negative; that is to say, at least one of the real plates must lie beyond the focus of the telescope where it will not intercept the light-beam.

It is therefore necessary either to complicate the system of correction, or to drop one of the conditions (1) to (3). The original Ross correctors (1935) abandoned the condition that spherical error should vanish, and the design in effect traded spherical aberration for coma. Wynne (1949) has shown that the amount of residual spherical aberration in a system field-corrected by an essentially thin lens is determined uniquely by the gross geometry. For practicable spacing of the corrector from the focus, it is of the order of one half the error which the primary mirror would have if it were spherical instead of figured. Since the original Ross corrector was used quite happily by astronomers, it is a little surprising that a spherical error of 1″ arc at prime focus was at one stage raised as a serious objection to figuring the 100-inch Isaac Newton primary for Ritchey-Chrétien working.

In order to obtain anastigmatic performance, it is necessary to add further degrees of freedom. The most obvious method is to use three aspheric plates, and systems of this kind have been described by Meinel (1953).

A second type of solution may be introduced by reference to the Maksutov (1944) telescope configuration. It is well known that the Schmidt telescope simultaneously cancels all three Seidel errors by means of a figured plate placed at the centre of curvature of the spherical primary mirror. It does so essentially because the system has circular symmetry, apart from small foreshortening effects of the nearly-plane corrector plate. However, the length of a Schmidt is twice the focal length, and this is sometimes inconvenient. The Maksutov configuration avoids this objection by using as a corrector a deep meniscus, both surfaces of which are concentric with the main mirror, placed in the parallel beam near to the position of the focal surface. It acts as a weakly diverging lens because the path of the light rays in the refracting medium is longer at the outside of the lens because of greater obliquity to the surfaces. This is a small effect, which it may indeed be advantageous to remove by slight modifications in curvature, and the principal effect is to place a virtual corrector plate at the common centre of curvature. The system is thus made to behave as if there were an optical element outside the physical tube of the telescope. The arrangement is highly successful in small and medium sizes, but it becomes difficult to manufacture or support the deep meniscus in very large sizes.

In a similar way, a refracting surface of a field-corrector convex towards the primary mirror can give rise to a virtual aspheric plate in the physically inacessible space beyond the focal surface. From the Wynne theorem already quoted, it follows that exploitation of this principle requires the field-corrector to have essential thickness, which in turn raises the well-known difficulty of obtaining simultaneous correction for longitudinal and lateral colour error. However, a number of excellent solutions have been successively developed, notably by Wynne (1967, 1968). Aspheric figurings permit the number of surfaces to be reduced, but the final performance is rather similar for both aspheric and spherically figured correctors.

The performance attained is typically an image spread having 1″ arc diameter at off-axis angles of up to $\frac{1}{2}°$. Meritorious though this performance is, the field area amounts to, at most, a few per cent of that attainable by a Schmidt telescope, and moreover the Schmidt gives image spreads of only 0·4″ arc diameter over a field 5° across at f/3·5, and about 1″ arc diameter over a 6° field at f/2·5. Moreover these images are perfectly symmetrical, which is not the case for systems employing field-correctors.

An important practical consideration is that the surfaces of refracting elements having converging power are necessarily relatively steep, and consequently the tolerance of centring is extremely fine, at most a few tens of microns. These relatively steep surfaces, in contrast to reflecting surfaces of similar power, also give rise to relatively large amounts of higher-order aberrations even when the design is carefully balanced.

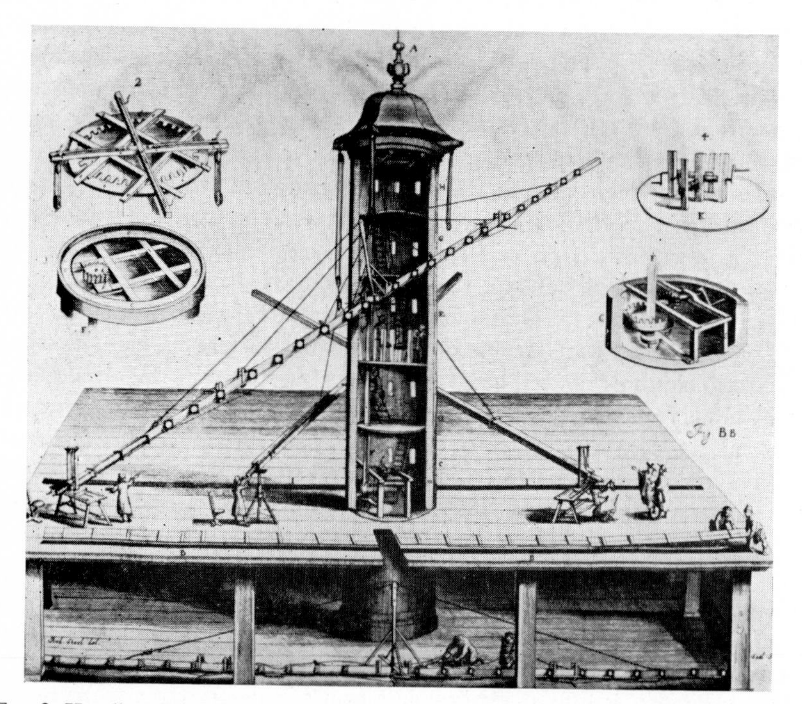

FIG. 2. Hevelius' 150-foot telescope. *By courtesy of Ronan Picture Library and Royal Astronomical Society.*

Copyright free. Available in 'Source Book in Astronomy' by H. Shapley and H. E. Howard, published by McGraw-Hill Book Company, 1929, and reproduced on p. 111 (Chapter: Bradley—'From Bradley's "Miscellaneous Works",' edited by S. P. Rigaud, 1831).

The work of the designers of field-correcting systems has been extremely important in enabling useful astronomical work to be carried out. Nevertheless they were presented by astronomers, as part of the terms of reference outside their own control, with telescope systems which were not, as we have seen, inherently well-suited to field-imaging. It is as if their efforts had made it possible to use a siding built on extremely rough ground where trains frequently became derailed, but clearly the correct route for the main-line railway does not lie in this direction.

Much of the difficulty arises essentially from confusion over what is meant by the power of a telescope. In the seventeenth century, it was taken as axiomatic that the power of a telescope was measured by its length, and some ludicrous extremes were consequently resorted to (see Figure 2). More recently, it has come to be assumed that the power of a telescope is measured by the diameter of its clear aperture. We shall discuss this more systematically in Section 3.2, but it may suffice to point out here that this assumption is reasonable enough in relation to flux collection but not field-imaging. The importance of field-imaging methods lies, as we shall see in more detail below, in their enormous efficiency and effectiveness in collecting astronomical information. It is sometimes objected that the majority of objects studied in astronomy appear essentially as point-objects, and are therefore not susceptible to study by field-imaging. The fallacy in this argument is that although individually the objects may appear as points, the system of such objects, for example stars or quasars, is extended over virtually the whole sky (Fellgett 1964, 1969).

In field-imaging, the capacity of a telescope to collect information about the sky is evidently proportional both to the square of the linear aperture and to the square of the diameter of the angular field. On this basis, a Schmidt of 20-inch aperture is at least as powerful as a 100-inch Ritchey-Chrétien, for example, working over a field 1° across and even so probably with lower accuracy. It will be shown in Section 3.3.2 that not only are the powers of these two telescopes equivalent in an abstract sense, but they are able to record the same information in the same time and down to the same limiting magnitude.

The objections sometimes raised against the Schmidt telescope, namely its comparatively long tube and the difficulty of obtaining a refracting disc of optical quality for the corrector plate, would indeed be apposite in relation to Schmidt telescopes of very large aperture, for example 200-inch diameter. However, even a comparatively small Schmidt can record in a few minutes images of anything from 10 to 100 thousand stars, and these images are susceptible to accurate measurement for position and brightness. This information capacity is already devastingly high in relation to current means of measurement and reduction of the resultant data, and there is accordingly no need to build very large Schmidts at the present time. The cost of a 20-inch telescope is in any event, even if it has a long tube, much lower than that of a 100-inch. There is no insuperable difficulty in obtaining refracting discs for use with mirrors up to at least 100 inches in diameter; for example, one such disc was procured for the

Isaac Newton at a time when a Schmidt design had been adopted for this telescope, and so far as the author is aware this disc is still extant (Woolley 1962).

2. SYSTEMS ANALYSIS AND DESIGN

The starting point of the previous section was the problem of optical design of astronomical telescopes as often presented by astronomers to opticians. The discussion which next follows is directed towards a systematic discussion of the basic problem of telescope design, namely to give the greatest effectiveness and economy in astronomical observation. This formulation makes it at once evident that the problem is too narrow as stated. The telescope is one out of many sub-systems all of which contribute to the effectiveness of the overall observational system. It is this overall efficiency which must be maximized; the design of any one sub-system cannot be considered in isolation from the rest, and it is improper to maximize the performance of any one component without regard to the effectiveness of the system as a whole.

In the days when visual observation was predominant, the telescope was the only sub-system susceptible to design. There was indeed a degree of control over the observer, in respect of his selection, training, and the use made of him. There was not in those days much opportunity to choose the geographical site of the telescope, observatories usually being set up close to established institutions of learning; consequently there was little control over the atmosphere through which the observations were made. There was therefore a measure of appropriateness in treating the design of the telescope itself as the principal part of the problem, since the other sub-systems were more or less standardized and their properties well known to everyone working in the field.

It is far otherwise today, when a large number of sub-systems are involved, some of considerable complexity, which may differ from one installation to another and in accordance with the purpose of the work being undertaken. Some of the more important of these sub-systems are listed below, and their properties and interaction are discussed.

Under these new conditions, it is no longer appropriate to propose or talk about 'an x-hundred inch telescope', where x takes a value such as 1, 1·5, or 2·3, without defining carefully the purpose of the installation, the other sub-systems with which the telescope will interact, and the overall systems design. This new approach accords well with what has already been customary in radio-astronomy, and is also notable in the Australian work in solar physics, where the design has been directed to carefully specified objectives. Adequate specification is not synonymous with narrowness, but is equally applicable to purposes of great generality.

A notable example in radio-astronomy is the development of aperture synthesis from the initial starting points of single-dish radio telescopes on the one hand, and interferometers on the other. These are both in themselves undesirable extremes, and the aperture-synthesis radio telescope greatly out-performs either.

For the purpose of systems design, it is convenient to classify methods of astronomical observation into field-imaging, flux-collecting, and point imaging. In the first of these, it is required to image an extended field in the sky with the greatest possible precision. In flux-collecting, the need is simply to collect as much radiation as possible from the object to be studied, so that this can be fed to suitable analysing equipment. The multiplex interferometric spectrometry of P. and J. Connes (1966, 1969) is a notable recent example, in which they have attained resolution and accuracy in the infrared spectra of Mars and Venus not inferior to that of standard atlases of the solar spectrum. By point-imaging is meant the class of observations in which precise imaging is required only of a single object which may be set on the optic axis of the telescope. A large amount of astronomical work is at present done in this way; but the discussion in Section 4 will suggest that this is largely predicated by the availability of telescopes suited to this mode, and does not represent optimal systems design.

Field-imaging methods have a special significance because they are capable, as we have already seen, of enormous throughput of information. Evidently a field-imaging method which is capable of giving measurements simultaneously on 10,000, or even only 100, objects at a time is likely to contribute the majority of the astronomical information at our disposal, even if individually it works considerably less efficiently than one-star-at-a-time point-imaging methods. It is accordingly these field imaging methods which deserve the closest attention in design, and on which a major part of the astronomical budget should be spent.

3. DISCUSSION OF SUB-SYSTEMS

3.1. *Atmosphere*

3.1.1. *Seeing and Transparency.* The properties of the atmosphere which affect astronomical observations are transparency and seeing. The meaning of transparency is obvious. Due to Rayleigh scattering, augmented by any impurities or haze that may be present, the light from a celestial object is attenuated in its passage through the atmosphere by an amount of the order of one magnitude at the zenith, increasing roughly as the secant of the zenith distance. For specialized purposes, for example observations of the solar corona, interference by this scattered light is important. However, these applications are likely to be supplanted, or at least greatly supplemented, by observations from outside the atmosphere.

The loss of light is of some importance in itself, and the variability of the attenuation is often the principal limitation to accuracy of brightness measurement of objects which are not observed simultaneously, as for example in conventional one-star-at-a-time photo-electric photometry. American astronomers habitually speak of 'the zenith extinction for the night'. Observers used to the more variable European skies, expressed by the jingle

> Photometric sky, photometric sky,
> Not too wet and not too dry

tend to regard this as a cruel joke.

The term 'seeing' has a more specialized meaning, and is sometimes confused by non-astronomers with transparency. It correctly refers to the optical effects of inhomogeneities of refraction of the atmosphere. The wave-fronts of starlight are substantially plane on entering the atmosphere, but become corrugated in passing through an inhomogeneous layer. If these corrugated wave-fronts propagate in undisturbed air, the corrugations become partly transmuted into variations of intensity over the wave-front, and a statistical balance between these two kinds of disturbance is built up in a calculable way (Booker, Ratcliffe & Shinn 1950). It is convenient to refer to the two effects respectively as phase-scintillation and amplitude scintillation (Fellgett 1956). Only phase-scintillation is present immediately below a disturbed layer, for example at the telescope if the disturbances are due to inhomogeneity in the first few hundred metres above the ground. Intensity scintillation builds up to its full amount over a drift-distance which is typically of the order of 10 kilometres.

Amplitude and phase-scintillation are similar to placing a screen of variable absorption or delay over the aperture of the telescope, and consequently both effects degrade the resolution. Even under the best conditions, the image given by a telescope is seldom less than 1″ arc in diameter, corresponding to the diffraction-limited resolution of a telescope of about 10 cm aperture. The human eye can, however, exploit favourable statistical fluctuations when the disturbances are momentarily unusually small, and in this way can utilize the diffraction-limited resolution of telescopes up to about 50 cm diameter. This facility is not, of course, available in ordinary photography.

Another important effect of astronomical seeing is on the efficiency of collection of radiation. As we shall see (Equation 10) seeing can be as important as aperture in determining the effectiveness of a stellar spectrograph.

3.1.2. *Site Testing and Selection.* There is, of course, at present no means of controlling the seeing at a particular site, but it is possible to choose sites having favourable seeing. This is mainly a function of altitude and latitude, with modifications by the local topography. Altitude is favourable in reducing the amount of atmosphere that has to be traversed by the light, and also sometimes in other ways. For example, being on a mountain may place the observatory above a temperature inversion or trade-wind layer. High seeing is adversely affected by temperate-zone depressions and fronts, and present evidence suggests that the bands at latitude $\pm 20°$ are particularly favourable. Surprisingly little systematic work appears to have been done on the effects of local topography. It seems to be very advantageous to raise the telescope 100 m or so above ground level, but no systematic investigation of the variation of inhomogeneity with height is known to the author prior to the work of Coulman (1969). An astronomical tradition asserts that seeing is better when the surrounding ground is covered by trees or bushes. This is certainly true for daylight observations, as in solar physics, but appears doubtful at night. Any camper knows that it tends to be warmer under vegetation, and presumably this warm air convects upward throughout the night. The author's observations suggest that a sea cape facing the prevailing wind can be favourable.

In view of the enormous importance of seeing and transparency on the effectiveness of a telescope, it has to be admitted that our present methods of site testing are inadequate, at least in respect to seeing. The statistics of the auto-correlation function of the complex amplitude representing the wave-front should be determined up to distances equal to the aperture of the proposed telescope. Few if any attempts have been made to measure this quantity interferometrically, and reliance has mainly been placed on attempts to extrapolate on doubtful theoretical bases from results obtained with small instruments. In terms of cost-effectiveness, it would be desirable to pay much greater attention to this aspect of systems design.

3.2. *Telescopes*

3.2.1. *Point Imaging.* Many existing large telescopes have a primary mirror figured to a paraboloid, and are used either at prime focus or in a Cassegrainian arrangement. Such systems cannot be regarded as stigmatic except essentially on the optic axis (Fellgett 1968*b*, 1969). For example, the coma wave-aberration of an f/3 paraboloid reaches the traditional figuring tolerance of $1/10\lambda$ at only $\sim 1''$ arc fiom the axis! Such telescopes are accordingly suited mainly to point-imaging work. Of work of this kind, spectrography is currently the most significant in terms of telescope-hours. Indeed large telescopes all over the world probably spend more time on spectrography than on all other programmes put together. It is accordingly appropriate to discuss here some relationships concerning the cost-effectiveness of slit spectrography; we shall return to a discussion of those aspects particularly concerned with photographic emulsions in Section 3.3.3, and compare this method with objective dispersion in Section 4.1.

If the slit of a grating spectrograph is opened until nearly all the light is admitted, the maximum resolution r that can be attained is

$$r \lessdot r_0 = \frac{b}{A\alpha} \tag{7}$$

where α is the diameter of the seeing-disc of the star, A the diameter of aperture of the telescope and b the optical delay between extreme beams striking the grating (Fellgett 1969). Under these conditions a mean signal-to-rms-noise ratio R can be attained, in each of n resolved spectral elements in the spectrum of a star giving Q photons per unit time and area, for an exposure time t given by

$$Qt = \frac{4nR^2}{\pi \epsilon_e A^2} \qquad (r \lessdot r_0) \tag{8}$$

where the meaning of ϵ_e is given in Section 3.3.1. If $n \sim r$, as is true over about an octave of spectrum,

$$Qt \sim \frac{4rR^2}{\pi \epsilon_e A^2} \qquad (r \lessdot r_0). \tag{9}$$

The limiting resolution r_0 depends, according to Equation (1), only on the relative sizes of telescope and spectrograph, measured respectively by A and b; interestingly no other dimensions are significant. For resolutions above r_0, the slit jaws must be closed so that approximately r_0/r of the incident light is admitted, giving

$$Qt \sim \frac{4\alpha r^2 R^2}{\pi \epsilon_e A b} \qquad (r > r_0) \tag{10}$$

Increasing the aperture of the telescope merely increases the length of slit illuminated, giving a gain in light proportional simply to A. These relations are illustrated graphically in Figure 3 in which the relative cost of the telescope has also been indicated on the assumption that it is proportional to the cube of the linear size. This accords with the engineering rule of thumb that a structure of a given kind costs approximately a constant amount per tonne.

This diagram raises serious questions about the cost-effectiveness of large point-imaging telescopes used in this mode. In the region of Equation (10) it is as profitable to increase b as A, and even in the more favourable region of Equation (9) telescope cost increases more rapidly than the advantage gained in Qt. This suggests that a multiplicity of small telescopes should be used, rather than one large expensive one, the number being chosen to balance capital against operating costs. Fastie (1967) has proposed to use a tesserated mirror in just this manner, the tesserae sharing a single spectrometer. Another effective method is discussed in Section 4.1.

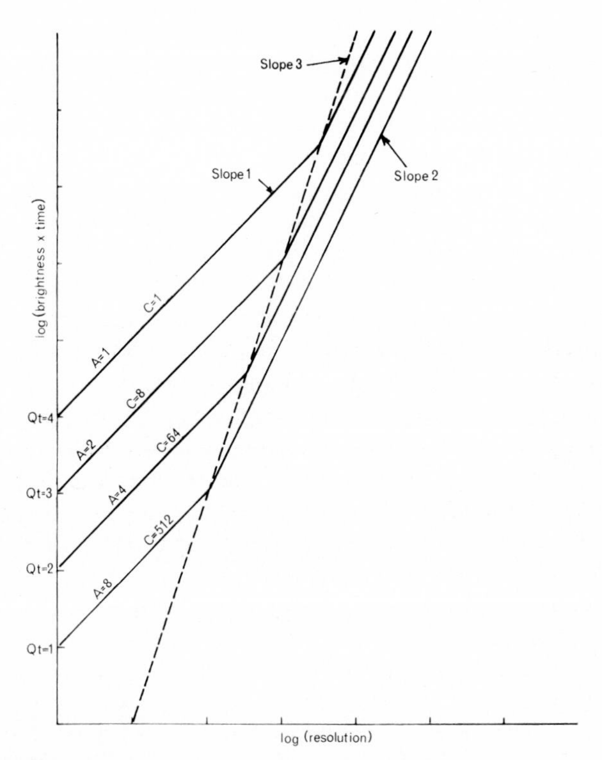

FIG. 3. Necessary exposure Qt as a function of spectrographic resolution for different aperture A and relative cost C of telescope.

In order to image accurately at any point of its field, a point-imaging telescope must be constructed with high precision, and this inevitably results in high cost. It appears doubtful whether the additional cost of attempting field-imaging is really worth-while, since none of the resultant systems approach anywhere near the performance of a Schmidt, as we have seen. There is indeed the need for a moderate number of fairly large point-imaging telescopes in the world, for those problems which can only be handled by this mode of working. However analysis indicates that many problems could be handled either by cheaper flux-collecting methods or by the much more efficient field-imaging methods, and that the predominance of the point-imaging approach may be a habit engendered by the fact that most of the large telescopes in the world today are essentially suited only to this mode.

3.2.2. *Flux-Collecting*. Point-imaging has been described first, since it corresponds most closely to the traditional design of astronomical telescope. However this is a logical inversion, since most applications of such telescopes, such as spectrography, are usually expressed in terms of the amount of radiation that can be collected, and it is necessary to state additional constraints in order to establish the necessity of collecting this radiation into a very small image.

With respect to spectrography itself, the major such constraint is the very low luminosity (product of image-area and solid-angle accepted) of conventional grating spectrographs. Other reasons include the necessity to separate the radiation to be studied from an interfering background, for example light of the night sky or light from a brighter neighbouring star. In special cases, for example in planetary investigations, it may be desired to analyse separately the radiation coming from different parts of the object. Nevertheless the luminosity constraint is often the crucial one, and when it is removed flux-collecting methods can be used in a wide range of problems. For example, in the planetary infrared spectrography of P. and J. Connes (1966, 1969) relaxed imaging tolerances are acceptable because the luminosity of the interferometric spectrometer is fully adequate to accept even a relatively broad image.

The relaxed demands on accuracy of construction react favourably on cost and lead-time. A flux-collector of 3 m aperture may be ranked as comparatively small, and one of 10 m aperture can be constructed much more cheaply than a major astronomical telescope of point-imaging design. An aperture of 25 m, similar to that of a successful class of radio-telescope, appears perfectly practicable (Horn-d'Arturo 1965, Fellgett 1967*a*, Mertz 1969).

The problem has been simplified in recent years by the development of new material and structural methods. New layouts have been proposed which are adapted to the sole task of feeding the radiation to fixed analysing equipment, without the constraints imposed by the multifarious demands on conventional telescopes. These matters have been discussed elsewhere, and reference may be made particularly to the work of Neugebauer, Martz & Leighton (1965) and of P. Connes (1969).

A structure proposed by the present author makes a complete break with conventional axes of rotation, and uses instead a set of six hydraulic legs the lengths of which are computer-controlled so as to enable the telescope to be pointed in any direction, while maintaining rotation about a fixed point where a servo-controlled mirror can direct the radiation into the analysing apparatus. It will be seen from Figure 4 that in this configuration the 'dish' forms the major part of the structure, which is evidently economically desirable.

FIG. 4. Model study of flux collector using six servo-controlled hydraulic actuators to give rotation about a fixed point. The reflecting structure is well supported and conventional axes are not needed.

3.2.3. *Field-imaging*. The tolerances on image quality for field-imaging applications are often under-estimated. The basic problem in the reduction of field photographs is that measurements can be made in two spatial coordinates x and y, and also of a brightness parameter which may be designated z. It is required to convert the vector (x, y, z) into the conventional astronomical measures (α, δ, m), in which α is the right ascension, δ the declination, and m the magnitude, upon some chosen system. The required transformation is in general non-linear and non-diagonal. It is customary, though logically it makes no ultimate difference, to adopt some convenient interpolation formula which approximately represents the transformation, and then to perform the actual calculations on the residuals from this assumed formula. If the three-dimensional residual corresponding to any star under measurement were known, (α, δ, m) could be calculated from the interpolation formula. The problem is therefore reduced to one of estimating this residual. This has to be done by calculating the residual for 'standard' stars the positions or brightnesses of which are assumed known. From the (x, y, z) measurement of these stars, the residual corresponding to them is known immediately, and the residual at the programme stars must then be estimated by interpolation. It will be noticed that this formulation is more general than the traditional concepts of plate-constants, magnitude equation, etc., which indeed it includes as special cases.

In the required transformation process, it is difficult or impracticable to extend the interpolations over regions wider than those within which the residuals may be regarded as statistically stationary. Systematic changes are caused by variations in image quality over the three-dimensional space $\langle x, y, z \rangle$. Regions within which the variation in image quality may be neglected, for a particular purpose, are called isoplanatic. A sufficient number of standards must be present within each such region to establish the transformation between (x, y, z) and (α, δ, m). Accordingly the number of standards depends inversely on the square of the linear dimensions of the isoplanatism regions. Since the establishment of standards is in general laborious, this fact is important in the overall systems productivity. It clearly favours wide-field systems such as the Schmidt in which the image quality is very uniform.

It is known experimentally that a variation of 10 per cent in the image spread has an appreciable effect on the photometric behaviour of the images. Since a telescope would normally be placed in a site where the seeing would be as good as 1″ arc, at least sometimes, this implies a tolerance on image spread of 0″·1 arc in the definition of an isoplanatism region for photometry.

The situation with respect to unsymmetrical aberrations, whether of design, construction or maladjustment, is even more demanding. It was emphasised by Danjon (1946) that once an image becomes unsymmetrical it no longer possesses a unique centre. Atkinson (1947, 1955, 1960, 1961, 1967) has emphasized the impossibility of exact adjustment of an astronomical instrument unless adequate criteria of adjustment are available. In the presence of unsymmetrical imaging such criteria are hard to find since, for example, the direction defined by auto-collimation is not necessarily that which pertains to the imaging of a star. In the presence of these difficulties, positional measures may be affected by up to the amount of the unsymmetrical image-spreads present. This argument is not basically affected when this asymmetry is swamped by larger amounts of symmetrical aberration; if an image is visibly unsymmetrical it is certainly unreliable, but the converse is not true.

Following the author's semi-qualitative arguments, Linfoot and Redman (1968) performed model calculations for positional shifts caused by the interaction of symmetrical and asymmetrical aberrations with photographic non-linearity. They concluded that a total coma spread as high as 0″·75 arc may be tolerable for traditional astrometric methods. However the systematic errors found in the international programme of solar parallax observations of Eros at its 1930–31 opposition (Jones 1941, Atkinson 1969), show that such methods are unsafe. M. Flavell and the author (unpublished) found by a semi-analytic method that shifts of 10 per cent of the coma present could occur even with quite mild non-linearity. The limit of accuracy of current astrometry is about 0″·01 in the mean over many plates (about 3,000 plates were used in the Eros work), and Linfoot and Redman adopt 0″·003 as a target giving scope for improvement. Their curves show that with a photographic gamma of 2·3 and 0″·75 of coma it is unsafe to measure stars differing by more than about 2^m, and sometimes much less, unless considerable sky-fog is present, and defocusing and astigmatism are less than 0″·5. We actually need to compare stars over a range of at least 10^m, wish to use high-gamma plates with clean background in the interests of noise-equivalent efficiency (see Section 3.3), and require freedom from the variation of shift with respect to focus, seeing, and other aberrations, as well as the variation of emulsion properties across the plate. Accordingly a much tighter coma tolerance must be imposed, and 0″·1 is suggested as a minimum requirement.

It may be concluded that symmetrical spread diameters not exceeding 0″·4 arc, and asymmetries of less than 0″·1 arc, are already necessary for the best astronomical work, and that further improvements to the limit of optical feasibility can be usefully exploited to give improved accuracy of measurement.

The above arguments serve further to emphasize the superiority of the Schmidt system over other exisiting arrangements. It is particularly important that not only is the theoretical performance of the Schmidt very high, but that practicable means are available for maintaining it in good adjustment. This is something which we do not at present know how to do with, for example, Schmidt-Cassegrain systems in general.

Although the Ritchey-Chrétien has only about 5 per cent of the field-imaging power of a Schmidt for equal aperture, Bowen (1966) has argued that a field 1° across is about as large as it is practicable to use, mainly because of differential refraction over the field. However, he assumes exposure times of several hours. This represents assumptions about the use of photographic emulsions which belong to an earlier stage of understanding, see Section 3.3, and we now know that it is more advantageous to use relatively short exposures and to average over several photographs, preferably by averaging the measurements made on them by automatic measuring engines.

One disadvantage of the Schmidt is that, apart from the choice of focal ratio, Schmidts have geometrical similarity and no freedom exists in the design of the relationship of overall length to system focal length. Indeed the practicable choice of focal ratios is itself rather restricted, values in the region of f/3 being optimal for most purposes.

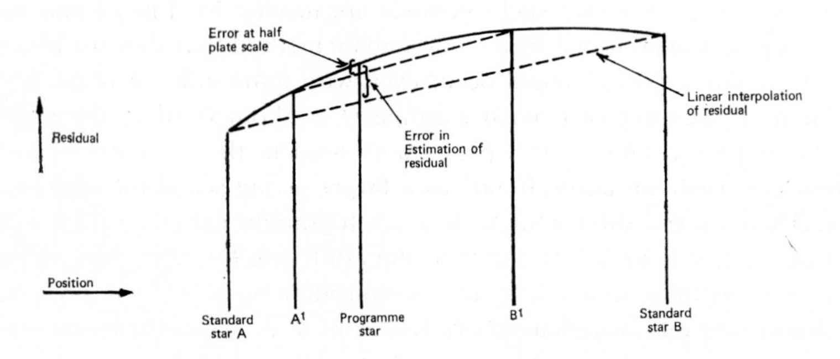

FIG. 5. Effect of plate scale on residual in position (see text).

Fortunately the resulting system focal length is usually quite convenient. Astronomers accustomed to systems which cannot give tolerable image quality at shorter ratios than say f/8 sometimes complain of the small plate-scale of the Schmidt. However this factor is actually favourable to isoplanatism with respect to the inevitable statistical variations in the properties of the photographic emulsion. For photometry this is obvious, and has been confirmed by the results obtained by Reddish (1966) on the Edinburgh Schmidt in which he has been able by photographic observations to correct photo-electric magnitude sequences which had been regarded as standards. For positional measurements, however, the matter bears a little more discussion.

Suppose that there are two standard stars A and B (Figure 5) and that a programme star lies somewhere between them. In the absence of further standards, the best that can be done is to interpolate the residual linearly between the two standards. The leading error in the estimated residual is then that due to curvature of the (unknown) true interpolation, and therefore proportional to the square of distance. Since the systematic part of the residual in position is caused by emulsion shifts, it depends on position on the plate rather than on the sky. Halving the scale of distance thus reduces by a factor of four the leading term in the error of the interpolation. Accordingly the accuracy of positional measures may actually be improved by reduction of plate-scale, in the same way as occurs for photometric measures. The physical reason is that the emulsion shifts at the programme stars are better correlated with those at the standards if they are close together on the plate. Against this, of course, a given error on a small-scale plate corresponds to a larger error projected back onto the sky, and more detailed calculations show that the variation of overall accuracy with plate-scale is fairly small within practicable limits.

Nevertheless there are occasions when a wider choice of system focal length would be desirable, either by the choice of system or by the choice of parameters of a given arrangement. A useful candidate is the monocentric Schmidt (Wayman 1950). This system is often dismissed with the remark that the aberrations are considerably larger than for a classical Schmidt of the same primary focal ratio (Bowen 1967). However it may often be more significant that the aberrations are in fact slightly smaller than for a classical Schmidt having the same system focal ratio and requiring the same size of dome.

The Cassegrain-like system with a single aspheric plate in the converging beam near to the position of the primary, due to Gascoigne (1960) and Schulte (1966), also appears particularly promising. Inspection of Schulte's spot diagrams suggests that the leading aberration is chromatic difference in focus. It would be interesting to investigate the image performance if this variation were designed out. Furthermore, it would be desirable to omit the field-flattener used by Schulte. It is not conducive to isoplanatism to have a refracting surface very close to the focus, and experience with Schmidts has shown that the inconveniences and disadvantages of bending plates to a spherical curvature are remarkably small.

The three-reflection arrangement described by Rumsey (1969) has focal ratio, field size and performance which fit nicely into the gap between Schmidts and Ritchey-Chrétiens, and the special advantages of strictly flat field and all-reflecting optics. The proposal of Brown (1969) gives almost the performance of a classical Schmidt but with reduced tube length. It could be a means of rescuing parabolic telescopes at the cost of re-figuring and a minimum of mechanical modification. Both these authors have addressed themselves to the important requirement that it must be possible to keep the system correctly adjusted in use (van Breda 1969).

3.3. *Photographic Emulsion*

The particular significance of the photographic sub-system lies in its enormous multiplicity. A photographic plate some 20 cm across, and costing a few decipounds, represents some 10^8 independent photo-receptors; one of the cheapest purchases on record. Although the efficiency is not quite as high (see below) as for some photo-electric devices, it must be remembered that a factor of 10 in multiplicity increases the rate at which astronomical information is gained by exactly the same amount as does the same factor in efficiency. It is then not hard to understand why more astronomical information has been secured by photographic methods than in any other way, and this is particularly true in relation to field-imaging. It is accordingly especially important to use photographic emulsions in the most efficient manner.

3.3.1. *Responsivity and Detectivity.* The traditional approach to photographic sensitivity was in terms of tonal reproduction in graphic art. It was therefore directed to the measurement of density as a function of exposure, as in the 'H & D' characteristic. In relation to the plate as a scientific receptor, this corresponds to the aspect of sensitivity which is now called responsivity; that is to say, the gross relation of input to output. The more significant aspect of sensitivity is termed detectivity, and concerns the signal-to-noise ratio which can be obtained under given conditions.

If N photons are available in an observation, it is possible in principle to measure intensity with a ratio of signal to rms noise $R = \sqrt{N}$. If in fact Q photons are necessary to attain this signal-to-noise ratio, then the efficiency of use of photons may conveniently be defined by the relationship

$$R^2 = N = \epsilon_e \times Q \tag{11}$$

in which the efficiency ϵ_e may be called the noise-equivalent quantum efficiency. If the measurement was performed using area a of a photographic emulsion, then the noise-equivalent quantum storage per unit area is defined as N/a. (Fellgett 1958; Jones 1953, 1958, 1959).

Measurement shows that the peak noise-equivalent quantum efficiency of the best emulsions is near to 1 per cent, considerably higher than has been estimated by some authors in the past. This maximum often occurs at small photographic densities, for example 0·2 to 0·3, and at considerably below the exposure which maximizes N. Astronomically this implies that if we are short of photons, which is nearly always the case, it is more efficient to make a larger number of short exposures which maximize ϵ_e, rather than continue a single exposure until the faintest limiting magnitude is attained, a condition which evidently corresponds to maximizing N. The averaging of results over a number of plates, which was difficult or tedious in the past, has now become much easier with the development of modern automatic measuring means.

The measurements also show that the sensitivity, as measured by ϵ_e, can be nearly as high for fine-grained emulsions as for those that are traditionally regarded as 'fast'. Consequently, suitably chosen fine-grain emulsions can be used in telescopes without loss of real efficiency. This gives freedom which is equivalent, in most significant ways, to changing the focal length of the telescope at will.

3.3.2. *Application to Direct Imaging.* As an example of the application of the averaging method, consider a Schmidt of 60-cm aperture working over a field 4° in diameter. If 16 repeated plates are taken of a given region of sky, the errors will be reduced to one quarter of those for a single plate. This result may be compared with that obtainable by a Ritchey-Chrétien of 2·5-m aperture working over a field 1° across. This telescope will require approximately 16 separate exposures to cover the same area of sky as the Schmidt, but if the focal ratios of the two instruments were the same the area occupied by each star-image on a single plate taken with the Ritchey-Chrétien is 16 times greater than for the Schmidt. The noise-equivalent photon storage is accordingly higher by the same factor, which again leads to an improvement by a factor of 4 in the statistical accuracy attained. Thus a series of 16 plates taken with the one instrument will yield the same accuracy, of which limiting magnitude is a special case, as a series of 16 plates taken with the other and covering the same total area of the sky. The time of exposure of a single plate is the same for both instruments if the focal ratios are the same, and in general it is easy to see that the exposures are proportional to the square of signal-to-noise ratio for either instrument. The cost-effectiveness of the small Schmidt is accordingly greater than for the larger and more expensive Ritchey-Chrétien (Fellgett 1964).

3.3.3. *Application to Spectrography.* The application of these principles to spectrography raises several interesting issues. The early dominance of responsivity ideas has led to the use of very 'fast' spectrograph cameras of very short focal ratio at great trouble and expense, with the result that the spectrum must then be widened in order to attain a useful signal-to-noise ratio. This is done as an afterthought rather than being properly incorporated into the overall design. A more logical approach is to design the spectrograph so that each resolved element of spectrum is focused onto an area having the required noise-equivalent storage. If a relatively light exposure maximizing ϵ_e is employed, the required area is larger than if the exposure is continued until N is maximized. This tends to obviate the need to focus onto very small areas by means of ultra-short focal ratio cameras. A further fallacy is to suppose that the slit of the spectrograph must be focused onto a resolution-length of the emulsion. This is indeed the smallest scale on which it can usefully be imaged, but there is no reason why a larger scale cannot be usefully employed as part of the deliberate choice of an area giving the required equivalent storage. Furthermore, the resolution length is customarily assumed to correspond to the conventional 'ultimate resolution' of 50 lines/mm for fast emulsions. However, this ultimate resolution corresponds to the finest line spacing that can be recognized visually in a standard test-pattern, and evidently this is a matter both of the photographic transfer function and the signal-to-noise ratio. The actual grains in even a coarse emulsion are much smaller than 1/50th mm, and measures of photographic transfer function in recent years have shown that the fall-off at higher spatial frequencies is determined more by the density and thickness of the emulsion than by the actual size of grain; a so-called coarse emulsion may indeed have just as much resolution as a considerably finer one. Thus the traditionally assumed 50 lines/mm has very little connection with the actual resolution of an emulsion.

Spectrographs can be improved with respect to convenience, cost and performance by taking account of these principles. The unnecessary loss of luminosity-matching by trailing a star on the spectrograph slit can be avoided by image slicers and in analogous ways. Nevertheless the difficulty remains (see Figure 3) that the luminosity of a high resolution grating spectrograph falls short of what is often necessary in order to match the star image. The difficulty is obviously even greater with respect to extended sources. There is an urgent need for a new type of spectrometer which would combine the high luminosity of interferometric instruments with the high multiplicity of the spectrographic method. The germ of a suitable method may be hidden in the principles of multiplexing (Fellgett 1951, 1957, 1957), of the SISAM spectrometer (Connes 1957) or the *Spectromètre à Grilles* (Girard 1963), but nobody has so far hit on the idea which is needed. This could be a very profitable field for inventive attention.

3.4. *Photoelectric Devices*

It has been said, perhaps a little unkindly, that the photomultiplier is for those who must attain fast time-resolution, or for those who are bad at electronics. Certainly the manufacture of photomultipliers has been largely dominated by the needs of nuclear physicists, and the characteristics of the photomultipliers commercially available are not well matched to astronomical needs. For the light-levels encountered in astronomy, very low dark emission is required. This is favoured by small sizes of photocathode, and by the avoidance of high gains which may lead to ion feedback and multiple pulsing. However, the multiplication per stage should be high, in the interests of a narrow pulse-height distribution.

If measurements are based on the total final anode current, the statistical fluctuations are enhanced by the difference in weighting given to different photo-electrons according to the size of cascade which they produce in the multiplier section; this is equivalent to a loss by about a factor of 2 in telescope time, or reducing a 3·5-m telescope to 2·5-m aperture. If on the other hand a pulse-counting method is used, all pulses which pass the discriminator are accorded the same weight, but with the penalty that some pulses are lost completely. When the discriminator level is set for optimum discrimination between light and dark counts, having regard to the brightness being measured, the results for the pulse-counting method do not seem to be very different from those which can be attained under good conditions by direct integration of the anode current. Pulse counting does, however, appear to give rather more stable operation, and has the advantage of giving the measurement directly in digital form. The channel photomultiplier gives a pulse height distribution which is so good that either pulse-counting or current-integrating methods should be equivalently near-perfect, and other forms of multiplier should probably be regarded as obsolescent for astronomical purposes.

All difficulties relating to pulse-height variations can of course be avoided by having no multiplier section; that is to say, by using a vacuum photo-diode instead of a photomultiplier at all. The techniques required are not difficult to those familiar with high-insulation and electrometer amplifier methods; for example, it is necessary to enclose the anode circuit of the photocell in a mildly evacuated enclosure. A particularly advantageous method is to charge the photocell before the exposure, normally by earthing the anode while the cathode is held at a controlled negative potential, and then to leave the cell completely isolated during the exposure. The photo-current charges the anode away from earth potential, and the integral of this current is measured by momentarily connecting the anode, by a mechanical switch, to the input of the amplifier at the end of the exposure. This free-anode method (Fellgett 1955, 1968) prevents the input current of the amplifier from mixing with the photo-current, and thus degrading the signal-to-noise ratio, during the exposure. Calculation shows that an rms charge error of better than 100 electrons is attainable. In order to attain a photometric accuracy of 1 per cent rms, at least 10^4 photo-electrons must be collected, and the uncertainty in this number is itself 100 electrons. Accordingly, detector noise will degrade the performance only slightly in photometry of this or higher accuracy. The receptor of a free-anode photometer can be cheap and compact, and in addition a large number of these units can time-share the amplifier and recording equipment. It is accordingly a method which seems well adapted to photometry of high multiplicity using field-imagers (see Section 4.3).

It may be noted that there is no such thing as 'dc' methods in photometry, since the zero-point of the equipment must be determined at some time, and in any case the sky background is unknown and fluctuating. All methods are essentially 'ac' and involve chopping between *sky* and *star* + *sky*. Moreover all these methods use phase-sensitive rectification, whether this is performed by hardware or by numerical subtraction of the *sky* reading from that for *sky* + *star*. All discussions of the relative merits of 'ac', 'dc', 'psr' methods are accordingly based on a misapprehension; one can only discuss the relative advantages of different methods of 'ac psr' observations.

The high efficiency and accurately quantitative response of the photo-cells can be exploited in some observations where, at first sight, imaging devices appear to be necessary. For example, informational analysis (Fellgett 1955*b*) shows that this is so for the measurement of radial velocity, leading to what may be termed a radial velocity photometer. This method has been used by Griffin (1968) with great success. He is able to obtain first quality radial velocities (± 1 Km/s) down to 9^m using a telescope of only 1 m aperture and observing an average of six stars per hour.

Where imaging is essential, the higher efficiency of photo-electric image tubes has to be set against the higher multiplicity of photographic methods. Moreover, as we have seen, the peak efficiencies of photographic emulsions are not far short of those which are practicable in a photo-electric device having the complexity of an image tube, with the compromises in processing which this necessarily implies. The technology of image tubes for astronomy is currently developing rapidly, and the properties of such tubes have been extensively treated elsewhere. Accordingly, comment will here be confined to emphasizing the need to take proper account of these characteristics in the systems design of the observing complex. Unfortunately astronomers have not omitted to design 'fast' spectrograph cameras for image tubes, although their properties are quite different from those of photographic emulsions and extreme minification of the image is seldom advantageous.

3.5. *Automatic Measuring Engines for Direct Photographs*

Reference has already been made to the need to match the large information-collecting power of the Schmidt by equally powerful means of measuring the information recorded on the Schmidt photographs. Fortunately modern photo-electric and control techniques are well adapted to the construction of machines for this purpose, and a number of projects of this kind are in progress around the world. Comments will here be confined to the General Automatic Luminosity and *X-Y* measuring engine GALAXY. This project was initiated at the Cambridge Observatories, moved with the author to the Royal Observatory, Edinburgh, and is currently the subject of a contract between Science Research Council and Hilger Electronics Limited. Acceptance trials of the first machine are now well advanced.

The GALAXY concept differs from that of other machines in a number of important respects. It was recognized from the beginning that the required rate of measurement made it impracticable to rely on the visual judgement of a human observer as a matter of routine. Accordingly the recognition of star images is performed automatically during a preliminary search phase, in which the plate is scanned by a combination of mechanical motion and a cathode-ray-tube flying-spot scanner. The patch of light which explores the plate is arranged to approximate to an optimal distribution for estimating the probability that a given clump of grains is due to a star image. More precisely, a wave-form is generated which approximates to the ratio of likelihood of the observed grain distribution on the two hypotheses that a star-image either is, or is not, present.

This emphasis on statistical optimization with respect to the photon-grain interaction in the original photograph is continued in the measurement phase. Each star image discovered in the search-phase, or a selection of these determined by computer program at the will of the astronomer, is explored by a spirally scanned cathode-ray-tube and statistical weights are applied to the different annulae, as a function of radius in the star image, so as to nearly maximize the statistical weight of the measurement with respect to two coordinates of position and one of brightness. An application of the theory of Woodward and Davies (Woodward 1953) shows that it is necessary to iterate the choice of these weighting functions successively with respect to both position and brightness (Fellgett 1965). That is to say, the star-image must be centred with all attainable accuracy in order to correctly determine the weighting functions which optimize the measurement of brightness, and the brightness must be determined with all attainable accuracy in order to optimize the weighting functions with respect to measurement of position. Careful attention to this requirement enables the statistical weight of the final measurement of brightness to be some two to three times greater than for a conventional iris photometer. The accuracy of positional settings is better than that of a skilled human observer only in regard to consistency and freedom from fatigue, since it is known that such an observer attains substantially the full accuracy which is theoretically possible.

Astronomers have hitherto tended to regard a measuring method as satisfactory if it gives repeatable readings on any one plate. This is necessary but not sufficient; the attention given in GALAXY to grain statistics may save the equivalent of as much as a factor of 2 in telescope aperture.

3.6. *Computers*

It is no good collecting large amounts of information and converting this information into digital measurements unless the measures can be adequately reduced and used by the astronomer. Traditional paper, pencil and desk-calculator methods are evidently totally inadequate to the output of a Schmidt and GALAXY measuring system; but fortunately modern electronic digital computers are highly suitable and fully adequate in speed. Indeed their power and versatility is so great that the only discussion needed is of the ways in which the astronomer can best exploit this great power.

The raw digital data fed into the computer will be reduced, by the methods that have been outlined in Section 3.2.3, to data giving the right ascension, declination and magnitude of the programme stars with respect to the systems defined by stars specified to be standards of position or brightness. In the old days, the data at this stage would have been regarded as fully reduced, and would have been published in catalogue form. With the very large data rates now available, this is neither necessary nor desirable. The data are considered by the astronomer as having been freed from the accidental parameters of the particular exposures from which they were derived, and reduced to a standard system. In other respects they are regarded as raw data which it is necessary to interrogate by program in order to answer specific astronomical questions.

In the formulation of these questions, the astronomer must take care to avoid being overwhelmed by excessive amounts of information. Thus he will normally ask for individual information about only those stars which belong to rare classes, and will be concerned with statistics of commoner types of stars. The amount of statistical information needed to define the structure of our own galaxy in all directions, to varying line-of-sight depths, and with respect to the numerous significant classifications of stars, is large even by the standards of modern data-processing. In this one problem alone, there is certainly work to do for many years ahead.

3.7. *The Astronomer*

In a significant respect the astronomer himself is the most important sub-system of all, since without him no useful interpretation of the astronomical observations can be made. It is just as important to optimize his integration into the overall system as that of any other sub-system.

Before the present century, the role of the astronomer was closely constrained by the nature of the observations, which were visual. Only the human eye had the sensitivity needed in an astronomical detector, and only the astronomer could interpret what his eye saw. Consequently it was vitally important for him to spend long nights at the telescope, even though this meant cold, discomfort and irregular hours for the other aspects of his work. The advent first of photography, and later of other man-made receptors of adequate sensitivity, has meant that the objectivity of the physical response of these detectors is preferred to visual estimates, except for some very special purposes. It is accordingly pertinent to ask whether the place of the astronomer is now at the telescope at all, except during experimental and setting-up stages. Since the beginning of the photographic era in astronomy it has traditionally been argued that the personal presence of the astronomer has been needed to provide good guiding of astronomical photographs. This argument fails to explain why special skill in guiding should necessarily be associated with knowledge of the theory of astronomy. In more recent years, technological developments have enabled automatic photo-electric guiders to out-perform the human observer, and there appears to be no reason why they should not replace him entirely in this task. The more recent argument is that the astronomer is needed to make tactical decisions according to the state of seeing, transparency, etc., which develops during the night. Given objective measures of these factors, however, it appears that the consequent decisions can be rationally based upon them, and consequently can be delegated to a computer program or a suitably programmed night assistant.

Whoever may actually be needed at the telescope during the night, there is now ordinarily no reason for him to endure the cold of an astronomical dome, where in any case his body-heat tends to spoil the seeing. Modern control methods make it possible and desirable to perform most operations from a warmed control room, which can if necessary overlook the telescope through a double-glazed panel. This arrangement was used on the twin 16-inch automatic photometric telescope set up at Royal Observatory, Edinburgh, during the author's appointment there (Reddish 1966). The success of the Image-isocon camera tube now gives a remote observer access to the image-surface of the telescope with sensitivity similar to that of the eye. Even a single sensitive photocell, combined with a programmed pattern of precise off-sets, is a surer method of field identification than eye comparisons with a finding chart.

The versatility of the small modern on-line digital computer makes it well-adapted to the control of an astronomical telescope, either alone or in partnership with a human operator. Most of the data defining the program of observations can conveniently be fed to the computer through a punched paper-tape reader. Catalogue information can be stored semi-permanently in the computer backing store, together with information about the characteristics of the telescope, and accumulated experience of refraction, flexure, etc. A convenient and natural means of interaction with the operator is by a conversational graphic display. The computer can in this way display star charts, measured colour-magnitude arrays to date, etc., and the operator can intervene by pointing to stars or other representative points and typing an instruction code. For example he can in this way indicate the order in which a group of stars are to be measured. Sensitive areas of the display can give directly additional command facilities, and numerical information can be displayed on request. It turns out that the overall console system can be cheaper in this way than by providing a multitude of special-purpose control knobs and display indicators for quantities such as right ascension, declination, hour angle, zenith distance, etc., often in units of time, decimal or hexagesimal angle, radians, etc.

4. SOME APPLICATIONS OF FIELD-IMAGING

In the foregoing discussion, the argument has returned us again and again to the great power and information capacity of field-imaging methods. It is accordingly pertinent to enquire how much astronomy can be done in this economical way, and it does indeed appear that many problems are susceptible to this approach which are traditionally done by laborious one-star-at-a-time methods.

4.1. *Objective Dispersion*

In Section 3.2.1 the cost-effectiveness of slit spectrometry by large telescopes was called into question, because the cost increases more steeply than the resultant gain, even in the more effective region where the majority of light from the star enters the slit. Moreover it is just in this region that a conventional spectrograph shows greatest logical redundancy of components. If the starlight does not fall on the jaws of the slit, the slit is to this extent redundant. The purpose of the collimator is to render the light parallel, but parallel light was already available before the wave-fronts entered the telescope. The spectrograph camera focuses the dispersed radiation onto the receptor, but a telescope is itself a device for focusing. Accordingly there is much to be said for placing the means of dispersion over the objective, and dispensing with a conventional spectrograph altogether. This immediately converts the spectrography to a field-imaging procedure, giving an advantage in output proportional to the high multiplicity which can be attained.

Objective dispersion was indeed the method by which extensive catalogues of stellar spectral type were originally produced. Objective prisms and gratings have been used successfully in the measurement of radial velocity, traditionally a stronghold of the slit spectrograph, by Fehrenbach (1955), Fehrenbach and Duflot (1956) and van Breda (1964). Good guiding is necessary, and one is dependent for resolution, instead of just efficiency, on good seeing. These requirements are probably not too objectionable for many applications.

In Section 3.2.1 it was pointed out that the effective size of grating in a slit spectrograph depends on the delay between extreme beams. If an objective grating is used in the first order, which is usually convenient, it is easy to see that the resolution attainable in given seeing depends only on the total number of lines of the grating across the aperture, and if this is held constant is independent of whether the grating is placed over the main aperture or in a re-imaged parallel beam elsewhere in the system. Gratings in which the line structure consists of variation in transmission necessarily give rise to symmetrical diffraction on either side of the zero order. If this variation is sinusoidal, orders $+1$, 0, and -1 are present. The separation of corresponding lines in the ± 1 orders was used by van Breda in his measurement of radial velocity, following a suggestion made by the author. Although this is useful, the maximum efficiency that can be attained with absorbing gratings is rather low. Better efficiencies are possible with gratings in which the line structure consists of a suitable variation in refractive index, and efficiencies approaching 100 per cent are theoretically possible for gratings having variation of refractive index in depth. Promising methods of constructing such gratings holographically are being developed, and some have been described at the present conference (see also Burch & Palmer (1961)).

The disadvantages of objective dispersion are that spectra of different stars may overlap, and the sky background is always superposed on the spectra. Overlapping can indeed be a nuisance in a crowded field, but sky background is less harmful than it at first appears. Because of photographic saturation, it does limit the exposure which can be given with a 'fast' emulsion to a few minutes. However, as was seen in Section 3.3.1, fine-grain emulsions can be used with little loss of efficiency, or indeed a number of plates with coarse-grain emulsions can be averaged. To consider the effect quantitatively, suppose that it is required to attain a signal-to-noise ratio R in each spectral element. Then it is necessary to expose each area corresponding to a spectral element until the storage is $N = R^2$ (see Equation (11)). The statistical fluctuation within each such area is then $\sqrt{N}$. If an rms accuracy of 1 per cent is required, $R = 100$ and $\sqrt{N} = 100$. We know, however, that stars near to the limiting magnitude of a reasonably exposed direct photograph can be measured to about $0^m{\cdot}1$, corresponding to about $R = 10$. The statistical fluctuation due to the background is thus approximately a tenth of that due to an element of spectrum in which 1 per cent accuracy can be attained, and is effectively negligible.

4.2. *A possible use for Old Parabolic Telescopes*

The opening section of this paper referred to the difficulties of attaining reasonable performance from a reflecting telescope once the primary has been parabolized. The view was taken in Section 3 that it is better to consign such telescopes to point-imaging, for which they are inherently suitable, rather than to elaborate means of making them into relatively ineffectual field-imagers. However, the parabola possesses the interesting property that if the focused light is rendered parallel by a second paraboloid, confocal with the first, the resultant afocal system is free from Seidel aberration (Schwarzchild 1905; Linfoot 1955). Some recent literature has thrown doubt on this conclusion (Meaburn 1968) and I am grateful to Professor H. H. Hopkins for confirming by direct calculation the correctness of the classical result.

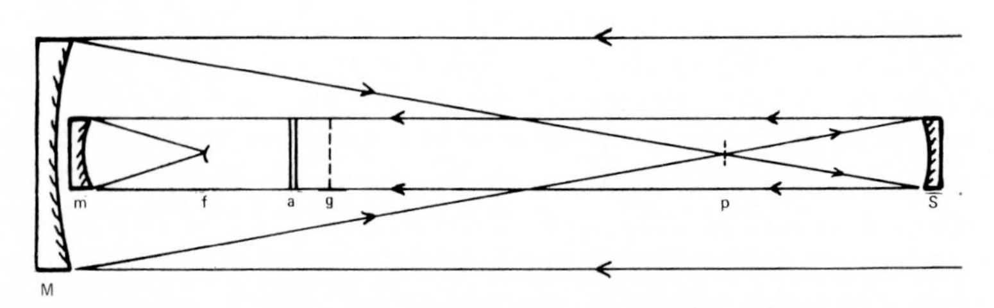

FIG. 6. A possible method of using a parabolic telescope for field-imaging spectrography. *M*, primary mirror; *S*, secondary mirror; *p*, perforated screen; *a*, aspheric plate; *m*, Schmidt mirror; *f*, photographic plate; *g*, dispersing element.

A possible system exploiting this property is shown in Figure 6. The starlight is first focused onto a screen, placed at the traditional prime focus, in which apertures are cut to select the light of the desired stars. This radiation is then rendered parallel by a secondary paraboloid, a dispersing element is placed in this parallel beam, and the spectra are finally focused by means of a camera which may conveniently consist of a small Schmidt telescope. The advantages of the system are that overlapping is controllable, and the sky background is reduced by a factor of approximately $\alpha n/\beta$, where β is the diameter of the prime-focus aperture, α the diameter of the seeing disc, and n the number of resolved spectral elements. The dispersing element needs to be only approximately 1/5th of the size of the primary mirror: the number of lines should be the same as for a primary-aperture grating, as explained in Section 4.1.

The useful field is limited essentially by coma of the primary. An aperture 10″ arc across is commonly used for photometry under fairly poor observing conditions. If this size is adopted, the maximum coma spread should not exceed about 5″ arc, giving a field radius of about 4′ for an f/3 primary. This small field, despite the fact that the primary imaging is required to be good enough only to select and not to image programme stars, is itself an indictment of the performance of a paraboloid.

4.3. *Schmidt Photometry and Astrometry*

Reference has already been made to the success of Schmidt telescopes in astrometry (Dixon 1962, 1963), despite the supposed disadvantage of short focal length. The measurement of stellar parallax has hitherto been the provence of one-star-at-a-time methods using long focus refractors. An element of the GALAXY project was that field-imaging methods of great multiplicity are really essential in parallax work if one is to have sufficient standards for adequate control of systematic error, and to be able to build adequate statistical weight against random errors without prohibitive labour. The Cambridge Schmidt is now being used in this way by Argue (1969).

It has become almost an article of faith with astronomers that photo-electric magnitudes are inherently better than those obtained photographically. There is an element of truth in his, inasmuch as the photographic plate, unlike a photocell, is not re-usable and therefore cannot be calibrated before or after use. Nevertheless one-star-at-a-time photo-electric photometry is inherently susceptible to accidental disturbances, whether of cloud, seeing, misguiding, or equipment malfunction, affecting any one observation. We saw in Section 3.2.3 that photographic photometry with a carefully tuned-up Schmidt is actually able to correct errors in published photo-electric magnitude sequences. This result is a consequence of the excellent isoplanatism of the Schmidt, and this includes the isoplanatism with respect to variation in emulsion sensitivity arising from the comparatively short focal length of the instrument.

4.4. *Field-Imaging Photoelectric Photometry*

It is reasonable to enquire whether the advantages of multiplicity, automatic compensation for transparency changes, and safety from malfunction affecting the measurement of a particular star, available in Schmidt photographic photometry, cannot be realized in some manner which also uses the high efficiency and possibility of precise calibration of photo-electric methods. It appears indeed that field-imaging photo-electric photometry is possible according to the following procedure.

A photograph is taken of the star-field to be studied, and from this a mask is made having apertures (of normal photometric size) admitting the light of the stars to be studied. A photo-electric photometer is then placed behind each of these apertures, the aperture plate inserted in the Schmidt, and simultaneous readings obtained on all the selected stars. The compactness, low cost, and possibility of sharing recording equipment, of the free anode photometer make it particularly suited to this application. To calculate roughly the optimum number of receptors, let the cost of a telescope of aperture A equipped with m photometers be

$$C = C_1 A^{\alpha} + C_2 m \tag{12}$$

where C_1, C_2, and α are constants. The speed of measurement is

$$S = S_1 A^{\beta} m \tag{13}$$

where S_1 and β are constants. By Lagrange's method or otherwise, the speed is maximized subject to the constraint of constant cost C when

$$C_2 n = (\alpha/\beta) C_1 A^{\alpha}. \tag{14}$$

Using Equation (12), this means that the cost of photometers should be α/β times that of the telescope. For example, if 250,000 currency units are available, 100,000 should be spent on the telescope and 150,000 on photometers, assuming $\alpha = 3$, $\beta = 2$. Although more exact calculations taking setting costs into account will modify the optimum, several hundred receptors will be called for in a typical case. The traditional single photometer on a 100-, 150- or 200-in telescope is evidently very far from optimum.

Because of the high multiplicity, it is economical to continue the exposure over a long period, with of course the usual alternation to determine sky background. Consequently the effective limiting magnitude will be much fainter than would normally be expected for photo-electric photometry on a telescope of given size. This is analogous to the gain in limiting magnitude for multiple re-photographing discussed in Section 3.3.2. For example with 100 photometers, it would be economically practicable to do photometry of $0^{m}\!\cdot\!01$ rms statistical accuracy down to $18^{m}\!\cdot\!5$ in a good sky with a 40-cm aperture telescope, or to 21^{m} at $0^{m}\!\cdot\!1$ rms. These estimates are based on the measured performance of the Edinburgh Twin 40-cm telescope (Reddish 1966) and assume a sky brightness of $21^{m}\!\cdot\!5$ per square arc second.

5. CONCLUSIONS

An outstanding example of excellent systems design in a flux collecting application has been given by Connes (1969). Every element of his method is directed towards optimizing the overall systems performance. Not only are the details of the apparatus carefully adapted to this end, but each of the principal characteristics of the method is essential to the results which have been obtained.

The method has been called MIF spectrometry, indicating that it is Multiplex, Interferometric, and Fourier. The interferometric attribute gives the high luminosity necessary for flux-collecting to be feasible (Jacquinot 1954). The Fourier attribute enables a wide spectral range to be covered, in contrast for example to the classical Fabry-Perot interferometer, and also gives the orthogonality of modulation of each spectral element which is exploited in the multiplexing. The multiplex attribute essentially enables the whole spectrum to be observed under signal-to-noise limited conditions in the same time as would be necessary for a single spectral element in a scanning spectrometer (Fellgett 1951, 1957*a*, 1957*b*, 1967). The multiplexing of one million resolved elements is now feasible, and will almost certainly be realized before the present communication is published. The corresponding multiplex gain means that even if a non-multiplex spectrometer had the wide spectral range of the Fourier method, and the large luminosity of interferometry, it would require 3,000 years of nightly observations to secure the results which the multiplex spectrometer can attain in one single night.

The potential power of field-imaging methods, which has so far been partially realized in practice, has been emphasized throughout this discussion. Astronomical knowledge will be enormously increased if the systems design of field-imaging methods can be made as good as has been achieved in the cited flux-collecting example.

REFERENCES

Argue, A. N., 1969, *Q. Jl. Roy. Astron. Soc.*, **10**, *Proceedings of the Cambridge Observatories*, 134.
Atkinson, R. d'E., 1947, *M. N. Roy. Astron. Soc.*, **107**, 291.
—— 1955, *M. N. Roy. Astron. Soc.*, **115**, 427.
—— 1955, *Observatory*, **75**, 239.
—— 1960, *M. N. Roy. Astron. Soc.*, **120**, 505.
—— 1961, *Roy. Obs. Bull.*, No. 34.
—— 1967, *Astrom. J.*, **73**, 60.
—— 1969, Astrometric Conference, Charlottesville, also colloquium, Cambridge University, *c*.1955.
Booker, H. G., Ratcliffe, J. A. and Shinn, D. H., 1950, *Phil. Trans. Roy. Soc.*, **242**, 579.
Bowen, I. S., 1967, *Q. Jl. ast. Soc.*, **8**, 9.
—— 1967, *Ann. Rev. Astronomy and Astrophysics*, **5**, 45.
van Breda, I. G., 1964, Determination of Stellar Radial Velocities using an Objective Diffraction Grating, Thesis, University of Cambridge.
—— 1969, *M. N. Roy. Astron. Soc.*, **144**, 73.
Brown, D., This volume, p. 521.
Burch, C. R., 1942, *M. N. Roy. Astron. Soc.*, **102**, 159.
Burch, J. M. and Palmer, D. A., 1961, *Optica Acta*, **8**, 73.
Connes, P., 1957, *Optica Acta*, **4**, 136.
—— 1966,
—— 1969, This Volume, p. 1.
Coulman, C. E., 1969, This Volume, p. 528.
Dixon, M. E., 1962, *M. N. Astron. Soc. S.A.*, **21**, 180.
—— 1963, Astronomy with a Schmidt Camera, Thesis, University of Cambridge.
—— 1963, *M. N. Astron. Soc. S.A.*, **22**, 6 and 32.
Danjon, A., 1946, *Bull. Astron.*, **13**, 1.
Fastie, W., 1967, *App. Optics.*, **6**, 397.
Fellgett, P. B., 1951, Doctoral thesis, University of Cambridge.
—— 1955*a*, International Astronomical Union General Assembly, Dublin.
—— 1955*b*, *Optica Acta*, **2**, 9.
—— 1956, *Astronomical Optics*, Z. Kopal, Ed., North Holland, Amsterdam.
—— 1957*a*, Les Progres Recents en Spectroscopie Interferentialle, CNRS, Bellevue.
—— 1957*b*, *ibid.*

—— 1964, *The Observatory*, **84,** 216.
—— 1967*a*, *J. de Phys.*, **28,** (Coll. C2-165, Supp. to 3–4).
—— 1967*b*, *Science Journal*, April, p. 58.
—— 1968*a*, Symposium on Instrumentation for Large Optical Telescopes (S.R.C.), Cambridge, 12–13 August.
—— 1968*b*, *J. Sci. Instrum.*, Series 2, **1,** 1141.
—— 1965, Report dated 20 October.
—— 1969, *Phil. Trans. Roy. Soc.* A, **264,** 309.
—— Also unpublished internal reports 1955–1968 (GALAXY).
FEHRENBACH, C., 1955, *J. des Observateurs*, **38,** 165.
FEHRENBACH, C. and BARBIER, M., 1955, *ibid*, **38,** 180.
FEHRENBACH, C. and DUFLOT, M., 1955, *ibid*, **38,** 176.
—— 1956, *ibid*, **39,** 53 and 104.
GASCOIGNE, S. C. B., 1960.
—— 1965, *The Observatory*, **85,** 79.
—— 1968, *Q. Jl. Roy. Astron. Soc.*, **9,** 98.
GIRARD, A., 1963, *J. de Physique*, **24,** 139.
GRIFFIN, R., 1968, Symposium on Instrumentation for Large Telescopes, Cambridge, 12–13 August.
HORN-D'ARTURO, G., 1965, *Pub. Osserv. Astron. Univ. Bologna*, **9,** 1.
JACQUINOT, P., 1954, 17² Congres du GAMS, Paris.
JONES, R. C., 1953, *Advances in Electronics*, **5,** 1.
—— 1958, *Photographic Science and Engineering*, **2,** 57.
—— 1959, *Advances in Electron Physics*, **11,** 87.
JONES, Sir H. SPENCER, 1941, *Mem. Roy. Astrom. Soc.*, **66,** 11.
LINFOOT, E. H., 1955, *Recent Advances in Optics*, Oxford University Press, London.
LINFOOT, E. H. and REDMAN, R. O., 1968, *M. N. Roy. Astron. Soc.*, **139,** 347.
MAKSUTOV, D. D., 1944, *J. Opt. Soc. Amer.*, **34,** 270.
MEABURN, J., 1968, *Astrophys. and Space Sci.*, **1,** 166.
MEINEL, A. B., 1953, *Astrophys. J.*, **118,** 335.
MERTZ, L., 1969, This Volume, p. 507.
NEUGEBAUER, G., MARTZ and LEIGHTON, 1965.
REDDISH, V. C., 1966, *Sky and Telescope*, **32,** No. 3.
ROSS, F. E., 1935, *Astrophys. J.*, **81,** 156.
SCHULTE, D. H., 1966, *Applied Optics*, **5,** 309.
SCHWARZSCHILD, K., 1905, *Astr. Mitt. K. Stemw. Gottingen*, No. 10.
WAYMAN, P. A., 1950, *Proc. Phys. Soc.*, B., **63,** 553.
WOODWARD, P. M., 1953, *Probability and Information Theory with Applications to Radar*, Pergamon Press, London.
WOOLLEY, R. V. D'R., 1962, *Q. Jl. Roy. Astron. Soc.*, **3,** 249.
WYNNE, C. G., 1949, *Proc. Phys. Soc.*, B, **62,** 772.
—— 1967, *Applied Optics*, **6,** 1227.
—— 1968, *Astrophys. J.*, **152,** 675.

Design for a Giant Telescope

Lawrence Mertz

Smithsonian Institution, Astrophysical Observatory,
Cambridge, Massachusetts 02138

Abstract—The advent of Fourier spectrometry has eliminated a major deterrent to development of giant telescopes, since spectrometers need no longer be correspondingly large. (This is achieved by the absence of an entrance slit.) Nevertheless, optical quality must be maintained to approximately the 'seeing' blur for high efficiency.

Since economic factors preclude scaling up conventional designs, a new basic design is proposed. Restricting the purpose of the telescope to collecting light for spectrometry, rather than giving pictorial images, permits unorthodox configurations. The proposed design, an optical analog to the Arecibo radio telescope, has four basic characteristics: (1) tessellated primary; (2) spherical primary; (3) alt-azimuth steering; (4) stationary primary.

Numerous mechanical features as well as the optical characteristics of the design are presented.

Before Fourier spectrometry was successfully demonstrated, projects to develop giant optical telescopes were rather pointless. No versatile instruments, other than interferometers, are able to cope with the necessarily large images emanating from a giant telescope. The aim of this study has been to develop details of a potential design for a giant telescope tailored for Fourier spectrometry.

It is necessary to emphasize first of all that the light-bucket concept is unacceptable. It remains important to reduce the image to minimum size in order to use the smallest possible infrared detector and also to exclude as many deleterious night-sky photons as possible. These considerations lead to a figure of merit for the telescope of D/θ, where D is the aperture diameter and θ is the angular blur. Hence, the goal to strive for is approximately 1-arcsec blur so that the instrument is limited by atmospheric 'seeing' rather than by telescope quality.

Since scaling up conventional telescope designs would be exorbitantly expensive, we must resort to unorthodox configurations. The proposed design shown in Figure 1 is an optical analog of the Arecibo radio telescope and has four basic characteristics:

1. *Tessellated Primary.* Manufacturing and handling a monolithic primary mirror are hopeless tasks. Optimum mirror size in terms of area

versus cost turns out to be about 0·5 to 0·75-m diameter, consonant with numerous tessellations.

2. *Spherical Primary.* Mass production of the tessellations requires a spherical figure. Furthermore, the ease of fabricating, aligning, and testing spherical surfaces is unexcelled.

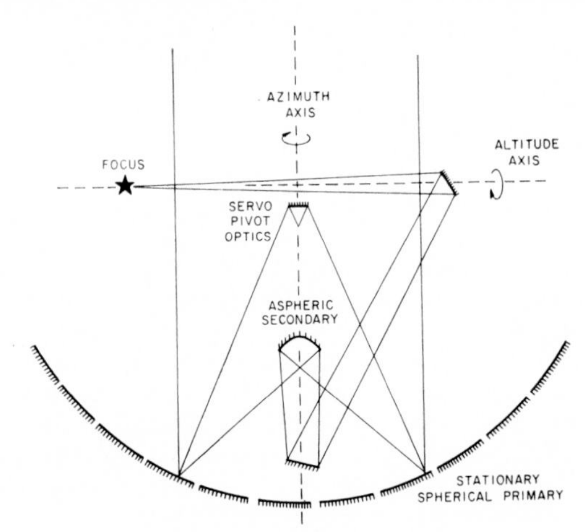

FIG. 1. Optical design of the telescope.

3. *Alt-azimuth Steering.* Radio telescopes have clearly demonstrated the economic virtues of alt-azimuth mounts. The requisite computer control is both inexpensive and convenient. The main disadvantage is the added difficulty of offset guiding; however, this is not deemed overwhelming.

4. *Stationary Primary.* The Arecibo concept is attractive in that it is virtually immune to gravitational-flexure problems. It has the disadvantage of requiring considerably more primary mirror surface than the effective aperture area, amounting to a factor of about 6 for our design. Our expectation is that economical mass production of the tessellations will sufficiently reduce the disadvantage. The alternative is individual servo-control of the tessellations.

The mechanical aspects of the mount are not shown in Figure 1. Apart from the primary mirror, all the remaining components are rigidly attached to one another. The altitude axis is conventional, and its bearings are attached on an annular frame, which is supported by air-pads. The azimuth axis (effective pivot location) is defined by a servo-control, whereby the image of a small light source is constrained on an adjacent four-quadrant detector. The bisector of the light source and the detector would thus lie at the centre of the primary mirror. Since that physical location would obstruct the stellar rays on the way to the focus, a small flat mirror is employed to fold the centre down out of the way. Direct definition of the azimuth axis by means of optics is far more precise than could be obtained from any large-diameter mechanical bearing.

The aspheric secondary mirror of the telescope presents an appreciable problem. A computer programme was written to calculate (with no approximations) the family of curves shown in Figure 2. Just which curve to employ is an intuitive choice, balancing secondary size and coma. The

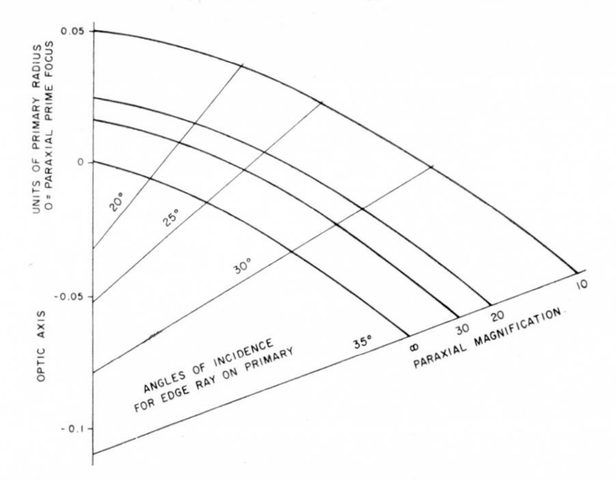

FIG. 2. Calculated aspheric secondary curves.

coma is so severe that zonal magnifications differ by considerable factors, and a small hole placed at the focus projects as a truncated cone on the celestial sphere. The family of cones is depicted in Figure 3, where the

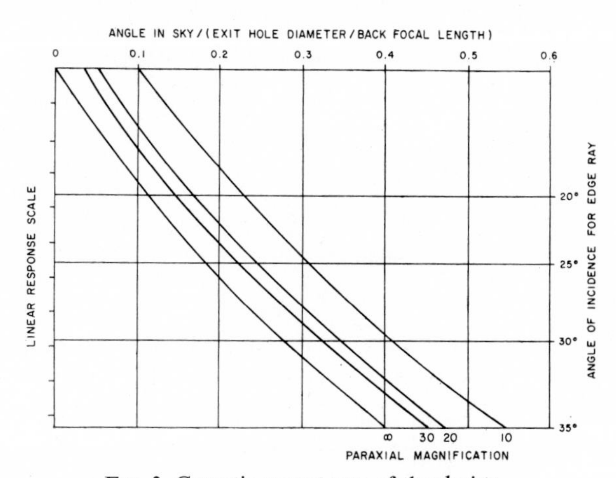

FIG. 3. Comatic acceptance of the design.

base of the response depends on the size of the secondary mirror (right-hand ordinate scale) and the cone surface accords with the paraxial magnification of the secondary mirror.

Figure 4 is composed of two graphs that show the trade-offs of secondary diameter, coma, sky coverage, and requisite primary-mirror area. The small circles represent a reasonable operating point that has been selected.

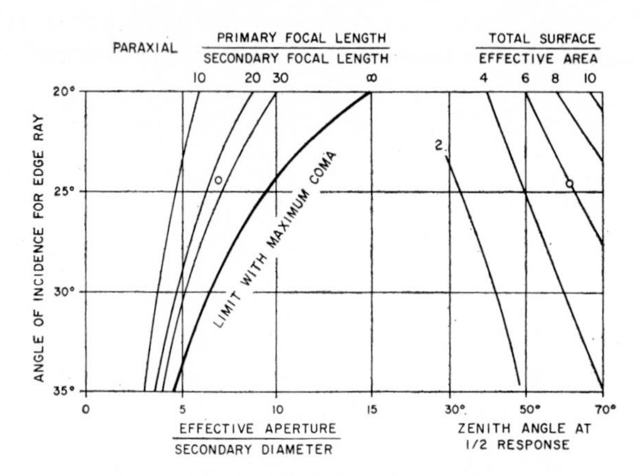

FIG. 4. Graphic aids in the choice of a secondary.

The fabrication of such a grossly aspheric surface is not expected to be easy. The simple linkage shown in Figure 5 has been computed to generate the required curve. It is expected that this linkage will greatly facilitate

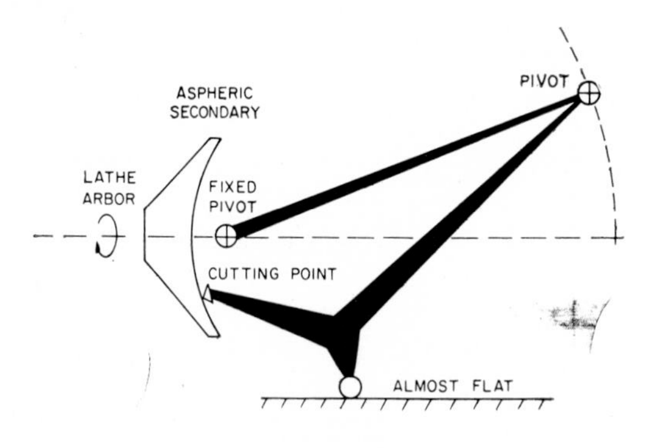

FIG. 5. Linkage to generate the aspheric secondary.

fabrication. Not only is the cam surface almost flat, but also the generated curve is insensitive to errors on that cam.

The primary tessellations will be fabricated by means of replication techniques similar to those used for diffraction gratings. A pilot project is now under way at Photo Tech, Inc., to accomplish this technique on cast aluminum substrates.

Mechanical support of the primary will be on a 'hierarchic' geodesic frame (the more basic frame members will be stronger and heavier). Alignment procedures for the individual tessellations will make use of the servo-pivot optics. A small portable vibrator will be used to induce nutating tremor in a tessellation. The modulated portion of the servo-signal is thus attributed to that particular tessellation. Manual alignment can then cancel out the servo-signal.

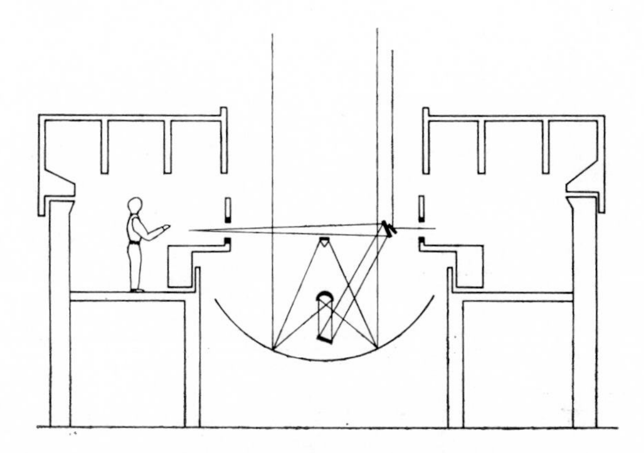

FIG. 6. Section of proposed 2-m prototype.

Table 1 gives the parameters proposed for a 2-m prototype telescope as well as those for a 15-m telescope, which is the ultimate goal. Apart from

TABLE 1. Recommended design parameters.

	Prototype	Goal
Effective aperture (metres)	2	15
Zenith angle at 1/2 response	60°	60°
Primary radius (metres)	2·5	19
Tessellations { number	100	2400
Tessellations { size (metres)	0·5	0·75
Secondary diameter (metres)	0·3	2·2
Focal convergence	f/20	f/20
Paraxial focal length (metres)	120	1050
'Scale' at 1/2 response (arcsec/mm)	6	0·6
Building diameter (metres)	12	60

size, the major structural difference between the two is that in the case of the smaller telescope the observer walks round to follow the azimuth,

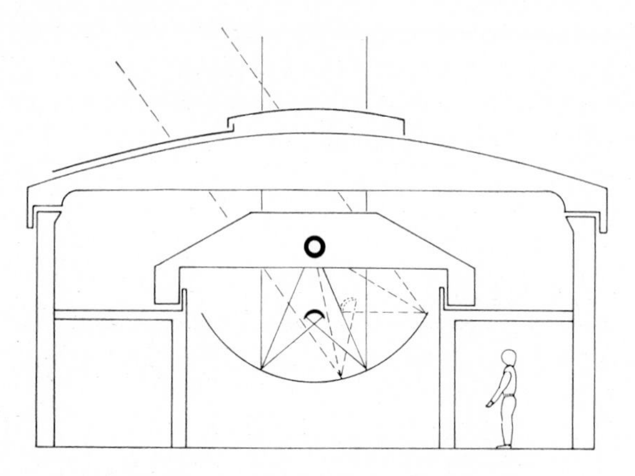

FIG. 7. Section of proposed 2-m prototype.

while in the large telescope the observing rooms, as well as several other rooms, are slung as an integral part of the dome, which is slaved to the

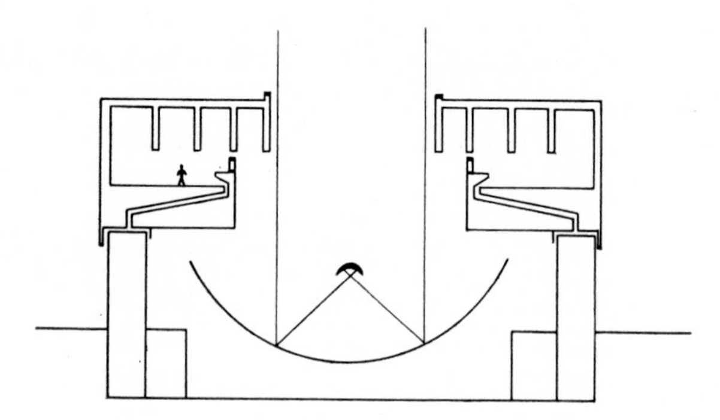

FIG. 8. Section of proposed 15-m telescope.

telescope azimuth. Sketches of these telescopes appear in Figures 6 to 8, and a photograph of a model of the giant is shown in Figure 9.

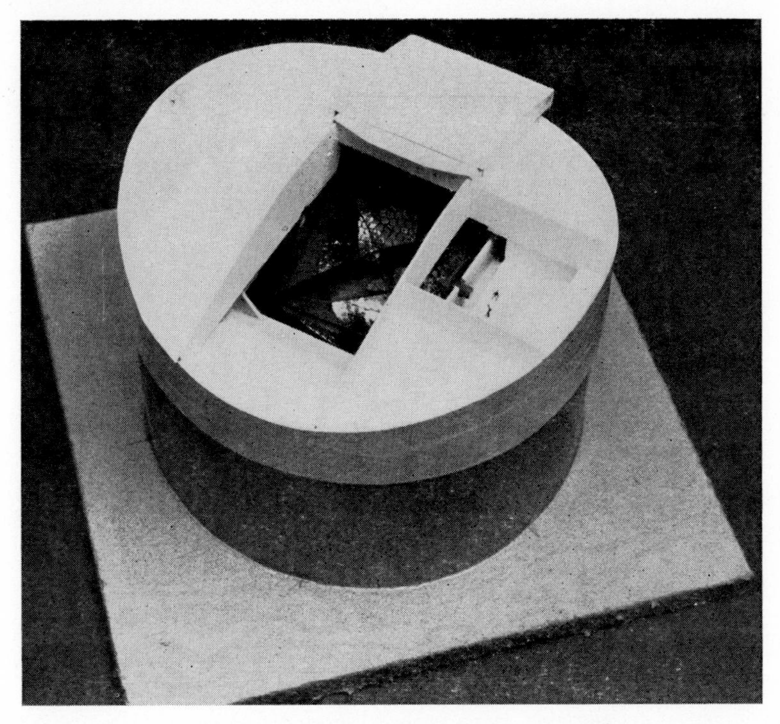

FIG. 9. Photograph of a model of the giant telescope.

ACKNOWLEDGEMENTS

Most of the ideas incorporated in this design are not novel, and I should like to acknowledge the suggestions contributed by Drs. A. Meinel, R. Miller, G. Fiocco, and N. Carleton. The main new feature is the integration of a number of ideas into a positive detailed design, which I feel is thoroughly practicable.

A Compact Three-Reflection Astronomical Camera

N. J. Rumsey

Physics and Engineering Laboratory, D.S.I.R.,
Lower Hutt, New Zealand

Abstract—Of the many systems of three mirrors that could possibly be used as astronomical cameras, one has certain unique advantages. This uses concave primary and tertiary mirrors that together form a continuous reflecting surface on a single disc of glass—though the equations for the profiles of the two mirrors are different. The secondary mirror is convex. To a first approximation, the profiles of all three mirrors are hyperbolic, with eccentricities increasing from the primary to the tertiary.

Advantages of the system are:

1. The tertiary mirror remains permanently in perfect alignment with the primary.
2. There are no supports for the tertiary crossing the light path and diffracting light.
3. Spherical aberration, coma, astigmatism and field curvature can all be controlled, so that the image can be made flat as well as sharply defined.
4. The image can be placed in a relatively convenient position, say, just outside the secondary mirror (which must then be perforated).
5. The system is compact, the overall length being little more than one-third of the focal length.

Schwarzschild (1905) showed that an astronomical camera, well corrected for spherical aberration, coma, astigmation and field curvature, could be constructed from just two aspheric mirrors. However, the primary mirror was necessarily convex, the separation between the mirrors twice the focal length, and the diameter of the secondary more than twice the clear aperture of the system (see Figure 1).

As a more practicable arrangement, Schwarzschild favoured a system in which astigmation and Petzval curvature were balanced against each other to give a flat mean field. The secondary mirror was now smaller than the primary, but the separation was still rather large, about 1·25 times the focal length (see Figure 2). The system has never found much favour among astronomers.

If a pure mirror system is to be more compact, and still is to give an image that is flat and sharply defined, it must use at least three mirrors. Paul (1935) suggested such an arrangement of three mirrors (see Figure 3) in a paper describing many systems for astronomical photography; and, independently, Lagrula (1942) proposed a rather similar system (see Figure 4). These systems can be reasonably compact. In Lagrula's case, the distance of the photographic plate from the pole of the primary is 60 per cent of the focal length.

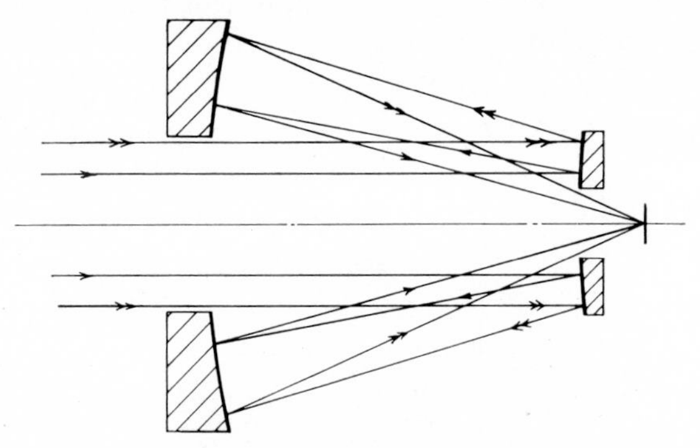

FIG. 1. Schwarzschild's well-corrected system.

However, these systems suffer from certain disadvantages. First, there must be supports for the tertiary mirror. In the simplest arrangement, these cross the light path twice. Thus, for oblique pencils where the tertiary supports cannot lie in the shadow of the secondary supports there will be, altogether, three times as much light diffracted by the supports as in a

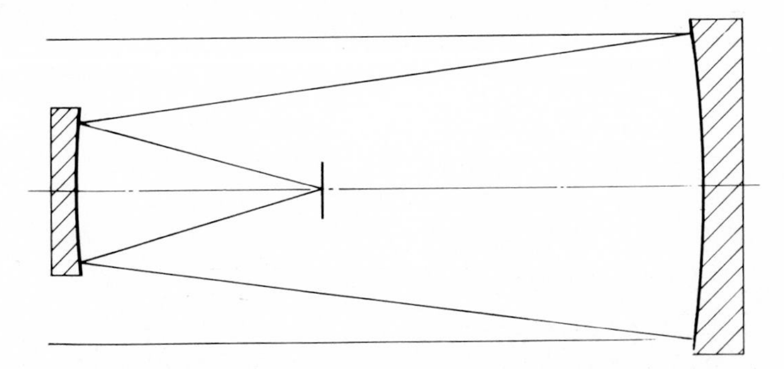

FIG. 2. Schwarzschild's camera with astigmatic flattening of the field.

two-mirror system. Second, the mechanical problem of maintaining alignment of the three mirrors as the instrument changes orientation is likely to be severe. Third, it seemed to me probable that a system of three mirrors met by the light in the order concave, concave, convex, would be more difficult to correct for high order aberrations than one with the more symmetrical arrangement: concave, convex, concave.

By adopting the latter arrangement, we find that it is easy to remove the first two difficulties also. If we insist that the tertiary mirror should be placed very close to the primary, it can be supported from behind the primary so that its supports do not cross the light path. Better still, the primary and tertiary mirrors can be made on the same disc. Then the tertiary does not have to be supported separately, and necessarily remains permanently aligned with the primary.

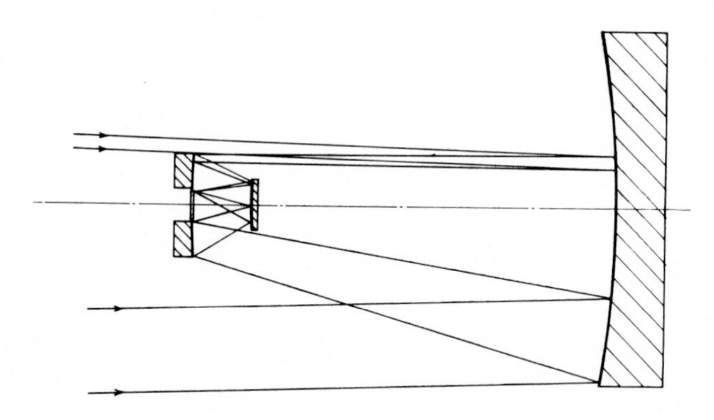

FIG. 3. Paul's three-mirror camera.

One particular system of this type has a unique combination of advantageous features. In this system, the primary and tertiary mirrors are not only made on the same disc, but their profiles are made to agree perfectly in position and slope where they meet (see Figure 5).

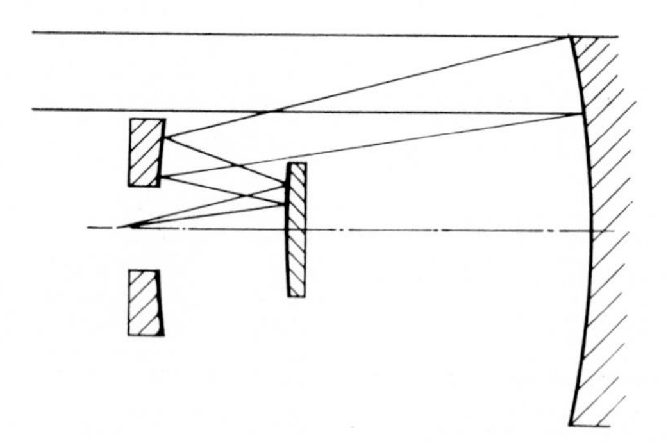

FIG. 4. Lagrula's three-mirror camera.

In order to see that it is possible to impose this restriction on the design without sacrificing control of the important aberrations, let us count the degrees of freedom available to the designer, and the number of conditions he must satisfy; let us ignore asphericities of high degree on the one hand,

and high order aberrations on the other. Then the designer has the following degrees of freedom available: the three paraxial curvatures and the three eccentricities of the three mirrors, and two axial separations (from primary to secondary and from secondary to tertiary), i.e. eight degrees of freedom altogether. With these he must achieve a prescribed focal length and a convenient location for the image plane, also control four primary

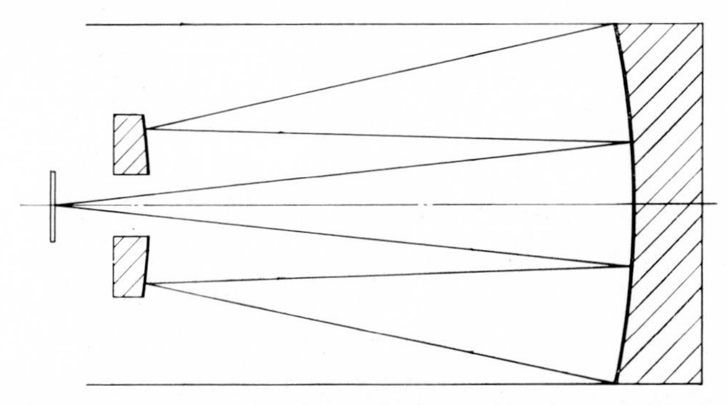

FIG. 5. New compact camera using three reflections, with primary and tertiary mirrors adjusted to form a continuous surface on a single disc.

aberrations: spherical aberration, coma, astigmatism and Petzval curvature, i.e. six conditions so far that he must satisfy. This leaves the two extra degrees of freedom necessary to match the primary and tertiary in position and slope where they meet.

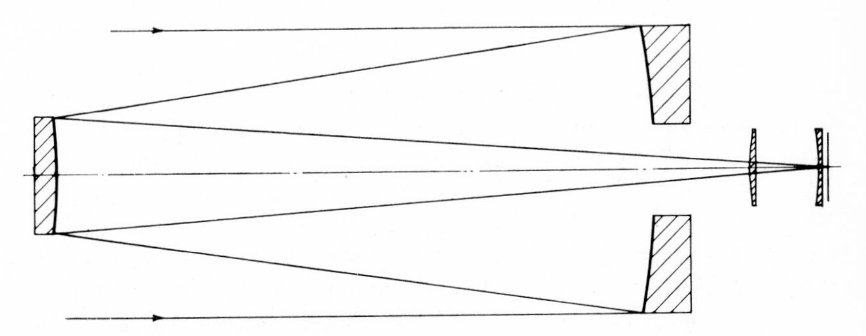

FIG. 6. Schulte's Ritchey-Chrétien design, with refracting components as suggested by Gascoigne.

If, in a system of this kind, the image plane is placed in a readily accessible position just outside the secondary mirror, the distance of the image from the pole of the tertiary mirror is one third of the focal length, while the axial separation of secondary and tertiary mirrors is 28 per cent of the focal length. If the diameter of the secondary mirror is made half that of the primary, the diameter of the substantially unvignetted image is about one sixth that of the primary mirror, irrespective of the focal length of the system. Thus, for a system of given aperture, the angular field and the focal length are inversely related. For a system of relative aperture f/5, the angular field is 2°.

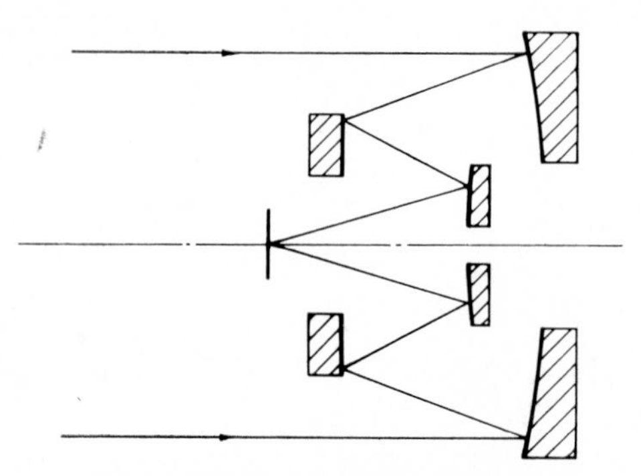

FIG. 7. Meinel's three-mirror camera.

It is natural to compare this new system with Ritchey-Chrétien systems on the one hand, and with Schmidt cameras on the other. Ritchey-Chrétien systems use only two mirrors, and can be made more compact than the new system; but without the help of auxiliary refracting components, their angular field is limited to less than 1°. With the help of refracting components, such as have been suggested by Violette (1922), Gascoigne (1965, 1966), Köhler (1966), Wynne (1965), and Schulte (1966*a*) the field can be increased. Schulte (1966*a*, *c*) has designed an f/7·5 Ritchey-Chrétien system with two auxiliary refracting components (see Figure 6) that gives star images of maximum dimension not more than half a second of arc over the range of wavelengths 340 to 660 nm at the edge of a field of $1\frac{1}{2}$° diameter. However, by a very small margin, his design is *less* compact than the new system using three reflections. If one were to insist on using the new system at the same relative aperture, the field would be limited by vignetting to a slightly smaller diameter (unless one tolerated a significantly larger central obstruction), but the definition of the star images would be distinctly better. On the other hand, if a relative aperture of f/5 is acceptable for the new system, the field can be of 2° diameter, at the edge of which the star images would be decidedly smaller than a quarter of a second of arc.

The Schmidt camera gives good definition and freedom from vignetting over a relatively extended field, often more than 5° diameter; but it requires a tube length that is twice the focal length, and a mirror of diameter larger than the clear aperture (usually by a factor of 1·5). The new three-reflection system, with a tube length only one third of the focal length, thus has an advantage over the Schmidt by a factor of six for this feature; while its primary mirror is no larger than the clear aperture. Thus it would be feasible to construct the new system with a larger aperture and a much longer focal length than for the largest feasible Schmidt. Since the faintest star that can be photographed becomes fainter as the focal length of the camera is increased, irrespective of aperture, astronomers may well prefer to use the new system with a relatively large focal length together with a modest field, rather than with a shorter focal length for the sake of the wider field that is then possible. The relative aperture of f/5 and field of 2° diameter mentioned already may well turn out to be an optimum compromise for a large instrument of this type. In the present state of knowledge, a system of aperture 3 m and focal length 15 m, in a tube of overall length about 6 m, with a field of 2°, would be a straightforward engineering proposition and an exceptionally powerful instrument.

The profiles of the mirrors for the new system are all hyperbolic, to a first approximation. The focal length of the primary is about half that of the complete system, and the departure of its shape from a paraboloid is about twice that of a typical Ritchey-Chrétien primary. This is a very advantageous shape for use with suitably redesigned prime focus correctors of the kinds suggested by Gascoigne (1965), Baranne (1966), Köhler (1966), Wynne (1965), and Schulte (1966*b*, *c*) to work with Ritchey-Chrétien primaries. In particular, with the new primary, the simple aplanatizing plate suggested by Gascoigne (1965, 1966) would give a field of about 1·4 times the diameter of that obtainable with a Ritchey-Chrétien primary of the same relative aperture. The other correctors would also benefit to some extent. Again, it is always possible to design secondary mirrors to provide Cassegrain and coudé foci free from spherical aberration wherever they may be wanted for photometry and spectroscopy; so it appears that if the new system were required to be the basic arrangement of a multipurpose instrument, it would be even more advantageous for this than the other systems already in use.

Historical Note

The author described the new system to Professor S. C. B. Gascoigne some time ago, who in turn mentioned it to Dr. A. B. Meinel. Dr. Meinel showed a diagram of such a system, except that he did not join the primary and tertiary mirrors, at the I.A.U. Symposium on the Construction of Large Telescopes (Meinel 1966), remarking that he did not know what name to attach to it. Dr. Meinel (1967) has himself proposed and examined another version of the three-mirror system, in which he makes the secondary paraxially flat and the tertiary convex (see Figure 7). Compared with the system recommended above, Meinel's system suffers from the disadvantages of a larger central obstruction, and a tertiary mirror that has to be made separately from the primary and subsequently mounted in accurate alignment with it.

REFERENCES

Baranne, A., 1966, *The Construction of Large Telescopes*, I.A.U. Symposium No. 27, D. L. Crawford (Ed.), Academic Press, London and New York, p. 22.
Gascoigne, S. C. B., 1965, *The Observatory*, **85,** 79.
—— 1966, I.A.U. Symposium No. 27, p. 36.
Köhler, H., 1966, I.A.U. Symposium No. 27, p. 9.
Lagrula, J., 1942, *Cahiers de Physique*, **8,** 43.
Meinel, A. B., 1966, I.A.U. Symposium No. 27, p. 29.
—— 1967, *J. Soc. Mot. Pict. Telev. Engrs.*, **76,** 201.
Paul, M., 1935, *Rev. d'Opt.*, **14,** 169.
Schulte, D. H., 1966*a*, *Appl. Opt.*, **5,** 309.
—— 1966*b*, *ibid.*, p. 313.
—— 1966*c*, I.A.U. Symposium No. 27, p. 32.
Schwarzschild, K., 1905, *Astron. Mitt. Kön. Sternw. Göttingen*, No. 10.
Violette, H., 1922, *Rev. d'Opt.*, **1,** 397.
Wynne, C. G., 1965, *Appl. Opt.*, **4,** 1185.

A Family of Short-Tube Wide-Field Catadioptric Telescopes

D. S. Brown

Abstract—The Schmidt telescope has made a great contribution to astronomical knowledge because of its relatively wide field of good definition. The main disadvantage of the Schmidt is its very large overall length which adds significantly to the cost of mounting and housing the instrument, and leads to flexure problems in large-sized instruments.

The optical systems described are a family of single-mirror short-tube anastigmats resembling the Wright camera, but having superior performance. The optical performance of these systems can be described to a reasonable approximation by simple expressions which are given, and the problem of designing a wide-field telescope to meet specific performance requirements at minimum cost is considered.

The problems of optical construction and alignment for these systems and for a classical Schmidt are discussed.

INTRODUCTION

In recent years the optical design of medium and large size astronomical telescopes has almost invariably followed a single pattern, the Ritchie-Chrétien Cassegrain arrangement, with a focal ratio of about f/8 for direct photography of faint objects, a prime focus station with a ratio of about f/3 and a coude station of about f/30, with additional Cassegrain stations for spectroscopic and photometric use. The advantages of this type of telescope are not in question, but the maximum field size attainable, though large by comparison with that of a paraboloid and Ross-type corrector, is not large enough to permit the surveying of large areas of sky. The classical Schmidt has adequate field size for this purpose but suffers from the mechanical handicap of its very long tube, which because of the cost of the necessary building and dome restricts the focal ratio to rather fast systems with significant field and chromatic aberrations, limited by short exposures. In the George Darwin Lecture, I. S. Bowen gave an admirably concise account of the astronomical reasoning leading to this situation.

The maximum exposure time that can be usefully employed for wide field photography is set by differential refraction. Figure 1 shows the relation for maximum image spread $\frac{1}{2}$ arc sec., the exposure times being the mean of the values given by Bowen for declinations $+60°$, $+30°$, $0°$, $-30°$. For a field of 5° diameter the maximum exposure is about one hour and with modern emulsions such as Kodak 103a—O this indicates a focal ratio of about f/5. Figure 1 also shows the relation between limiting exposure and focal ratio, and by joining the two curves by vertical lines of constant exposure the optimum focal ratio for any particular field diameter is easily established. Clearly, calculations of this sort are approximate and dependent on the validity of the photographic emulsion data

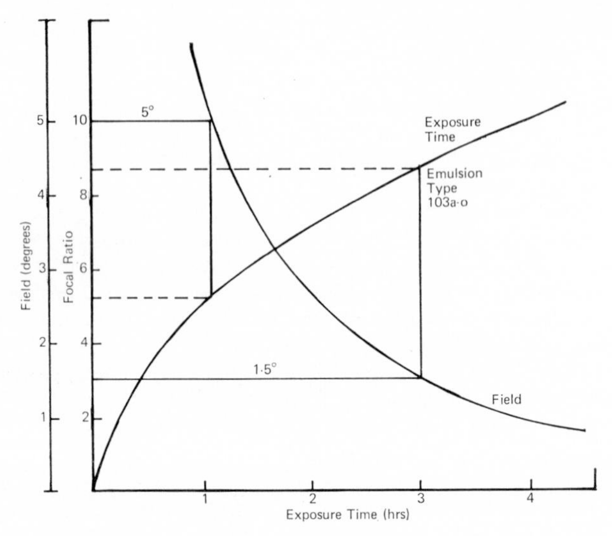

FIG. 1. Relationship between maximum useful field, maximum exposure time (103 a–o) and focal ratio.

during the useful life of the instrument. However, there can be little doubt that there is a real need for an optical system capable of giving good, wide field performance but having a tube length less than that of the classical Schmidt.

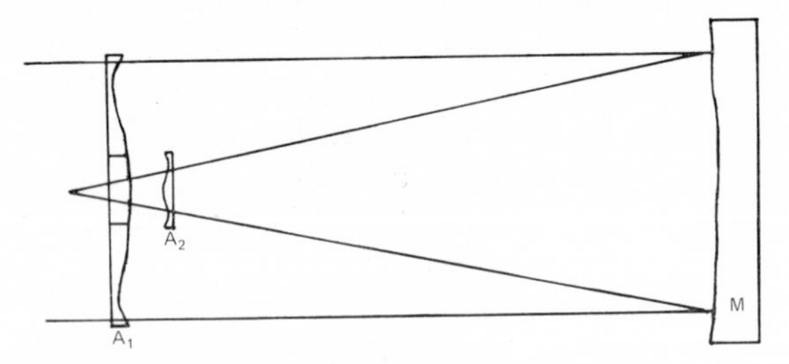

FIG. 2. Short-tube anastigmat optical system.

Optical System

To correct the primary aberrations of a spherical mirror by means of a set of aspheric plates it will generally be necessary to use three plates. If the mirror is itself made aspheric only two aspheric plates are needed. The problem remaining is to locate these in the most favourable positions and to determine the optimum aspheric profiles. It is physically convenient to locate a small corrector plate close to the focal surface in the converging beam and a larger one near the aperture stop of the system at a position as close to the centre of curvature as possible, which for a short tube

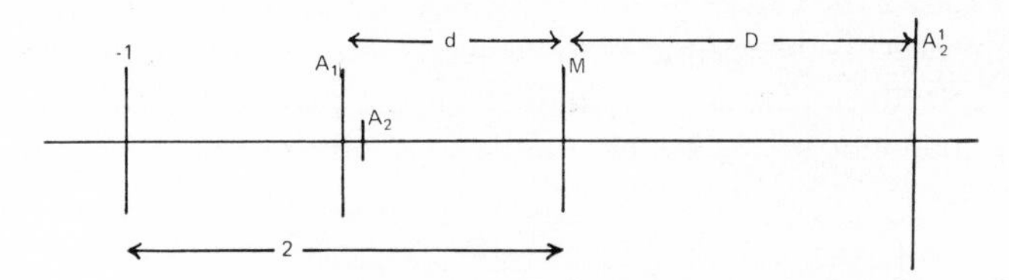

FIG. 3. Plate diagram for short-tube anastigmat.

instrument means at a distance from the mirror approximately equal to its focal length. Figure 2 shows the layout of the system and Figure 3 the plate diagram imaged into object space. If we adopt the convention that the power of the equivalent Schmidt corrector plate is equal to one, and focal length also equal to one, then the powers of the two refracting plates are given by:

$$A_1 = (2D + 4)/(d^2 + dD) \tag{1}$$

$$A_2 = (4 - 2d)/(D^2 + Dd), \tag{2}$$

and the power of the mirror asphericity by:

$$A_m = 1 - (A_1 + A_2) \tag{3}$$

Chromatic Errors

The main residual aberrations in an instrument of moderate focal ratio are likely to be the chromatic effects due to the two aspheric plates. These are dependent on the powers of the plates (and on the distance from the focal surface in the case of the second plate). Figure 4 shows the variation of power, A_1, of the large plate with separation from the mirror, for four different values of D. The case $D = \infty$ is essentially the limiting case of the Wright camera, and we have the expected result $A_1 = 2$ at $d = 1$. The plate power decreases as d increases and since the overall length of the system cannot be made much greater than the focal length without increasing the building dimensions, the initial selection of $d = 1$ appears a good compromise. Since the large aspheric plate is at, or near the aperture stop of the system, the chromatic error introduced will be wavelength-dependent, spherical aberration, uniform over the field.

The value of A_2 given in Equation (2) is the power of the small aspheric plate over the on axis section only. For a system covering a field angle ϕ with focal ratio f, the power of the full-size plate is given by:

$$A_T = \phi^4 f^4 (D+1)^4 (4-2d)/(D^2+Dd) \tag{4}$$

which reduces for the case $d=1$ to:

$$A_T = 2\phi^4 f^4 (D+1)^3/D \tag{5}$$

Figure 5 shows the values of A_T for the case $d=1$, $\phi=0{\cdot}1$, $f=5$. Although the total power of the second plate depends strongly on D, the chromatic errors it introduces are less dependent on the plate position. When D is large the diameter of the incident cone is small and the plate introduces

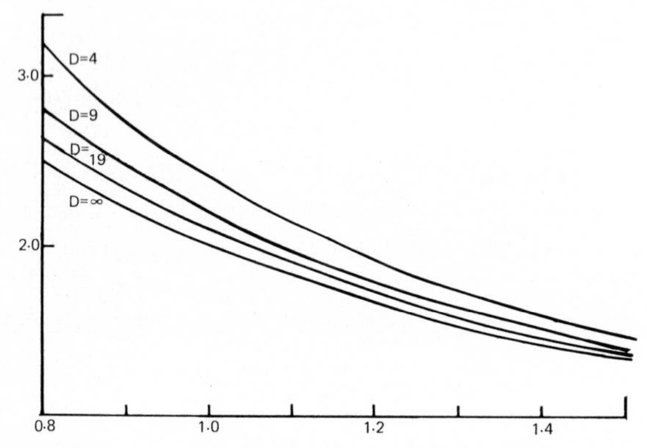

FIG. 4. Relationship between aspheric power of main corrector and aspheric plate positions.

almost pure astigmatism. For the condition D large, the rest of the optical system reduces to a Wright camera with field astigmatism independent of D, corrected by the aspheric plate. The chromatic astigmatism introduced by the aspheric plate is equal to the astigmatism of the equivalent Wright system divided by the Abbé number for the wavelength range considered. In addition to the chromatic astigmatism there will be a small transverse chromatic aberration which can be made zero for any selected field radius by selection of a suitable quadratic term in the plate profile, and a very small chromatic coma. If desired, this last aberration may be reduced to zero by selection of a suitable position for the aperture stop of the system. The transverse chromatic aberration is proportional to A_T and is reduced as D decreases, but only at the expense of some increase in A_1.

Higher Order Aberrations

To obtain optimum performance in a system with three aspheric surfaces it is possible to introduce higher order terms in the aspheric profiles to correct higher order aberrations which may remain when only r^2 and r^4 terms are used. For example, the introduction of an r^6 term into each profile makes it possible to correct spherical aberration of this order and also one astigmatic and one comatic aberration of the same order. The main field error contribution of the large aspheri plate is a higher astigmatism, whilst the fact that the mirror asphericity is located on a curved surface is likely to introduce a high-order comatic term. It seems probable that both these effects can be corrected by the introduction of suitable terms in the aspheric profiles, although no detailed calculations have yet been made.

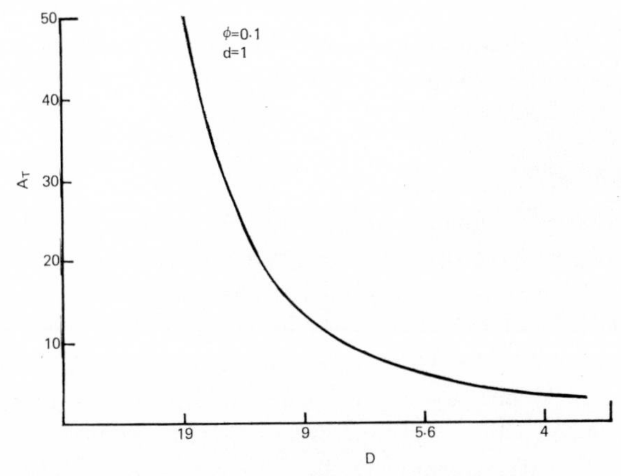

FIG. 5. Relationship between total power and position of small corrector plate.

Meniscus Lens Systems

Clearly either of the two aspheric corrector plates may be replaced by meniscus lenses, and provided that the mean centre of curvature is used to describe the position of the meniscus, the relations previously given remain valid for these systems. Figure 4 shows that there is some advantage to be gained by making $d > 1$, which is possible in the meniscus case, and the reduced power of the corrector results in a reduction of the chromatic errors introduced by it quite apart from the lower inherent chromatic error of the meniscus. The use of a meniscus lens in the converging beam tends to introduce transverse chromatic aberration which can usually be corrected, but only by moving the centres of curvature of the two surfaces so far apart that the simple relationships given are no longer valid.

Adjustment

One of the main practical advantages of the classical Schmidt is that the adjustment of the optical system is relatively easy to carry out. The adjustment of a system with three aspheric surfaces might be expected to be much more difficult, but in this case the two aspheric plates are placed so close together that the smaller plate can conveniently be mounted from the larger, and with modern manufacturing methods there is no difficulty in ensuring that the two plates remain in their correct relative positions. The only adjustments needed are to square on the aspheric plate assembly, the mirror and the photographic plate. Inaccurate squaring of the mirror produces axial coma, and since the focal surface is easily accessible this may be observed visually. When any axial coma has been removed, the presence of axial astigmatism indicates lack of squareness of the aspheric plate assembly. When mirror and plate assembly are aligned the plate-holder may be aligned by focus measurements at several positions near the edges of the field. In practice the performance is little affected by mis-alignment of the aspheric plate assembly, whilst mirror and plate-holder alignment are critical, as with the classical Schmidt.

Total System Cost

The total cost of producing an operational astronomical instrument is largely controlled by the optical design. This is because the optical design controls the physical size of the instrument and its associated building while the cost of optical components represents only a small part of the whole. In most telescope designs the cost of the optical components is between 10 and 30 per cent of the instrument cost. The cost, including building and dome, will be perhaps two or three times the instrument cost though this can vary enormously due to differences in labour costs in different locations. The total cost of the installation will therefore depend mainly on dome diameter and will rise approximately as $(\text{diameter})^{2\cdot5}$. In considering the relative cost of alternative telescope designs it is not

TABLE 1. Relative Cost of Construction and Operation.

	Ritchie Chrétien	Classical Schmidt	Short-tube	
Construction	1	67	12	Equal focal
Operation	28	1	1	length
Construction	1	20	3·6	Equal
Operation	11	1	1	aperture

possible to find a universally valid basis for comparison. For photographic systems one might require the systems to be capable of recording equally faint objects, in a single exposure, regardless of exposure time, which implies systems of equal focal length. Alternatively we may consider systems of equal aperture, which would imply a reduced ability to detect faint objects in the faster system unless high efficiency slower plates become available or superposition of multiple exposures is used. Table 1 gives the relative capital and operating costs for Ritchey Chrétien (f/8, $1\frac{1}{2}°$ field), classical Schmidt (f/5, 5° field), and short tube (f/5, 5° field) systems for both alternatives.

These cost figures assume that the instruments are used exclusively for wide-field photography, and would be significantly altered if allowance were made for spectrographic use, but such allowance is hardly possible without detailed consideration of the observational programmes to be carried out.

CONCLUSIONS

The short-tube systems described have more powerful aspheric correctors and hence larger chromatic errors than a classical Schmidt of the same focal ratio, but at the focal ratios that would be most effective for wide-field photography the performance is quite adequate. The capital cost of the instrument and building is significantly higher than for a Ritchey Chrétien telescope, but much less than that of an equivalent Schmidt. For survey-type photography the operational cost is much lower than that of the Ritchey Chrétien telescope.

The author would like to thank the members of the Optical Laboratories of Grubb Parsons & Co. Ltd., for their help in producing this paper and in the construction of the prototype instrument, and the Directors of the Reyrolle-Parsons Group for permission to publish this paper. The optical designs described are the subject of a patent application.

Quantitative Treatment of 'Seeing' in Systems Design of Solar Astronomical Telescopes

C. E. Coulman

CSIRO Division of Physics,
National Standards Laboratory,
Sydney, Australia, 2008

Abstract—The modulation transfer function of a telescope which views a celestial object through the Earth's atmosphere may be calculated from the properties of the instrument and certain statistics of the fluctuating atmospheric temperature field.

It is shown that the size of telescope may be matched to expected atmospheric 'seeing' conditions at a site on the basis of simple micrometeorological measurements. The improvement in optical performance and increase in utilization which result from raising a telescope above the ground on a tower are also shown to be capable of quantitative estimation. Some examples of the performance and cost of construction of telescope installations at an actual observatory are given.

INTRODUCTION

Astronomers have long been aware that the image-transmitting properties of the Earth's atmosphere often limit the resolution of detail in telescopic images of celestial objects. However, only recently has quantitative computation of the optical performance of a complete system comprising telescope and atmospheric propagation path become possible. In this short paper we shall consider the methods which are available for the purpose, the range of conditions for which they have been experimentally substantiated, and the extent of the relevant micrometeorological data available.

As examples of the utilization of this approach we shall show how the effect of the atmosphere on the performance of a solar telescope may be calculated and how this estimated performance may be improved by raising the telescope above ground level on a tower.

Optical Image Transmission through the Atmosphere

As an electromagnetic wave propagates in the atmosphere it exhibits amplitude and phase fluctuations which depend upon the fluctuations of refractive index of the air. In the image-plane of a telescope which receives such a wave, we may observe corresponding spatial and temporal fluctuations in the light distribution, and hence in the image quality. In practice, image-detection occupies some finite time and measurement or specification of the quality of an image involves consideration of the effects of time-averaging. It is convenient to specify the image quality of a system which comprises an image-forming instrument and an atmospheric propagation path by its optical transfer function (contracted to O.T.F.). For an image averaged over time t' the modulus of the O.T.F., or modulation transfer function (M.T.F.), may be written $\langle M(f) \rangle_{t'}$, where f is spatial frequency. Hufnagel and Stanley (1964) have shown theoretically that for an image averaged over a long time

$$\langle M(f) \rangle_{t' \to \infty} = A(f) \exp\left[-\tfrac{1}{2} D_w(r)\right] \tag{1}$$

Spatial frequency f in the image plane is related to r, a correlation distance, in the entrance-pupil plane by

$$f = r/\lambda F,$$

where F is the focal length of the imaging system and λ the wavelength of light. The intrinsic O.T.F. of the optical system is $A(f)$ and $D_m(r)$ is a statistical measure of the amplitude and phase fluctuations of the wave received at its entrance pupil. It is called the wave structure function* (Fried 1966) and may be expressed as

$$D_w(r) = 2{\cdot}91 \left(\frac{2\pi}{\lambda}\right)^2 r^{5/3} \int_0^{x_1} C_n^2(x) \mathrm{d}x, \tag{3}$$

for propagation of an initially plane wave from $x = 0$ to $x = x_1$ (Tatarski 1959) in a locally homogeneous, isotropic medium.

For such a medium the structure function of a passive variable, such as refractive index n, may be written as

$$D_n(r) = C_n^2 r^{2/3}, \tag{4}$$

following Kolmogorov (1941).

The O.T.F. of a system comprising a telescope and a horizontal propagation path above level, uniform land has been investigated experimentally as a function of the structure coefficient C_T of the atmospheric temperature

* A structure function is a statistical second moment defined, in one dimension and for a property T, by

$$D_T(r) = \langle [T(x) - T(x+r)]^2 \rangle,$$

where r is a span along coordinate x and the sharp brackets denote the averaging operation. It is convenient to express $D_T(r)$ in the form

$$D_T(r) = C_T^2 \phi(r),$$

where C_T is called the structure coefficient of T and ϕ is some function of r which characterizes the statistical properties of the field of variable T.

field and the range x by Coulman (1965 and 1966). Analysis of some 500 experimental measurements has yielded the relationship

$$\langle M(f)\rangle = A(f)\exp\left[-\text{const.}\,f^{1\cdot 65}C_T^{\,2\cdot 00}x^{1\cdot 06}\right], \tag{5}$$

which closely resembles the equation obtained when the expression for $D_w(r)$ in Equation (3) is substituted into Equation (1). The approximately

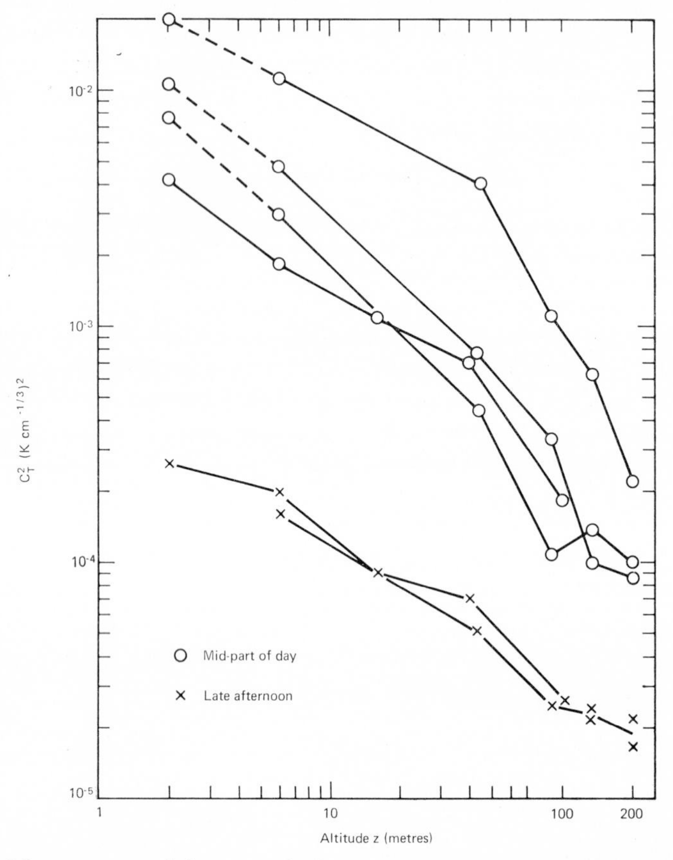

FIG. 1. Measurements of the atmospheric temperature structure coefficient $C_T{}^2$ have been made on clear summer days over uniform level grasslands with instruments suspended from captive balloons. These results refer to average conditions in the mid-parts and afternoons of such days.

linear relationship of the optical refractive index of air to the air-temperature is expressed by the constant of proportionality in Equation (5).

Expressions for the M.T.F. for an image averaged over a very short time $\langle M(f)\rangle_{t'\neq\infty}$ have been given by Fried (1966) but these have not yet been verified experimentally, although some discussion of the practical effects of varying image-averaging time has been given by Coulman (1968).

Observational Data on Atmospheric Temperature Structure

The foregoing experimental and theoretical relationships provide a basis for calculating the O.T.F. associated with image propagation through a locally homogeneous, isotropic atmosphere provided that the distribution of C_T^2 along the propagation path is known.

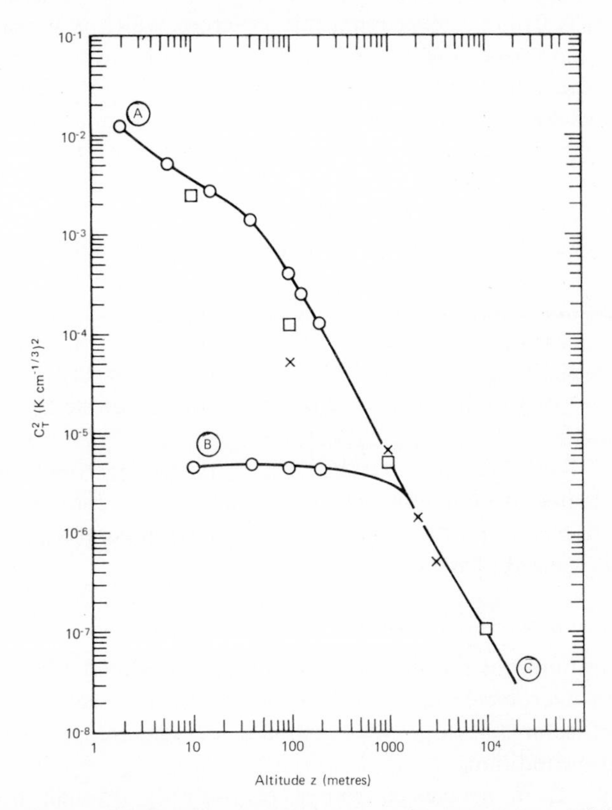

FIG. 2. Variation of C_T^2 with altitude z. Curve AC refers to average conditions during the mid-part of clear summer days. Curve BC refers to conditions in the temperature-quiescent downdraughts between convective plumes. The origins of data points are: ○ Coulman (1969); □ Hufnagel (1966); × Koprov and Tsvang (1966).

The variation of C_T^2 with altitude above level, uniform grassland has recently been investigated with rapid-response thermometers suspended from captive balloons (Coulman 1969). Some of the results of this work are shown in Figure 1. In Figure 2 values of C_T^2 obtained during the mid-parts of clear summer days have been averaged and are denoted by circles on curve AC. The only other comparable measurements available were made by Koprov and Tsvang (1966) with an instrumented aircraft, but some indirect estimates of C_T^2 have also been given by Hufnagel (1966).

By combining all of these data a reasonable estimate of the daytime C_T^2-profile up to about 20 km altitude may be made, as seen in Figure 2, curve *AC*.

It has been shown that short periods of exceptionally good 'seeing' conditions are sometimes observed as a consequence of thermal structure in the lower layers of the atmosphere (Coulman & Hall 1967). Furthermore, there is strong evidence that this structure, which is associated with convective motions of the air, persists to heights of a kilometre or so and facilitates the intermittent use of a large telescope for high-resolution observations by day. Measurements of C_T^2 in the temperature-quiescent downdraughts associated with this convective structure have yielded the results shown by circles on curve *BC*, Figure 2. This subject is dealt with in greater detail by Coulman (1969).

In the systems design of solar telescopes precise knowledge of the C_T^2-profile is essential in the lower troposphere where, under normal daytime conditions, a large contribution to the wave structure function $D_w(r)$, and hence to astronomical 'seeing', arises. The portion of the curve relating to altitudes above about 3 km in Figure 2 must be regarded as a tentative estimate and its use in calculating the performance of stellar telescopes, for example, must be correspondingly cautious until direct measurements of C_T^2 have been made at high altitudes. In particular, it is thought that layers of temperature-inhomogeneous air sometimes exist in the upper troposphere but no direct measurements of the temperature structure of such layers have yet been made.

Applications to the Systems Design of Solar Telescopes

As an example of the use of the approach outlined above we shall employ the micrometeorological data of the previous section to estimate the size of solar telescope which should be selected to match given atmospheric conditions.

For light from a celestial source propagating through the earth's atmosphere, we may rewrite Equation (3) as:

$$D_w(r) = 2{\cdot}91\left(\frac{2\pi}{\lambda}\right)^2 r^{5/3} \sec\omega \int_{z_T}^{z_m} C_n^2(z)\,\mathrm{d}z, \tag{6}$$

where z is the vertical coordinate with origin at the ground, ω is the angle between zenith and the viewing direction, and we neglect effects due to curvature of the earth. An image is assumed to be formed by a diffraction-limited, axially symmetric optical system situated at height z_T, and z_m is the height above which the value of C_n^2 becomes negligibly small. On the basis of present knowledge, Figure 2, we believe that no serious error is involved if we take $z_m = 20$ km.

We utilize Equations (5) and (6), together with a well-known relationship between C_n and C_T (Coulman 1969), to calculate the M.T.F. for the time-averaged image $\langle M(f) \rangle$ produced by an 11 cm-diameter telescope mounted at height $z_T = 10$ m, viewing zenith through an atmosphere whose C_T^2-profile is given by curve *AC* in Figure 2. The M.T.F. of this composite system is denoted by the full line (1) in Figure 3, whereas the broken line represents the M.T.F. of this diffraction-limited telescope in a completely uniform atmosphere.

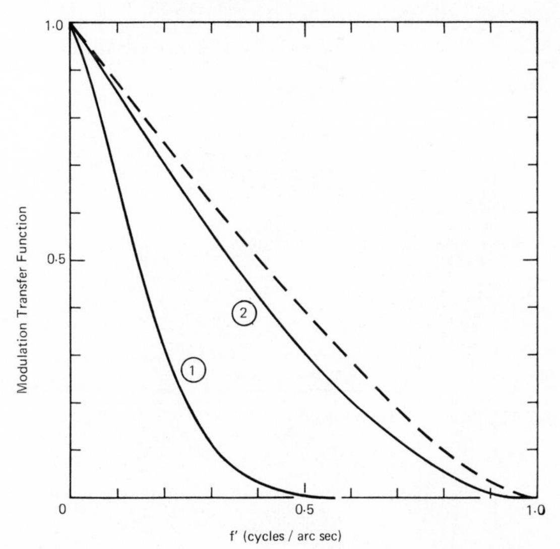

FIG. 3. The long-time-average MTF for a system comprising an 11 cm-diameter telescope and a propagation path through an atmosphere having C_T^2–profile represented by curve *AC* of Figure 2 is here denoted by curve (1). Curve (2) represents the MTF of this system under especially favourable conditions when the C_T^2-profile is as shown by curve *BC*, Figure 2. The telescope is assumed to be diffraction-limited, with MTF as denoted by the broken line, and mounted at $z_T = 10$ m above ground. Wavelength $\lambda = 0{\cdot}55 \times 10^{-4}$ cm.

Spatial frequency f' is expressed in cycles per unit angle of field to render the results independent of focal length F of the imaging system.

$$f' = Ff. \tag{7}$$

This is common practice in astronomical optics; the resolution cut-off limit of an instrument is customarily quoted in arc sec.

Better optical performance can be achieved with this system if a time-averaged image is obtained during one of the periods of exceptionally favourable conditions referred to above (p. 532), and characterized by the C_T^2-profile of curve *BC*, Figure 2. The M.T.F. of the system is then as shown by curve (2), Figure 3.

If similar calculations are performanced for a larger telescope, of 30 cm-diameter, it will be seen from curve (1), Figure 4, that under average mid-day conditions the performance is not significantly better than that of the smaller telescope referred to in Figure 3. The larger instrument is therefore atmosphere-limited. However, during periods of exceptionally favourable conditions, characterized by the C_T^2-profile of curve *BC*, Figure 2, the M.T.F. of the composite system is represented by curve (2), Figure 4. Much of the potential of the larger telescope for resolving fine detail can then be utilized. Further improvement is possible by raising the telescope to greater height z_T. and by the use of short image-averaging time (Coulman 1969).

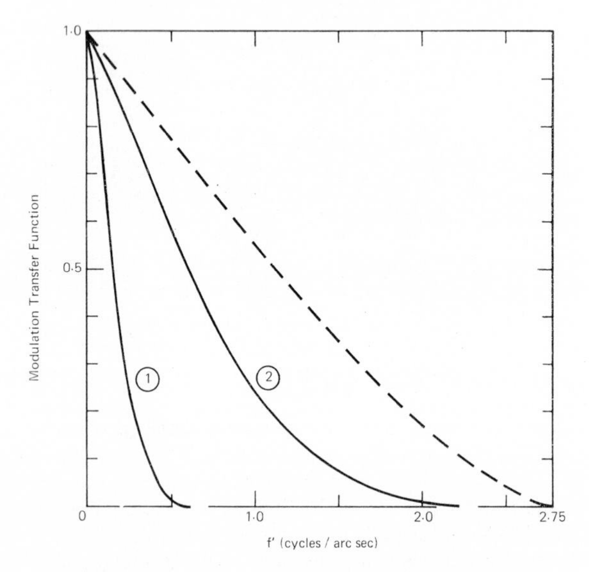

FIG. 4. Similar to Figure 3 but for a telescope of 30 cm-diameter. Note that the abscissa scale is different from that of Figure 3 but that the resolution cut-off limit of curve (1) is again close to $f' = 0{\cdot}5$ cycle/arc sec. The system under these conditions is atmosphere-limited.

As a further example of the use of the methods described in this paper, the improvement in performance of a telescope which may be obtained by raising it above ground level on a tower will be calculated. Figure 5 shows the M.T.F. of a telescope of 11 cm-diameter viewing zenith at mid-day from heights of 2, 10, 30 and 100 m above level uniform grassland; the data from Figure 2, curve *AC* are used in these calculations. The improvement in the M.T.F. which results from eliminating the effects of the lower layers of air, where C_T^2 attains large values, is evident.

Discussion of Some Actual Observatory Installations

The economic aspect of systems design of astronomical instruments and installations invariably becomes a matter of major importance in the construction or extension of observatories. The methods outlined in this paper and elsewhere (Coulman 1969) facilitate the calculation of changes

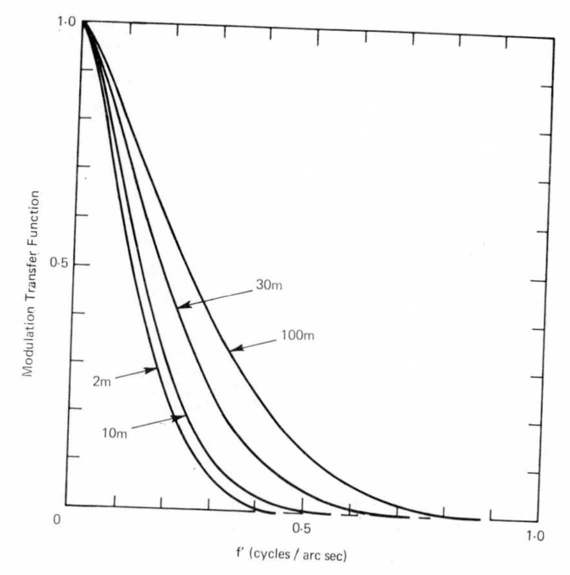

FIG. 5. The MTF is shown for an 11 cm-diameter telescope mounted at 2, 10, 30 and 100 m above ground and used under conditions when the $C_T{}^2$-profile is as given by curve *AC* of Figure 2. Note the improvement in performance as the telescope is raised above the lowest layers of air where $C_T{}^2$ attains large values.

in performance and utilization which result from changes in major design parameters. It would be extremely useful if such variations in performance could be related to cost of construction in some general way, but it seems that the only reliable guide to cost is obtainable from the prices tendered for actual projects. It is interesting to consider the development of the CSIRO Division of Physics' Optical Solar Observatory at Culgoora, N.S.W., Australia, as an example in this connection.

The first instrument installed there was a solar flare-patrol telescope, of effective aperture 10 cm, which is required to operate in continuous daily duty using an image exposure time of about 0·05 sec. The example cited in Figure 3, curve (1), is roughly comparable with this instrument. It may be judged that a reasonable compromise between cost and performance has been made by mounting this flare-patrol telescope at 6 m above ground on a structure which cost approximately $US 9000 in 1964. This sum covered the cost of construction of a cast concrete pier to support the telescope and a mechanically retractable protective canopy (instead of the more usual dome) supported by a steel framework. The telescope, instrumentation and control cabin walls, fittings and services are not included.

Subsequently, a 30 cm-aperture chromospheric telescope was installed at 20 m above ground on a steel structure which cost about $US 44,000 in 1966. Again this sum covered the cost of construction of the tower and protective canopy but excluded such items as the telescope, instrumentation and services. Full details are given by Loughhead, Bray, Tappere and Winter (1968). For much of the research work for which this instrument is intended it is acceptable to confine operation to the especially favourable conditions referred to above (p. 532). The expected performance should be slightly better than indicated by Figure 4, curve (2) because of the greater height ($z_T = 20$ m) and there is some evidence that this is so (Loughhead *et al.* 1968).

CONCLUSION

It has been shown above that the optical performance of a telescope viewing a celestial object through the earth's daytime atmosphere may be estimated quantitatively, in conditions of locally isotropic, homogeneous turbulence, from measurements of the fluctuating atmospheric temperature field. It is likely that such measurements can be conducted more cheaply than can the traditional optical site testing and they can be utilized quantitatively to optimize some aspects of the design of an observatory from both economic and performance viewpoints.

In principle there is no reason why the approach outlined in this paper should not be applied to all the circumstances of interest to astronomers. However, in many such circumstances, it appears that homogeneous, isotropic turbulence does not exist and the form of the structure function of refractive index or temperature cannot be given in the simple form of Equation (4). Some meteorological measurements of temperature fluctuations in stable thermal equilibrium at night over land by Okamoto and Webb (1969) and observations of the temperature structure in oceanic air-streams by Coulman and Shaw (1969) support this view.

There is thus a clear need for further micrometeorological research in order to extend the application of present methods for the quantitative treatment of 'seeing' to all those circumstances which interest the designers of astronomical telescope systems.

REFERENCES

Coulman, C. E., 1965, *J. Opt. Soc. Am.*, **55,** 806.
—— 1966, *J. Opt. Soc. Am.*, **56,** 1232.
—— 1968, *J. Opt. Soc. Am.*, **58,** 1668.
—— 1969, *Solar Phys.*, **7,** 152.
Coulman, C. E. and Hall, D. N., 1967, *Appl. Opt.*, **6,** 497.
Coulman, C. E. and Shaw, D. E., 1969, Colloq. on the Spectra of Meteorological Variables (URSI), Stockholm, 9–19 June 1969.
Fried, D. L., 1966, *J. Opt. Soc. Am.*, **56,** 1372.
Hufnagel, R. E. and Stanley, N. R., 1964, *J. Opt. Soc. Am.*, **54,** 52.
Hufnagel, R. E., 1966, Restoration of Atmospherically Degraded Images, Woods Hole Summer Study, Washington, D.C.
Kolmogorov, A. N., 1941, *Dok. Akad. Nauk S.S.S.R.*, **30,** 301. [Translated 1961: *Turbulence*, Friedlander, S. K. and Topper, L., (eds.) Interscience, New York.]
Koprov, V. M. and Tsvang, L. R., 1966, *Izv. Atmos. i Okean. Fiz.*, **2,** 1142.
Loughhead, R. E., Bray, R. J., Tappere, E. J. and Winter, J. G., 1968, *Solar Phys,*. **4,** 185.
Okamoto, M. and Webb, E. K., 1969, in press.
Tatarski, V. I., 1959, *Wave Propagation in a Turbulent Medium.* [Translated R. Silverman, 1961, McGraw Hill, New York.]

The Effects of Atmospheric Scattering and Absorption on the Performance of Optical Sensors

L. G. Mundie* and H. H. Bailey*

The RAND Corporation,
Santa Monica, California

Abstract—The apparent radiance N of a source of inherent (zero-range) radiance N_0 as viewed through a scattering atmosphere, is comprised of transmitted radiance and path luminance, and is given by the relation $N = N_0T + N_q(1 - T)$, where T is the atmospheric transmittance and N_q, the 'equilibrium' radiance, is approximated by that of the horizon sky measured at an appropriate azimuth. The performance of photographic systems and vision at high light levels is measured directly by transmitted contrast. With the aid of the above relation, the contrast transmittance τ_c is shown to be $\tau_c = T[T + K(1 - T)]^{-1}$, where K, the sky/ground ratio, is the ratio of N_q to the scene radiance. When detecting targets against the horizon sky the contrast transmittance is equal to T. Low light level photoelectric sensors are fundamentally limited by photon noise, their limiting S/N being proportional to the quantity $(N_t - N)/(N_t + N_b)^{\frac{1}{2}}$, where N, and N_b represent, respectively, the radiance of the target and the background. The transmittance τ_{ph} of this function is given by the relation $\tau_{ph} = T[T + K(1 - T)]^{\frac{1}{2}}$. A similar factor applies to visual systems at low ($<10^{-1}$ cd/m^2) luminance levels. The S/N transmittance with gated viewing systems (also assumed to be limited photon noise) is equal to T, while with passive infrared sensors, which are essentially unaffected by path radiance, the corresponding factor is T_{IR}, the infrared transmittance. The latter sensors are usually considerably less degraded in performance by a scattering atmosphere than are visual or photographic systems, while the degradation suffered by photon-noise-limited systems such as image intensifiers or low-light-level television equipment lies between these extremes.

INTRODUCTION

The scattering and absorption of electromagnetic radiation by the atmosphere have been studied extensively (Middleton 1958; Duntley 1948; Cambridge Research Laboratory 1965; Duntley 1964). This paper is concerned with the effects of atmospheric absorption and scattering on the performance of optical systems designed to assist in the detection of targets against terrestrial backgrounds; atmospheric refraction, including the effect of turbulence on 'seeing' conditions, is ignored. The sensors

* This research is supported by the United States Air Force under Project RAND—Contract No. F44620–67–C–0045. Views or conclusions contained in this study should not be interpreted as representing the official opinion or policy of the United States Air Force.

considered are the human eye, photoelectric detection systems, gated viewers, and passive infrared scanners. The performance of each of these sensors is examined as a function of the apparent brightness or radiance of the target and background, and the 'transmittance' of the appropriate function through the atmosphere is expressed in terms of the energy transmittance (indicated by the visibility range) and the sky/ground brightness ratio.

The Effects of Atmospheric Scattering on Apparent Radiance

The apparent radiance N_R of a surface of inherent (zero-range) radiance N_0 viewed through a scattering atmosphere having a transmittance T is equal to the sum of the transmitted radiance N_0T and the path luminance, which can be expressed as $N_q(1-T)$,* so that

$$N_R = N_0T + N_q(1-T) \tag{1}$$

where N_q is the apparent radiance of an infinite (equilibrium) path identical with the actual path as regards particle characteristics and illumination. Practically, N_q is approximately equal to the radiance of the horizon sky viewed at such an azimuth that the line of sight is inclined to incident sunlight at the same angle as is the actual viewing path. (In some geometrical configurations no such azimuth exists; however, the definition still stands.)

Visual and Photographic Systems

The performance degradation suffered by visual and photographic systems operating through a scattering atmosphere can be described by stating the contrast transmittance, which can be derived as follows. The inherent contrast, C_0, is defined by the realtion

$$C_0 = (B_t - B_b)/B_b \tag{2}$$

where B_t and B_b represent, respectively, the inherent luminance of target and background. The apparent contrast C_R, at range R is found, with the aid of Equation (1) (expressed in terms of luminance) to be:

$$C_R = C_0T[T + K(1-T)]^{-1} \tag{3}$$

where $K = N_q/N_b$. Since, as mentioned above, N_q is (conceptually, at least) approximated by the radiance of the horizon sky observed at an appropriate azimuth, K is termed the 'sky/ground ratio'.

* See, for example, Middleton (1958).

The contrast transmittance, T_c, defined as the ratio of C_R to C_0, is thus given by:

$$T_c = T[T + K(1 - T)]^{-1} \tag{4}$$

T_c is plotted against K in Figure 1 for three values of T.

The ranges of K values commonly encountered over three representative types of terrain* are indicated at the top of the Figure.

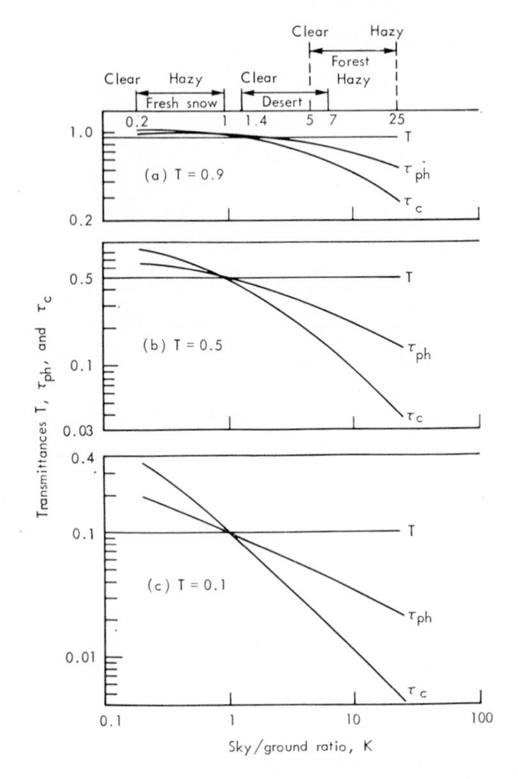

FIG. 1. Transmittances T, τ_{ph}, and τ_c versus sky/ground ratio for three values of T.

Image Intensifiers

Electro-optical image intensifiers, whether arranged for direct viewing (scotoscopes) or coupled into a television camera (low-light-level television) are fundamentally limited by photon noise, i.e., random fluctuations in the density of the arriving stream of photons, at low levels of natural

* See Middleton (1958) p. 73. Note that the term 'hazy' has been substituted for 'overcast', since the authors consider that Equation (3) is more generally valid in hazy conditions than under a stratified overcast.

illumination. Practical devices may have a higher quantum detection efficiency than the eye, and they provide enough gain to permit the final output to be viewed photopically. But, of course, they cannot increase the information available at the entrance pupil of the device, which is fundamentally limited by the photon noise in the arriving radiation. They may, in fact, introduce additional noise; but any such contributions are ignored here. It is now a straightforward matter to analyse the inherent signal-to-noise ratio and the degradation in this ratio caused by the atmosphere.

Returning to the radiance notation and expressing the fluctuation in the signal $(N_t - N_b)$ by the form $\sqrt{(N_t + N_b)}$,* one can write for $(S/N)_0$, the signal-to-noise ratio in the absence of an intervening atmosphere,

$$(S/N)_0 = A \mid N_t - N_b \mid (N_t + N_b)^{-\frac{1}{2}},$$

where A is a constant. When the atmosphere is present we have, with the aid of Equation (1):

$$(S/N)_R = (S/N)_0 T[T + K(1 - T)]^{-\frac{1}{2}}$$

where K is the ratio of the sky radiance to the average of N_t and N_b. The ratio of (S/N). to $(S/N)_R$, denoted by the symbol τ_{ph}, is given by

$$\tau_{ph} = T[T + K(1 + K(1 - T)]^{-\frac{1}{2}} \qquad (5)$$

τ_{ph} is actually the transmittance of the ratio of the difference between the apparent radiances of target and background to the square root of the sum of these radiances. Since the S/N ratio of a sensor limited by photon noise is proportional to this function, we will refer loosely to τ_{ph} as the 'transmittance' of this S/N ratio, while recognizing that the S/N is not, of course, transmitted.† It is seen that the best possible signal-to-noise ratio transmittance, which is realized in the absence of path radiance $(K = 0)$, is equal to the square root of the energy transmittance. With path radiance present, the S/N transmittance is reduced further, but this time only by the square root of the same factor that reduces contrast. Figure 1 shows the variation of τ_{ph} with T and K. It may be noted that τ_{ph} is the geometric mean of T and τ_c for any given value of K.

* The arriving photons are describable by a Poisson distribution. The standard deviation during some time interval is $\sqrt{N}$, where N is the mean number of photons arriving during that time. The standard deviation of the difference signal $(N_t - N_b)$ is the root-sum-square of the individual standard deviations.

† A further comment on terminology: Since radiance transmittance, contrast transmittance, τ_c, and S/N transmittance, τ_{ph} all involve path luminance, they do not represent transmittances in the usual sense. The authors would like to suggest the term 'transferance' to represent these concepts.

Vision at Low Light Levels

At low light levels (background luminance values below 10^{-1} cd/m^2) the eye also appears to be limited by photon noise, since it obeys the de Vries-Rose law (Rose 1948) which states that $(B_t - B_b)_l/B_b$ is independent of B_b, where $(B_t - B_b)_l$ is the liminal or threshold value of $B_t - B_b$.* The performance of the human eye at low light levels thus appears to be limited by the same quantity that determines the S/N ratio of photon-noise-limited sensors. Accordingly the degradation due to the atmosphere in the performance of the human eye at background luminance values below 0·1 cd/m^2 is determined by the transmittance of this quantity, given by Equation (5). Typical values can be read from the τ_{ph} curves of Figure 1.

Passive Infrared Sensors

In the infrared spectral region attenuation arises from molecular absorption as well as from scattering. In terrestrial reconnaissance this lost radiation will be replaced by an approximately equal quantity of emitted and scattered radiation when the absorbing atmosphere is at the same temperature as the scene. As a result the detector noise is, in this case, unaltered by atmospheric effects, so that S/N is reduced directly in proportion to the infrared transmittance, T_{IR}. Accordingly, the S/N transmittance in the infrared is generally given simply by T_{IR}. The same conclusion can be reached by noting that the value of K, the sky/ground ratio, is unity in the passive infrared case.

Laser Scanners and other Gated-Viewing Systems

The performance of active optical reconnaissance systems is limited by back-scatter, as is discussed at length elsewhere (Steingold & Strauch 1968). With such systems the back-scatter can be reduced by spatial separation of source and receiver or by range gating. If back-scatter is completely removed, the atmosphere will degrade performance only through beam attenuation. Since two-way transmission is involved, both $(N_t - N_b)$ and $(N_t + N_b)$ will be decreased by the factor T^2. Assuming photon noise limitation, the S/N will therefore be reduced by a factor T due to the atmosphere. The S/N transmittance is indicated by the corresponding curves in Figure 1. Note that in Figure 1, T (being independent of the sky/ground ratio) is indicated by horizontal lines. If the noise is predominantly internal detector noise, as it would be for long wavelength infrared laser systems, the S/N is reduced by a factor T^2 due to the atmosphere but is still independent of K.

* The spatial resolution of the normal eye is sufficiently reduced at low light levels to result in a photon-noise-limited discrimination of grey levels.

Detection of Targets against the Horizon Sky

Consider an airborne target of inherent radiance N_t which is observed from the ground against a sky background of radiance N_s. In accordance with Equation (1) the apparent target radiance is given by $N_tT+N_q(1-T)$; the signal associated with the target will then be proportional to the radiance difference $|N_tT+N_q(1-T)-N_s|$. When the target is near the horizon, N_q is approximately equal to N_s, whereupon this expression reduces to $|N_t-N_s|T$. The effect of the atmosphere intervening between the target and the observer is thus to reduce the signal by the factor T. Since the background radiance is independent of the target range, the transmitted contrast will thus be degraded by approximately the factor T due to atmospheric scattering. Following similar reasoning, and invoking Kirchoff's law, the same relation can be shown to hold in the detection of thermal targets against the sky through an isothermal absorbing atmosphere using a passive infrared sensor.

One interesting consequence of this, in the visual case, is that the performance of ground-based aircraft spotters is (usually) less degraded by the intervening atmosphere than is the simultaneous performance of observers in the spotted aircraft when they are detecting ground targets in the neighbourhood of the spotters. In cases when the spotter is looking into the horizon sky, his advantage can be made explicit; it is T/τ_c, a factor which (as has been shown) is greater than unity whenever K is greater than unity. In comparing the actual performance of the two observers, other factors must of course be taken into account. These include, for example, the inherent contrasts available in the two cases (aircraft against the sky versus target against the terrain background), the relative target size, the search area, the proximity of the sun to the line of sight,* and so forth.

Variation in Transmittance with Range

The effect of the above considerations on the practical performance of sensors is shown in Figures 2 and 3, in which T, τ_{ph}, and τ_c are plotted against R/V_2, the ratio of range to visibility with atmospheric visibility ranges, V_2, of 2, 10, and 50 mi. These figures apply, respectively, to the cases in which K is equal to 5 and 25.

* The airborne observer often has the option of flying so as to keep the sun at his back, which not only reduces path luminance for him (since rearward scattering is less than forward scattering) and increases his ground visibility, but also forces the spotter to look nearly into the sun. As a consequence, while the spotter in this case observes little or no change in contrast transmission (since T is independent of direction) he does suffer from 'sun-blinding', i.e., a reduction in resolution or visual acuity accompanying his accommodation to very high light levels. Because of this last effect, the air-borne observer may actually realize a significant advantage.

CONCLUSIONS

The performance degradation due to atmospheric effects is seen from the above to vary markedly between the various sensor types. It is most severe with daytime visual and photographic systems, whose performance

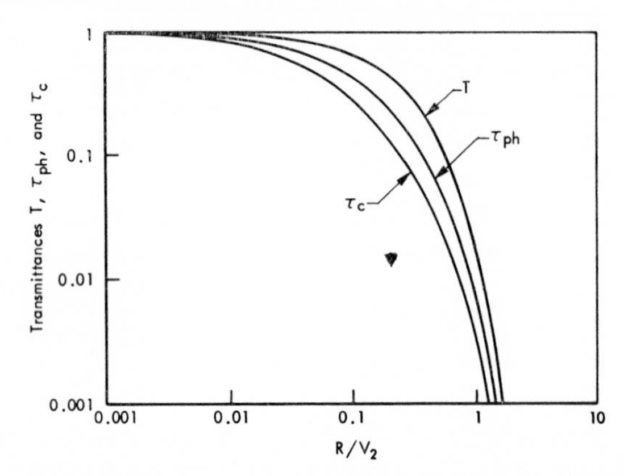

FIG. 2. Transmittances T, τ_{ph}, and τ_c versus R/V_2 (sky/ground ratio, $K=5$)

is measured directly by transmitted contrast. The performance of low-light-level systems, which are limited by photon noise, is less severely degraded. Infrared and gated viewing systems are still less degraded,

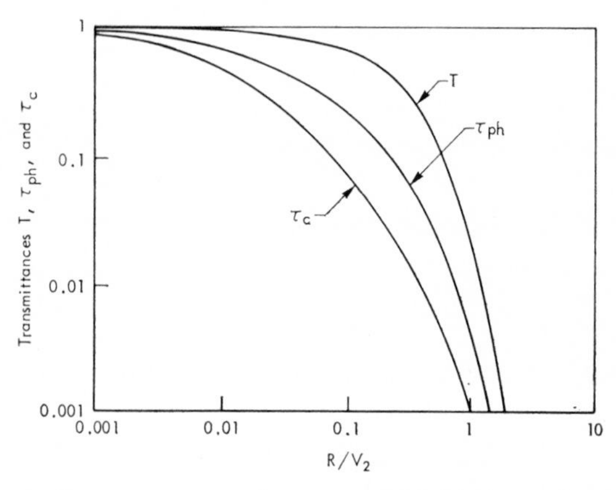

FIG. 3. Transmittances T, τ_{ph}, and τ_c versus R/V_2 (sky/ground ratio, $K=25$).

because the attenuation coefficient is usually smaller in this region, and because their performance is essentially independent of path radiance.

A typical example will illustrate the above conclusions and the use of the various transmittance factors discussed in this paper. Consider the case in which observations are to be made over a three-mile path under conditions for which $K=25$, when the visibility is 10 miles. Assume the humidity to be, alternatively, 'dry' (containing 11·1 mm H_2O/mi) or

TABLE 1. Atmospheric transmittance factors.

Sensor type		Performance transmittance factors*	
		Determining factor	Value (%)
Vision (day)		τ_c	1·75
Vision (low light)		τ_{ph}	7·4
Photoelectric (low light)		τ_{ph}	7·4
Gated viewer		T	30
Passive infrared (3·2–4·8 μm)	dry	T_{IR}	53
	humid	T_{IR}	34
Passive infrared (8–13 μm)	dry	T_{IR}	64
	humid	T_{IR}	18

* See text for assumed conditions.

'humid' (43·9 mm/mi). Under these conditions the performance transmittance for the various sensors is compared in Table 1, in which data for the sensors utilizing reflected radiation were taken directly from Figure 3, while that for the infrared sensors (the transmittance averaged over the two most commonly used atmospheric windows) was deduced from Taylor and Yates (1957) and Yates and Taylor (1960).

REFERENCES

Cambridge Research Laboratory, U.S. Air Force, *Handbook of Geophysics and Space Environments*, Cambridge, Mass., 1965.

DUNTLEY, S. Q., 1948, *J. Opt. Soc. Amer.*, **38,** 179.

DUNTLEY, S. Q., *et al.*, 1964, *Appl. Opt.*, **3,** 549.

MIDDLETON, W. E. K., 1958, *Vision Through the Atmosphere*, University of Toronto Press, Toronto.

ROSE, A., 1948, *J. Opt. Soc. Amer.*, **38,** 196.

STEINGOLD, H. and STRAUCH, R. E., 1968, *Back-scatter Limitations in Active Night-Vision Systems*, The RAND Corporation, RM-5442-PR.

TAYLOR, J. H. and YATES, H. W., 1957, *J. Opt. Soc. Am.*, **47,** 223.

YATES, H. W. and TAYLOR, J. H., 1960, *Infrared Transmission of the Atmosphere* Naval Research Laboratory, Report No. 5453.

New Data on the Properties of Infrared-Transmitting Optical Materials

Stanley S. Ballard

University of Florida,
Gainesville, Florida, U.S.A.

Abstract—The development of new optical materials for the infrared spectral region continues, although more slowly than a decade ago. The accumulation of data on the physical and chemical properties of the new materials that are coming into use also continues, but at a discouragingly slow pace. There are special needs for physical measurements made at high and at low temperatures.

New values of thermal expansion coefficients determined in the author's laboratory in the range from room temperature to as low as 16°K for some materials are reported; these supplement earlier data taken for the same materials in the temperature range 0–100°C.

Reference is made to two new collections and collations of data on the properties of optical materials: One is to appear in the third edition of the *American Institute of Physics Handbook*, and the other in the new *Handbook of Optics* being prepared under the auspices of the Optical Society of America.

NEW OPTICAL MATERIALS

Following World War II there was brisk activity in the development of commercially available optical materials for infrared instrumentation. These activities have been summarized in a number of review articles, including one presented by this author at the ICO-sponsored Conference on Photographic and Spectroscopic Optics held in Japan in 1964 (Ballard 1965). The pace of development of these materials has since slowed somewhat, so that in the intervening five years only a few specific infrared optical materials have been developed, although of course the new solid-state-type materials such as the III-V and II-VI compounds are always of interest and possible utilization in this spectral region.

The new materials that have appeared include Irtran-6, a micro-crystalline aggregate of cadmium telluride produced by the Eastman Kodak Company. It has attractive physical and chemical properties and transmits to beyond 28 μ. Its refractive index is around 2·7, which means that reflection losses are high unless suitable reflection-reducing films are applied to the surfaces of the element.

The firm of Texas Instruments, Inc., has pursued the development of infrared-transmitting glasses of the non-oxide chalcogenide type. Their glass No. 1173 has been available for several years. Its chemical composition is $Ge_{28}Sb_{12}Se_{60}$, and it transmits in the infrared to about 15 μ. Their newer glass No. 20, $Ge_{33}Se_{55}As_{12}$, has a higher softening point of 470°C as compared to 370°C for glass 1173.

THERMAL EXPANSION DATA

Measurements of the coefficient of linear thermal expansion of optical materials have been made for some time in the author's laboratory. Samples as short as 1/2 inch are used. The earlier work covered the temperature range 0–100°C, and employed an interferometric-type dilatometer apparatus in which the expansion of the sample was measured against that of fused silica (Ballard, Browder, Kaylor & Streete 1968). The method utilizes the simple fringe pattern resulting from a thin air-wedge formed between two fused-quartz (fused-silica) optical flats; the top flat is supported by the sample by means of a fused-quartz screw. The expansion of the material is determined by relating its change of length to the measured change of fringe spacing, as temperature is changed. The fringe spacing is measured by a travelling microscope suspended above the expansion cell. Thermal expansion and other data on both new and older optical materials have been collated by Ballard and Browder (1966).

An apparatus was subsequently developed for measuring the thermal expansion of small samples of optical materials in the range from room temperature down to about 16°K. It employs a 3-terminal capacitance-type dilatometer in which the samples are measured against the thermal expansion of oxygen-free, low-conductivity copper. A minute change of the length of a sample produces a change of the spacing of a parallel-plate capacitor with guard ring; the resulting change of capacitance is measured with a highly sensitive bridge. The expansion coefficient is then determined by relating the change of capacitance to the change of dimensions of the sample. Data in this low-temperature region have recently been reported for the six hot-pressed, compacted Irtran materials and glass 1173 (Browder & Ballard 1969). These data are summarized in Figure 1. The referenced article shows that good agreement was obtained between the measurements conducted in our laboratory and those made by other investigators using different techniques and of course different samples.

DATA COMPILATIONS

For several years, the most quoted compilations of the physical properties of infrared-transmitting optical materials have been the reports prepared by the author and two colleagues (Ballard, McCarthy & Wolfe 1959, 1961). These data were largely included in the second edition of the

American Institute of Physics Handbook, published in 1963 by the McGraw-Hill Book Company. However, only the data on refractive index, transmission, and absorption of the infrared materials are grouped in Sections 6*b* and 6*c*; data on the mechanical, thermal, chemical, and other properties

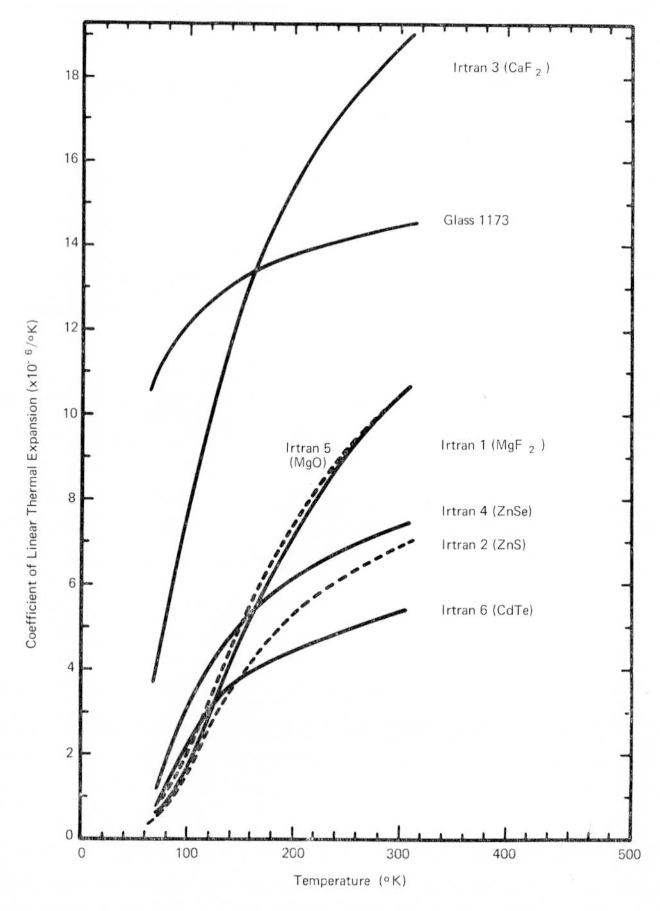

FIG. 1. Coefficient of linear thermal expansion versus temperature, for an infrared-transmitting glass (Texas Instruments, Inc.) and six Irtran samples (Eastman Kodak Company). The basic materials from which the several hot-pressed, microcrystalline Irtran materials were made are given in parentheses.

of these same materials are located in others of the eight sections of the handbook, especially Section 2, Mechanics; and Section 4, Heat. (Section 6 is devoted to Optics).

A more recent compilation of the properties of optical materials is given in Chapter 8 of *Handbook of Military Irfrared Technology*, edited by William L. Wolfe and published in 1965 by the U.S. Government Printing Office. The chapter concludes with a list of 79 references.

The *American Institute of Physics Handbook*, first published in 1957, was prepared under the direction of Dwight E. Gray as coordinating editor, and eight section editors; there were 125 individual contributors to the second edition. This popular handbook is now being revised a second time, and the third edition should appear in 1970. Dr. Gray is again the Coordinating Editor, and Bruce H. Billings continues as section editor for Optics. The present author, with the assistance of colleagues at the University of Florida, is revising Subsection 6*b*, Refractive Index of Special Crystals and Certain Glasses; and 6*c*, Transmission and Absorption of Special Crystals and Certain Glasses. In preparing for this revision he has directed a literature search going back to January 1959 and hence overlapping the literature search conducted for the second edition of 1963. Data have been noted on some 500 crystals and glasses, less than 400 of which are non-vitreous materials. The materials for which 18 or more references have been found are SiO_2, Ge, Si, LiF, CaF_2, and BaF_2. There are many other materials for which 10–17 references were found. The total number of possibly useful articles is around 450. At that, only the readily available journals have been searched, namely the *Journal of the Optical Society of America*, *Applied Optics*, *Optics and Spectroscopy*, *Infrared Physics*, and *Optica Acta*. Many more references could be found on the semi-conductor materials if a search were made in the journals reporting research in solid state physics, but we do not feel that this step is appropriate in the present instance.

The *Handbook of Optics* is soon to be published by the McGraw-Hill Book Company for the Optical Society of America. There had been discussion for some years about the need for a handbook of optics which would provide a large amount of accurate, reliable, and up-to-date information in a condensed, easy-to-use form. As it is now being developed, it will contain basic definitions and formulas, but not the derivation of formulas. Extensive and detailed tables of data such as optical glass characteristics and black-body radiation data will be condensed or presented in graphical form. Written material will be kept to a minimum and will be used to explain the use of tables, graphs, and formulae, and to clarify definitions. The Optical Society of America authorized this project, which was undertaken in August 1967 with Walter G. Driscoll as editor-in-chief and William Vaughan as associate editor. There are 18 chapters, each with a contributor (author), and one or more readers who serve as active critics of the prepared material.

For the purposes of the present paper it can be noted that Chapter VII is entitled 'Properties of Optical Materials'. The principal contributor is William L. Wolfe, who was one of the present author's colleagues in the preparation of Sections 6*b* and 6*c* in the second edition of the *American Institute of Physics Handbook*. As a matter of fact, it was decided between Mr. Wolfe and the author that the former would take the primary responsibility for the optical materials presentation in the new OSA *Handbook of Optics*, while the latter would be responsible for the revision of somewhat similar material in the *American Institute of Physics Handbook*. It is expected that the same numbers (data) will thus appear in both compilations.

The present time schedule for the *Handbook of Optics* indicates that all chapters should be submitted during this summer or early fall. An *ad hoc* committee has been appointed by the Board of Directors of the Optical Society of America to make an overall review of the *Handbook* manuscript in its final form, and the writer is a member of this four-man committee. It will be the duty of this group to look for evenness and parallelism of presentation in the several chapters and for the desired full but not overlapping coverage of the many sub-fields of optics.

These documentation projects of the American Institute of Physics and the Optical Society of America forecast a more satisfactory situation for data compilation in optical materials, and in other branches of optics and physics, than has heretofore existed. One who has been in this general field for some years realizes keenly, however, that the mere compilation of data is not enough. More emphasis must be placed on the acquisition of new data through making additional, often difficult measurements. I refer especially to determining the several important physical properties of optical materials both at high and at low temperatures. It is discouraging to realize how few quantitative data are available, even on common optical materials, at temperatures much above 100°C, and at temperatures much below 0°C. To be sure, these measurements are in general much more difficult than those made at room temperatures. But their potential value is such that investigators should press with vigour toward obtaining them.

REFERENCES

BALLARD, S. S., MCCARTHY, KATHRYN A. and WOLFE, W. L., 1959, *Optical Materials for Infrared Instrumentation*, IRIA State-of-the-Art Report No. 2389-11-S, Ann Arbor, University of Michigan.

—— 1961, *Optical Materials for Infrared Instrumentation*: *Supplement*, IRIA State-of-the-Art Report No. 2389-11-S_1, Ann Arbor, University of Michigan.

BALLARD, S. S., 1965, *Japan. J. Appl. Phys.*, **4**, Suppl. 1, 23.

BALLARD, S. S. and BROWDER, J. S., 1966, *Appl. Opt.*, **5**, 1873.

BALLARD, S. S., BROWDER, J. S., KAYLOR, H. M. and STREETE, J. L., 1968, *J. Opt. Soc. Amer.*, **58**, 155.

BROWDER, J. S. and BALLARD, S. S., 1969, *Appl. Opt.*, **8**, 793.

List of Delegates

Australia	
Coulman, C. E.	C.S.I.R.O., Chippendale
Kobler, H.	C.S.I.R.O., Chippendale
Simpson, A. M.	University of Queensland, Brisbane
Steel, W. H.	C.S.I.R.O., Chippendale
Waterworth, M. D.	University of Tasmania, Hobart
Austria	
Gabler, F.	Klostergasse 37, Vienna
Reuschal, A.	36–36 Gobergasse, Vienna
Schindl, K. P.	Amalienstrasse 34/1/3, Vienna
Belgium	
Collard, R.	Belgian Committee for Optics, Liége.
de Clerck, G.	Glaverbel Laboratoire, Gilly
Malaise, D.	33 rue de Batty, Cointe-Schlessin
Monfils, —	Institut d'Astrophysique, Cointe-Schlessin
Canada	
Baird, K. M.	National Research Council, Ottawa
Chassé, Y.	University Laval, Quebec
Lansraux, G.	National Research Council, Ottawa
China	
Billings, B.	American Embassy, Taipei
Czechoslovakia	
Cuchy, Z.	Monokrystaly, Preperska Ul.
Havelka, B.	Palacky University, Olomouc
Keprt, E.	Palacky University, Olomouc
Knittle, Z.	Meopta, Prerov.
Miler, M.	Institute of Radio Engineering and Electronics, Prague
Nabelek, B.	Physical Institute of the Czechoslovakian Academy of Sciences
Polasek, J.	Staropramenna 16, Prague
East Germany	
Schwider, J.	Arndtstrasse 42, DDR 1199 Berlin
France	
Albe, F.	I.S.L., St. Louis
Arnulf, A.	Institut d'Optique, Paris
Belvaux, Y.	Institut d'Optique, Orsay
Bergeron, C.	Institut d'Optique, Orsay
Bernstein, L.	Institut d'Optique, Orsay
Berny, F. M. M.	Institut d'Optique, Orsay
Blaise, J.	13 rue Eliane, 92 Mendon
Body, Y.	Laboratoire d'Optique, Paris
Bouchareine, P. M.	14 allée Descartes, 91 Orsay
Bourgeon, M.-H.	3 ville Floré, Paris
Brethon, A.	10 alleé de la Résidence 94, Fresnes
Bulabois, J.	University of Besançon
Claire, J. J.	Laboratoire d'Optique, Paris
Connes, P.	C.R.N.S., Bellevue
Corno, J.	28 Domaine de Chateau-Gaillard, 94 Maisons-Alfort

France—continued

CUISENIER, M.	Observatoire de Mendon
DEBRUS, S.	Laboratoire d'Optique, Paris
DETAILLE, M.	14 rue du Chateau d'If, Marseille
DUDERMEL, M.	47 rue Victor Hugo, 91 Plaiseau
DUPOISET, H.	118 ave. St. Exupery, 92 Antony
ENARD, D.	36 ave du Président Wilson, le Pecq
FERRAY, M.	Institut d'Optique, Paris
FLEURY, —	Institut d'Optique, Paris
GUELACHVILI	28 boulevard St. Michel, 91 Etampes
IMBERT, C.	59 ave de la Source, 94 Nogent
IMBERT, M.	15 rue des Peupliers, 91 Longjumeau
LAFAIT, J.	7 rue Christian Dewet, Paris
LANDRAUD, A.	Institut d'Optique, Paris
LEBLANC, M.	le Paradis, 91 Longpost s/Orge
LEVY, M.	Institut d'Optique, Paris
Le. LUYER, —	C.E.R.C.O., Courbevoie
LOUISNARD, N.	O.N.E.R.A., Chatillon
LOWENTHAL, S.	Institut d'Optique, Orsay
MAILLARD, J. P.	78 Trappes.
MARÉCHAL, A.	Institut d Optique, Paris
MARIOGE, J.	15 rue Petratque, Paris
MAY, M.	Laboratoire d'Optique, Paris
MICHEL, J. J.	2 rue P. J. Redoute, 92 Mendon le Forêt
MONDAL, P.	Laboratoire d'Optique, Paris
NOMARSKI, G.	Institut d'Optique, Paris
OLIVIE, M. M.	Ecole Polytechnic, Paris
PETIT, R.	Laboratoire d'Optique, Marseille
PIEUCHARD, G.	Jobin & Yvon, Arcueil
PINARD, J.	18 rue Mozart, 78 Foutenay le Fleury
POUEY, M.	C.N.R.S., Bellevue
POULEAU,	2 chemin de Billere, Billere-Pau
RAGOUT, J.	1 rue Martin, Bougival
ROBLIN, G.	Institut d'Optique, Paris
RODOLFO, R.	128 rue Mouffetard, Paris
ROUSSEAU, M.	Institut d'Optique, Paris
ROUSSEAU, P.	Institut d'Optique, Paris
ROUSSEL, P. C. H.	S.E.M.S., Clamart
SAMUEL, F.	Huet, Paris
SIMON, J. F.	2 rue Nelaton, Paris
SLANSKY, S.	Institut d'Optique, Paris
SMIGIELSKI, P.	I.S.L., St. Louis
SPITZ, E.	C.S.F., Corbeville par Orsay
TERRIEN, J. C.	Pavillon de Breteil, 92 Sèvres
VEDEL, F. J.	299 E2 ave de St. Julien, Marseille
VEDEL, M. E.	299 E2 ave de St. Julien, Marseille
VIENOT, J. Ch.	University of Besançon

Germany

FRANKE, G. G.	633 Wetzlar, Laufdorferweg 2
GLATZEL, E.	Carl Zeiss, Oberkochen
HAINA, D.	6100 Darmstadt, Hochschulstr. 2
HAMMERSCHLAG, R. H.	Ohmstrasse 33, 33 Braunschweig
HERTEL, J.	Optisches Institut der Technischen Universitat, Berlin
HOCK, F.	633 Wetzlar, Norgenweide 32
KOHLER, H.	Sauerbruchstrasse 6, 792 Heidenheim, Brenz
KRAUTTER, M.	Sauerbrichstrasse 18, 7082 Oberkochen
KROSS, J.	655 Bad Kreuznach, Postfach 947
LANZL, F.	I. Physik Institut Technischen Hocheschule, 61 Darmstadt
MUEHSCHLAG, R.	655 Bad Kreuznach, Postfach 947
ROSENBRUCH, K. J.	Wohlerstrasse 1, 33 Braunschweig
SCHEDEWIE, —	703 Boblingen, Heusteigstrasse 34
SCHOBER, H.	Institut für Med. Optik., Barbarastrasse, 8 München 13

Germany—continued

WAIDELICH, W.	6100 Darmstadt, Hochschulstrasse 2
WILSON, R. N.	Carl Zeiss, 7082 Oberkochen/Wurtt.
WOELTCHE, W.	655 Bad Kreuznach, Postfach 947

Hungary

HAHN, E.	Research Institute for Telecommunications, Budapest
KANTOR, K.	Central Research Institute for Physics, Budapest

India

RAO, V. V.	Regional Engineering College, Andhra Pradesh

Israel

GELD, A.	Ministry of Defence, Tel-Aviv

Ialy

ATZENI, C.	Consiglio Nazionale delle Ricerche, Firenze
CEPPATELLI, G. G.	Observatory of Florence, Firenze
CHECCACCI, V. R.	Istituto di Ricerce sulle onde Ellettromagnetiche, Firence
FIORENTINE, A.	Laboratorio di Neurofisiologia, Piza
GELLI, G.	Officine Galileo spa, Firenze
OTTAVIANI, M. C.	Istituto di Fisica Superiore, via Panciatichi 56, Firenze
PANTANI, L.	Istituto di Ricerca sulle onde Elettromagnetiche, Firenze
RIGHINI, A.	Osservatorio Astrofisico di'Arcetri, Firenze
RIGHINI, G.	Osservatorio Astrofisico di Arcetri, Firenze
RIGHINI, G. C.	Istituto di Fisica, Superiore, Firenze
SCANDONE, F.	Officine Galileo spa, Firenze
SCEEL, G.	Officine Galileo spa, Firenze
TORALDO di FRANCIA, G.	Istituto di Ricerca sulle onde Elettromagnetiche, Firenze

Japan

ISSHIKI, M.	Nippon Kogaku, K.K., Tokyo
KINISITA, K.	Gakoshoin University, Tokyo
NAKANO, T.	Nippon Kogaku, K.K., Tokyo
OSE, T.	University of Tokyo, Tokyo
SATO, T.	Institute of Technology, Tokyo
YOSHINAGA, H.	University of Osaka

Netherlands

AALBERS, A. D.	Jacob van Campenweg 31, Rotterdam
BECKER, J.	N.V. Optische Industrie "de Oude Delft", Delft
BEERNINK, G. J.	Institute of Applied Physics TNO-TH, Delft
BOERSMA, S. L.	"de Oude Delft", Delft
BOUWHUIS, G.	Philips Research Laboratory, Eindhoven
DANDLIKER, G.	Philips Research Laboratory, Eindhoven
GODFROY, H.	Köninklijke Shell Laboratory, Badhuisweg 3
FREIE, H. G.	N.V. Oprische Industrie "de Oude Delft", van Miereveltlaan 9, Delft
KINGMA, R. V.	"de OudeDelft", Delft
KRAMER, P.	Philips Research Laboratory, Eindhoven
LAMBERTS, C. W.	Institute of Applied Physics, P.O. Box 155, Delft
de LANG, H.	Technische Hogeschool, Lorentzweg 1, Delft
PEEK, T. H.	Philips Research Laboratory, Eindhoven
SNOEK, C.	Zeeman Laboratorium, Plantage Muidergracht 4, Amsterdam
VAN DEELEN, W.	Schollevaarstraat 96, Maassluis
VAN DE STADT	Sterrewacht "Sonnenborgh," Utrecht
VAN DEN BRINK, H. G.	17 van Renevveg, Eindhoven
VELZEL, C. H. F.	Philips Research Laboratory, Eindhoven
VOS, J. F.	Laboratory for Technology, Delft

New Zealand
RUMSEY, N. J. — D.S.I.R., Lower Hutt

Poland
GAJ, M. — Institut fur Technische Physik, Wroctaw
SKALINSKI, T. — Institute of Physics, Warsaw

Spain
BELLANATO, J. — Instituto de Optica, Madrid
CASAS, J. — University of Zaragoza
CATALINA, F. — Instituto de Optica, Madrid
GARCIA, O. — Instituto de Optica, Madrid
IGLESIAS, L. — Instituto de Optica, Madrid
RICO, F. — Instituto de Optica, Madrid
TINAUT, D. — Instituto de Optica, Madrid
YZUEL, M. J. — University of Zaragoza

S. Africa
VAN ROOYEN, E. — National Physical Research Laboratory, Pretoria

Sweden
BACK, A. — Jungner Instrument A.B., Stockholm
DONNE, W. — L.K.B. Produkter A.B., Bromma
GUSTAFSSON, S. — Chelmers Institute of Technology, Gothenburg
INGELSTAM, E. — The Institute of Optical Research, Stockholm
JACOBSSON, R. — AGA Aktiebolag Div. 10, Lidingo
MOLLER, B. — Jungner Instrument A.B., Stockholm
NISS, E. — AGA AB Div. EGO, Lidingo
STENSLAND, L. — Institute of Optical Research, Stockholm
VOGL, G. — Jungner Instrument A.B., Lidingo
WAHREN, P. — Frithofsvagen 10, Djursholm
WALLES, S. — AGA AB Div. EGO, Lidingo
WALLIN, L. E. — Chelmers Institute of Technology, Gothenburg.

Switzerland
GREENAWAY, D. D. — Laboratoires R.C.A., Zurich
GUGGER, W. — Paillard S. A., Orbe
ITEN, P. D. — B.B.C., Daetwil
LOTMAR, W. — Swiss Office of Weights & Measures, Wabern, Bern
MERCIER, R. — Ecole Polytechnique, Lausanne
TIZIANI, H. — Institut für Technishe Physik, Zurich

United Kingdom
ALABASTER, P. S. — University of Manchester
ARCHBOLD, E. — National Physical Laboratory, Teddington
ASHTON, A. — Rank Precision Industries, Leicester
ASPDEN, J. — Unilever, Isleworth, Middlesex
BACH, A. G. — 99 Hartle Down, Purley, Surrey
Bailey, W. — Fermain, 60 Down Road, Portishead, Bristol
BAKER, L. R. — SIRA, South Hill, Chislehurst, Kent
BARTON, N. P. — F.V.R.D.E., Chertsey, Surrey
BATEMAN, D. E — Royal Aircraft Establishment, Farnborough
BEACH, A D. — Atomic Weapons Research Establishment, Aldermaston
BEESLEY, M. J. — Services Electronics Laboratory, Baldock, Herts
BENNETT, S. J. — 82 Cross Deep, Twickenham, Middlesex
BENTLEY, J. M. — Rank Precision Industries Ltd., Leeds 8
BIRCH, K. G. — Optical Metrology Division, National Physical Laboratory, Teddington
BLANDFORD, B. — University of Reading
BOTTOMLEY, S. C. — Hilger & Watts, London
BOULTON, G. B. — Imperial College, London
BRADDICK, H. J. J. — University of Manchester, Manchester 13
BROWN, B. S. — Atomic Weapons Research Establishment, Aldermaston
BROWN, D. S. — Grubb Parsons, Newcastle upon Tyne 6

United Kingdom—continued

BURCH, J. M.	National Physical Laboratory, Teddington
BUSSELLE, F. J.	25 Wealdon Close, Hildenborough, Kent
BUTLER, C.	Optical Measuring Tools Ltd., Maidenhead, Berks.
CHUBB, T. W.	Royal Aircraft Establishment, Farnborough
CLARK, A. D.	57 Broadwood Avenue, Ruislip, Middlesex
CLARKE, D.	University of Glasgow
CLARKE, J. A.	Mullard Research Laboratory, Redhill, Surrey
CLARKE, R. W. G.	Mullard Research Laboratory, Redhill, Surrey
COLEMAN, K.	Atomic Weapons Research Establishment, Aldermaston
COOK, G. H.	Rank Precision Industries, Leicester
COOK, R. W. E.	National Physical Laboratory, Teddington
COOPER, P. R.	W. Watson & Sons, West End Lane, Barnet, Herts.
CRUIKSHANK, M. A.	Degenhardt & Co. Ltd., 31–36 Foley Street, London
DAINTY, J. C.	Imperial College, London
DAVIDSON, J. N.	Royal Army Radar Development Establishment, Sevenoaks, Kent
DAVIES B. H.	3 Roebuck Court, Turpin's Rise, Stevenage, Herts.
DAY, D. J.	Wray Optical Works Ltd., Bromley, Kent
DITCHBURN, R. W.	University of Reading
DREWITT, J. M.	W. Watson & Sons, Barnet, Herts.
DYSON, J.	National Physical Laboratory, Teddington
EDGAR, R. F.	Imperial College, London
ENNOS, A. E.	National Physical Laboratory, Teddington
FELLGETT, P. B.	University of Reading
FREEMAN, D.	Imperial College, London
FREEMAN, G. C. H.	National Physical Laboratory, Teddington
FREEMAN, M. H.	Pilkington Perkin-Elmer, St. Asaph., N. Wales
GATES, J. W. C.	National Physical Laboratory, Teddington
GILLINGHAM, P. R.	Pateena, Pevensey, Sussex
GOULD, J. A.	Royal Military College of Science, Faringdon, Berkshire
GRAINGER, J. F.	University of Manchester
GRANT, R. E.	R. &. J. Beck Ltd., Watford, Herts.
GREEN, F. S.	National Physical Laboratory, Teddington
GREENLAND, K. M.	Scientific Instruments Research Association, Chislehurst
GRIGGIN, W. G.	Astrophysics Research Unit, Culham Laboratory, Abingdon, Berks
GROVE, F. J.	Pilkington Brothers Ltd., Ormskirk, Lancs.
HABELL, K. M.	National Physical Laboratory, Teddington
HAIG, N. D.	'Gartlans', Norwood Lane, Medham, Near Gravesend, Kent
HALL, R. G. N.	National Physical Laboratory, Teddington
HARMER, C. F. W.	Royal Greenwich Observatory, Herstmonceux, Hailsham
HARMER, D. L.	Royal Greenwich Observatory, Herstmonceux, Hailsham
HARPER, D. W.	Pilkington Brothers Ltd., Ormskirk, Lancs.
HEATHCOTE, A.	Imperial Chemical Industries, Macklesfield, Cheshire
HEAVENS, O. S.	University of York
HERCHER,	National Physical Laboratory, Teddington
HOME-DICKSON, J.	22 Westminster Mansions, Great Smith Street, London S.W.1
HOPKINS, H. H.	University of Reading
HORNE, D. F.	Hilger & Watts Ltd., Loughton, Essex
JAMES, J. F.	The Schuster Laboratory, Manchester University
JAMIESON, T. H.	Barr & Stroud Ltd., Anniesland, Glasgow W.3
JERRARD, H. G.	University of Southampton
JONES, G. F.	R. & J. Beck Ltd., Greycaine Road, Watford, Herts.
KELLY, R. A.	27 Gordano Gardens, Easton-in-Gordano, BS20 OPD, Somerset
KENWORTHY, J. G.	Imperial Chemicals Industries, Runcorn, Cheshire
KEYTE, G. D.	Royal Aircraft Establishment, Farnborough
KIDGER, M. J.	Imperial College, London
KING, R. J.	National Physical Laboratory, Teddington

United Kingdom—continued

Leendertz, J. A.	University of Technology, Loughborough, Leicestershire
Leifer, I.	University of Aston, Birmingham
Liddell, H.	Queen Mary College, London E.1
Lïng, R. S.	13 Grayshott Close, Winchester, Hampshire
Lomas, G. M.	Pilkington Brothers Ltd., Ormskirk, Lancashire
Lothian, G. F.	University of Exeter
Lunn, C. H.	Atomic Weapons Research Establishment, Aldermaston Berkshire
Luxmore, A.	University College of Swansea, Glamorgan
Marchant, E.	Central Unit for Scientific Photography, Royal Aircraft Establishment, Farnborough
Mason, V. A.	Pilkington Brothers Ltd. Ormskirk, Lancashire
Milward, R.	Research & Industrial Instrument Co., Worsley Bridge Road, London, S.E.26
Minakovic, B.	Department of Engineering Science, Parks Road, Oxford
Minns, T. A.	W. Watsons & Sons Ltd., Barnet, Hertfordshire
Mitchell, C. J.	Queen's University Belfast
McKinnon Marshall, P.	Physics Department, Heriot Watt University, Edinburgh
McNaughton, L.	Schuster Laboratory, Manchester University
Noda, Mrs	18 Suffolk Road, Barnes, London, S.W.13
North, J. C.	Post Office Research Station, Dollis Hill, London, N.W.2
Palmer, E. W.	National Physical Laboratory, Teddington
Palmer, J. M.	Rank Precision Industries, 98 St. Pancras Way, London N.W.1
Payne, B. O.	Vickers Ltd., Haxby Road, York
Poole, J.	'Brightwell', Chelsfield Lane, Orpington
Porteous, R. L.	Hilger Electronics (Scotland) Ltd., Corn Exchange, Dalkeith
Redman, J. D.	Atomic Weapons Research Establishment, Aldermaston
Reid, A. M.	Pilkington Brothers Ltd., Ormskirk, Lancs.
Reid, C. D.	Atomic Weapons Research Establishment, Aldermaston
Ring, J.	Infra-red Astronomy Group, Imperial College, London
Roberts, —	University of York
Rogers, G. L.	University of Aston, Birmingham
Rout, R. J.	Atomic Weapons Research Establishment, Aldermaston
Rowley, R. W. C.	National Physical Laboratory, Teddington
Rudd, M. J.	Cavendish Laboratory, Free School Lane, Cambridge
Scholfield, J.	Imperial College, London
Seeley, J. S.	University of Reading
Shorter, G. B.	Optics Technology, Iliffe House, 32 High Street, Guildford
Smethe, H. J.	Schuster Laboratory, University of Manchester
Smith, F. H.	137 Foxley Lane, Purley, Surrey
Smith, R. W.	Imperial College, London
Stephens, C. L.	Infra-red Astronomy Group, Imperial College, London
Stephens, N.	University of Aston, Birmingham
Sternberg, R. S.	Schuster Laboratory, Manchester University
Stewart, D. J.	K. S. Paul & Associates, 52 Oaklands Drive, Wokingham Berkshire
Stewart, W. J.	Hatherleigh, Tower Road, Hindhead, Surrey
Tanner, L. H.	Queen's University, Belfast
Tarrant, A. W. S.	University of Surrey, Guildford
Thetford, A.	University of Reading
Thomas, B. R.	Atomic Weapons Research Establishment, Aldermaston
Thomson, W. C. M.	Royal Aircraft Establishment, Farnborough
Thorpe, L.	Research & Development Instrument Company, Worsley Bridge Road, London S.E.26
Tickner, A. S.	R. & J. Beck Ltd., Bushey Mill Lane, Watford, Hertfordshire
Trevelyan, B.	F.V.R.D.E., Chertsey, Surrey

United Kingdom—continued

TUCK, M. J.	Pilkington Perkin Elmer, St. Asaph, N. Wales
TURNER, W.	Rutherford Laboratory, Chilton, Didcot, Berkshire
TURPIN, J.	Atomic Weapons Research Establishment, Aldermaston
URWIN, F. W.	Pilkington Perkin Elmer, St. Asaph, N. Wales
VANN, M. A.	Guided Weapons Division, British Aircraft Corporation Ltd., Filton, Bristol
VERRILL, J. F.	National Physical Laboratory, Teddington
WALL, M. R.	Atomic Weapons Research Establishment, Aldermaston Berkshire
WATRASIEWICZ, B.	Guided Weapons Division, British Aircraft Corporation Ltd., Filton, Bristol.
WATTON, R.	Royal Radar Establishment, Malvern.
WATTS, J. K.	Imperial College, London
WELFORD, W. T.,	Imperial College, London
WEST, P.	SIRA, South Hill, Chislehurst, Kent
WILLIAMS, D. C.	National Physical Laboratory, Teddington
WILLIAMS, T. L.	SIRA, Chislehurst, Kent
WITHRINGTON, R. J.	Imperial College, London
WORRALL, M. J.	Rank Precision Industries, SIRA, South Hill, Chislehurst, Kent
WRIGHT, W. D.	Imperial College, London
WYNNE, C. G.	Imperial College London
YEE, H.	British Aircraft Corporation Ltd., Gunnel Wood Road Stevenage, Hertfordshire

United States

BALLARD, S. S.	Department of Physics, University of Florida, Florida
BARNES, W. P. (Jr.)	Itek Corporation, Building 10, 3rd Avenue, Burlington, Massachusetts 01803
BRYNGDAHL, O.	I. B. M. Research Laboratory, San José, California
DRUMMETER, L. F. (Jr.)	5503 Belfast Drive SE, Washington D.C. 20022
DUTTON, D. E.	Institute of Optics, University of Rochester, New York 14627
GOODMAN, J. W.	Department of Electrical Engineering, Stanford University, California
GUSTAFSON, D. E.	Optical Research Associates, 550 North Rosemead Boulevard, Pasadena, California
HEIMER, R. J.	Fairchild, 621 Hawaii Street, El Segundo, California 90245
HOOD, J. M.	U.S. Naval Electronics Laboratory, San Diego California 92151
HUPPI, E. R.	14 Alonquin Road, Chelmsford, Massachusetts
HYDE, W. Lewis	New York University, New York 10453
JONES, R. C.	Polaroid Corporation, Cambridge, Massachusetts 02139
KING, W. B.	Perkin Elmer Corporation, Norwalk, Connecticut 06852
KINGSLAKE, R.	56 Westland Avenue, Rochester, New York
KINGSLAKE, H. (Mrs.)	56 Westland Avenue, Rochester, New York
KOESTER, C. J.	American Optical Corporation, P.O. Box 187, Framinham Centre, Massachusetts 010701
LANGENBECK, P.	Perkin Elmer Corporation, Norwalk, Connecticut 08652
MEIRON, J.	Perkin Elmer Corporation, Norwalk, Connecticut 06852
MERTZ, L.	25 Hayes Lane, Lexington, Massachusetts 02173
MOSS, T.	The Ealing Corporation, 2225 Massachusetts Avenue, Cambridge, Massachusetts 02140
MUNDIE, L. G.	The Rand Corporation, 1700 Main Street, Santa Monica California
MACGOVERN, A. J	Itek Corporation, 10 Maguire Road, Lexington, Massachusetts
NICHOLSON, S.	2124 Linda Flora Drive, Los Angeles, California 90024

United States—continued

OFFNER, A.	Perkin Elmer Corporation, Norwalk, Connecticut 06852
RHODES, M. B.	Chemistry Department, University of Massachusetts, Massachusetts 01002
SCHAWLOW, A. L.	Physics Department, Stanford University, California 94305
SCOTT, B.	387 Harbor Road, Erie, Pennsylvania 16511
SCOTT, R. M.	Perkin Elmer Corporation, Norwalk, Connecticut 06852
SHANNON, R. R.	Optical Sciences, University of Arizona, Tucson, Arizona 85721
SMITH, F. DOW	Itek Corporation, 10 Maguire Road, Lexington, Massachusetts 02173
STRAUB, H. W.	United States Department of Commerce, Washington Science Center, Rockville, MD 20852
STROKE, G. W.	Electro-Optical Sciences Center, State University of New York, Stony Brook, New York 11790
THOMPSON, B. J.	The Institute of Optics, University of Rochester, Rochester, New York
VANASSE, G. A.	71 Old Stage Road, Chelmsford, Massachusetts 01701
VAN LIGTON, R. F.	Pleasant Street, Connecticut P.O. Box 187, Framington Center, Massachusetts 01701
DE VEER, J. D.	American Optical Corporation, P.O. Box 187, Framington Center Massachusetts 01701
WILCZYNSKI, J.	P.O Box 218, Yorktown Heights, New York 10598

U.S.S.R.

KANDYBA, W. W.	Committee for Standards, Measure and Measuring Instruments, Moscow
KARTASHEV, A. J.	Committee for Standards, Measure and Measuring Instruments, Moscow
SEMENOV, A. W.	Committee of Science and Technique, Moscow
SUKHODREV, N.	P. N. Zebeder Institute, Moscow
TROFOMOV, J. J.	Committee of Science and Technique, Moscow

INDEX